Smart Nanomaterials for Infectious Diseases

Nanoscience & Nanotechnology Series

Titles in the series:

1: Nanotubes and Nanowires
2: Fullerenes: Principles and Applications
3: Nanocharacterisation
4: Atom Resolved Surface Reactions: Nanocatalysis
5: Biomimetic Nanoceramics in Clinical Use: From Materials to Applications
6: Nanofluidics: Nanoscience and Nanotechnology
7: Bionanodesign: Following Nature's Touch
8: Nano-Society: Pushing the Boundaries of Technology
9: Polymer-based Nanostructures: Medical Applications
10: Metallic and Molecular Interactions in Nanometer Layers, Pores and Particles: New Findings at the Yoctolitre Level
11: Nanocasting: A Versatile Strategy for Creating Nanostructured Porous Materials
12: Titanate and Titania Nanotubes: Synthesis, Properties and Applications
13: Raman Spectroscopy, Fullerenes and Nanotechnology
14: Nanotechnologies in Food
15: Unravelling Single Cell Genomics: Micro and Nanotools
16: Polymer Nanocomposites by Emulsion and Suspension
17: Phage Nanobiotechnology
18: Nanotubes and Nanowires, 2nd Edition
19: Nanostructured Catalysts: Transition Metal Oxides
20: Fullerenes: Principles and Applications, 2nd Edition
21: Biological Interactions with Surface Charge Biomaterials
22: Nanoporous Gold: From an Ancient Technology to a High-Tech Material
23: Nanoparticles in Anti-Microbial Materials: Use and Characterisation

24: Manipulation of Nanoscale Materials: An Introduction to Nanoarchitectonics
25: Towards Efficient Designing of Safe Nanomaterials: Innovative Merge of Computational Approaches and Experimental Techniques
26: Polymer–Graphene Nanocomposites
27: Carbon Nanotube-Polymer Composites
28: Nanoscience for the Conservation of Works of Art
29: Polymer Nanofibers: Building Blocks for Nanotechnology
30: Artificial Cilia
31: Nanodiamond
32: Nanofabrication and its Application in Renewable Energy
33: Semiconductor Quantum Dots: Organometallic and Inorganic Synthesis
34: Soft Nanoparticles for Biomedical Applications
35: Hierarchical Nanostructures for Energy Devices
36: Microfluidics for Medical Applications
37: Nanocharacterisation, 2nd Edition
38: Thermometry at the Nanoscale: Techniques and Selected Applications
39: Nanoceramics in Clinical Use: From Materials to Applications, 2nd Edition
40: Near-infrared Nanomaterials: Preparation, Bioimaging and Therapy Applications
41: Nanofluidics, 2nd Edition
42: Nanotechnologies in Food, 2nd Edition
43: ZnO Nanostructures: Fabrication and Applications
44: Diatom Nanotechnology: Progress and Emerging Applications
45: Nanostructured Materials for Type III Photovoltaics
46: Chemically Derived Graphene: Functionalization, Properties and Applications
47: Graphene-based Membranes for Mass Transport Applications
48: Carbon Nanostructures for Biomedical Applications
49: Surface Chemistry of Colloidal Nanocrystals
50: Reducing Agents in Colloidal Nanoparticle Synthesis
51: Carbon Nitride Nanostructures for Sustainable Energy Production and Environmental Remediation
52: Nanotubes and Nanowires, 3rd Edition
53: Bionanodesign: Old Forms for New Functions, 2nd Edition
54: Photothermal Nanomaterials
55: Concepts and Design of Materials Nanoarchitectonics
56: Anticorrosive Nanomaterials: Future Perspectives
57: Sustainable Nanotechnology

58: Anisotropic 2D Materials and Devices
59: Nanotechnology for Diabetes Management
60: Graphene-based Nanocomposite Sensors: Detection to Diagnosis
61: Nanoparticles as Sustainable Environmental Remediation Agents
62: Smart Nanomaterials for Infectious Diseases

How to obtain future titles on publication:
A standing order plan is available for this series. A standing order will bring delivery of each new volume immediately on publication.

For further information please contact:
Book Sales Department, Royal Society of Chemistry, Thomas Graham House, Science Park, Milton Road, Cambridge CB4 0WF, UK
Telephone: +44 (0)1223 420066, Fax: +44 (0)1223 420247
Email: booksales@rsc.org
Visit our website at books.rsc.org

Smart Nanomaterials for Infectious Diseases

Edited by

Suvardhan Kanchi
CHRIST (Deemed to be University), India
Email: ksuvardhan@gmail.com

Nidhi Sharotri
Sri Sai University, India
Email: nidhisliet11@gmail.com

Rajasekhar Chokkareddy
Aditya College of Engineering, India
Email: chokkareddys@gmail.com

Deepali Sharma
Toronto Research Chemicals, Canada
Email: dpschem@gmail.com

and

Falah H. Hussein
University of Babylon, Iraq
Email: abohasan_hilla@yahoo.com

ROYAL SOCIETY OF CHEMISTRY

Nanoscience & Nanotechnology Series No. 62

Print ISBN: 978-1-83767-082-6
PDF ISBN: 978-1-83767-281-3
EPUB ISBN: 978-1-83767-282-0
Print ISSN: 1757-7136
Electronic ISSN: 1757-7144

A catalogue record for this book is available from the British Library

The Royal Society of Chemistry is a charity, registered in England and Wales, Number 207890, and a company incorporated in England by Royal Charter (Registered No. RC000524), registered office: Burlington House, Piccadilly, London W1J 0BA, UK, Telephone: +44 (0)20 7437 8656.

For further information see our web site at www.rsc.org

Printed in the United Kingdom by CPI Group (UK) Ltd, Croydon, CR0 4YY, UK

Preface

In recent years, the world has witnessed unprecedented challenges posed by infectious diseases. The rapid spread of pathogens and the emergence of drug-resistant strains have pointed towards the urgent need for innovative solutions in the field of medicine. As researchers and scientists delve deeper into the realm of nanotechnology, new avenues of combating infectious diseases are being unveiled. This book, *Smart Nanomaterials for Infectious Diseases*, explores the cutting-edge advances in nanotechnology and its promising potential in revolutionizing the way we approach and manage infectious illnesses.

The journey through the pages of this book will take you on an enlightening tour of nanomaterials and their applications in combating infectious diseases. We have carefully curated a collection of chapters that not only discuss the theoretical aspects but also present practical approaches to deal with infectious diseases.

The exploration begins with "Carbon-reinforced Polymer Nanocomposites Against Infectious Diseases," which sets the foundation for understanding the role of nanocomposites in disease management. From there, we delve into "Nanostrategies for Infectious Pulmonary Diseases: Current Progress and Future Prospects," addressing a critical aspect of respiratory infections and the possibilities of nanomedicine.

"Biofilm-associated Infections and Their Management" presents novel strategies for tackling the stubborn problem of biofilms, and "Green Nanomaterials Zinc Oxide and Chitosan for Antimicrobial Activity Against Oral Pathogens" explores eco-friendly alternatives for oral health.

The book ventures further into "Biopolymer-based Drug Delivery Systems: Synthesis Strategies, Functionalization, and Applications," offering insights into sophisticated delivery systems, and "Mechanism of Silver and Gold

Nanoscience & Nanotechnology Series No. 62
Smart Nanomaterials for Infectious Diseases
Edited by Suvardhan Kanchi, Nidhi Sharotri, Rajasekhar Chokkareddy, Deepali Sharma and Falah H. Hussein

Published by the Royal Society of Chemistry, www.rsc.org

Nanoparticles Against Infectious Disease-causing Pathogens," shedding light on the fascinating antimicrobial properties of noble metal nanoparticles.

As we progress, the book also addresses the critical issue of "Nanotoxicology and Its Remediation," ensuring the safety of nanomaterials, and "Biocompatibility and Functionalization of Sustainable Nanomaterials," enhancing their potential applications.

In the face of specific infectious diseases, "Nanomaterial-based Narrowband UVB Light-emitting Devices for Phototherapy Treatment of Infectious Skin Diseases" presents a breakthrough in skin therapy, while "Bioengineered Nanomaterials: Recent Trends and Bottlenecks in the Management of Infectious Diseases" explores avenues to manage infectious diseases.

The ongoing challenges posed by SARS-CoV-2 are addressed in the chapters "*Ex Vivo* Immunization: A Strategy for Immunization Against SARS-CoV-2" and "Neuroinvasion by SARS-CoV-2 Influences the Prognosis of Neurological Disorders," offering valuable insights into potential solutions.

Amidst these scientific endeavors, it is crucial to acknowledge the societal aspects. "Socioeconomics: Involving Communities in the Fight Against Infectious Illnesses for Stronger Health Systems" explores the collective role in battling diseases.

The book culminates with two essential aspects: "Sustainable Nanomaterials for Mosquito-based Infectious Disease Control" and "Sustainable Nanomaterials and Diagnosis for Arboviral Infections," highlighting eco-friendly solutions for disease management.

Lastly, we offer "A Brief Appraisal of Nanomaterials for COVID-19," a succinct summary of the role that nanomaterials played in the fight against the pandemic.

We are confident that the contents of this book will serve as a valuable resource for researchers, practitioners, and enthusiasts alike, fostering interdisciplinary collaborations to combat infectious diseases effectively. Our sincere appreciation goes to the contributors who have shared their expertise and knowledge to make this endeavor possible.

Embrace the knowledge presented within these pages and join us in shaping a healthier, more resilient future through smart nanomaterials.

Suvardhan Kanchi
Nidhi Sharotri
Rajasekhar Chokkareddy
Deepali Sharma
Falah H. Hussein

Contents

Nanoscience & Nanotechnology Series No. 62
Smart Nanomaterials for Infectious Diseases
Edited by Suvardhan Kanchi, Nidhi Sharotri, Rajasekhar Chokkareddy, Deepali Sharma and Falah H. Hussein

Published by the Royal Society of Chemistry, www.rsc.org

CHAPTER 1

Carbon-reinforced Polymer Nanocomposites Against Infectious Diseases

BISWAJIT PARHI, DEBASRITA BHARATIYA, ANURADHA BISWAL AND SARAT K. SWAIN*

Department of Chemistry, Veer Surendra Sai University of Technology, Burla, Sambalpur, Odisha 768018, India
*Email: skswain_chem@vssut.ac.in

1.1 Introduction

In the present era, everyone is aware of the importance of the terms 'nano' and 'the nano effect' owing to the abundant adoption of this technology in the development of many uses with different real-time applications, such as nano-SIM cards, nanocars, nanodevices, nanomedicine, nanoscaffolds, *etc.*, with their outstanding mechanical, thermal, electrical, magnetic, and optical properties.[1] The use of nanotechnology is extensively seen in the electronics, food, biomedical, and packaging industries owing to their interesting morphology, high tensile strength, biocompatibility, low toxicity, nanosize, and large surface area. The development of nanomaterials and their derivatives has advanced their use predominantly to solve biomedical and health-related issues. Research on the development of smart and low-cost materials has been prioritized. For the growth of society through nanoscience and nanotechnology, advances in the preparation of new sustainable nanocomposite materials are essential.

Nanoscience & Nanotechnology Series No. 62
Smart Nanomaterials for Infectious Diseases
Edited by Suvardhan Kanchi, Nidhi Sharotri, Rajasekhar Chokkareddy, Deepali Sharma and Falah H. Hussein

Published by the Royal Society of Chemistry, www.rsc.org

Nanocomposites are materials that have different dimensions in which at least one phase must be in the nano range. Nanocomposites are composed of two parts, namely a filler part and the matrix. The various types of filler materials used for the preparation of nanomaterials are nanotubes (NTs), metal oxide and ceramic nanoparticles (NPs), and polymeric nanomaterials. Depending on the type of matrix, nanocomposites are broadly classified into three types, namely ceramic matrix nanocomposites, metal matrix nanocomposites, and polymer matrix nanocomposites. All of these nanocomposite materials possess unique thermal, mechanical, and pharmacological properties with different morphologies that differ from those of their bulk substances. The enhanced properties at the nano level encouraged applications in different areas such as packaging, electronics, and biomedical uses. Different types of nanocomposites and their synthesis routes and applications are shown in Figure 1.1.

Polymer matrix nanocomposites have been shown to be more effective in biomedical applications owing to their high biocompatibility and high pharmacological activity. Researchers all over the world are continuously involved in developing new composite materials from biopolymers, which have been proved to be more sustainable for use in the biomedical field. Widely used materials include natural and synthetic polymer matrix nanocomposites and polymers reinforced with carbon nanotubes (CNTs), graphite, graphene and inorganic NPs, focusing on light weight, easy processing, recyclability, oxygen and moisture permeability, solvent resistance, thermal stability, barrier properties, pharmaceutical features, and high strength, which inspired many scientists to work on treatments against infections caused by different agents.[2,3] In addition to these properties, the high conductivity, effective adsorption, and high thermal stability of such composites attracted attention towards their application as materials useful in defence,

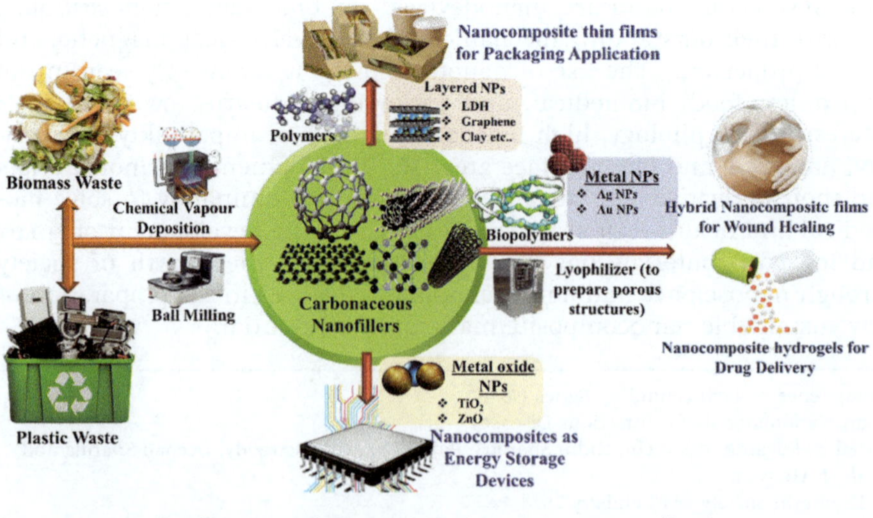

Figure 1.1 Synthesis routes of nanocomposites and some applications.

electronics, sensor, and biomedical areas.[4] The phenomenon of the dispersion of nanofillers on different matrices shows the reinforcement that is extensively observed in the case of polymer nanocomposites bearing superior properties. The multipurpose uses of the polymer matrix are illustrated in Figure 1.2.

One of the greatest challenges that the whole world is facing, along with biomedical science, is the problem of tackling the concurrent occurrence of a pandemic due to parasitic pathogen infection. An infection is defined as the attack of a pathogenic parasite on healthy host tissue through toxin production with replication of the pathogen cells. This infection causes serious illness in the host body and in many cases efficiently kills the host if the immune system is unable to fight back properly. The host body for infection can be a plant, animal, or human depending on the nature of the pathogen involved. Plants are least susceptible to pathogenic infection whereas the trend of infection is high in the case of animals and humans. Many pathogens are found in nature that can cause infection in both animals and humans. The transmissibility factor of pathogens helps them to cause severe, deadly pandemics such as COVID-19. The majority of infections are caused by bacteria and viruses, and some infections are caused by other pathogens such as fungi, protozoa, worms, and prion proteins. Traditional research has been ongoing aimed at the development of vaccines and drugs to control pathogen infections. A few of the drugs and vaccines have been registered for use in society for specific infection control. However, in many cases, there are no suitable drug candidates available to combat infection. For two decades, with the rapid increase in sustainable nano-biomaterials development, there has been bright hope in biomedical science regarding the availability of infection-resistant drugs and vaccines.

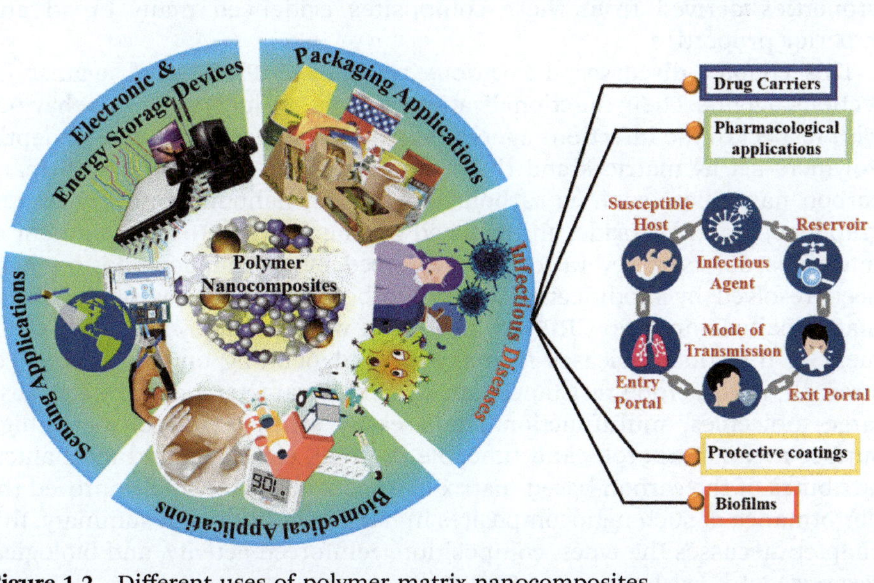

Figure 1.2 Different uses of polymer matrix nanocomposites.

Direct and indirect transmission of the viruses, bacteria, worms, prion proteins, protozoa, and fungi from the outside to the inside of living cells poses serious threats to humans, sometimes even taking lives. The spread and transmission of infected microbes can occur by coughing, sneezing, and direct and indirect touching, and also by air- and water-borne microbes. The combating effect of nanocomposites has increased substantially in the biomedical field by fighting many serious issues, including working against many agents such as bacteria, viruses, worms, *etc.* To counter these infectious agents and to achieve their complete eradication, the development of drug delivery modes, scaffolds, catheters, and stents, tissue engineering, and anti-infectious drugs has been advancing, and polymer matrix nanocomposites have come into the picture.[5,6]

Carbon-reinforced polymer nanocomposites (CRPNCs) are one category of sustainable nanomaterials that have attracted attention in the field of material science and technology owing to safety aspects, quality, renewability, low cost, ready availability, high surface area, conductivity, thermal stability, and porosity, planar structure, regular framework, nanosize and thickness, flat surface, and good mechanical strength.[7] In the current scenario, CRPNCs have shown high efficacy against infectious diseases. These sustainable nanomaterials are made up of a polymer matrix reinforced with carbon-based materials such as CNTs, fullerenes, graphene, nanodiamond, carbon dots, *etc.*[8] In some infections caused by bacteria, fungi, and protozoa, the reinforced composite material efficiently destroys the cell wall of the pathogen with the production of reactive oxygen species (ROS), which leads to the death of the pathogen and the curing of the disease. In the case of viral infections, CRPNCs suitably enhance the production of interferons for the protection of healthy cells.[9,10] The mixed properties derived from these composites evidenced many broad and superior properties.

This chapter discusses the various types of CRPNCs used against infectious agents. Their functionalization, factors, influences, and behaviour with respect to the infectious agents causing disease are explained in depth. Polymers act as matrices and the reinforcement is carried out with many carbon nanofillers such as carbon fibre, CNTs, nanodiamond, fullerenes, graphene, graphene oxide, and reduced graphene oxide. In the treatment of infectious diseases, several challenges faced by the nanocomposites have been resolved by modification of the embedded properties of the filler–matrix nanocomposite. CRPNCs have been widely employed for the treatment of infectious diseases in many areas depending on the mechanical strength, antimicrobe resistance, antioxidant parameters, *etc.* The low cost, large molecules, multifunctional and easily synthesized polymers, high surface area, and porous and tuneable thermal, electrical, and mechanical attributes of the carbon-based matrix have contributed to and improved the performance of such nanocomposites in harsh conditions. In summary, this chapter discusses the types, composition, reinforced activity, and biological response and development of CRPNCs.

1.2 Categories of Infectious Agents

Living beings, especially humans, are quite hostile to microbes causing infections and their related health threats. The agents for infectious disease are microbes, worms, and a few pathogenic proteins. Most infections are the result of the pathogenic activities of microorganisms and easy transmission *via* air, water, and soil. The variable shapes and sizes affect the infection rates and fatalities among living beings. A detailed discussion of several types of infection-causing microbes is presented below. Figure 1.3 illustrates different types of infectious agents and the infections they cause.

1.2.1 Bacteria as Infectious Agents

Bacteria are unicellular organisms that have a cell wall with other cellular materials such as cytoplasm, ribosomes, and nucleoids. Numerous types of bacteria are present on this green planet, and they are considered one of the first living organisms on Earth. Bacteria can be classified into different types from different aspects, but with regard to their infection-causing ability they can be classified into two types: symbiotic bacteria and parasitic bacteria. Symbiotic bacteria are conventionally non-pathogenic whereas parasitic bacteria are pathogenic. Many of the infections in animals and humans are caused by pathogenic bacteria and represent a global challenge. The rapid multiplication of bacteria cells in the host body leads to infection, causing serious illness. The major infectious diseases seen in humans include *Acinetobacter* infection, actinomycosis, *Arcanobacterium haemolyticum* infection, bacterial meningitis, bacterial vaginosis, cavities (dental caries), Carrion's disease, cat-scratch disease, chancroid, and cholera, whereas the major bacterial infectious diseases observed in both humans and animals

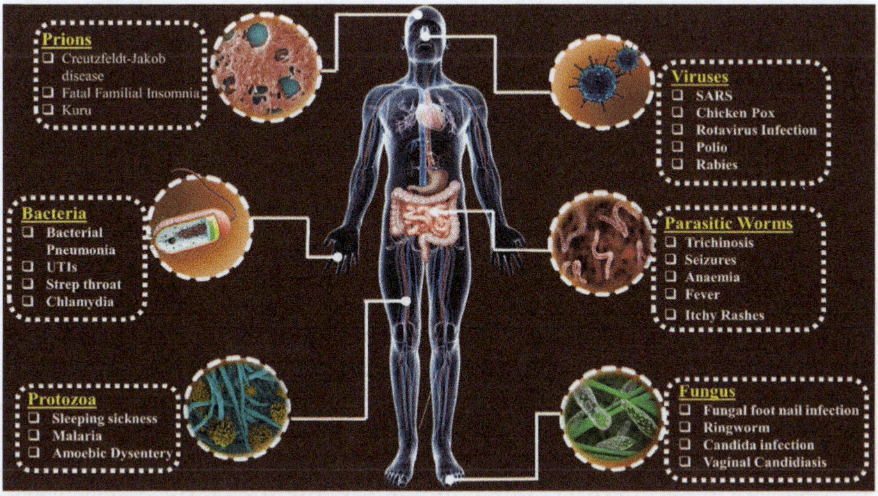

Figure 1.3 Different types of infectious agents that cause deadly diseases.

include anaplasmosis, anthrax, *Bacillus cereus*-caused illness, bacterial pneumonia, bartonellosis, brucellosis, bubonic plague, *Burkholderia* infection, Buruli ulcer, campylobacteriosis, cellulitis, chlamydia, *Chlamydia pneumoniae* infection, *Clostridioides difficile* colitis, and diphtheria.[11,12] Being ubiquitous, bacteria play a significant role in the maintenance of a healthy life. As bacterial infections and their transmissions are common and occur by infected guests reaching susceptible hosts from the environment through air, water, food, and physical contact, they are also easy to treat compared with viral and fungal infections. To inhibit bacterial growth, antibiotics and antibacterial agents have been increasingly used depending on the chemical structure, target site, and types of bactericidal agents.[13]

1.2.2 Fungi as Infectious Agents

Fungal infection is one of the most common infections observed in living cells in the form of yeasts or moulds. They grow like plant species. The primary agents responsible for these infections are the genera *Candida*, *Cryptococcus*, and *Aspergillus*, all of which cause serious infections. The reason lies with the cell itself or is caused by host factors. They can affect the cells and tissues of the living being with varying infection rates. The aggravated cells multiply to large numbers in a very short time and diagnosis can be difficult, leading to high fatality rates.[14] At the present time, fungal infections caused by SARS-CoV-2 are suspected to be due to superinfected agents affecting the lungs, kidneys, heart, and eye tissues.[15,16] To face such challenges, appropriate measures should be taken, such as the recording of fungal distribution and drug resistance capacity, and consideration of risk factors. Fungal disease is very chronic once a suitable host has been found, and can be life threatening if it is not treated early.

1.2.3 Viruses as Infectious Agents

Viruses can replicate inside the host itself if there is a chance of a growth-permissible environment. Common viral diseases include AIDS, hepatitis, COVID-19, chickenpox, smallpox, cold sores, colds, flu, *etc.*, caused by direct contact or intake of the virus into the body. Viruses exist in many shapes and sizes, including helical, complex, simple, cylindrical, icosahedral, and envelope types. They are made up of nucleotides derived from genes, a protein for coating, and phospholipids to act as an envelope outside the cell for their survival purposes. They can affect other infected and non-infected agents including humans, plants, animals, bacteria, and fungi. The study of and research on these agents have been quite difficult owing to their small size and fast growth rate, affecting major organs of living beings. Recent viral infections include the dangerous COVID-19, which created havoc in the world by causing a life-threatening epidemic. This RNA virus causes severe pulmonary damage over time by mutation transmitting from one host to another.[17] The infection was difficult to control because of its unknown structure and the rate of invasion was too fast. Effective interventions and research to reduce viral infections are predominantly based on the use of vaccines,[18] drugs, and systematic treatments.

1.2.4 Protozoa as Infectious Agents

Infections due to protozoa are caused by *Plasmodium* agents targeting tissue damage, leading to diseases that mainly affect the respiratory tract. Their size lies in the micrometre range. Protozoa such as *Entamoeba*, *Acanthamoeba*, *Balamuthia*, *Leishmania*, *Trypanosoma*, *Trichomonas*, *Lophomonas*, *Cryptosporidium*, *Cyclospora*, *Toxoplasma*, *Plasmodium*, *Enterocytozoon*, and *Balantidium*, cause dangerous clinical conditions such as AIDS, malign haemopathies, giardia, toxoplasmosis, *etc.*[19] Diseases such as malaria and chagas disease are mainly dominated by protozoa infections and cause the highest mortality due to protozoa infection. The immunological response of the parasitic host antigen and cytokine profiles gives the infection profile of damaged tissue. The detection, finding, and treatment of these parasitic protozoa these days depend on methodology, availability of commercial kits, and modern molecular biological techniques.[20]

1.2.5 Worms as Infectious Agents

Worms resemble helminths with varying micron sizes. Parasitic worms are spread and transmitted to the living body by drinking polluted water, contact with infected soil, eating undercooked and date-expired food, and mosquito bites. The common diseases mainly caused by worms are ascariasis, schistosomiasis, amoebiasis, cryptosporidiosis, and filariasis.[21] The worms affect the intestines and stomach of humans and reduce the nutritional value of the intake of food and water.[22] To face this challenge, treatment to eradicate the worms should be carried out by restoring the required energy, proteins, and micronutrients.

1.2.6 Prion Proteins as Infectious Agents

Prion diseases caused by prion proteins are extremely dangerous as they affect the delicate parts of humans such as neurological tissues. The prion proteins first malfunction and then form aggregates called prions that cause infectious brain disease, and as this affects the neurological path the condition is very difficult to treat. The diseases caused by prion proteins are Creutzfeldt–Jakob disease (CJD), Gerstmann–Sträussler–Scheinker syndrome, and fatal insomnia.[23] The chemical modification of prion proteins might be one of the reasons for this infective nature.[24] Painful and incurable diseases are not easy to treat but the symptoms may be eased with supportive care. Early measures should be taken to control prion-causing diseases to avoid severe chronic health issues.

1.3 Types of Carbon-reinforced Polymer Nanocomposites (CRPNCs) and Their Use in Controlling Infectious Diseases

To combat these infectious agents and to reduce the fatality rate, it is wise to use suitable drugs with formulated anti-infective resistance and supportive care. To simplify these issues further, reinforced polymers came into the picture with an additive manufacturing process that enhanced the biological and pharmaceutical performance.[25] The wide range of beneficial properties and the existence of various forms of carbon have challenged many issues such as improper alignment, formation of cracks and voids, and unsuccessful adhesion. The properties of CRPNCs depend on the dimensions of the carbon fillers. For a better understanding, different types of carbon-based fillers with various dimensions are illustrated in Figure 1.4.

1.3.1 Carbon Fibre-reinforced Polymer Nanocomposites (CFRPNCs)

Carbon fibres are broadly obtained naturally from plant sources such as rice husk, palm, banana, jute, cotton, coir, *etc.*, and artificially from graphene–basalt, glass, *etc.* Polymers such as epoxies, polyethylene (PE), polyesters, and polyacrylonitrile (PAN) act as templates or matrices for the development of CFRPNCs. These fibres have been adopted for several uses as a consequence of their light weight, availability, low cost, natural origin, biodegradability, low toxicity, high strength, and durability.[26] The reinforcement of high-strength carbon fibres as a filler of the chemical environment of the polymer matrix contributes strengthened and useful CFRPNCs by polymerization, *in situ*,

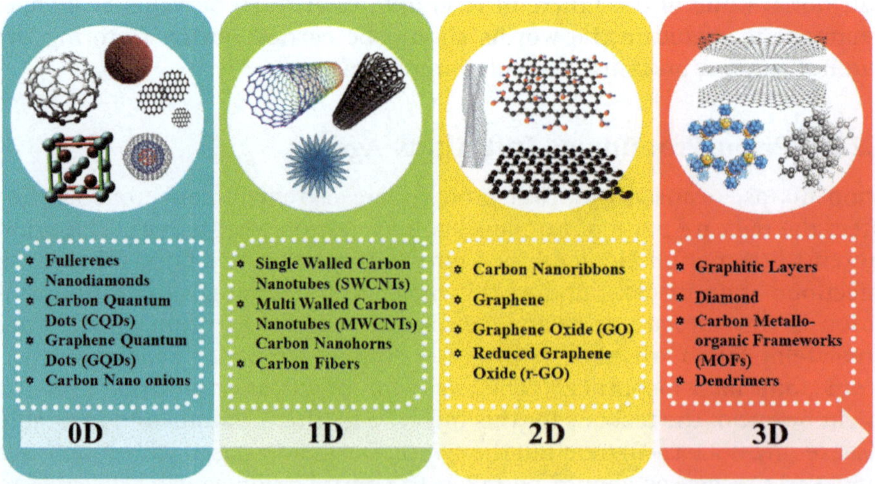

Figure 1.4 Different types of carbon-based fillers with various dimensions.

one-pot, and sol–gel synthetic routes. The fibre content, morphology, size, shape, composition, orientation, and preparation method impact the quality of CFRPNCs. The extraordinary multifunctional nature, biocompatibility, zwitterionic effect, anticorrosive properties, high flexural strength, and stiffness damping of these nanocomposites enhanced their importance in biomedical fields to establish the optimum characteristics of the material for a desired application.

Matharu *et al.* designed graphene oxide (GO)–poly(methyl 2-methylpropenoate) (PMMA) CFRPNCs *via* a three-step synthesis. These ~1.4 mm sized 8 wt% CFRPNCs show effective antibacterial properties against *Escherichia coli* bacteria with strong bioactivities of 46–85% compared with than GO–poly(vinyl alcohol) (PVA) fibres due to the presence of the non-water-soluble polymer PVA.[27] The incorporation of the naturally derived polymeric fibre chitosan (CS) in a poly(butylene adipate-*co*-terephthalate) (PBAT) matrix prepared using the solution casting technique showed outstanding activity against fungal and bacterial growth. The biocompatibility, biodegradability, low cost, and hydrophilic nature of CS led to the easy and effective incorporation of NPs in the polymer matrix. The interaction of the amine ($-NH_2$) groups of the CS polymer creates leakage in the fungal and bacterial cell walls by removing the proteinaceous and other intracellular constituents of the bacteria. These CFRPNCs show effective antifungal and antibacterial activity against the fungi *Aspergillus niger* and *Mucor rouxii*, and Gram-positive *Staphylococcus aureus* and *Bacillus subtilis* and Gram-negative *Salmonella enteritidis* and *E. coli* bacteria. The ductility reduction, high water permeability, and better strength improved the anti-infective resistance focusing on strong H-bonding interactions between the hydroxyl (–OH) and $-NH_2$ groups of CS and the carbonyl moieties of the polymer. The study suggested that prepared CFRPNCs can act as potential agents for the inhibition of fungal and bacterial growth and can be used in antimicrobial packaging systems for food and medical applications.[28] de Faria *et al.* functionalized GO–Ag–CS-embedded poly(lactide-*co*-glycolide) (PLGA) CFRPNC, which prevented bacterial colonization on its surfaces. The 356 nm diameter fibre nanocomposites were formed by the interlinking of the carboxyl (–COOH) group of GO and the $-NH_2$ group of the polymer. The direct interaction of CFRPNC with the bacterial cell wall showed an efficacy of 99% and 76% inactivation growth rate for Gram-negative (*E. coli* and *Pseudomonas aeruginosa*) and Gram-positive (*S. aureus*) bacteria by an ROS mechanism.[29]

The risk of infection post-surgery always scares the patient. To avoid this problem to some extent, the development of a graphene fibre-based polycaprolactone (PCL) matrix CFRPNC was achieved by melt spinning. The release of the antibacterial agent chlorhexidine (CHX) works successfully with controlled and sustainable kinetics. The increased mechanical strength of CFRPNCs due to the amount of graphene present in it showed elastic modulus and tensile strength increments of +175% and +115%, respectively. Even the trapping of the required drug and its release occurred with the porous graphene filler in the nanocomposite, causing an improved anti-infective environment for the patient.[30] The biodegradability, size, cylindrical and round morphology, drug release capacity of the fibres, and suitable chemical interactions between carbon filler and polymer matrix led to an effective inhibition effect on the growth of infection-causing agents.

1.3.2 Carbon Nanotube-reinforced Polymer Nanocomposites (CNTRPNCs)

A rolled graphene sheet closed by hemifullerenes is called a carbon nanotube (CNT). CNT-reinforced polymer nanocomposites (CNTRPNCs) possess good physical and pharmacological properties. With excellent physical properties such as electrical, mechanical, magnetic, thermal, and optical properties, these polymer nanocomposites have been used for antimicrobial applications, drug delivery, nerve cell generation, and catalysts.[31] The size, thickness, functionalization, and easy production make CNTs very useful for several applications. A CNTRPNC facilitates the interaction between the infected cell and itself and protects itself from the adsorption of non-specific proteins. CNT-based polymer nanocomposites can prevent bacterial growth and biofilm adhesion due to the functionalization, CNT length, electronic structure facilitating ionic conductivity, limited hydrophobic surface, and the presence of a large surface area.[32]

Joo *et al.* prepared a useful CNTRPNC by using CNT as filler and poly[2-(dimethylamino)ethyl methacrylate] (PDMAEMA) as a matrix *via* grafting and atom transfer radical polymerization (ATRP). This CNT-based CNTRPNC showed an outstanding antimicrobial effect against yeast, and an antibacterial effect against Gram-positive *S. aureus* and Gram-negative *E. coli*. The highest viability loss was observed for Gram-negative *E. coli* bacteria with the highest sheet resistance of $<9.68 \times 10^3$ Ω per square. The solubility, stimuli responsiveness, and quaternary ion formation of this CNT-based polymer nanocomposite contribute to the effective inhibition of microbial growth.[33] The long, thin, and round CNTs induce antibacterial and antifungal activities by breaking the infected cells. CNT-based CS reduces the enzyme activities of infected fungal and bacterial cells and facilitates DNA breaking.[34] It has been demonstrated that CNTRPNCs show fascinating anti-infective mechanisms towards fungi, bacteria, and viruses such as *Candida albicans, E. coli, B. subtilis, P. aeruginosa*, and *S. aureus* by following the steps (i) interaction of CNTRPNCs with the infective microbial surface, (ii) destruction of protein and DNA, and (iii) ROS formation.[35] The peptide-based CNTRPNCs, otherwise called host defence peptides (HDPs), also kill the infection-causing microorganisms through multiple steps such as alteration of the cell membrane, membrane permeabilization, activation of autolysin, inhibition of biomolecule synthesis, inhibition of certain enzymes, induction of cell and membrane damage, DNA destruction, protein dysfunction, and release of secondary products, such as ROS.[36]

The functionalization of hydrazide and hydrazone derivatives, epoxides, and hydroxyl and amine groups resulted in improved antimicrobial activities against the growth of *Enterococcus faecalis*, *Staphylococcus epidermidis*, *E. coli*, *Aspergillus niger*, *Cryptococcus neoformans*, and *Candida tropicalis* as judged by their higher inhibition zone diameters using the agar well diffusion method.[37] Liu *et al.* fabricated a CNT-based CRPNC as a poly(lactic acid) (PLA)–CNTs–CS composite and studied its antimicrobial behaviour. It was reported that an increase in the CS content in the composite led to the enhancement of the mechanical properties, solubility, shelf life, swelling ratio of the fibre films, and antimicrobial activity of the composite fibre films against *Botrytis cinerea*, *Rhizopus*, *S. aureus*, and *E. coli*, making it useful for fruit and vegetable preservation.[38] Xiang *et al.* discussed the activity of CNTs for biomedical and therapeutic purposes. The CNT-based CRPNCs show reduced neuron damage in the brain and spinal cord owing to the presence of accumulated prion proteins in the human body. The prion protein formation is caused by the accumulation of misfolded proteins, oxidative stress, apoptosis, inflammation, perturbation of vesicle trafficking, and synapse dysfunction.[39] The structural modification, multifunctionalized moieties, divergent synthesis procedure, composition, and reduced self-aggregation originated from powerful van der Waals interactions, and greater interaction between the MWCNTs and microorganisms led to an enhanced anti-infective mechanism against the growth of harmful microorganisms. The easy preparation and fabrication of CNT-based RPNCs can improve the microbial growth inhibition rate and provide a sustained eradication procedure to fight recurrent and deadly infection-causing microbes such as bacteria, worms, protozoa, *etc.*

1.3.3 Nanodiamond-reinforced Polymer Nanocomposites (NDRPNCs)

Nanodiamonds (NDs) are a type of carbon-based nanomaterial used in different applications such as bioimaging, biomedical devices, and implants. The surface composition, biocompatibility, enough acid anhydride groups, small size, and presence of a partially oxidized and negative surface of NDs made them potent antimicrobial agents with a new nanomaterial. Generally, they are composed of a non-toxic sp^3-carbon diamond core and a graphitized outer layer containing sp^2-hybridized partially oxidized carbon, which allow the unique morphologies and functionalization through electrostatic interactions among the groups present.[40] The PLGA–polypropylene–PLA-embedded NDRPNCs have shown great efficacy in biomedical use. The ease of surface functionalization and fascinating mechanical and interesting surface morphology has been widely used to reinforce biopolymers and the participation of hydrophilic groups led to the creation of good antibacterial activities with the presence of water-soluble antimicrobial drugs.[41] ND-reinforced PMMA–PLA nanocomposites possess very good antibacterial activity depending on the weight percentage of doping of NDs in the composite, leading to good doping and depositional capacity.[42] Xie *et al.* synthesized silicone–ND–polydimethylsiloxane–polyurea nanocomposites and studied their antifouling activity. These NDRPNCs are non-toxic and environmentally friendly. The presence of perfluoro and polysiloxane over the top surface of the CRPNC improved the self-healing properties and performance, showing an effective reduction in and control over antimicrobial growth on its surface.[43] The incorporation of NDs in the polymer nanocomposites greatly impacted the reduction of infective agent growth due to the interesting silicone modification in the nanocomposite, good adhesion and toughness, presence of hydrogen bonds, size effects, surface area, and carbon-based hybridizations. The biocide-free, eco-friendly nature, high strength, and oxygen abundance nature of the ND-incorporated polymer nanocomposites provide strong evidence for utilization against antimicrobial growth.

1.3.4 Fullerene-reinforced Polymer Nanocomposites (FRPNCs)

Being a carbon allotrope, closed or semi-closed mesh-like structured fullerenes are interconnected by double and single bonds and are useful for biomedical purposes. They possess sp^2- and sp^3-hybridized carbon atoms having a high transport charge ability and electron affinity. The production of peroxides (hydrogen peroxide, H_2O_2), superoxide ($^{\bullet}O_2^{-}$), singlet oxygen ($^1O_2$), and hydroxyl radicals ($^{\bullet}OH$) also enhances the inhibition of microbial growth depending on the electrostatic interaction. The positive surface charge of fullerenes and the functional groups in the polymer in the CRPNCs enhance the efficacy against Gram-negative *E. coli* bacteria. Fullerene-reinforced polymer nanocomposites (FRPNCs) are mainly useful for bone tissue engineering and tissue generation and antimicrobial growth. The FRPNCs have been shown to be effective against bovine leukaemia virus (BLV) and *Lactobacillus rhamnosus* and *Lactobacillus acidophilus* bacteria.[44] The incorporation of fullerenes in a natural polymer such as CS and cyclodextrin polysaccharides to give cyclodextrin–CS–fullerene nanocomposite CRPNCs showed significant antibacterial action against *C. albicans*, *S. aureus*, and *E. coli*.[45] The activity, membrane permeability, elasticity, mesh-type structure, interesting size, and photodynamic surface possessed by fullerene CRPNCs are very useful properties for antimicrobial performance.

1.3.5 Carbon Dot-reinforced Polymer Nanocomposites (CDRPNCs)

Carbon dots (CDs) have been extensively used in numerous energy storage and biomedical applications depending on their quantum size, size effect, irregular surface, surface defects, and fluorescent properties. ROS production depends on the temperature effect and photothermal agents are responsible for the effective generation of reactive oxygen species. The fluorescent-type quasi-spherical CDs show high water solubility and high stability, are environmentally benign, have good biocompatibility, are cost-effective, and have a spherical structure with an average size of 10 nm. As CDs are among the best photothermal agents, they show significant antibacterial resistance following the photodynamic inactivation mechanism in Gram-positive (*S. aureus*) and Gram-negative (*E. coli*), *Klebsiella pneumoniae*, and *P. aeruginosa* bacterial species and against multidrug-resistant pathogens.[46] The strategic development of CD incorporation in polymer nanocomposites (CDRPNCs) creates more opportunities for various applications depending on the high abundance of CDs as the precursor, low cost, easy synthesis methods, large surface area, and interesting morphology. The CDRPNCs are produced by sol–gel and green chemistry methods, hydrothermal processing, ultrasonic wave use, and microwave radiation and oxidation methods.[47] Wu *et al.* discussed some CDPRNCs for application in photodynamic antimicrobial chemotherapy (PACT) using multidrug resistance (MDR) concepts. To combat and prevent the MDR effect without any side effects, these CRPNCs seem to be most useful in exhibiting powerful photodynamic effects. These can kill both bacterial and cancer cells under visible light irradiation, hence presenting themselves as strong candidates for application against serious infective agents.[48]

The viruses from the plague, chickenpox, Zika, swine flu, Ebola, and SARS-Cov-19 have caused havoc in the daily routine of humans, and have taken billions of lives over the last few centuries. Although they are inactive and incapable of growing on their own, they cause life-threatening diseases once they get a chance in a suitable environment. They are otherwise called parasites. Currently, the world is still facing the SARS-CoV-2 attacks that began in 2019, and to combat this deadly pandemic CDs and their derived carbon quantum dots (CQDs) incorporated in polymer nanocomposites have contributed extensively. They act as effective antiviral agents owing to their biocompatibility, small size, good sustainability, environmental friendliness, and potential candidature. The encapsulation of antiviral drugs into CQD–DNA CDRPNCs through electrostatic interactions and adsorption helps with easy, sustainable, and effective controlled drug delivery. These CDRPNCs also affect the viral growth in living organisms occurring in the following steps: viral entry–uncoating, their replication, and release in cell lines. This growth is prevented by the presence of CQDs in the CDRPNCs depending on their composition, chemical moieties, topography, surface morphology, and hydrogen and covalent bondings.[49] Maruthapandi *et al.* designed an interesting CD-initiated polypyrrole (PPY)–CuO composite *via* a one-step sonochemical method and studied its antibacterial behaviour against *E. coli* and *S. aureus*. The best antibacterial action of these CDRPNCs was observed at a concentration of 1 $mg\,mL^{-1}$ of the composite involving 0.234 $mg\,mL^{-1}$ of CuO, which destroyed the bacterial cell wall after ROS generation. The ROS production in this case also depends on the synergistic effect between the polymer and CuONPs. Even the aromatic ring and amine chemical moiety generated from PPY also participated in the bacterial DNA damage with the production of an electrostatic interaction leading to electrolyte loss from the bacterial cell wall, causing its death.[50]

1.3.6 Graphene-reinforced Polymer Nanocomposites (GRPNCs)

Two-dimensional, single-sheet, sp^2-hybridized graphene has been extensively used as an important precursor for derived species such as GO and reduced GO (rGO). There are many resources available in nature that can be extracted by mechanical separation, chemical vapour deposition, exfoliation, *etc.* The honeycomb-shaped graphene sheet itself possesses antibacterial properties, which are enhanced when incorporated with a bioactive polymer matrix. The use of graphene increased after its mixing with polymers, which enhanced the conductive nature of graphene-reinforced polymer nanocomposites (GRPNCs). These have been widely used in pharmaceutical, sensor, and energy storage device applications.[51] The antimicrobial properties of graphene are followed by three important mechanisms: membrane stress, oxidative stress, and electron transfer pathways. Graphene–poly(*N*-vinylcarbazole) (PVK) underwent van der Waals interactions *via* physical mixing and blending with the chemical reaction between these two precursors and showed efficacy against numerous bacteria such as *B. subtilis*, *E. coli*, *Rhodococcus opacus*, and *Cupriavidus metallidurans*. Chemical interaction develops between the π–π stacking of graphene and the aromatic rings of PVK.[52] The zwitterionic effect, production of quaternary ammonium salts, grafting, and multifunctional, polyionic complex formation of polymer nanocomposites facilitate the inhibition of microbial growth on any surface. There are several GRPNCs, such as graphene-reinforced and Ni–P-incorporated polyethylenimine, graphene–PMMA, graphene–PLA, graphene–polyindole, and graphene–polyurethanes, that can eliminate 99.99% of bacterial growth due to their efficient dispersion rate, controlled flow direction, and limited agglomeration effect, depending on their synthesis routes.[53] Mousavi *et al.* discussed graphene-incorporated poly(ethylene glycol) (PEG) and its probable pH-controlled drug release properties with antiviral and antibacterial agents residing in it, prepared by condensation polymerization, chemical vapour deposition, and exfoliation methods.[54] Joshi *et al.* designed electrochemical immunosensors to detect the presence of the influenza virus H1N1, which is a worldwide threat, by using thermally decomposed rGO and shellac. The high stability and reproducibility due to the presence of the phenolic –OH moiety made this method effective for the inhibition of microbial growth. The limits of detection of H1N1 in phosphate-buffered saline and saliva samples were found to be 26 and 33 PFU mL^{-1}, making these GRPNCs successful immune sensors.[55]

1.3.7 Graphene Oxide (GO)-reinforced Polymer Nanocomposites (GORPNCs)

GO with a hexagonal structure is the most common form. The morphology of GO looks like a transparent paper-like structure. The synthesis routes such as epitaxial growth, layer-by-layer assembly, chemical and mechanical exfoliation, and laser-assisted approaches generated an improved version of graphene as GO. GO materials are versatile to use for different purposes as they are oxygen-rich, cost-effective, and easy to prepare and handle. However, the underlying disadvantages such as limited-scale preparation, high production cost, low production yield, and toxicity led scientists to hybridize GO with other nanocomposites such as metal oxides and polymers to improve such issues. It has been reported that when different forms (graphite, graphite oxide, GO, and rGO) were tested, GO showed the lowest toxicity, and exhibited significant microbial inhibition of >80% for planktonic microbial cells, *C. metallidurans*, *B. subtilis*, and *E. coli*.[54] The toxicity and low-scale production of GO-based material can be improved by the development of new GO-reinforced polymer nanocomposites (GORPNCs) with superior composite properties. The other reasons for enhanced antimicrobial activity lie in the size dependency attribute of GO, the smaller sheet size prepared *via* sonication, the oxidative stress caused by the high defect density of GO, and the contact-based interaction between the infected cell and itself.[56] The excellent catalytic, thermal, mechanical, and electrical properties, renewability, biocompatibility, large surface area, small size, and easy preparation of such GORPNCs are favourable attributes.

Kumar *et al.* discussed in detail several GO-functionalized polymers such as PVA–CS–GO against *E. coli* and *B. subtilis via* an Agar Diffusion Assay (ADA) evaluation method with an inhibition effect up to 1.25 and 1.40 mm, PDMS–GO–DMA against *E. coli* and *S. aureus* through a plate count method with a degree of inhibition of ~40%, 2 wt% GO–CS against *E. coli* and *S. aureus via* ADA, and 10 nm sized GO@PEG@AgNPs against *E. coli* and *S. aureus*, which demonstrated that GO@PEG@AgNPs have enhanced antibacterial properties.[57] The outbreak during the recent COVID-19 pandemic caused by the coronavirus was reduced with the practical application of GORPNCs. The use of surface coating strategies with biopolymers such as amino acids, peptides, and naturally rich GO became an important pathway for a sustainable solution to control such dangerous diseases. Technically graphene and its derivatives have the strength to fight against Ebola, dengue, Zika, human coronavirus 229E, and African swine flu by inhibiting their growth through cell wall penetration, ROS and agglomerate formation, and negatively charged interactions with positively charged capsids, but the coordination of GO along with the polymer makes it more effective for antiviral activity. The trapped pores and hydrophobic groups in the polymer make this composite more useful for coronavirus treatment by controlling the growth of viruses.[58]

Cheong *et al.* synthesized the GO-reinforced polymer nanocomposites (GORPNCs) GO–PEG and expanded GO (EGO)–PEG *via* a coupling reaction with new amide bond formation. An antifungal study was performed for *C. albicans* using the *in vitro* cut-well method, showing extensive shrinkage and porosity deformations. The participation of Zn and Cu metals at different weight percentages in these GORPNCs exhibited effective antifungal behaviour in broth dilution assays. A 30 : 70 combination Cu in GO–PEG showed a minimum inhibitory concentration (MIC_{50}) of 185–225 $\mu m\ mL^{-1}$.

The viability results indicated that GO–PEG appeared to be as good an antifungal agent as CuNPs, which can be explained by the high colloidal stability with CuNP concentration increment.[59] The complex preparation of GO–CS GORPNCs has limited demand but still has been extensively used in wound healing treatment depending on the permeability to bacteria, high surface area, prosody-driven drug delivery path, and high exudate absorption capacity.[60] Similar work by Khan *et al.* was performed using numerous bacteria, fungi, and wound-healing treatments. They designed GO and its derivative-based nanocomposites to study the pharmaceutical activities of bacteria (*P. aeruginosa* and *S. aureus*) and fungi (*Saccharomyces cerevisiae* and *Candida utilis*) and the interesting results are dependent on the laser-activated modified surface of GO.[61,62] There are many nerve diseases involving the presence of prion proteins, which cause spongiform encephalopathy with an abnormal mass that often leads to fatal brain infection. This disease is not curable but can be controlled to some extent by the use of GO–PMMA GORPNCs.[63] Their outstanding thermal and biomedical properties, unique surface, and planar structure have led to a great demand for their substantial development as antibacterial and antimicrobial agents influenced by their interesting functionalities and rich oxygen groups.

1.3.8 Reduced Graphene Oxide-reinforced Polymer Nanocomposites (rGORPNCs)

Improved rGO is prepared from GO by chemical reduction in the presence of numerous reducing agents at different temperatures. It is composed of mixed sp^2 and sp^3 carbons and also planar geometries. The catalytic and mechanical activities of rGO have resulted in important biomedical uses.[64] The high surface area, the possibility of unsaturated bond formation, environmentally benign properties, hydrogen–electrostatic interactions, and π–π stacking provide a good platform for conjugation with polymer nanocomposites to produce better pharmaceutical agents such as rGO-reinforced polymer nanocomposites (rGORPNCs). Ouyang *et al.* designed water-soluble poly-L-lysine–reduced graphene oxide–copper nanoparticle (PLL–rGO–CuNP) nanocomposites and studied their antibacterial activity against Gram-negative *E. coli* and Gram-positive *S. aureus*. The composite showed additive antibacterial activity, and the CuNPs on PLL–rGO were more stable than those on polyvinylpyrrolidone, resulting in a long-term additive antibacterial effect depending on the presence of multifunctional rGO and the band gap of Cu metal in the composites.[65]

Studies have suggested that the biocidal activity of rGO is higher than those of GO and graphene, especially for Gram-negative *E. coli* bacteria with an outer membrane, whereas it was less effective towards Gram-positive *S. aureus* without an outer membrane. The toxicity in the case of rGO is also greater owing to the presence of sharp edges during the contact interaction.[66] Vatandost *et al.* designed natural-based rGORPNCs derived from rGO and polyphenol-containing green tea extract and studied the pharmaceutical

Table 1.1 List of CRPNCs, their biomedical uses, test methods, and factors responsible for antimicrobial effects.

Type of CRPNC	Biomedical uses	Target infective agents	Method	Factors responsible for the enhancement of properties to act against diseases	Ref.
Polymer PLA–PLGA	Wound suturing, fixing ligaments, surgical implants	Bacteria	Agar plate diffusion	Network and crosslinked structure	26
Nanosheet GO carbon fibres	Bacteria control, drug delivery, tissue engineering	*E. coli* bacteria	Flow cytometry–colony counting	Concentration of GO	27
CS–PBAT	Packaging of food	*S. aureus*, *B. subtilis*, *S. enteritidis*, and *E. coli* bacteria	Broth dilution	Presence of strong H-bonding interactions	28
PLGA–CS	Antimicrobial coating	*E. coli*, *P. aeruginosa*, and *S. aureus* bacteria	Luria–Bertani broth dilution	Covalent binding, cost-effective, scalable, ability to hinder microbial proliferation	29
CNT–PDMAEMA	Antibacterial film	*S. aureus* and *E. coli* bacteria	Agar plate diffusion	Surface composition, solubility, stimuli, responsiveness, and quaternary ion formation	33
CNT–CS	Antimicrobial templates	*A. niger*, *C. neoformans*, *C. tropicalis*, and *E. faecalis* fungi and *S. epidermidis* and *E. coli* bacteria	Agar well diffusion	Structural modification and multifunctionalized moieties	37
CNT–PLA–CS	Fruit and vegetable storage	*E. coli*, *S. aureus*, and *B. cinerea* bacteria and *Rhizopus* fungus	Colony-forming count	CS concentration and non-uniform distribution	38

Table 1.1 (*Continued*)

Type of CRPNC	Biomedical uses	Target infective agents	Method	Factors responsible for the enhancement of properties to act against diseases	Ref.
CNT–polymer	Drug delivery and tissue regeneration	Prion	Low-temperature sterilization technology	Structural modification, multifunctionalized moieties, divergent synthesis procedure, composition, and reduced self-aggregation	39
Fullerene–polymer	Immunosensor	Bovine leukaemia virus	Tissue culture	Activity, membrane permeability, elasticity, interesting size and shape	44
rGO–polymer	Biosensor	Influenza H1N1	PBS–saliva samples	High stability and reproducibility due to the presence of phenolic –OH moiety	55
GO–polymer	Antiviral coating	Ebola, dengue, Zika, human coronavirus 229E	Blood, mucous, saliva	Cell wall penetration, ROS and agglomerate formation, negatively charged interaction with positively charged capsid	58
GO@polyurethane–siloxane	Antimicrobial film	*E. coli* and *S. aureus* bacteria and *C. albicans* fungus	Colony-forming count	Incorporation of GO nanoplatelets	68

properties. It exhibits a strong bacterial growth resistance for *E. coli* and *S. aureus* with more antioxidant activity.[67] The rapid and easy preparation of rGORPNCPs through several simple synthesis techniques with specific monitoring and recognition capabilities provides a fruitful strategy in the fight against infections from bacteria, fungi, viruses, and other agents.[68] Table 1.1 lists different types of CRPNCs, their biomedical uses, and factors responsible for eradicating infectious diseases.

1.4 Future Perspectives

In view of the present health scenario of the globe, it can be readily hypothesized that the major worldwide health alert in the future will be infections arising as a result of antimicrobial resistivity. The major challenge will be to design sustainable drugs or vaccines that will suppress the antimicrobial resistivity to tackle infectious diseases caused by parasitic microorganisms as discussed above. The proper selection of carbon-based filler materials with a bioactive polymer matrix with additional biocomponents will be highly effective in developing natural or modified carbon-reinforced hybrid polymer composites for dealing with infectious diseases by suppressing antimicrobial resistivity.[69–72] However, some challenges that are yet to be solved can be overcome by a better understanding of the science behind carbon fillers, polymer matrix, their combination, incorporation of other natural molecules, their interaction, and morphology. The mechanisms, pathways, and the participation of CRPNCs in the inhibition of microbe growth by damaging the microbe cell wall and DNA destruction are shown in Figure 1.5.

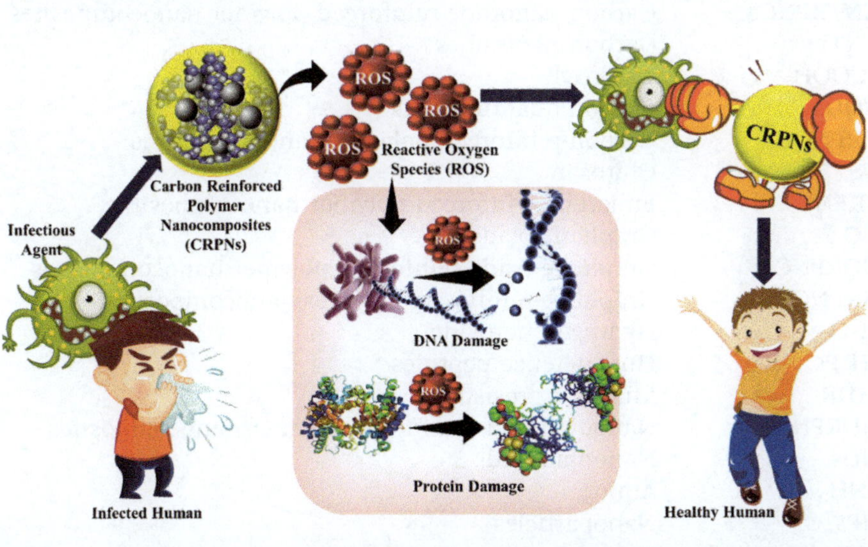

Figure 1.5 Mechanisms and participation of CRPNCs that promote the inhibition of microbe growth.

1.5 Concluding Remarks

In this chapter, a summary of different types of carbon-reinforced polymer nanocomposites and their experimental activity against various infectious diseases is presented. Infection-causing agents such as bacteria, viruses, protozoa, fungi, helminths, and prion proteins are discussed with their host attack and cell damage mechanisms. Different studies have been reported for the preparation of types of CRPNCs and their activity against model pathogens. The chemical structure, nanosize, quantum effect, multifunctional attributes, stability, thermal, mechanical, and electrical behaviour, and interesting morphology have led to the adoption of CRPNCs for different pharmaceutical purposes in recent years. The excellent stability, sensitivity, and accuracy of these nanocomposites have influenced their activity against these life-threatening diseases. Exceptionally, the low-cost preparation methods, superior activities, and environmentally benign and in a few cases green synthetic methods of CRPNCs have motivated the use of these materials against the growth of infective agents and led to a reduction in fatalities.

Abbreviations

ATRP	Atom transfer radical polymerization
BLV	Bovine leukaemia virus
CDRPNCs	Carbon dot-reinforced polymer nanocomposites
CDs	Carbon dots
CFRPNCs	Carbon fibre-reinforced polymer nanocomposites
CJD	Creutzfeldt–Jakob disease
CNTRPNCs	Carbon nanotube-reinforced polymer nanocomposites
CNTs	Carbon nanotubes
–COOH	Carboxyl
CQDs	Carbon quantum dots
CRPNCs	Carbon-reinforced polymer nanocomposites
CS	Chitosan
FRPNCs	Fullerene-reinforced polymer nanocomposites
GO	Graphene oxide
GORPNCs	Graphene oxide-reinforced polymer nanocomposites
GRPNCs	Graphene-reinforced polymer nanocomposites
H_2O_2	Hydrogen peroxide
HDPs	Host defence peptides
MDR	Multidrug resistance
NDRPNCs	Nanodiamond-reinforced polymer nanocomposites
NDs	Nanodiamonds
$-NH_2$	Amine
NPs	Nanoparticles
NTs	Nanotubes
–OH	Hydroxyl

$^{\bullet}OH$	Hydroxyl radical
$^{\bullet}O_2-$	Superoxide
$^1O_2$	Singlet oxygen
PACT	Photodynamic antimicrobial chemotherapy
PAN	Polyacrylonitrile
PBAT	Poly(butylene adipate-*co*-terephthalate)
PCL	Polycaprolactone
PDMAEMA	Poly[2-(dimethylamino)ethyl methacrylate]
PE	Polyethylene
PEG	Poly(ethylene glycol)
PLA	Poly(lactic acid)
PLGA	Poly(lactide-*co*-glycolide)
PLL	Poly-L-lysine
PMMA	Poly(methyl 2-methylpropenoate)
PPY	Polypyrrole
PVA	Poly(vinyl alcohol)
PVK	Poly(*N*-vinylcarbazole)
rGO	Reduced graphene oxide
rGORPNCs	Reduced graphene oxide-reinforced polymer nanocomposites
ROS	Reactive oxygen species

Conflicts of Interest

The authors declare that there is no conflict of interest in publishing this chapter.

Acknowledgements

The authors acknowledge research support grants awarded by University Grants Commission (UGC) India and a DS Kothari Post-Doctoral fellowship to Dr Biswajit Parhi. Dr Debasrita Bharatiya acknowledges Veer Surendra Sai University of Technology (VSSUT), Burla, India, for its support with research facilities as a Research Associate (RA).

References

1. A. Hiremath, A. A. Murthy, S. Thipperudrappa and K. N. Bharath, *Cogent Eng.*, 2021, **8**(1), 1991229.
2. A. Kausar, *J. Plast. Film Sheeting*, 2020, **36**, 94.
3. R. Gogoi, A. K. Maurya and G. Manik, *Compos., Part C: Open Access*, 2022, **8**, 100279.
4. M. M. Harussani, S. M. Sapuan, G. Nadeem, T. Rafin and W. Kirubaanand, *Def. Technol.*, 2022, **18**, 1281.
5. A. P. Mouritz, J. Galos, D. P. Linklater, R. B. Ladani, E. Kandare, R. J. Crawford and E. P. Ivanova, *Nano Sel.*, 2021, **2**, 2061.

6. H. Hara and S. Sakaguchi, *Int. J. Mol. Sci.*, 2021, **22**, 12439.
7. S. H. Kamarudin, M. Rayung, F. Abu, S. Ahmad, F. Fadil, A. A. Karim, M. N. Norizan, N. Sarifuddin, M. S. Z. M. Desa, M. S. M. Basri, H. Samsudin and L. C. Abdullah, *Polymers*, 2022, **14**, 174.
8. I. O. Oladele, T. F. Omotosho and A. A. Adediran, *Int. J. Polym. Sci.*, 2020, 8834518.
9. J. V. Lim, S. T. Bee, L. T. Sin, C. T. Ratnam and Z. A. A. Hamid, *Polymers*, 2021, **13**, 3547.
10. X. Li, W. Xu, Y. Xin, J. Yuan, Y. Ji, S. Chu, J. Liu and Q. Luo, *Polymers*, 2021, **13**, 513.
11. A. Reta, A. B. Kifilie and A. Mengist, *Adv. Prev. Med.*, 2019, 4380309.
12. P. Manohar, B. Loh, R. Nachimuthu and X. Hua, *Front. Med.*, 2020, **7**, 420.
13. S. Doron and S. L. Gorbach, in *International encyclopedia of public health*, 2008, p. 273.
14. Y. Li, Y. Gao, X. Niu, Y. Wu, Y. Du, Y. Yang, R. Qi, H. Chen, X. Gao, B. Song and X. Guan, *Front. Cell. Infect. Microbiol.*, 2020, **10**, 553648.
15. A. A. Shishido, M. Mathew and J. W. Baddley, *Curr. Fungal Infect. Rep.*, 2022, **16**, 87.
16. T. Marinelli and S. van Hal, *Curr. Fungal Infect. Rep.*, 2022, **16**, 75.
17. C. R. Triggle, D. Bansal, H. Ding, M. M. Islam, E. A. B. A. Farag, H. A. Hadi and A. A. Sultan, *Front. Immunol.*, 2021, **12**, 631139.
18. M. Javanian, M. Barary, S. Ghebrehewet, V. Koppolu, V. K. R. Vasigala and S. Ebrahimpour, *J. Med. Virol.*, 2021, **93**, 4638.
19. R. Martínez-Girón, J. G. Esteban, A. Ribas and L. Doganci, *Eur. Respir. J.*, 2008, **32**, 1354.
20. L. Frías, D. Leles and A. Araújo, *Mem. Inst. Oswaldo Cruz*, 2013, **108**, 1.
21. A. Hall, G. Hewitt, V. Tuffrey and N. De Silva, *Matern. Child Nutr.*, 2008, **4**, 118.
22. B. H. Kumar, K. Jain and R. Jain, *Med. J. Armed Forces India*, 2014, **70**, 144.
23. H. Hara, J. Chida, K. Uchiyama, A. D. Pasiana, E. Takahashi, H. Kido and S. Sakaguchi, *Sci. Rep.*, 2021, **11**, 10109.
24. I. V. Baskakov and L. Breydo, *Biochim. Biophys. Acta, Mol. Basis Dis.*, 2007, **1772**, 692.
25. V. K. Srivastava, P. K. Jain, P. Kumar, A. Pegoretti and C. R. Bowen, *J. Mater. Eng. Perform.*, 2020, **29**, 4162.
26. D. K. Rajak, D. D. Pagar, P. L. Menezes and E. Linul, *Polymers*, 2019, **11**, 1667.
27. R. K. Matharu, T. A. Tabish, T. Trakoolwilaiwan, J. Mansfield, J. Moger, T. Wu, C. Lourenço, B. Chen, L. Ciric, I. P. Parkin and M. Edirisinghe, *J. Colloid Interface Sci.*, 2020, **571**, 239.
28. A. M. Díez-Pascual and A. L. Díez-Vicente, *RSC Adv.*, 2015, **5**, 93095.
29. A. F. de Faria, F. Perreault, E. Shaulsky, L. H. Arias Chavez and M. Elimelech, *ACS Appl. Mater. Interfaces*, 2015, 7, 12751.
30. R. Scaffaro, A. Maio, L. Botta, E. F. Gulino and D. Gulli, *Eur. Polym. J.*, 2019, **110**, 221.

31. L. Giraud, A. Tourrette and E. Flahaut, *Carbon*, 2021, **182**, 463.
32. R. Teixeira-Santos, M. Gomes, L. C. Gomes and F. J. Mergulhão, *iScience*, 2021, **24**, 102001.
33. Y. T. Joo, K. H. Jung, M. J. Kim and Y. Kim, *J. Appl. Polym. Sci.*, 2013, **127**, 1508.
34. C. I. Idumah, M. Zurina, J. Ogbu, J. U. Ndem and E. C. Igba, *Compos. Interfaces*, 2020, **27**, 1.
35. M. A. Saleemi, Y. L. Kong, P. V. C. Yong and E. H. Wong, *Adv. Pharm. Bull.*, 2022, **12**, 449.
36. Z. Zhu, Z. Wang, S. Li and X. Yuan, *J. Biomed. Mater. Res., Part A*, 2019, **107**, 445.
37. N. A. Mohamed and N. A. Abd El-Ghany, *Int. J. Biol. Macromol.*, 2018, **115**, 651.
38. Y. Liu, S. Wang, W. Lan and W. Qin, *Int. J. Biol. Macromol.*, 2019, **121**, 1329.
39. C. Xiang, Y. Zhang, W. Guo and X. J. Liang, *Acta Pharm. Sin. B*, 2020, **10**, 239.
40. J. Wehling, R. Dringen, R. N. Zare, M. Maas and K. Rezwan, *ACS Nano*, 2014, **8**, 6475.
41. A. Rehman, S. Houshyar and X. Wang, *J. Biomed. Mater. Res., Part A*, 2020, **108**, 906.
42. L. Bacakova, A. Broz, J. Liskova, L. Stankova, S. Potocky and A. Kromka, *Diamond and Carbon Composites and Nanocomposites*, 2016, p. 59.
43. Q. Xie, C. Liu, X. Lin, C. Ma and G. Zhang, *ACS Appl. Polym. Mater.*, 2020, **2**, 3181.
44. I. M. Lukavenko, V. V. Andryushchenko and O. V. Yazykov, The laser-induced coagulation method of biological tissues, *Springer Proceedings in Physics*, 2020, vol. 244, p. 121.
45. A. M. Díez-Pascual, *Int. J. Mol. Sci.*, 2021, **22**, 10511.
46. M. Ghirardello, J. Ramos-Soriano and M. C. Galan, *Nanomaterials*, 2021, **11**, 1877.
47. Z. Feng, K. H. Adolfsson, Y. Xu, H. Fang, M. Hakkarainen and M. Wu, *Sustainable Mater. Technol.*, 2021, **29**, e00304.
48. X. Wu, K. Abbas, Y. Yang, Z. Li, A. C. Tedesco and H. Bi, *Pharmaceuticals*, 2022, **15**, 487.
49. Y. Xue, C. Liu, G. Andrews, J. Wang and Y. Ge, *Nano Convergence*, 2022, **9**, 15.
50. M. Maruthapandi, A. P. Nagvenkar, I. Perelshtein and A. Gedanken, *ACS Appl. Polym. Mater.*, 2019, **1**, 1181.
51. A. Qadir, T. K. Le, M. Malik, K. A. Amedome Min-Dianey, I. Saeed, Y. Yu, J. R. Choi and P. V. Pham, *RSC Adv.*, 2021, **11**, 23860.
52. K. Hossain, M. Rafatullah, S. Z. Abbas, A. Ahmad, N. Ismail and A. Y. Maruthi, *Antimicrobial activity of graphene-based nanomaterials*, Elsevier Inc., 2019.
53. S. Pandit, K. Gaska, R. Kádár and I. Mijakovic, *Chem. Phys. Chem.*, 2021, **22**, 250.

54. S. M. Mousavi, F. W. Low, S. A. Hashemi, C. W. Lai, Y. Ghasemi, S. Soroshnia, A. Savardashtaki, A. Babapoor, N. Pynadathu Rumjit, S. M. Goh, N. Amin and S. K. Tiong, *Artif. Cells, Nanomed., Biotechnol.*, 2020, **48**, 1189.
55. S. R. Joshi, A. Sharma, G. H. Kim and J. Jang, *Mater. Sci. Eng., C*, 2020, **108**, 110465.
56. F. Perreault, A. Fonseca de Faria, S. Nejati and M. Elimelech, *ACS Nano*, 2015, **9**, 7226.
57. P. Kumar, P. Huo, R. Zhang and B. Liu, *Nanomaterials*, 2019, **9**, 737.
58. K. Shirvanimoghaddam, M. K. Akbari, R. Yadav, A. K. Al-Tamimi and M. Naebe, *APL Mater.*, 2021, **9**, 031112.
59. Y. K. Cheong, M. P. Arce, A. Benito, D. Chen, N. L. Crisóstomo, L. V. Kerai, G. Rodríguez, J. L. Valverde, M. Vadalia, A. Cerpa-Naranjo and G. Ren, *Nanomaterials*, 2020, **10**, 819.
60. W. Feng and Z. Wang, *iScience*, 2022, **25**, 103629.
61. M. Shahnawaz Khan, H. N. Abdelhamid and H. F. Wu, *Colloids Surf., B*, 2015, **127**, 281.
62. L. Shi, J. Chen, L. Teng, L. Wang, G. Zhu, S. Liu, Z. Luo, X. Shi, Y. Wang and L. Ren, *Small*, 2016, **12**, 4165.
63. M. Barani, M. Mukhtar, A. Rahdar, G. Sargazi, A. Thysiadou and G. Z. Kyzas, *Molecules*, 2021, **26**, 186.
64. M. H. Kahsay, N. Belachew, A. Tadesse and K. Basavaiah, *RSC Adv.*, 2020, **10**, 34916.
65. Y. Ouyang, X. Cai, Q. S. Shi, L. Liu, D. Wan, S. Tan and Y. Ouyang, *Colloids Surf., B*, 2013, **107**, 107–114.
66. A. M. Díez-Pascual and J. A. Luceño-Sánchez, *Polymers*, 2021, **13**, 2105.
67. E. Vatandost, A. G. H. Saraei, F. Chekin, S. N. Raeisi and S. A. Shahidi, *ChemistrySelect*, 2020, **5**, 10401–10406.
68. H. Yu, W. Guo, X. Lu, H. Xu, Q. Yang, J. Tan and W. Zhang, *Food Control*, 2021, **127**, 108117.
69. B. Parhi, D. Bharatiya and S. K. Swain, *Saudi Pharm. J.*, 2020, **28**, 1719.
70. D. Bharatiya, S. Patra, B. Parhi and S. K. Swain, *Int. J. Polym. Mater. Polym. Biomater.*, 2021, **1**, 1.
71. D. Bharatiya, B. Parhi and S. K. Swain, *J. Appl. Polym. Sci.*, 2023, **140**, e53389.
72. B. Parhi, D. Bharatiya and S. K. Swain, *J. Appl. Polym. Sci.*, 2022, **139**, 52116.

CHAPTER 2

Nanostrategies for Infectious Pulmonary Diseases: Current Progress and Future Prospects

SAMEER NADAF,*[a] POPAT KUMBHAR,[b]
DURGACHARAN BHAGWAT[c] AND SHAILENDRA GURAV*[d]

[a] Bharati Vidyapeeth College of Pharmacy, Palus, Maharashtra 416310, India; [b] Tatyasaheb Kore College of Pharmacy, Warananagar, Taluka Panhala, District Kolhapur, Maharashtra 416113, India; [c] Bharati Vidyapeeth College of Pharmacy, Kolhapur, Maharashtra 416013, India; [d] Department of Pharmacognosy, Goa College of Pharmacy, Goa University, Panaji, Goa 403001, India
*Emails: sam.nadaf@rediffmail.com; shailendra.gurav@nic.in

2.1 Introduction to Infectious Diseases

Microbial diseases are spreading throughout the world at an alarming rate, and their prevalence will only continue to rise. Colonialism, enslavement, and wars all contributed to the global spread of deadly diseases in early modern periods, with disastrous effects. Prior to the development of vaccines, some common diseases such as tuberculosis, poliomyelitis, smallpox, and diphtheria caused multiple deaths. Animal diseases such as rinderpest (cattle plague), now eradicated, proliferated widely across trade routes and with roving armies, having catastrophic effects on livestock and reliant large populations.[1,2] Myriad severe infectious disease outbreaks have been observed in the 21st century. The COVID-19 pandemic is one of several diseases that have wreaked havoc on people's lives and way of life all across

Nanoscience & Nanotechnology Series No. 62
Smart Nanomaterials for Infectious Diseases
Edited by Suvardhan Kanchi, Nidhi Sharotri, Rajasekhar Chokkareddy, Deepali Sharma and Falah H. Hussein

Published by the Royal Society of Chemistry, www.rsc.org

the world. Noticeable morbidity and mortality have been observed in the last two decades as a result of several outbreaks including COVID-19, swine flu, Ebola virus, and Zika virus. Each of these outbreaks crossed international boundaries to infect humans in nations all over the world.[2]

Since the 1950s, emerging and re-emerging infectious illnesses have plagued healthcare systems, and evolving pathogens are increasingly viewed as a significant microbiological problem for public health.[3] With increased cleanliness and the creation of antimicrobials and vaccinations, infectious disease prevention, control, and even elimination have advanced significantly. Infectious diseases continue to be the largest cause of disease burden worldwide, with high rates of morbidity and mortality, particularly in poorer nations. In addition, over the last 30 years, there have been threats of newer diseases resulting from the growth and adaptation of bacteria, and also the resurgence of old diseases as a result of increased antibiotic resistance and the capability to spread to new geographic areas.[4]

In the course of an infection, pathogens engage closely with infected cells in a number of different ways. The equilibrium between bacterial virulence or infectivity and host immunity determines an organism's ultimate pathogenicity. Phagocytic cells and a healthy immune system both contribute to the host's ability to withstand infections. Once bacteria have entered tissues and come into contact with macrophages, an inflammatory response occurs as a result of signalling eicosanoids and complement mechanisms.[5] The prevention of postoperative infections is always crucial, but it is a complicated issue that necessitates a variety of preventive measures. These recommendations are an important tool for safeguarding patients.[6]

Bacterial infections that are developing or redeveloping pose a major risk to both human and animal health. Extracellular bacteria live freely, whereas intracellular bacteria proliferate inside eukaryotic host cells. Extracellular bacteria usually leave the body after being taken up by phagocytes such as macrophages (M), neutrophils, and dendritic cells (DCs) and being broken down in lysosomal compartments within these cells. Nevertheless, these bacteria have the potential to be harmful, for instance by releasing toxic substances or by excessively inducing inflammatory responses that harm tissues and cells. These intracellular bacteria, which cause significant morbidity and mortality in humans, include *Salmonella enterica*, *Escherichia coli*, *Staphylococcus aureus*, *Rickettsia massiliae*, *Chlamydia* sp., *Brucella abortus*, *Mycobacterium tuberculosis*, and *Listeria monocytogenes*. *Mycobacterium*, *Brucella*, methicillin-resistant *Staphylococcus aureus* (MRSA), *Shigella*, *Listeria*, and *Salmonella* are just a few of the pathogens that can invade and remain in mammalian host cells, particularly macrophages, where they multiply and create a repository, leading to persistent and recurring infections.[7,8] Several strategies are used by intracellular bacteria to avoid phagolysosomal destruction, allowing them to live and grow inside host cells. The bacteria proliferate either freely in the cytosol or in specific cellular compartments, depending on the type.[7,8]

Owing to their low toxicity, infections with intracellular bacteria take longer to develop. Unfortunately, these organisms live inside cells, making it difficult to treat completely the infections that they cause. These characteristics make intracellular bacterial infections more likely than acute infections to result in chronic or recurrent illness. Cell-mediated immunity is crucial for battling intracellular bacterial infections. Those subjects with impaired immune systems who are unable to mount cell-mediated immune responses are more susceptible to contracting an infection from these bacteria.[9,10] Optimal protective immunity typically comprises a blend of many immune system components to coordinate cellular responses that result in protection against intracellular microorganisms. To be more precise, to resist facultative intracellular bacterial infections, acquired cell-mediated immunity and T lymphocyte activation of macrophages are necessary.[7,10,11]

The World Health Organization (WHO) ranks respiratory infectious diseases first in terms of the number of years lost due to death or disability as a result of disease burden. After heart and encephalic vascular diseases, lower respiratory tract infections are the third largest cause of death globally, but in nations with low economic standing they rank first.[12] This chapter overviews respiratory infections and illuminates several barriers of bacterial infections. The features and benefits of nanosized materials for diagnosing various lung diseases are analysed. The current trends of different nanosystems such as lipid-based vesicles, polymeric micelles, solid lipid nanoparticles (NPs), nanostructured lipid carriers, dendrimers, *etc.*, for treating several life-threatening lung infections, including COVID-19, are reviewed and discussed with emphasis on their advantages and limitations.

2.2 Pulmonary Infectious Diseases

Acute respiratory infections have been cited as a leading factor in ambulatory consultations, hospitalizations, and fatalities in numerous regional and global estimations. Bronchitis, bronchiolitis, and pneumonia are lower respiratory tract illnesses that are the second most common reason for antibiotic prescriptions. Most of these diseases are brought on by viruses, and their typical natural course is recovery on its own.[13] Viral, bacterial, fungal, and parasitic infections of the parenchyma can result in lung infections.

Across the world, several hundred million children and adults suffer from lung disorders brought on by microbial infections. Antibiotic resistance is mostly caused through the treatment of lung infections in Western cultures. Microorganisms in the environment provide a persistent threat to the lungs and airways. An adult who is sedentary can breathe in more than 10 000 L of air each day, which can cover 40–80 m^2 of lung surface. This air can contain up to 100 000 bacteria per litre in urban areas, some of which can enter the thoracic airways and beyond. The gut, in comparison, has a surface area of 30 m^2, and stomach acid restricts microbial ingression. As a result, it is not unexpected that lung infections most frequently impact people's health.[14]

Approximately 4 million people die each year from respiratory infections, which are the main cause of death in underdeveloped nations. Pneumonia, which is typically linked to a bacterial infection in children under the age of 5 years, is responsible for 1.3 million fatalities per year and 18% of all-cause mortality. More adults die from pneumonia than from HIV or malaria.[14–16]

2.2.1 Bacterial Infections

Owing to their limited ability to remove germs from their respiratory tracts, patients with chronic respiratory disease are more likely to develop persistent, incapacitating respiratory infections. Because resistant biofilms frequently form, such infections can turn into chronic, lifelong problems that are challenging to treat with antibiotics. The microbial populations in the upper and lower respiratory tracts alter as various respiratory disorders develop, frequently becoming less diverse and dysbiotic and associated with increasing patient morbidity. A common set of respiratory pathogens, including *Haemophilus influenzae*, *Streptococcus pneumoniae*, *S. aureus*, *Pseudomonas aeruginosa*, and *Moraxella catarrhalis*, are present in people with chronic respiratory disease.[17–19] Bacterial infections have been linked to serious, potentially fatal lung illnesses (see Figure 2.1).

The lower respiratory tract illness pneumonia is brought on by the body's inability to remove microorganisms from the lower airways and alveoli. The production of cytokines and local inflammatory markers leads to the formation of white blood cells and fluid congestion, which results in pus, further damaging the lungs.[20] Pneumonia is mainly caused by *S. pneumoniae*.

One of the main pathogens responsible for one of the most deadly infectious diseases, tuberculosis, is *Mycobacterium tuberculosis*. It mostly affects the lungs and it occurs due to the adhesion of *M. tuberculosis*, which is found in aerosol droplets, to the surfaces of the lung alveoli.[21] The proteobacteria *Moraxella* spp. and *Neisseria* spp. can also less frequently lead to acute lung infections. Together with *M. catarrhalis* and *S. pneumoniae*, unencapsulated non-typeable *Haemophilus influenzae* (NTHI) strains are the primary causes of recurrent airway infection in chronic obstructive pulmonary disease (COPD).[22] It is possible that *Chlamydia* or *Mycoplasma pneumoniae* infection occurs before the onset of asthma, according to some data.[23]

The lungs, digestive system, and other body organs are mainly impacted by cystic fibrosis (CF), a genetic disorder. Common human infections including *S. aureus* and *H. influenzae* as well as various opportunistic pathogens, the most significant of which is *P. aeruginosa*, are among the bacterial species most frequently linked to respiratory tract infection in CF.[24,25]

Non-CF bronchiectasis-related infections are a large category of illnesses that cause respiratory tract damage, increased mucus discharges, and infections that permanently widen the airways. Infections associated with bronchiectasis that are unrelated to CF have been linked to pathogens such as *Veilonella* spp., *Prevotella* spp., and *Neisseria* spp.[26]

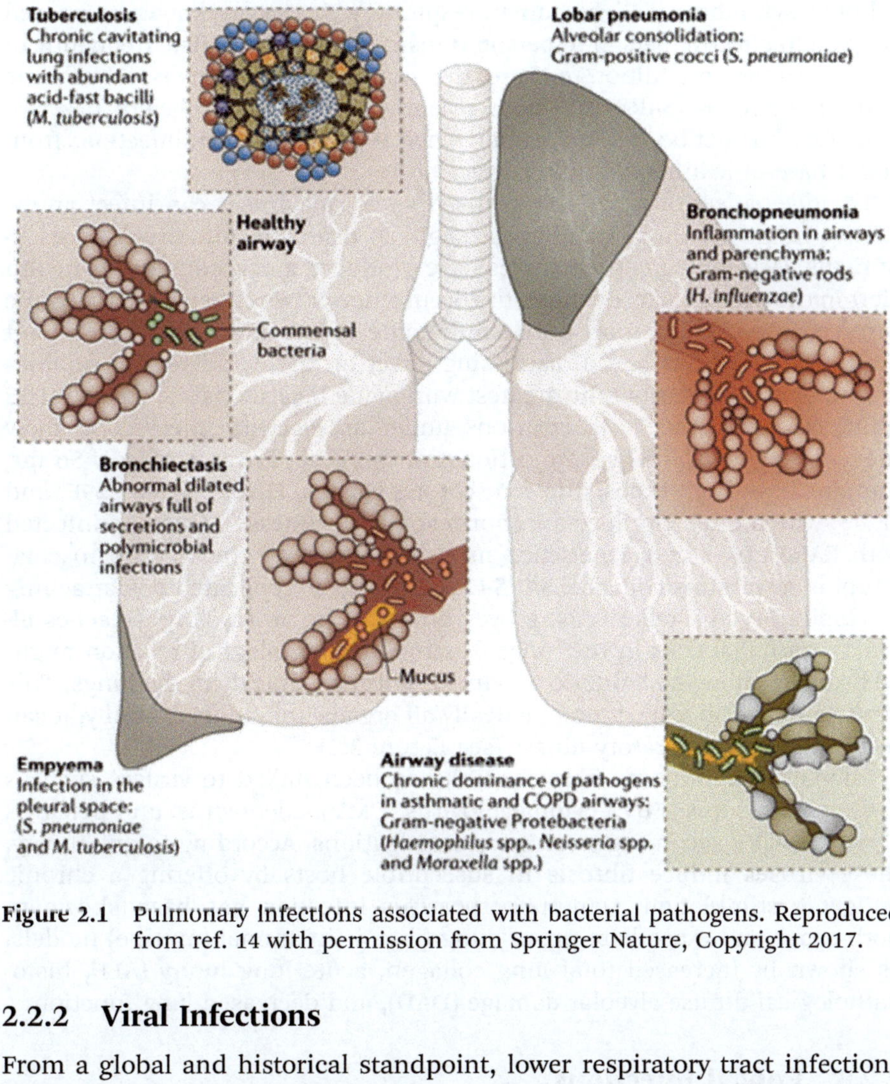

Figure 2.1 Pulmonary infections associated with bacterial pathogens. Reproduced from ref. 14 with permission from Springer Nature, Copyright 2017.

2.2.2 Viral Infections

From a global and historical standpoint, lower respiratory tract infections have a larger scope and impact than any other infectious condition, and viral pneumonias have become established as some of the deadliest and most dramatic human diseases.[27] In the modern era, viral respiratory infections (VRIs) have attracted huge attention owing to the unexpected outbreak and global spread of COVID-19.[28] The most frequent cause of VRIs is a virus, specifically one of the Orthomyxoviridae and Paramyxoviridae, Picornaviridae, coronaviruses, or adenovirus families.[29] Viral infections are mainly observed in the upper airways, mainly the nose, paranasal sinuses, pharynx, and larynx. The influenza virus, adenovirus, rhinovirus, enterovirus, parainfluenza virus, respiratory syncytial virus, metapneumovirus, and coronavirus are among the pathogens that cause viral upper respiratory infections.[30] A syndrome of rhinovirus infections, often known as rhinosinusitis, is characterized by cold-like symptoms.[31] Eight human respiratory viruses are known to be well adapted to effective person-to-person transmission and circulate frequently in all age groups. In addition to these, the avian influenza virus H5N1 and the SARS coronavirus (SARS-CoV) have recently become public health hazards. SARS-CoV has not been in circulation since 2003, and human infections from H5N1 have only infrequently occurred.[32]

The diverse set of viruses recognized as coronaviruses can infect an extensive range of animals and cause slight to deadly respiratory diseases in humans. Developing coronaviruses have produced a new community health alert in the 21st century owing to the occurrence of two extremely pathogenic coronaviruses.[33] With more than 6 million deaths worldwide as of the end of 2019, COVID-19 has had a distressing effect on the world's demographics and currently represents the greatest worldwide health crisis since the 1918 influenza pandemic.[34] Observations under an electron microscope show the presence of spike-like projections on the coronavirus surfaces. So far, humans have been exposed to four corona viruses, HKU1, NL63, 229E, and OC43, which often produce a minor respiratory illness.[35] Humans infected with SARS-CoV-2 can experience mild symptoms all the way up to catastrophic respiratory collapse. SARS-CoV-2 begins to replicate upon attaching to respiratory epithelial cells, moves down to the airways, and reaches alveolar epithelial cells in the lungs. A strong immunological reaction might be brought on by the SARS-CoV-2 virus replicating quickly in the lungs. This virus can have an impact on practically all organs and, most critically, it can result in acute respiratory illness (see Figure 2.2).[36]

Idiopathic pulmonary fibrosis (IPF) has been linked to viruses such as Epstein–Barr virus (EBV), cytomegalovirus (CMV), adenovirus, and hepatitis C virus (HCV), according to several investigations. According to one theory, these viruses induce fibrosis in susceptible hosts by offering a chronic antigenic stimulation. Gamma-herpesvirus infection has been shown to both cause and exacerbate pre-existing fibrosis in animal (murine) models, as shown by increased total lung collagen, acute lung injury (ALI), histopathological diffuse alveolar damage (DAD), and decreased lung function.[37]

2.2.3 Fungal Infections

The fungi that cause pulmonary fungal infections can be endemic or opportunistic. The endemic mycoses are a varied group of fungal organisms that have a number of traits in common, such as the capacity to exhibit temperature dimorphism, the capacity to infect otherwise healthy humans, and the propensity to live in particular geographic areas. Children with weakened immune systems or changed microbiota or those with impaired integumentary barriers are frequently affected by fungi that cause opportunistic illnesses. However, the fungi that cause opportunistic infections can occasionally arise in healthy hosts.[38]

Pulmonary mycoses are categorized into three groups, as depicted in Figure 2.3.[39]

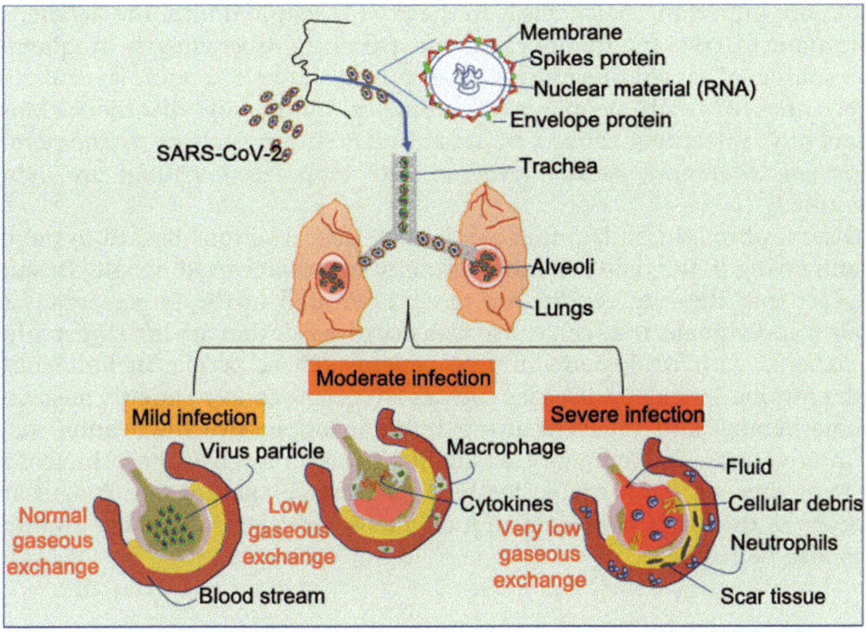

Figure 2.2 Effect of SARS-CoV-2 infection on alveolar cells depending on the severity of the infection. Reproduced from ref. 36, https://doi.org/10.3390/cancers12123822, under the terms of the CC BY 4.0 license https://creativecommons.org/licenses/by/4.0/.

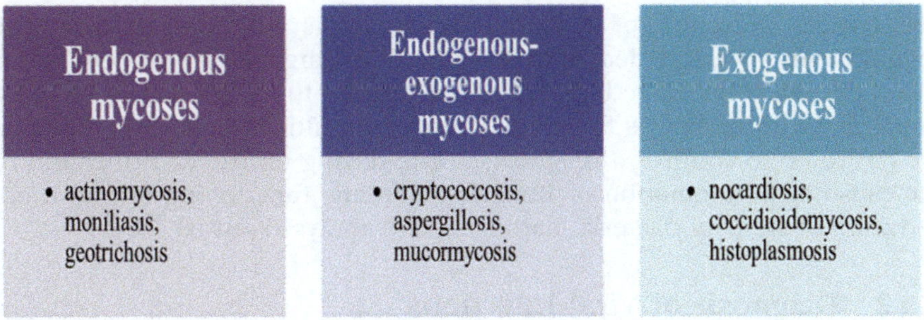

Figure 2.3 Types of pulmonary mycoses.

Based on environmental and other variables that support the growth of these organisms in the soil, the endemic mycoses are geographically restricted. Patients with histoplasmosis and blastomycosis are primarily found in the Mississippi and Ohio River valleys, whereas those with coccidioidomycosis are mostly found in the American southwest desert. Particularly in people with cellular immunological deficiency, cryptococcosis can also manifest as a pulmonary illness. As opportunistic disease causes in immunocompromised individuals, these mycoses are becoming more and more important, particularly in those with acquired immune deficiency syndrome (AIDS). In granulocytopenic patients, *Aspergillus* is frequently the source of a significant invasive fungal infection, and it can also infect the lungs in people with atopy or pulmonary disorders. More frequently, infections caused by less virulent fungi such as *Trichosporon*, *Fusarium*, *Alternaria*, *Pseudallescheria*, and dematiaceous fungi are being identified.[40]

Diseases brought on by fungi continue to pose a serious hazard to public health owing to the rising number of immunocompromised people. Despite the fact that these infections are rarely identified in the target organs of healthy individuals, they can cause invasive diseases that are life threatening in patients with weakened immune systems. These people include those with immunodeficiency diseases such as HIV/AIDS, cancer patients receiving chemotherapy, and those receiving immunosuppressive medication such as in bone marrow/stem cell transplants. Invasive mycoses have increased in frequency with infectious deaths as a result of pathogenic fungus infections in the lungs, especially in individuals with significant host immune response deficiencies.[41]

2.3 Diagnosis

Early diagnosis of diseases is imperative to control the spread of diseases and deadly outcomes thereof.

2.3.1 Diagnosis of Bacterial Infections

The correct detection of bacterial infections is essential to limit the requirement for unneeded antibiotics and to target the right course of treatment. Bacterial infections are mostly related to inflammation or some impairment of the body's function. Therefore, multiple diagnostic methods are desirable to confirm a diagnosis. Tests relying on the identification of the cause of inflammation or impaired systemic functioning and also observation of signs of bacterial pathogens are always essential.[42]

2.3.2 Diagnosis of Viral Infections

The majority of disorders that require medical care globally, *i.e.* ~60%, are viral infections. Unfortunately, there was a dearth of medical supplies during epidemics to treat all individuals with viral illnesses.[43] Correct viral infection identification improves the clinician's capacity to choose the best course of treatment for patients, monitor disease development, and avoid antibiotic overuse.[44] There are numerous molecular and immunological diagnostic techniques available to find human viral infections. Rapid viral identification in patients' samples is made possible by molecular diagnostic techniques, which are reasonably cheap, quite sensitive, and very targeted methods. Human viral infections have been detected and the epidemiology of these illnesses has been widely studied using immunologically based methods. In clinical samples, they can identify viral antigens or antiviral antibodies.[45]

2.3.3 Diagnosis of Fungal Infections

With a large worldwide illness burden, chronic pulmonary aspergillosis exacerbates pre-existing lung disorders, particularly those linked to cavities or bullae.[46] The use of rapid antigen detection methods in blood and bronchoalveolar lavage fluid has improved the diagnosis of invasive pulmonary aspergillosis, histoplasmosis, and blastomycosis.[47] Compared with culture or serum testing, *Aspergillus* antigen testing and polymerase chain reaction (PCR) testing are more subtle. A new lateral flow antigen testing instrument might offer quick bedside diagnosis of aspergillosis.[46]

2.4 Current Therapies

Depending on the severity of the illness and the situation, several treatment plans are advised. It is imperative to treat pathogenic bacteria in relation to serious illnesses.[48] Depending on the harshness of the infection and the type of bacteria involved, antibiotics are frequently the first line of treatment for bacterial infections. Patients are given prescriptions for drugs including tetracyclines, penicillins, and cephalosporins. Owing to a few clear benefits, inhalation treatments are appealing options for the treatment of respiratory illnesses.[49] Inhalation products containing tobramycin and dry powder are available to treat CF. Formulations including aztreonam and amikacin sulfate are also beneficial.[49]

Remdesivir, lopinavir–ritonavir, and interferons are just a few of the safe antiviral medications that have been used in the treatment of coronavirus infections such as severe acute respiratory syndrome (SARS) and Middle East respiratory syndrome (MERS). Remdesivir (GS-5734), a nucleoside analogue with antiviral activity against several viral families, including the Filoviridae, Pneumoviridae, Paramyxoviridae, and Coronaviridae, has been tested.[50]

The development of the novel azoles voriconazole and posaconazole, which have broad antifungal activity, has improved the treatment of angioinvasive moulds. Amphotericin B is used less frequently and, when it is, it is frequently administered as a lipid formulation to reduce toxicity. Because they interfere with the metabolism of the fungal cell wall, a structure not shared by human cells, the newest agents, the echinocandins, are particularly safe.[47]

2.5 Limitations of Current Therapies

The following factors have been found to be limiting factors for the effectiveness of current antibacterial and antiviral therapy over time.

2.5.1 Barriers

It can be challenging to administer medications to only the targeted tissues or cells. The biology underpinning the molecular, cellular, and tissue-level barriers that hinder the delivery process is a major factor in this.[51] The distribution of medications to the target site is impacted by a variety of biological obstacles (see Figure 2.4). Depending on the normal or pathological conditions in the respiratory tract, the characteristics of the barriers may change. It is essential to be aware of these biological barriers and the challenges they provide in order to improve drug delivery and, consequently, therapy.

Mucus acts as a natural protective barrier in the respiratory tract. The mucus can trap foreign objects and pathogens that are inhaled *via* steric and adhesive trapping, possibly followed by elimination *via* ciliary movement.[52] Hence barrier qualities need to be studied in order to direct medications to particular areas.

2.5.2 Drug Resistance

The use of antibiotics in treating, preventing, and controlling bacterial infections is a crucial part of modern medicine. Many strategies have been evolved by bacteria to counteract the effects of antibiotics. There are three primary ways by which bacteria resist the effects of antibiotics: (1) enzyme-based antibiotic hydrolysis, (2) modifications to porins that alter membrane permeability, and (3) efflux pump-mediated antibiotic outflow (Figure 2.5).[53] Moreover, bacteria alter the target's structure or reduce an antibiotic's affinity for attaching to the target, inhibiting attachment or binding.[54]

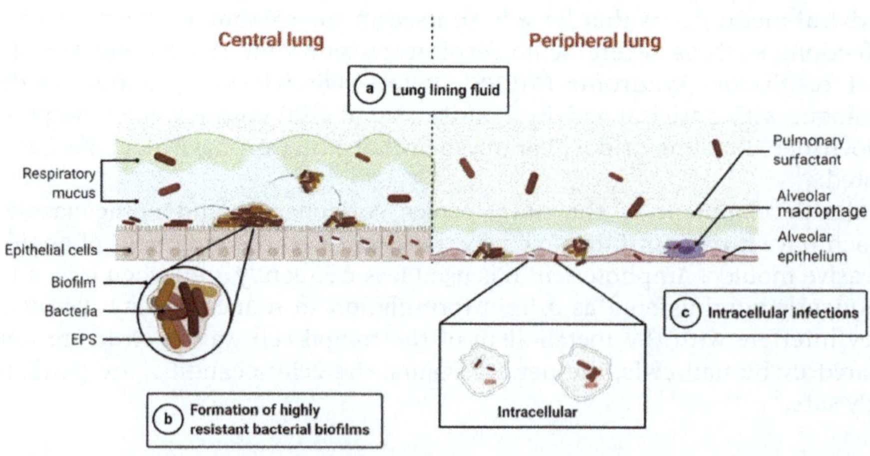

Figure 2.4 Illustration of biological barriers. (a) Lung lining fluid; (b) formation of highly resistant bacterial biofilms; (c) intracellular infections. Reproduced from ref. 26 with permission from Springer Nature, Copyright 2021.

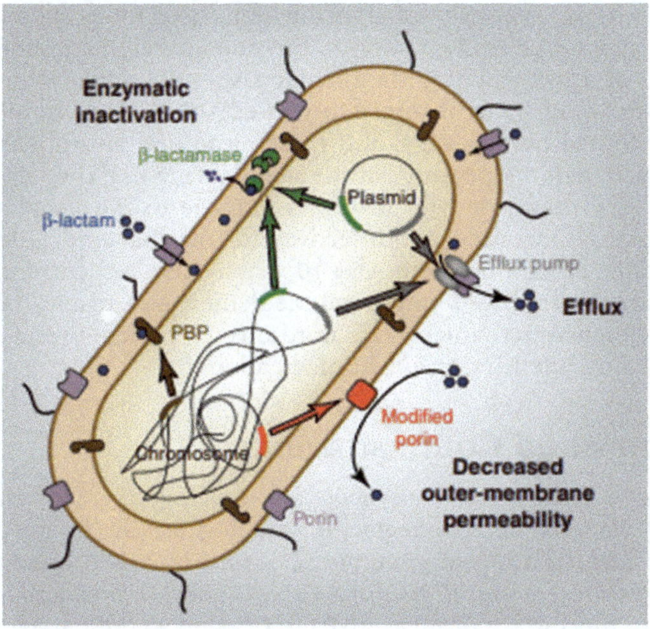

Figure 2.5 Primary mechanisms of β-lactam resistance in Enterobacteriaceae. Reproduced from ref. 53 with permission from Elsevier, Copyright 2012.

Moreover, the microbes create a special structure known as a biofilm in order to survive against adverse environmental factors such as recognized antibiotics. Because they are based on self-produced extracellular polymeric compounds, biofilms can withstand the body's immune response while being resistant to antibiotics, disinfectants, and phagocytosis [exopolysaccharides (EPS)]. The three-dimensional organization of bacteria in a biofilm, the geometry of the matrix, and the antibiotic resistance of the biofilm are all related.[55]

2.5.3 Other Limitations

The necessity to devise ways to dramatically increase the water solubility of inhaled anti-infectives is highlighted by the tiny volume of the pulmonary lining fluid, poor bioavailability of treatments, and high doses of inhaled anti-infectives.[56] Moreover, damage to healthy lung cells may result on repeated administration of high dosages of antibiotics without regulated release and without particular targeting to the site of the disease.[57] To keep drug levels beyond their minimum inhibitory concentrations (MICs) at the sites of infection, high dosages of the medication are necessary.[58] There have been cases of vaccine-induced increased susceptibility to virus infection. Thus, the development of coronavirus vaccines has been significantly hampered by vaccine-induced enhancement.[59]

The majority of antifungal azole medications used today are administered either orally or systemically. Owing to the non-availability of adequate concentrations of drugs reaching the infection site, the different drug delivery methods are failing to treat respiratory infections.[60] Because of their weak solubility, limited permeability, and rapid clearance, antifungal medicines may not reach their full potential at the site of infection if they are not able to penetrate. Also prevalent with azoles is damage to the heart and liver.[61] Although using inhalation anti-infectives to treat bacterial infections has been standard of care over the past 10 years, there are currently no completely established inhaled antifungal medicines. As a result, there are still numerous unanswered questions, unresolved problems, and difficulties with developing inhaled antifungal medications.[60]

2.6 Nanosized Contrast Agents

Precise bacterial detection is crucial for directing therapy, and contrast agents for bacterial infection-specific imaging are an interesting area.[62] Early-stage identification of pneumonia caused by MRSA remains a major clinical problem. Huo *et al.* conjugated anti-MRSA monoclonal antibodies (mAbs) onto the surface of Au@AgNPs to develop a contrast agent. *In vitro*, the targeting of the MRSA was enhanced 10.66-fold by these antibody-modified NPs (AAMA NPs). Rats having pneumonia brought on by various infections were used to gauge the *in vivo* target efficacy. These AAMA NPs displayed stronger computed tomography (CT) contrast enhancement [498 HU (Hounsfield units)], according to CT data, than did raw Au@Ag and Omnipaque (<100 HU). Furthermore, using spectroscopic CT, lesions marked by AAMA NPs might be differentiated from lung parenchyma.[63] Methoxy-PEG-SH (m-PEG)-modified Au@Ag core–shell NPs showed good CT attenuation efficiency and low cytotoxicity towards common bacteria such as *E. coli* and *S. aureus* (see Figure 2.6).[64]

2.7 Nanotechnology and Pulmonary Delivery

Antibiotic resistance is a current threat in medicine. In this respect, nanoparticulate systems have appeared as a worthwhile strategy to go beyond the restrictions of routine approaches and combat drug resistance.[65]

2.7.1 Lipid-based Vesicles

The delivery of drugs using various nanocarriers (NCs) has shown enormous potential in the treatment of a variety of respiratory infections. The benefits offered by these NCs include improved solubility, pulmonary accumulation, tissue penetration, stability of drugs, and reduced drug resistance and systemic toxicities of therapeutics. The various NCs, including polymeric, lipidic, and inorganic types, have shown great promise in the treatment of diverse respiratory infections (see Figure 2.7).

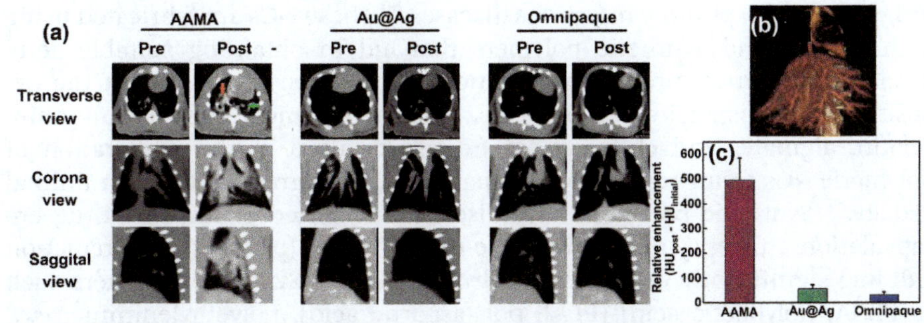

Figure 2.6 (a) *In vivo* CT findings for rats bearing MRSA pneumonia 4 h post-administration of AAMA NPs, Au@AgNPs, and Omnipaque in transverse, coronal, and sagittal views. (b) High-resolution 3D reconstruction image of rats post-injection of AAMA NPs. (c) Contrast enhancement in the form of HU values (Hounsfield units); all the data are averages of five randomly picked areas in the images. Reproduced from ref. 64 with permission from Elsevier, Copyright 2014.

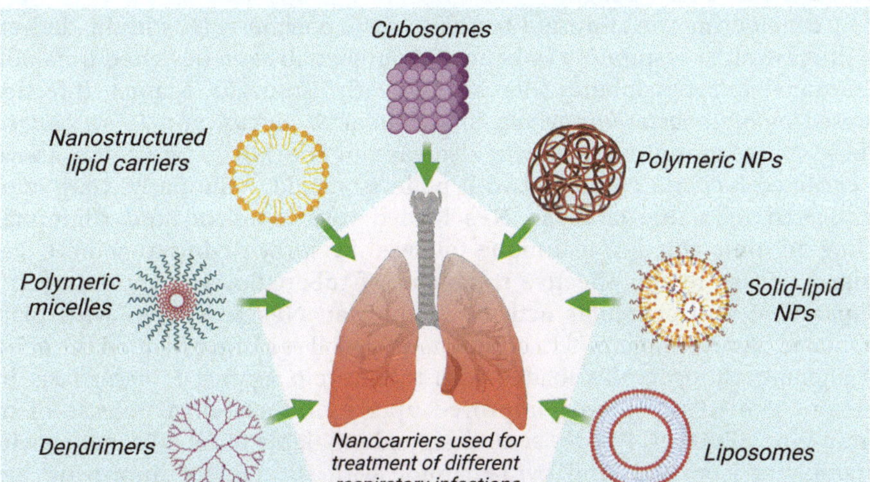

Figure 2.7 Nanoparticulate systems for the treatment of diverse respiratory infections.

2.7.2 Polymeric NCs

Polymeric NCs, including chitosan, poly(lactic-*co*-glycolic acid) (PLGA), polymeric micelles, *etc.*, have been found to be effective in the delivery of drugs to treat respiratory infectious diseases.[66] These NCs are fabricated using both natural and synthetic polymers that aid in obtaining suitable aerodynamic properties, prevent aggregation of particles, and augment the deposition and dispersion of particles. Natural polymers such as albumin, gelatin, alginate, and chitosan are extensively used in the preparation of polymeric NCs owing to their biocompatibility, biodegradability, and minimal toxicity.[67] Synthetic polymers have also been reported to improve drug encapsulation efficiency and control the cargo release (prolong the circulation half-life) significantly in respiratory drug delivery.[68] Synthetic polymers such as PLGA, poly(lactic acid) (PLA), poly(aspartic acid), polyethylenimine, *etc.*, have been widely employed in pulmonary drug delivery. Polyethylenimine has been reported to be highly beneficial in pulmonary drug delivery owing to its potency in binding to the negatively charged mucin, stimulating the process of endocytosis.[69] The surface modification of these polymeric NCs can target the drug to the infected lung sites *via* the protection of encapsulated drugs. Moreover, the charge (cationic or neutral) present on the surface of polymeric NCs is capable of adhering to the mucus layer of the lungs, resulting in sustained release of the drug.[70,71]

By considering the aforesaid benefits of the polymeric NCs in the delivery of drugs against respiratory infections, Pourshahab *et al.* designed inhalable chitosan–tripolyphosphate NPs loaded with isoniazid against infection caused by *Mycobacterium avium*, intracellular *S. aureus*, and *P. aeruginosa*. These NPs showed a significant decrease in the MIC against the above-mentioned bacteria compared with plain isoniazid.[72] Similarly, Garg *et al.* fabricated inhalable chitosan NPs loaded with isoniazid and rifampicin using an ionic gelation technique followed by spray drying to achieve sustained release in the effective treatment of tuberculosis. They observed a remarkable antitubercular activity of chitosan NPs compared with plain isoniazid and rifampicin.[73] In another study, Hill *et al.* investigated the effect of alginate–chitosan NPs loaded with tobramycin against *P. aeruginosa* infection. The NPs displayed improved uptake and sustained release of tobramycin.[74] Further, Pandey *et al.* designed inhalable PLGA NPs loaded with rifampicin, isoniazid, and pyrazinamide with the aim of improving bioavailability and therapeutic efficacy against *M. tuberculosis*. They observed a significant increase in the mean residence time and elimination half-life of the drugs compared with oral administration of the plain drugs, resulting in improved relative bioavailability.[75]

2.7.3 Polymeric Micelles

Polymeric micelles are amphiphilic polymer-based nanoscale vehicles that can self-aggregate in water when their critical micelle concentration (CMC) is exceeded. The polymeric micelles are composed of a hydrophobic corona core that can encapsulate the hydrophobic drugs and a hydrophilic corona that provides colloidal stability to the micelles. The nano size of polymeric micelles offers improvements in the solubility, lung deposition, and macrophage uptake of the drug. The surface decoration of polymeric micelles can aid in the improvement of their characteristics, site-targeted delivery, and the achievement of stimuli-responsive release of the drug.[76,77] In view of this, Grotz *et al.* fabricated Soluplus micelles loaded with rifampicin for pulmonary delivery against *M. tuberculosis*. The micelles displayed a 14-fold increase in the solubility and a 2.5-fold improvement in the *in vitro* antimicrobial activity of rifampicin. Notably, the micelles showed considerable rifampicin accumulation in the lungs of rats for more than 24 h.[78]

2.7.4 Dendrimers

Dendrimers are globular macromolecules with a core nucleus and several radiating branches.[79,80] They are made up of multiple customizable surface functionalities and have an exact, uniform-sized structure.[81,82] The types of polymers employed in the fabrication of dendrimers include melamine, polypropylenimine, polyesteramine and polyamidoamine (PAMAM). The tremendous versatility of dendrimers allows the customization of their physicochemical characteristics, biodistribution, and tissue or organ targeting. They can be altered in any of their molecular elements and functionalized in a variety of ways.[83,84] For instance, Bellini *et al.* prepared PAMAM dendrimers loaded with rifampicin for the treatment of tuberculosis. These dendrimers demonstrated considerable stability and rapid release in an acidic medium, revealing their better targeting against *Mycobacterium*.[85]

2.7.5 Liposomes

Liposomes have a bilayer structure of lipids and are commonly used as NCs for pulmonary drug delivery. Liposomes are heterogeneous drug delivery systems comprised of an aqueous core enclosed by a lipid bilayer.[86] Distearoylphosphatidylcholine, dipalmitoylphosphatidylcholine, and distearoylphosphoethanolamine are commonly used phospholipids in the fabrication of liposomes. Both hydrophilic and hydrophobic types of drugs can be encapsulated in the liposomes either in the lipid membrane or in the aqueous core, depending on the solubility of the therapeutics. There are two methods for loading drugs into liposomes, passive loading and active loading. In the former method, the cargo and lipids are mixed and the cargo is loaded within the liposomes at the same time during the construction of liposomes. In the latter approach, the cargo is loaded within the preformed liposomes by means of pH gradients. The safety and controlled release of the cargo are the chief benefits of inhalable liposomes.[87,88] In addition, they offer delivery of a range of therapeutics in large volumes to the peripheral airways *via* medical nebulizers. Liposomes also improve the biopharmaceutical and pharmacokinetic properties and stability of therapeutics, and deposit medication *via* phagocytosis within alveolar macrophages to enhance the treatment of intracellular infections.[87–89]

By considering the above benefits, Castoldi *et al.* fabricated liposomes loaded with calcifediol for the treatment of pulmonary infection caused by *P. aeruginosa* locally. The inhalable liposomes demonstrated extensive deposition in the lungs and a reduction in the survival of bacteria.[90] In another interesting study, Blanchard *et al.* designed inhalable negatively charged liposomes encapsulated with ciprofloxacin for the treatment of *P. aeruginosa* and *M. avium* lung infections. The liposomes exhibited sustained release and a high lung clearance half-life of the cargo compared with the plain ciprofloxacin.[91] Further, Hamed *et al.* investigated the antibacterial potential of moxifloxacin-loaded nanoliposomes against *M. tuberculosis*. They observed significant lung deposition of inhalable dry nanoliposomes. The antitubercular activity of moxifloxacin was observed to be substantially higher following its incorporation into the liposomes. In addition, the liposomes displayed a greater macrophage uptake.[92]

2.7.6 Solid Lipid Nanoparticles (SLNs)

SLNs consist of a solid-state lipid-based matrix comprised of solid lipids stabilized by an emulsifier. SLNs are capable of loading a variety of drugs including both hydrophilic and hydrophobic types. SLNs can serve as an ideal candidate for the pulmonary delivery of drugs as they can be prepared without organic solvents. These NCs offer various benefits such as biodegradability, biocompatibility, simplicity in large-scale production, and protection from harsh environmental conditions. Deep lung deposition can be achieved significantly using SLNs *via* their aerosolization. However, the low drug loading capacity and initial burst cargo release are the chief drawbacks of SLNs.[93,94] SLNs represent a widely used platform in antibiotic delivery against pulmonary infections. For instance, Gaspar *et al.* prepared microencapsulated SLNs loaded with the antibiotic rifabutin to treat *M. tuberculosis*. These SLNs demonstrated initial burst release followed by sustained release of rifabutin for 24 h. These SLNs were reported to show substantial antimycobacterial activity in a murine mouse model infected with *M. tuberculosis*.[93] In another interesting study, Varshosaz *et al.* delivered amikacin using SLNs *via* the pulmonary route for the treatment of CF lung infections. They observed a reduction in the dosing frequency and side effects of amikacin following pulmonary delivery compared with intravenous delivery of the free drug owing to the increased lung deposition. In addition, the SLNs demonstrated sustained release of amikacin and enhanced patient compliance.[95]

2.7.7 Nanostructured Lipid Carriers (NLCs)

NLCs are lipid-based NCs comprised of solid and liquid lipids. NLCs are capable of overcoming the various limitations of SLNs. The advantages of NLCs include improved cargo solubility in lipids, higher drug loading capacity owing to the imperfect crystal structure, and controlled release.[96]

Additionally, enhanced permeability, targeting, stability, biocompatibility, non-immunogenicity, prolonged circulation, and residence time are other noteworthy benefits of NLCs.[97]

In view of these attributes, Almurshedi *et al.* designed ciprofloxacin-loaded NLCs for the treatment of non-CF bronchiectasis infection. The NLCs displayed sustained release (80%) in 10 h. Moreover, the maximum growth of inhibition was observed with NLCs compared with plain ciprofloxacin against *S. aureus*.[98] Similarly, Pardeike *et al.* investigated the potential of itraconazole-loaded NLCs against aspergillosis in falcons (pulmonary infection). The size of the NLCs was found to be in the nanometre range, revealing a better opportunity for penetration into the respiratory tract of falcons.[99]

2.7.8 Cubosomes

Two extensively convoluted non-intersecting water channels and a deformed lipid bilayer serve as the primary structural components of cubosomes, a dispersion of the bicontinuous cubic mesophase.[100] Cubosomes are ideal NCs for the effective encapsulation of lipophilic, hydrophilic, and amphiphilic drug molecules owing to their high lipidic content and large surface area.[101] Biodegradability, biocompatibility, bioadhesion, and extended cargo release ability are other noteworthy features of cubosomes. Moreover, the nano size of cubosomes can contribute to the deep lung deposition of therapeutics, thereby achieving better therapeutic effects and reduced systemic side effects.[102]

In view of the above attributes of cubosomes, Kazi and Dehghan prepared inhalable cubosomes loaded with nystatin for the management of pulmonary aspergillosis. The nystatin-loaded cubosomes displayed prolonged release for 48 h compared with the plain nystatin dispersion. Additionally, the antifungal activity of nystatin-loaded cubosomes was higher than that of the plain nystatin dispersion.[103]

2.7.9 Phytosomes

Phytosomes (phyto–phospholipid complexes) are vesicular systems generated by the formation of hydrogen bonds *via* the interactions amongst phytoactive components and hydrophilic parts of phospholipids. The method of combining plant medications or extracts with phospholipids has been shown to be one of the most effective and challenging ways to increase the bioavailability and beneficial viability of diverse insufficiently absorbed plant elements. This avenue combines phospholipid particles with phosphatidylcholine in their structure with standardized domestic concentrates and the specific bioactive component of plants to form complexes that increase layer penetrability and partition coefficient, augment bioavailability, and increase solubility and stability. Phytosomes have been reported to be promising carriers in the treatment of pulmonary infections. For instance,

Singh *et al.* developed gingerol-loaded phytosomes complexed with chitosan for the treatment of respiratory infections. The *in vitro* release of gingerol from the phytosomes was observed to be sustained (88% release) up to 24 h. The zone of inhibition observed with phytosomes was higher than that with plain gingerol against *S. aureus* and *E. coli*. Moreover, the $AUC_{0-\infty}$ of phytosomes was 20 times higher than that of plain gingerol after oral administration in albino rabbits.[104]

2.7.10 Inorganic (Metal-based) NCs

These NCs are mainly produced using pure metals such as gold, silver, titanium, *etc.*, or some metal oxides. These NCs have been found to be important for treating pulmonary infections owing to their plasmonic and magnetic properties. These systems are known to retard the release of different drugs from their formulations. They can be produced by diverse approaches including physical, chemical, and biological methods. Various factors such as the size, shape, capping, zeta potential, ligands, charge, *etc.* of these NCs affect the antimicrobial activities. These NCs can target the infected sites and reduce the dosage and toxicity of loaded antibiotics. In addition, biocompatibility, high stability, and resistance to microbial degradation are other significant properties of these NCs, which play a vital role in the treatment of pulmonary infections.[105,106]

Tenland *et al.* fabricated mesoporous silica NPs loaded with NZX (peptide) for the treatment of *M. tuberculosis*. These NPs were found to be taken up effectively by macrophages. Moreover, they demonstrated noticeably greater *M. tuberculosis* eradication than the free peptide murine TB model.[105] In another study, Miranda *et al.* achieved pulmonary delivery of an anti-TB (P3) drug *via* superparamagnetic iron oxide NPs in microparticles. These microparticles were observed to be promising for P3 entry into the lower airways and phagocytosis through alveolar macrophages. Moreover, the system displayed sustained cargo release and ameliorated anti-TB efficacy.[106]

2.8 Nanotechnology for COVID-19 Treatment

The COVID-19 pandemic posed unprecedented risks to the public's health. The rapid spread of the disease prompted calls for new methods of virus eradication. Nanotechnology has not only demonstrated a role in the treatment but has also shown great promise in diagnosis and prevention. Thus, by playing a crucial part in preventive, diagnostic, and therapeutic strategies for COVID-19 management, nanotechnology acts as a formidable tool with the potential to reduce infections.[107–109] The nano size of nanomaterials, which is comparable to the size of viruses, has contributed significant functional capabilities for battling viruses. Notably, the tiny size and the large surface-to-volume ratio of nanomaterials make them ideal carriers for therapeutics, augmenting site-specific delivery of cargo, and rapid virus detection *via* interaction between the analyte and sensor. Furthermore, the nanomaterials also confer benefits such as enhanced cargo solubility, biocompatibility, reduced dose and dose frequency, and decreased clearance speed, degradation, and toxicity of the drug. The surface engineering of the nanomaterials can improve their interactions with the cell receptors or virus particles that impede virus–cell interactions.[107–109]

Nanotechnology has also been shown to have potential applications in the delivery of vaccines. Different NP-based vaccines with adjustable loading of antigens serve as intrinsic adjuvants. These NP-based vaccines can target both innate and adaptive immune systems at the cellular level.[109]

Different types of nanomaterials such as organic NPs (polymeric, lipidic, and hybrid polymeric–lipidic NPs), inorganic NPs (including metal NPs such as copper, silver, and titanium oxide), and magnetic NPs have been used for the delivery of a variety of antiviral drugs in the successful treatment of COVID-19. However, the chief bottleneck in the use of inorganic NPs is their toxicity, which necessitates further modifications and in-depth investigations of the aforesaid NPs.[109]

2.9 Nanovaccines for Treating Lung Infections

Vaccines are an efficient immune system booster that assist in preventing the spread of various lung infections. Vaccines are composed of attenuated viruses or their pieces. Millions of individuals have benefited from vaccination in the fight against, control of, and eradication of pulmonary infectious illnesses. However, autoimmunity or severe allergic reactions, a return to a virulent state, manufacturing, and stability-related issues are the main drawbacks with conventional vaccines.[110]

Nanotechnology has taken on an important role in solving the aforesaid challenges. Nanotechnology has improved the specificity and effectiveness of vaccines against diverse types of pulmonary infections. Nanomaterials are reported to boost immune response and dendritic cell input owing to their nano size. These nanomaterials (polymeric, lipidic, and inorganic) can be used as immunogens or adjuvants to enhance and prolong the antigenic delivery. Polymeric NPs, polymeric micelles, dendrimers, liposomes, solid lipid NPs, nanosuspensions, nanocrystals, and nanoemulsions have been extensively employed as adjuvants.[110–112]

Different types of nanovaccines such as polymeric, lipidic, inorganic, nucleotide-based, and protein- or peptide-based nanovaccines have been explored in the treatment of various pulmonary infections. These nanovaccines stimulate a potent immune response that reduces the spread of disease and permits disease control or death. They are designed to allow the antigen to reside either within or outside the NPs, enabling effective dissemination and controlled release at the targeted site. These controlled releases of antigens avoid exaggerated reactions and remove the necessity for a booster dose. Notably, the nanovaccines have been reported to improve the cellular uptake and stability of antigens.[110–112]

A range of routes such as oral, rectal, vaginal, and intranasal have been found to have potential in the delivery of vaccines that offers systemic or mucosal immune protection. Recent research has also revealed that the alternative route of mucosal delivery shields the mucosal surfaces against the majority of infections. Nasal vaccination has shown a great ability to mobilize a particular immunoglobulin A (IgA), cytotoxic T cells (CTLs), and systemic antibody response in comparison with other mucosal methods.[110]

In view of this, McGill *et al.* investigated the potential of a mucosal polyanhydride NP-based nanovaccine containing post-fusion F and G glycoproteins against respiratory infection caused by the syncytial virus. They studied the efficacy of this nanovaccine in a neonatal calf model. Significant reductions in lung pathology, viral load, and virus shedding were observed following the administration of the nanovaccine.[111] In another interesting study, Zheng *et al.* designed an inhalable bionic-virus nanovaccine resembling the SARS-CoV-2 structure by employing liposomes as capsid structures and a "spike" as the receptor binding part. This nanovaccine induced significant mucosal immunity against respiratory infections.[112]

In vitro and *in vivo* performances of drug-loaded nanosystems against various respiratory infections are summarized in Table 2.1.

2.10 Current Challenges and Future Prospects

The aforementioned examples demonstrate a number of therapeutic approaches that have great promise for creating nanomedicines that can both treat and prevent infectious diseases. Nanotechnology solutions will need to overcome a number of manufacturing, financial, and regulatory obstacles to be as effective as possible. Initially, lowering the cost will be the greatest challenge in creating new treatments and systems based on nanotechnology.[113]

The main issue with nanoformulations is their liquid-state stability. Despite the fact that lyophilization has been shown to be an efficient way to resolve this problem, it has raised the cost of the formulations. However, most scientists are having difficulty in scaling up these nanoparticulate systems. Nanomedicines must be successfully translated, and this requires a network of stakeholders to come together. Academics, business and government investors, and firms that conduct contract manufacturing and research are all included in this. Moreover, the development of vaccines is being severely hampered by the variations of viruses. The major worries are undoubtedly the toxicity and environmental hazards.

Even though there are various disagreements, this technology offers great hope for the future. The majority of organizations are focusing on nanoparticulate systems as a promising strategy in response to the COVID-19 epidemic. The majority of future research will be focused on creating nanoparticulate systems to fight respiratory diseases. The employment of several cutting-edge nanoparticulate systems and nanodevices is anticipated to have a tremendously favourable effect on people's health. Nanomedicine now has greater prospects for use owing to ongoing developments in a

Table 2.1 *In vitro* and *in vivo* performance of drug-loaded nanosystems against various respiratory infections.

Type of nanocarrier	Drug	Performance: *In vitro*	Performance: *In vivo*	Ref.
Chitosan–tripolyphosphate NPs	Isoniazid	Significant reduction in MIC against *M. avium*, intracellular *S. aureus*, and *P. aeruginosa* compared with the plain drug	—	72
Chitosan NPs	Isoniazid and rifampicin	Significant antitubercular effect compared with the plain drug	—	73
Chitosan NPs	Itraconazole	Significant penetration and antibacterial effect against *Cryptococcus neoformans*	—	107
Alginate–chitosan NPs	Tobramycin	Sustained release and enhanced uptake of tobramycin from the NPs	—	74
PLGA NPs	Rifampicin, isoniazid, and pyrazinamide	Controlled release of drugs	Significant improvement on the absolute and relative bioavailability of all three drugs from the NPs in guinea pigs infected with *M. tuberculosis*	75
PLGA NPs	Lopinavir	Significant (100-fold) reductions in the expression of interleukin-6 and 10-fold in interleukin-1β were observed in macrophages	—	108
Polymeric micelles	Rifampicin	Enhanced solubility and antimicrobial activity	Significantly improved accumulation of rifampicin in the lungs	76
Dendrimers	Rifampicin	Improved stability and targeting	—	77
Liposomes	Calcifediol	Significant reduction in the survival of *P. aeruginosa*	—	90
Liposomes	Ciprofloxacin	Sustained release of ciprofloxacin	Enhanced lung clearance up to 12 h compared with plain ciprofloxacin (1 h)	91
Liposomes	Amphotericin B	Targeted (dectin-2 coated) liposomes demonstrated 10-fold more inhibition of *Aspergillus fumigatus* than non-targeted liposomes	Significant decrease in the lung fungal burden by targeted liposomes compared with non-targeted liposomes	109

Table 2.1 (*Continued*)

Type of nanocarrier	Drug	Performance		Ref.
		In vitro	*In vivo*	
Liposomes	Gentamicin	—	Increased half-life, volume of distribution, bronchoalveolar concentration of gentamicin from liposomes compared with plain gentamicin following intravenous and nebulized administration in foals	110
Liposomes	Remdesivir	Controlled release of remdesivir for 6 h	Substantially higher (100-fold) concentration and lung accumulation of remdesivir metabolite in the lung were observed via inhalation administration of liposomes *in vivo* in BALB/c mice	111
Nanoliposomes	Moxifloxacin	Significant anti-tubercular activity and macrophages uptake	Enhanced lung deposition of nanoliposomes in rats	92
Solid liposomes	Icariin	Sustained release of icariin	Significantly improved oral bioavailability of icariin from the liposomes compared with plain icariin *in vivo*	112
SLNs	Rifabutin	Initial burst release followed by sustained release of drug	Significant antimycobacterial activity in an *M. tuberculosis*-infected murine mice model	93
SLNs	Amikacin	Sustained release of amikacin	Enhanced lung deposition of amikacin through SLNs delivered *via* the pulmonary route in rats	95
NLCs	Ciprofloxacin	Sustained release and remarkable antimicrobial activity against *S. aureus*	—	98
Cubosomes	Nystatin	Sustained release and significant antifungal activity	—	103
Phytosomes	Gingerol	Sustained release and considerable antimicrobial activity against *S. aureus* and *E. coli*	Enhanced oral bioavailability of gingerol from phytosomes compared with plain gingerol in rabbits	104
Mesoporous silica	Peptide NZX	Significant growth inhibition of *M. tuberculosis* compared with the plain peptide	Remarkable *M. tuberculosis* eradication in a murine TB model with *M. tuberculosis* H37Rv compared with the plain peptide	105

number of medical specialties. Investigations are currently being conducted on its potential use in diagnostic and regenerative medicine. Diagnoses would be made more quickly, allowing the immediate treatment of sick cells before they spread to and harm other bodily areas.

2.11 Conclusion

Nanosystems hold great promise in the detection and treatment of pulmonary infections. Right now, the majority of researchers are working on the development of several nanoparticulate systems that will surely control the spread of pulmonary outbreaks. The development of nanovaccines is also an area in which the whole world is interested. Undoubtedly, in the near future, these nanovaccines will be the first priority of scientists to achieve immunity against life-threatening pulmonary diseases.

References

1. P. Roeder, J. Mariner and R. Kock, *Philos. Trans. R. Soc., B*, 2013, **368**, 20120139.
2. R. E. Baker, A. S. Mahmud, I. F. Miller, M. Rajeev, F. Rasambainarivo, B. L. Rice, S. Takahashi, A. J. Tatem, C. E. Wagner, L. F. Wang, A. Wesolowski and C. J. E. Metcalf, *Nat. Rev. Microbiol.*, 2022, **20**(4), 193.
3. M. Vouga and G. Greub, *Clin. Microbiol. Infect.*, 2016, **22**(1), 12.
4. T. D. Chugh, *J. Biosci.*, 2008, **33**(4), 549.
5. A. O. Azghani and C. A. Clark, in *Regulation of the Inflammatory Response in Health and Disease*, ed. A. O. Azghani and E. J. Miller, 2009, pp. 1–18.
6. N. M. Aljamali, Z. H. Al-zubaidy and A. H. Enad, *Trends Pharm. Nanotechnol.*, 2021, **3**(2), 13.
7. A. Ali, A. Waris, M. A. Khan, M. Asim, A. U. Khan, S. Khan and J. Zeb, *Life Sci.*, 2023, **314**, 121332.
8. A. Osterloh, *Vaccines*, 2022, **10**(5), 751.
9. T. W. Mak and M. E. Saunders, in *The Immune Response Basic and Clinical Principles*, Academic Press, Elsevier, 2006.
10. S. H. Kaufmann, *Annu. Rev. Immunol.*, 1993, **11**, 129.
11. A. Thakur, H. Mikkelsen and G. Jungersen, *J. Immunol Res.*, 2019, **2019**, 1356540.
12. L. A. Carvajal and C. P. Pérez, in *Pediatric Respiratory Diseases*, ed. P. Bertrand and I. Sánchez, Springer, Cham, 2020, pp. 263–272.
13. R. Cohen, F. Angoulvant, S. Biscardi, F. Madhi, F. Dubos and Y. Gillet, *Arch. Pediatr.*, 2017, **24**(12S), S17.
14. W. Cookson, M. Cox and M. Moffatt, *Nat. Rev. Microbiol.*, 2018, **16**, 111.
15. T. Ferkol and D. Schraufnagel, *Ann. Am. Thorac. Soc.*, 2014, **11**, 404.
16. C. L. F. Walker, I. Rudan, L. Liu, H. Nair, E. Theodoratou, Z. A. Bhutta, K. L. O'Brien, H. Campbell and R. E. Black, *Lancet*, 2013, **381**(9875), 1405.
17. V. R. Ramakrishnan, L. J. Hauser and D. N. Frank, *Head Neck Surg.*, 2016, **24**, 20.

18. W. H. Man, W. A. A. de Steenhuijsen Piters and D. Bogaert, *Nat. Rev. Microbiol.*, 2017, **15**, 259.
19. A. L. Welp and J. M. Bomberger, *Front. Cell. Infect. Microbiol.*, 2020, **10**, 213.
20. R. Rafeq and L. A. Igneri, *Emerg. Med. Clin. North Am.*, 2022, **40**(3), 503.
21. I. Smith, *Clin. Microbiol. Rev.*, 2003, **16**(3), 463.
22. S. Sethi and T. F. Murphy, *N. Engl. J. Med.*, 2008, **359**, 2355.
23. D. L. Hahn, R. W. Dodge and R. Golubjatnikov, *J. Am. Med. Assoc.*, 1991, **266**, 225.
24. J. J. Lipuma, *Clin. Microbiol. Rev.*, 2010, **23**(2), 299.
25. M. R. Kiedrowski and J. M. Bomberger, *Front. Immunol.*, 2018, **9**, 3067.
26. Z. Huang, S. N. Kłodzińska, F. Wan and H. M. Nielsen, *Drug Delivery Transl. Res.*, 2021, **11**, 1634.
27. S. R. Zaki and C. D. Paddock, in *Dail and Hammar's Pulmonary Pathology*, ed. J. F. Tomashefski, P. T. Cagle, C. F. Farver and A. E. Fraire, Springer, New York, NY, 2008, pp. 426–475.
28. D. Patchett, J. Yang, J. Northern, M. Salinas and B. A. Bauer, *Mayo. Clin. Proc. Innov. Qual. Outcomes*, 2021, **5**(2), 480.
29. H. R. van Doorn and H. Yu, in *Hunter's Tropical Medicine and Emerging Infectious Diseases*, ed. E. T. Ryan, D. R. Hill, T. Solomon, *et al.*, Elsevier, 2020, pp. 284–288.
30. N. Saleri and E. T. Ryan, in *Travel Medicine*, ed. J. S. Keystone, P. E. Kozarsky, B. A. Connor, *et al.*, Elsevier, 2019, pp. 527–537.
31. A. M. Fendrick, *Am. J. Ther.*, 2003, **10**(3), 193.
32. H. F. Boncristiani, M. F. Criado and E. Arruda, *Encycl. Microbiol.*, 2009, 500.
33. B. Hu, H. Guo, P. Zhou and Z. L. Shi, *Nat. Rev. Microbiol.*, 2021, **19**(3), 141. Erratum in, *Nat. Rev. Microbiol.*, 2022, **20**(5), 315.
34. M. Cascella, M. Rajnik, and A. Aleem, *et al.*, Features, Evaluation, and Treatment of Coronavirus (COVID-19), in *StatPearls [Internet]*, StatPearls Publishing, Treasure Island (FL), 2022, Available from: https://www.ncbi.nlm.nih.gov/books/NBK554776/ [Updated 2022 Oct 13].
35. T. Singhal, *Indian J. Pediatr.*, 2020, **87**(4), 281.
36. P. Sadhukhan, M. T. Ugurlu and M. O. Hoque, *Cancers*, 2020, **12**(12), 3822.
37. N. Azadeh, A. H. Limper, E. M. Carmona and J. H. Ryu, *Chest*, 2017, **152**(4), 842.
38. A. S. Miller and R. W. Wilmott, in *Kendig's Disorders of the Respiratory Tract in Children*, ed. R. W. Wilmott, R. Deterding, A. Li, F. Ratjen, P. Sly, H. J. Zar and A. Bush, Elsevier, 9th edn, 2019, pp. 507–527.
39. E. L. Keeney, *Calif. Med.*, 1954, **81**(6), 367.
40. L. J. Wheat, M. Goldman and G. Sarosi, *Semin. Respir. Infect.*, 2002, **17**(2), 158.
41. Z. Li, G. Lu and G. Meng, *Front. Immunol.*, 2019, **10**, 1524.
42. T. H. Boyles and S. Wasserman, *S. Afr. Med. J.*, 2015, **105**(5), 1–3.
43. J. Dronina, U. Samukaite-Bubniene and A. Ramanavicius, *J. Nanobiotechnol.*, 2021, **19**(1), 348.

44. M. Pretorius and M. Venter, in *Viral Infections in Children*, ed. R. Green, Springer, Cham, vol. 1, 2017, pp. 151–182.
45. D. H. Reta, T. S. Tessema, A. S. Ashenef, A. F. Desta, W. L. Labisso, S. T. Gizaw, S. M. Abay, D. S. Melka and F. A. Reta, *Int. J. Microbiol.*, 2020, **2020**, 8832728.
46. G. E. Hayes and D. W. Denning, *Curr. Opin. Pulm. Med.*, 2013, **19**(3), 259.
47. J. A. Smith and C. A. Kauffman, *Respirology*, 2012, **17**(6), 913.
48. T. Velkov, N. Abdul Rahim, Q. Zhou, H-K. Chan and J. Li, *Adv. Drug Delivery Rev.*, 2015, **85**, 65.
49. J. Li, H. Zheng and S. S. Y. Leung, *Pharm. Res.*, 2022, **40**(5), 1057.
50. D. R. Tompa, A. Immanuel, S. Srikanth and S. Kadhirvel, *Int. J. Biol. Macromol.*, 2021, **172**, 524.
51. Editorial: Paved ways through delivery barriers, *Nat. Biomed. Eng.*, 2018, **2**, 267.
52. D. J. Thornton and J. K. Sheehan, *Proc. Am. Thorac. Soc.*, 2004, **1**(1), 54.
53. P. Nordmann, L. Dortet and L. Poirel, *Trends Mol. Med.*, 2012, **18**(5), 263.
54. D. C. Domínguez and S. M. Meza-Rodriguez, in *Pharmaceuticals and Personal Care Products: Waste Management and Treatment Technology, Development of antimicrobial resistance: future challenges*, ed. M. N. V. Prasad, M. Vithanage and A. Kapley, Butterworth-Heinemann, 2019, pp. 383–408.
55. M. A. Rather, K. Gupta and M. Mandal, *Braz. J. Microbiol.*, 2021, **52**(4), 1701.
56. J. E. Hastedt, P. Bäckman, A. R. Clark, W. Doub, A. Hickey, G. Hochhaus, P. J. Kuehl, C-M. Lehr, P. Mauser, J. McConville, R. Niven, M. Sakagimi and J. G. Weers, *AAPS Open*, 2016, **2**, 1.
57. D. K. Ho, B. L. B. Nichols, K. J. Edgar, X. Murgia, B. Loretz and C. M. Lehr, *Eur. J. Pharm. Biopharm.*, 2019, **144**, 110.
58. Q. T. Zhou, S. S. Leung, P. Tang, T. Parumasivam, Z. H. Loh and H. K. Chan, *Adv. Drug Delivery Rev.*, 2015, **85**, 83.
59. W. Huisman, B. E. Martina, G. F. Rimmelzwaan, R. A. Gruters and A. D. Osterhaus, *Vaccine*, 2009, **27**(4), 505.
60. K. Ito, *Expert Opin. Drug Delivery*, 2022, **19**(7), 755.
61. L. L. Benitez and P. L. Carver, *Drugs*, 2019, **79**(8), 833.
62. Y. Zhang, M. Hao, L. Li, Q. Luo, S. Deng, Y. Yang, Y. Liu, W. Fang and E. Song, *Trends Anal. Chem.*, 2023, **159**, 116916.
63. D. Huo, J. Ding, Y. X. Cui, L. Y. Xia, H. Li, J. He, Z. Y. Zhou, H. W. Wang and Y. Hu, *Biomaterials*, 2014, **35**(25), 7032.
64. D. Huo, J. He, H. Li, H. Yu, T. Shi, Y. Feng, Z. Zhou and Y. Hu, *Colloids Surf., B*, 2014, **117**, 29.
65. F. Andrade, D. Rafael, M. Videira, D. Ferreira, A. Sosnik and B. Sarmento, *Adv. Drug Delivery Rev.*, 2013, **65**(13–14), 1816.
66. G. Sheng, N. Tian, H. Duan, Z. Sun and H. Chu, *Acta Mater. Med.*, 2022, **1**(3), 343.

67. M. Paranjpe and C. C. Müller-Goymann, *Int. J. Mol. Sci.*, 2014, **15**, 5852.
68. E. Rytting, J. Nguyen, X. Wang and T. Kissel, *Expert Opin. Drug Delivery*, 2008, **5**, 629.
69. S. Di Gioia and M. Conese, *Drug Des., Dev. Ther.*, 2009, **2**, 163.
70. N. Sanders, C. Rudolph, K. Braeckmans, S. C. De Smedt and J. Demeester, *Adv. Drug Delivery Rev.*, 2009, **61**, 115.
71. A. Salvati, C. Åberg, T. dos Santos, J. Varela, P. Pinto, I. Lynch and K. A. Dawson, *Nanomedicine*, 2011, **7**, 818.
72. P. S. Pourshahab, K. Gilani, E. Moazeni, H. Eslahi, M. R. Fazeli and H. Jamalifar, *J. Microencapsulation*, 2011, **28**(7), 605.
73. T. Garg, G. Rath and A. K. Goyal, *Artif. Cells, Nanomed., Biotechnol.*, 2016, **44**(3), 997.
74. M. Hill, M. Twigg, E. A. Sheridan, J. G. Hardy, J. S. Elborn, C. C. Taggart, C. J. Scott and M. E. Migaud, *Pharmaceutics*, 2019, **11**(8), 379.
75. R. Pandey, A. Sharma, A. Zahoor, S. Sharma, G. K. Khuller and B. Prasad, *J. Antimicrob. Chemother.*, 2003, **52**(6), 981.
76. E. Grotz, N. Tateosian, N. Amiano, M. Cagel, E. Bernabeu, D. A. Chiappetta and M. A. Moretton, *Pharm. Res.*, 2018, **35**, 213.
77. S. C. Owen, D. P. Y. Chan and M. S. Shoichet, *Nano Today*, 2012, **7**, 53.
78. E. Grotz, N. L. Tateosian, J. Salgueiro, E. Bernabeu, L. Gonzalez, M. L. Manca and N. Amiano, *J. Drug Delivery Sci. Technol.*, 2019, **53**, 101170.
79. E. R. Gillies and J. M. J. Frechet, *Drug Discovery Today*, 2005, **10**(1), 35.
80. H. Choudhury, N. F. B. Zakaria, P. A. B. Tilang, A. S. Tzeyung, M. Pandey, B. Chatterjee, N. A. Alhakamy, S. K. Bhattamishra, P. Kesharwani, B. Gorain and Md. Shadab, *J. Drug Delivery Sci. Technol.*, 2019, **54**, 101301.
81. M. Liu and J. M. J. Fréchet, *Pharm. Sci. Technol. Today*, 1999, **2**(10), 393.
82. S. Singh, D. Hassan, H. M. Aldawsari, N. Molugulu, R. Shukla and P. Kesharwani, *Drug Discovery Today*, 2020, **25**(1), 223.
83. P. K. Maiti, T. Çağin, G. Wang and W. A. Goddard, *Macromolecules*, 2004, **32**(16), 6236.
84. D. Kaur, K. Jain, N. K. Mehra, P. Kesharwani and N. K. Jain, *J. Nanopart. Res.*, 2016, **18**, 146.
85. R. G. Bellini, A. P. Guimarães, M. A. C. Pacheco, D. M. Dias, V. R. Furtado, R. B. De Alencastro and B. A. C. Horta, *J. Mol. Graphics Modell.*, 2015, **60**, 34.
86. L. Sercombe, T. Veerati, F. Moheimani, S. Y. Wu, A. K. Sood and S. Hu, *Front. Pharmacol.*, 2015, **6**, 286.
87. L. Corrie, R. Gundaram and L. Kukatil, *Recent Innovations Chem. Eng.*, 2021, **14**(4), 347.
88. P. Nakhaei, R. Margiana, D. O. Bokov, W. K. Abdelbasset, M. A. Jadidi Kouhbanani, R. S. Varma, F. Marofi, M. Jarahian and N. Beheshtkhoo, *Front. Bioeng. Biotechnol.*, 2021, **9**, 705886.
89. A. Elhissi, *Curr. Pharm. Des.*, 2017, **23**(3), 362.

90. A. Castoldi, C. Herr, J. Niederstraßer, H. I. Labouta, A. Melero, S. Gordon, N. Schneider-Daum, R. Bals and C-M. Lehr, *Eur. J. Pharm. Biopharm.*, 2017, **118**, 62.
91. J. D. Blanchard, V. Elias, D. Cipolla, I. Gonda and L. E. Bermudez, *Antimicrob. Agents Chemother.*, 2018, **62**, e00440-18.
92. A. Hamed, R. Osman, K. T. Al-Jamal, S. M. Holayel and A.-S. Geneidi, *J. Drug Delivery Sci. Technol.*, 2019, **51**, 513.
93. D. P. Gaspar, M. M. Gaspar, C. V. Eleuterio, A. Grenha, M. Blanco, L. M. D. Goncalves and P. Taboada, *Mol. Pharm.*, 2017, **14**(9), 2977.
94. G. Yoon, J. W. Park and I. S. Yoon, *J. Pharm. Invest.*, 2013, **43**(5), 353.
95. J. Varshosaz, S. Ghaffari, S. F. Mirshojaei, A. Jafarian, F. Atyabi, F. Kobarfard and S. Azarmi, *BioMed Res. Int.*, 2013, **2013**, 136859.
96. A. Beloqui, M. A. Solinis, A. Rodriguez-Gascon, A. J. Almeida and V. Preat, *Nanomedicine*, 2016, **12**(1), 143.
97. P. Kumbhar, A. Manjappa, R. Shah, N. K. Jha, S. K. Singh, K. Dua, J. Disouza and V. Patravale, *J. Control. Release*, 2022, **341**, 1–15.
98. A. S. Almurshedi, H. A. Aljunaidel, B. Alquadeib, B. N. Aldosari, I. M. Alfagih, S. S. Almarshidy, E. K. D. Eltahir and A. Z. Mohamoud, *Int. J. Nanomed.*, 2021, **16**, 2405.
99. J. Pardeike, S. Weber, H. P. Zarfl, M. Pagitz and A. Zimmer, *Eur. J. Pharm. Biopharm.*, 2016, **108**, 269.
100. E. Esposito, R. Cortesi, M. Drechsler, L. Paccamiccio, P. Mariani, C. Contado, E. Stellin, E. Menegatti, F. Bonina and C. Puglia, *Pharm. Res.*, 2005, **22**(12), 2163.
101. X. Shi, T. Peng, Y. Huang, L. Mei, Y. Gu, J. Huang, K. Han, G. Li, C. Hu, X. Pan and C. Wu, *Pharm. Dev. Technol.*, 2017, **22**(3), 322.
102. N. Islam and V. Ferro, *Nanoscale*, 2016, **8**(30), 14341.
103. M. Kazi and M. H. Dehghan, *Istanbul J. Pharm.*, 2020, **50**(3), 224.
104. R. P. Singh, H. V. Gangadharappa and K. Mruthunjaya, *Eur. J. Pharm. Sci.*, 2018, **122**, 214.
105. E. Tenland, A. Pochert, N. Krishnan, K. Umashankar Rao, S. Kalsum, K. Braun, I. Glegola-Madejska, M. Lerm, B. D. Robertson, M. Lindén and G. Godaly, *PLoS One*, 2019, **14**(2), e0212858.
106. M. S. Miranda, M. T. Rodrigues, R. M. A. Domingues, R. R. Costa, E. Paz, C. Rodríguez-Abreu, P. Freitas, B. G. Almeida, M. A. Carvalho, C. Gonçalves, C. M. Ferreira, E. Torrado, R. L. Reis, J. Pedrosa and M. E. Gomes, *Adv. Healthcare Mater.*, 2018, **7**(15), e1800124.
107. H. Rashidzadeh, H. Danafar, H. Rahimi, F. Mozafari, M. Salehiabar, M. A. Rahmati, S. Rahamooz-Haghighi, N. Mousazadeh, A. Mohammadi, Y. N. Ertas, A. Ramazani, I. Huseynova, R. Khalilov, S. Davaran, T. J. Webster, T. Kavetskyy, A. Eftekhari, H. Nosrati and M. Mirsaeidi, *Nanomedicine*, 2021, **16**(6), 497.
108. S. Ye, K. Shao, Z. Li, N. Guo, Y. Zuo, Q. Li, Z. Lu, L. Chen, Q. He and H. Han, *ACS Appl. Mater. Interfaces*, 2015, **7**(38), 21571.
109. F. Vahedifard and K. Chakravarthy, *Emergent Mater.*, 2021, **4**(1), 75.

110. G. Rodrigues, M. Gonçalves da Costa Sousa, D. C. da Silva, T. M. BertoRezende, P. C. de Morais and O. L. Franco, *ACS Appl. Bio. Mater.*, 2021, **4**(5), 3880.
111. J. L. McGill, S. M. Kelly, P. Kumar, S. Speckhart, S. L. Haughney, J. Henningson, B. Narasimhan and R. E. Sacco, *Sci. Rep.*, 2018, **8**(1), 3021.
112. B. Zheng, W. Peng, M. Guo, M. Huang, Y. Gu, T. Wang, G. Ni and D. Ming, *Chem. Eng. J.*, 2021, **418**, 129392.
113. A. R. Kirtane, M. Verma, P. Karandikar, J. Furin, R. Langer and G. Traverso, *Nat. Nanotechnol.*, 2021, **16**(4), 369.

CHAPTER 3

Biofilm-associated Infections and Their Management

MANU SHARMA,*[a] REETIKA RAWAT,[a] EASHA PANDEY,[a] SHEELU SHARMA,[b] VEERA SADHU[c] AND RAGHAVA REDDY KAKARLA*[d]

[a] Department of Pharmacy, Banasthali Vidyapith, Banasthali, Rajasthan 304022, India; [b] Pt. Deendayal Upadhayay National Institute for Persons with Physical Disabilities, Delhi University, New Delhi 110002, India; [c] School of Physical Sciences, Kakatiya Institute of Technology and Science (KITS), Warangal, Telangana 506015, India; [d] School of Chemical and Biomolecular Engineering, University of Sydney, Sydney, NSW 2006, Australia

*Emails: aasmanu2018@gmail.com; reddy.chem@gmail.com

3.1 Introduction

Around 80% of microbial infections are due to biofilm-forming microorganisms.[1] Infections associated with biofilms are broadly categorized as non-device and device associated. Vaginitis,[2] colitis,[3] conjunctivitis,[4] gum disease,[5] urethritis,[6] and middle ear infection[7] are commonly associated with biofilm-forming bacteria. Biofilms consist of an extracellular polymeric matrix (EPM), also known as glycocalyx, slime, or capsule. Various cations such as calcium and magnesium undergo association among polymeric strands of biofilms and give strength *via* crosslinking. This matrix works as a reinforced barrier between the surface and microbial cells.[8] Owing to hydrogen bonding, it can incorporate a large amount of water and remains hydrated. The hydrating property of the EPM enhances the

Nanoscience & Nanotechnology Series No. 62
Smart Nanomaterials for Infectious Diseases
Edited by Suvardhan Kanchi, Nidhi Sharotri, Rajasekhar Chokkareddy, Deepali Sharma and Falah H. Hussein

Published by the Royal Society of Chemistry, www.rsc.org

antimicrobial resistance of biofilms by inhibiting the mass transport of antimicrobials across the biofilm by directly binding to these agents. In addition, both hydrophilic and hydrophobic functional parts of the EPM can link with macromolecules such as DNA, proteins, lipids, and divalent cations. The quantity of EPM produced varies among different organisms and also with their age.[9]

Biofilm-associated infections commonly occur with medical implants. Biofilm growth depends on electrostatic interactions among various surfaces and bacteria. Sometimes, these interactions among flagella and fimbriae or pili of bacteria and proteins of the host result in the implanted device being completely hidden. This is mediated *via* various adhesins in microbes such as flagella. Biofilm development proceeds in a sequential manner that includes attachment, microcolony formation, and maturation resulting in dispersal.[10,11]

Various pathways are involved in making microbes in biofilms more susceptible to causing disease than planktonic microbes. Among them, the dispersal of a large number of cells from biofilms in the environment is a recognizable process through which bacteria in a biofilm start an infection on the associated surface. Second, phenotypically heterogeneous bacteria in the biofilm may survive and multiply. The third and most important pathway is the regulation of quorum-sensing networks that can regulate virulence processes in the bacteria, such as the regulation of endotoxins by Gram-negative bacteria enclosed in the biofilm.[12]

Biofilm formation can occur in both natural and synthetic environments.[13] Depending on the environment, biofilms vary in their nature. On medical devices, single or multiple biofilm-forming organisms show a high level of organization, resulting in single or multiple layers with three-dimensional structures.[14] The establishment of biofilms on devices raises the threat of infection, and sometimes causes morbidity and mortality.

3.2 Biofilm Formation

Biofilm development is an active process. Research in the field of environmental and industrial microbiology concluded that biofilm development is usually affected by the environment in which microbes live, the surface on which they will attach, and the space where they are acclimatized with water, lipids, albumin, EPM, or various other nutrients present in the environment. Biofilm development occurs in several steps. These include adherence of bacteria to the surface in a reversible manner followed by an irreversible association. Subsequently, a prominent biofilm is developed. The adherence to the surface depends on various factors such as bacterial species, the configuration of the cell surface, nature of the surface, accessibility of nutrients, communication among cells, and overall governing network (see Figure 3.1).

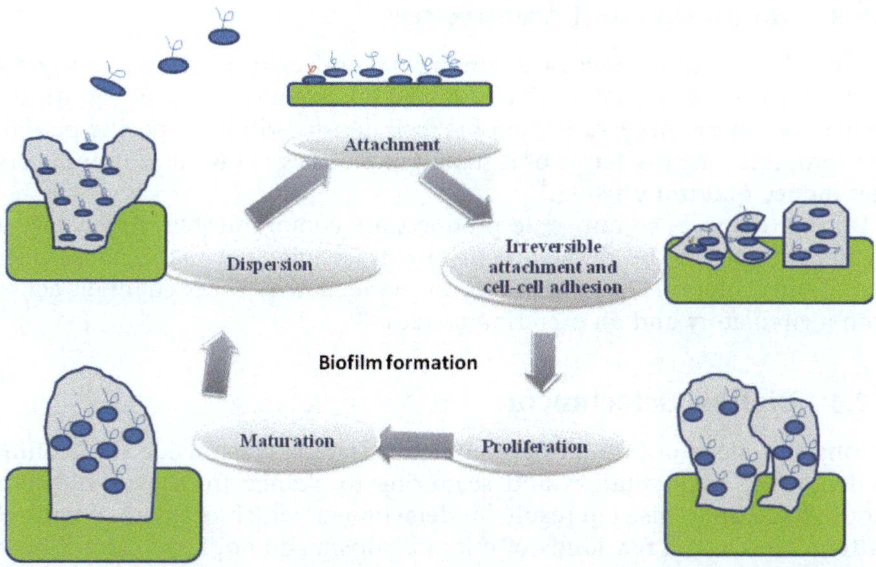

Figure 3.1 Biofilm life cycle. (1) Planktonic microscopic organisms adhere to the surface, but the association is reversible at this stage. (2) Subsequent bonding with irretrievable connection occurs. Microcolonies begin to form as the microorganisms go through cell division, allowing further attachment areas for the enlistment of various cells. (3) Extracellular polymeric substances (EPSs) are transmitted, which achieves a self-delivered network that shows the signs of the beginning of biofilm improvement. (4) The biofilm becomes completely grown, with cell-to-cell correspondence, through quorum sensing (QS), which oversees different physiological cycles. (5) The biofilm closes with a unit of small segments from the biofilm. These microscopic planktonic organisms could move and colonize various surfaces.

3.2.1 Attachment to the Surface

The initial phase of biofilm development includes attachment of microbial cells to the surface by pili or flagella, or sometimes connected through van der Waals forces, electrostatic interactions, *etc.* The hydrophobicity of the surface plays an important role in bacterial adhesion. Microorganisms adhere more to hydrophobic regions that are non-polar, such as Teflon, than hydrophilic and polar surfaces, such as metals and glass.[15]

3.2.2 Microcolony Formation

After the attachment of bacteria, microbial cells start to multiply and divide by chemical signaling within the extracellular polymeric substances, resulting in the development of microcolonies. A single bacterial colony in a biofilm consists of a diverse variety of microcommunities that harmonize with each other in various ways. Such harmonization plays a vital role in the distribution of important metabolic end products.[15]

3.2.3 Maturation and Architecture

Microbial cells convey messages among themselves by autoinducer signals after microcolony formation.[16] Cell-to-cell communication is a significant cycle to attain the necessary microbial cell density within a specific period. This promotes the discharge of signaling molecules known as autoinducers that induce quorum sensing.[17]

During this step, certain gene products are communicated, which play a significant role in the development of extracellular polymeric substances (EPSs) and evolution of interstitial voids in the matrix. These channels act as both a circulatory and an excretion system.[18]

3.2.4 Biofilm Detachment

During this stage of biofilm development, bacteria cells inside the biofilm undergo fast augmentation and scattering to change from a sessile to a motile structure. This step results in detachment, which occurs in a natural pattern. However, a few kinds of microorganisms do not form extracellular polysaccharides, and the bacterial cells scatter straightforwardly in the environment. Occasionally, mechanical stress may also be associated with this interaction.

Throughout the separation process, microbial networks inside the biofilm yield various saccharolytic enzymes. These enzymes assist in delivering a surface to organisms in another area for colonization. In addition, bacterial cells express the proteins connected with flagella development and allow microbes to move to another location. Separation of microbial cells and transfer to a pristine location help in the dissemination of diseases.[19]

3.3 Pathophysiology of Bacterial Biofilm Infections

The pathogenesis of bacterial infection increases due to the formation of a biofilm, as it protects the bacteria from the external environment. It has been reported that during the pathogenesis of bacterial infection, both innate and adaptive immunities become activated.[20]

A host immune response is crucial for the formation of biofilms and for their pathogenesis (see Figure 3.2). The biofilm releases cyclic dinucleotides (CDNs), which initiate an immune response *via* a stimulator of interferon genes (STING), resulting in activation of type 1 interferons (IFNs). Further, IFNs trigger helper T cells (TH cells) with neutrophils and macrophages along with the activation of natural killer cells (NK cells). On initiation, NK cells release interferon-gamma (INF-γ), which develops antibodies in company with more neutrophils and macrophages. However, the defensins barrier, antigen masking, and anticipation of phagocytosis inhibit the role of defensins by neutrophils, antibodies, and phagocytosis by macrophages, respectively.[21]

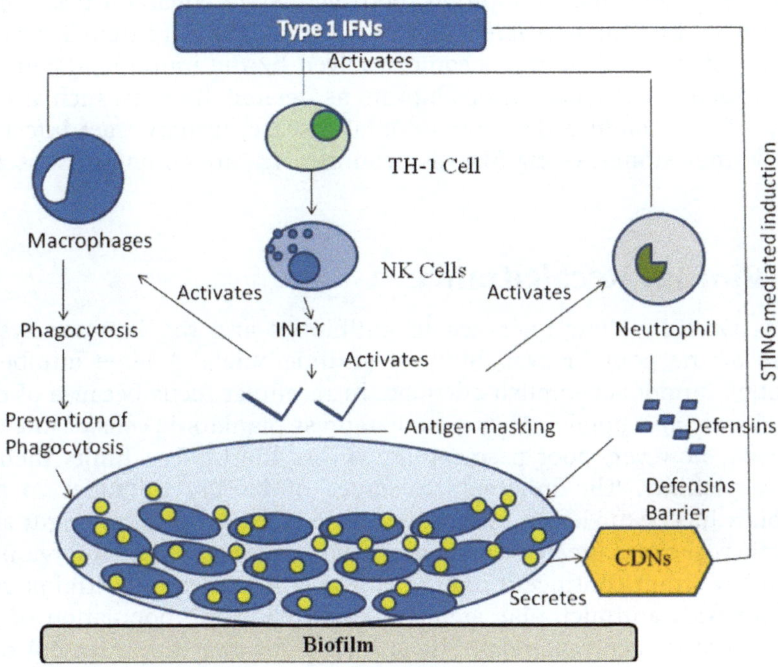

Figure 3.2 Pathogenesis of bacterial biofilm infection.

3.4 Biofilm-associated Infection

A huge number of microbes are known to form biofilms. Biofilm-related infections are difficult to analyze using normal culture methods, which may not precisely distinguish biofilm-related disease. Biofilm formation by microbes can be initiated at various sites in the human body. It can occur due to a reduced immune defense in an existing infection, or it can be acquired in hospital as biofilms survive over non-living foreign bodies.[22] Additionally, immunocompromised patients are more prone to bacterial biofilm infections. Patients fitted with medical devices such as implants are also more prone to such infections as the surface of the devices offers adhesion and colonization of microbes, leading to biofilm development.[23]

Biofilms are ubiquitous in nature and their formation represents a survival technique for bacteria in unfavorable conditions. While residing in the matrix of a biofilm, bacteria use various mechanisms to escape from the host defense systems. Initially, bacteria remain dormant and cause local tissue damage resulting in acute infection. Later, the bacteria become habituated to environmental anoxia and nutrient limitation, resulting in a reduced metabolic rate and cell division rate.[24] Moreover, owing to these changes, the bacteria become more resistant to antibiotic therapy by deactivating the antimicrobial targets. Although innate and acquired host immune responses become instigated in the course of biofilm infections, both fail to eradicate the biofilm pathogens. In addition, both accelerate collateral tissue damage.[25] Hence biofilm-associated diseases are usually persistent infections that grow slowly and hardly become resolved by the immune system and antimicrobial treatments. Various biofilm-associated diseases, such as otitis media, infective endocarditis, periodontal disease, urinary tract infections (UTIs), kidney stones, cystic fibrosis, wounds, *etc.*, are commonly observed globally.

3.5 Biofilm Recalcitrance

Biofilms typically show tolerance to antibiotics and the immune system because of the poor bioavailability of antimicrobials. A large number of antibiotics cannot accomplish adequate intracellular focus because of their poor solvency and uptake. They usually repress planktonic cells of infecting organisms; however, poor permeability across the biofilm limits their restorative viability. The inherent resistance of biofilm microbes to antimicrobials has been viewed as an overall peculiarity. A biofilm might show antibiotic resistance greater than that shown by planktonic microorganisms of a similar strain contingent upon a combination of various strains. After treatment with antimicrobial agents, a small enduring population of persistent microbes can repopulate the surface right away and turn out to be impermeable to additional treatment.[26] Biofilm recalcitrance is a multifactorial phenomenon.

3.5.1 Mechanisms of Recalcitrance

As soon as a biofilm is entrenched, it acts as a shield for bacteria. As a result, the bacteria become capable of surviving after several types of physiochemical aggression.[27] Moreover, the bacteria in a biofilm display a characteristic property of survival, *i.e.* recalcitrance. This is their capability to endure in the presence of antibiotics at higher concentrations, which subsequently results in treatment failure and recurrence of infection. However, microbes are fully susceptible to antimicrobials *in vitro* under planktonic conditions. Recalcitrance is a complex phenomenon that includes various tolerance and resistance mechanisms as described below (see Figure 3.3).[28]

3.5.2 Resistance

This is the capability of a microbe to grow in the presence of bacteriostatic or bactericidal antimicrobials.[29] Resistance is evaluated by estimating the minimum inhibitory concentration of the active constituent. Various mechanisms behind this process include antibiotic efflux, lower penetration of antibiotics, and enzyme activity that affects antimicrobial activity. These processes allow microbes to grow and avoid antibiotics. Moreover, the inheritance of resistance is genetically transferred from the precursor to the next generation, or it can be attained *via* horizontal gene transfer.[30]

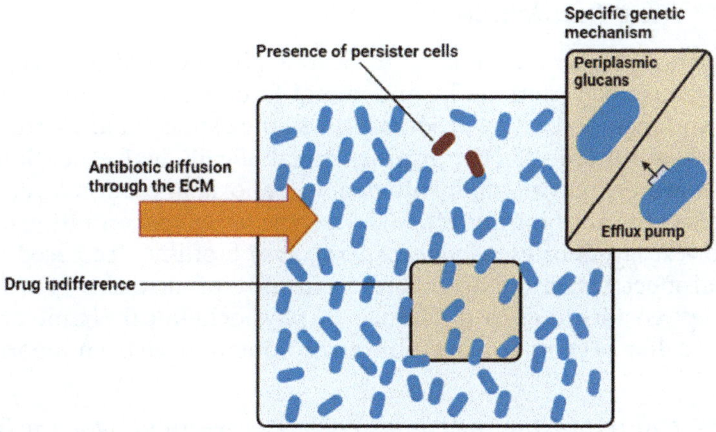

Figure 3.3 Principal mechanisms involved in the development of biofilm recalcitrance.

3.5.2.1 Tolerance

In contrast to resistance, tolerance is mostly associated with bactericidal antibiotics. According to the Clinical and Laboratory Standards Institute (CLSI), the lowest antibiotic concentration that facilitates the death of 99.9% of the bacterial population on exposure overnight is known as the minimal bactericidal concentration (MBC). Hence tolerance is explained as the lack of progress but the continuing existence of microbes in the presence of antimicrobials.[31]

Tolerance is of two kinds, physiological and phenotypic. Bacterial subpopulations present in biofilms exhibit wide variations in their metabolic rate. Subpopulations present on the inner layers of biofilms are physiologically active at a very slow rate, and usually no significant growth is observed. On the other hand, subpopulations present at the edge are highly metabolically active. Hence low metabolic rates of bacterial subpopulations affect the efficacy of all bactericidal antibiotics. Phenotypic tolerance results when the environment is unfavorable towards the action of antibiotics, resulting in a reduced eradication ability of antibiotics. However, phenotypic tolerance reverses rapidly as soon as bacteria achieve a growth-promoting medium.[32]

3.5.2.2 Antibiotic Penetration

Various reports have suggested that impaired antibiotic penetration is a reason behind biofilm recalcitrance. The mechanical and physiochemical characteristics of the bacterial extracellular matrix can lower the permeation of numerous compounds, including antibiotics and antiseptics.[33] Similarly, other studies also confirmed that positively charged aminoglycosides bind to a negatively charged extracellular polymeric matrix of biofilm. Thus, the permeation of aminoglycosides is hindered across the biofilm matrix at low drug concentrations. However, at higher drug concentrations, after saturation of surface binding a retardation of bacterial growth was observed.[34,35]

3.5.2.3 Drug Indifference

Many layers of biofilm are related to various physiochemical microenvironments because of variations in nutritional level, waste products, pH level, oxygen, and metabolized by-products across the extracellular matrix.[36] A few antibiotics are more effective against metabolically active developing microbes. However, nutritional insufficiency or anoxia in biofilm layers can adversely affect the antibiotic effect. Likewise, the lower concentration of oxygen in the lowest layers of *Pseudomonas aeruginosa* biofilms decreased the antimicrobial effect (bactericidal) of tobramycin and ciprofloxacin.[34] Lower effects of aminoglycosides due to difference in physiochemical characteristics of biofilm, *i.e.* low pH and an anaerobic environment, have been reported.[37]

3.5.2.4 Contributions of Genetically Determined Mechanisms

Various studies on microbial strains suggested that during biofilm development, efflux pumps eliminate antimicrobials from the bacterial cytoplasmic space and can induce biofilm-specific recalcitrance.[38] In a reported case, upregulation of efflux pumps such as AcrAB-TolC sequestrated antibiotics along with the toxins.[39] It has been reported that *Escherichia coli* and *P. aeruginosa* efflux pumps reduce antimicrobial actions. However, the preventive action of these pumps against antibiotics is mostly active at lower concentrations.[40]

3.5.2.5 Bacterial Persistence

Bacterial persistence is due to persister cells or persisters, which are an isogenic subpopulation of tolerant bacteria.[41] Persister cells are believed to arise *via* a phenotypic switch.[42] During the immense stress conditions, persister cells (dormant cells) that are highly recalcitrant to antimicrobial challenge form spontaneously within biofilm. Some studies have suggested that persisters are those bacterial cells that might have escaped antibiotic-induced programmed cell death (PCD). The sequence of such events involves the interaction of antibiotics to achieve their goal of contributing to microbial growth inhibition, whereas some bacterial cells survive due to the inactivation of PCD.[43] Various studies have shown the occurrence of highly tolerant persister cells in a biofilm of pathogens.[44–46]

3.6 Treatment of Biofilm-associated Infections

Bacterial biofilms cause a broad range of infections progressing to chronic tissue diseases due to the use of exogenous devices such as catheters and prosthetic joints. Most of the infections lead to the disturbance of normal protein synthesis, intense respiratory distress, impeded bilirubin metabolism, and many more issues in the host. Biofilm-related microorganisms like *Pseudomonas aeruginosa* mainly colonize in the lungs of cystic fibrosis patients, *Staphylococci* spp. colonizes over clinical devices, *Streptococci* spp. in dental plaque progenesis, *etc.*[47]

Biofilm treatment for device-related infections involves the removal of the device from the site, followed by treatment with antimicrobial agents. Systemic or extracellular infections can be easily dealt with effectively in comparison with intracellular or biofilm-mediated infections owing to the lower permeability across the biofilm, higher first-pass metabolism, or low bioavailability of the drug.[48,49]

3.6.1 Biofilm Eradication Agents (BEAs)

Treatment of biofilms involves biofilm eradication agents (BEAs). An assortment of promising BEAs and their activity, design, *etc.* are discussed below.

3.6.1.1 Antimicrobial Peptides (AMPs)

AMPs are universal agents used to control the formation and eradication of biofilms. Many plants, invertebrates, and animal species are common sources of AMPs.[50] AMPs differ enormously in size (between 5 and more than 90 amino acids) and molecular mass (1–5 kDa). Most of them are cationic, although anionic antimicrobial peptides are also available.[51] The activity of AMPs as antimicrobial agents is still not yet completely understood, but their mechanism of action is frequently connected with disordering of cytoplasmic membranes followed by cell lysis and hampering of the synthesis of cell walls, DNA, RNA, and proteins along with activation of autolysins that induce the death of microbes.[52] In addition, AMPs prevent the surface colonization of microbes and facilitate the disruption of the biofilm architecture.

3.6.1.2 Quaternary Ammonium Amphiphiles

These compounds have been widely used as bactericidal antibiotics for decades. Quaternary ammonium cations (QACs) have functional end hydroxyl groups on the surface, which can be altered to activate the –OH groups. The lipophilic alkyl chain of QACs facilitates their penetration into bacterial cell walls and membranes, leading to autolysis. QACs are toxic to eukaryotic cells. Researchers are continuously working to devise less toxic analogs of QACs.[53,54]

3.6.1.3 Antimicrobial Lipids

Antimicrobial lipids are single-chain lipid amphiphiles consisting of fatty acids and monoglycerides. They follow various mechanisms such as membrane permeability enhancement, cellular lysis, and damaging of electron transport chains. Various lipids such as monoglycerides (glycerol monolaurate) and fatty acids (docosahexaenoic acid and eicosapentaenoic acid) cause extensive damage to the outer biofilm membrane and have been found to be safe at concentrations up to 100–200 μM. Antimicrobial lipids have very low toxicity towards eukaryotic cells and promising biofilm eradication potential.[55]

3.6.1.4 Phenazines and Quinolines

Phenazines are redox-active secondary metabolites, normally developed *via* numerous microbial strains, *e.g.* *P. aeruginosa*,[56] *Pantoea agglomerans*,[57] *etc.* They are composed of a dibenzo annulated pyrazine moiety.[58] Halogenated phenazines have been found to be non-toxic to mammalian cells with a high potential to eradicate biofilms. They show inhibition of DNA synthesis *via* bacterial DNA cleavage in DNA–enzyme complexes, resulting in rapid bacterial death. Quinolines resemble phenazines in structure, but their activity is restricted to Gram-positive bacteria such as *Staphylococcus aureus*, *Staphylococcus epidermidis*, and *Enterococcus faecium*.

3.6.1.5 Nitric Oxide-releasing Antibiotics

Nitric oxide utilization in biofilm dispersion has shown success, although it is notably difficult to deal with nitric oxide and its organization and delivery to an objective site are frequently troublesome. Low non-toxic levels of nitric oxide may activate biofilm dispersal in various microbial species. This implies that nitric oxide in combination with the antimicrobial agent is able to destroy dispersed cells.[59]

3.6.1.6 Antibiotic Treatment Strategies for Combating Biofilm Infections

Biofilms composed of microorganisms intrinsically show greater tolerance to antimicrobial treatment than planktonic cells of that particular strain. On the basis of the results of antimicrobial susceptibility tests in clinical microbiology studies, therapeutic outcomes may not generally be obtained owing to the antimicrobial tolerance of biofilms. These include the following.

3.6.1.6.1 Utilizing Higher Antibiotic Concentrations. Topical administration facilitates higher concentrations by providing antibiotics straightforwardly to the infection site with low serum concentrations and associated difficulties. Topical antibiotics are useful when the target site is situated in an avascular area where the parenteral route fails to facilitate the drug. Moreover, this approach might diminish the possibilities of antibiotic resistance.[60]

3.6.1.6.2 Integrated Consecutive Antimicrobial Therapies. Antimicrobial integrated therapies are regularly utilized for preventing or delaying the initial stage of resistance to biofilm-related infections.[61] Biofilms are bacterial networks that show different primary regions and various metabolic states. Thus, combining a few agents (*e.g.* ciprofloxacin, tobramycin, or β-lactams) striking on metabolically activated layers with others (*e.g.* colistin), which particularly eradicates biofilm cells with lower metabolic activity, provides a coherent way to deal with establishing combination therapy.[62]

3.6.1.6.3 Adjuvants to Increase Antimicrobial Effects. Antibiotic adjuvants are widely employed nowadays to overcome antibiotic resistance. This strategy can increase the lifespan of these life-saving medications. These agents are synonymously known as resistance breakers or antibiotic potentiators.[63] They have low antibiotic potential, but co-administration with an antimicrobial helps in impeding antibiotic resistance. β-Lactamase inhibitors, efflux pump inhibitors, and outer membrane permeabilizers are commonly used antibiotic adjuvants to obstruct the mechanisms of antibiotic resistance. Adjuvant drug therapies also help in avoiding the need for efforts to develop challenging and expensive new therapies from the drug discovery point of view.[64]

3.7 Strategies for Biofilm Control

Among the constraints of various preventive approaches, the majority of them depend on the utilization of antibiotics. Hence numerous approaches have been adopted to recognize preventive systems in light of particular information available on bacterial adherence and biofilm arrangement (see Figure 3.4).

3.7.1 Inhibiting Microbial Adhesion

Biofilms cannot be formed without the adhesion of microbes on the surface. Hence, in order to hinder microbial adhesion, the underlying processes in biofilm arrangement must be targeted, as described below.

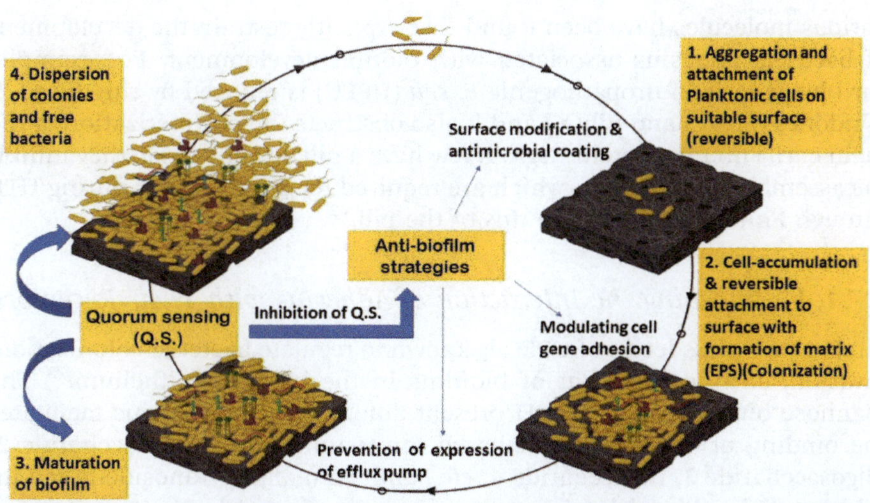

Figure 3.4 Various anti-biofilm strategies to prevent and eradicate biofilm formation.

3.7.1.1 Material Optimization, Surface Modifications, and Biosurfactants

The primary microbial attachment for the genesis of biofilms depends upon bacterial and surface factors. Metallic surfaces such as zirconium oxide and pure titanium revealed anti-biofilm adhesion properties.[65] Similarly, a silicone coating over a polymeric coat can considerably reduce bacterial attachment to the surface.[66] However, the method of biomaterial fabrication can also alter the roughness and physicochemical properties of the surface and consequently influence bacterial attachment.[67]

A further significant methodology for lessening bacterial attachment is to adjust the surface by safeguarding it using anti-adhesive molecules. For example, a peptide and its pegylated analog coated over Ti metal substantially impeded the adsorption of fibronectin and *S. aureus* attachment.[68] Similarly, biosurfactants, surface-active materials produced by many microbes, hinder the attachment of contenders by modifying surface properties such as wettability and surface charge and preventing the formation of bacterial bonds.[69,70]

3.7.1.2 Non-antibiotic Coatings

In view of limitations associated with antibiotic coatings, *e.g.* their impact is limited to non-resistant microscopic organisms, various groups have attempted to identify non-antibiotic coatings for hindering microbial attachment. The utilization of controlled-release hydrogel systems loaded with antibodies provides an opportunity to decrease bacterial attachment, improve bacterial eradication, and reduce the likelihood of infection.[71]

3.7.1.3 Inhibition of Production of Adhesins

Various molecules have been found that explicitly restrain the development of bacterial adhesins associated with biofilm development. For example, curli biogenesis in uropathogenic *E. coli* (UPEC) is reduced by ring-fused 2-pyridones (FN075 and BibC6) and it also obstructs the polymerization of the main curli subunit protein CsgA. A few have a pilicide effect, *i.e.* they inhibit the assembly of type 1 pili, which are required for pathogenesis during UTI, through FimH adhesin at the tips of the pili.[72]

3.7.1.4 Impeding the Interaction of Adhesins with Their Receptors

FimH type 1 pilus lectin of UPEC is known to regulate bacterial colonization, invasion, and development of biofilms in the bladder epithelium.[73] The mannose binding site of FimH present among all *E. coli* strains facilitates the binding of oligosaccharide-based FimH inhibitors (oligosaccharide 2, oligosaccharide 3, trisaccharide 4, *etc.*) and α-D-mannopyranoside-based inhibitors (*e.g.* thiazolylaminomannosides, neothiazolylaminomannosides, *etc.*). Thus, mannosides inhibit UTIs when administered prophylactically.[73]

3.7.1.5 *Use of Lactoferrin*

The use of lactoferrin is a successful approach to inhibiting biofilm development. Lactoferrin, an iron chelator, is known to impede the irrevocable attachment of *P. aeruginosa* to the surface by inhibiting the twitching motility of bacteria.[74]

3.7.1.6 *Cyclic Di-GMP Biosynthesis Inhibition*

Guanosine monophosphate (GMP) biosynthesis plays the chief role in the development of biofilms from planktonic form. Hence inhibition of GMP biosynthesis by diguanylate cyclase (DGC) may stop the development of biofilms and can keep the bacteria in a planktonic state.[75]

3.7.1.7 *Physical Approaches*

Many physical strategies to inhibit biofilm formation as a defensive procedure have been explored.[72] A modification of urethral catheters was introduced to convey current to electrodes present in the catheter tip, which yields ions of soluble salts, acting as a local biocide. The outcome of the study revealed a reduction in the bacterial burden in urine as acoustic waves, along with antimicrobials, helped in the eradication of biofilms.[76]

3.7.1.7.1 Confining Communication. The purpose of this strategy is to constrain biofilm development by modifying the primary connection of bacteria and progression of microcolonies. Quorum sensing (QS) is an important feature for biofilm communication, and interference with it can lead to the complete eradication of bacteria. Various molecules have been developed that inhibit QS signals to prevent the formation of a *P. aeruginosa* biofilm. The ability of azithromycin to interfere with *lasI*-mediated QS signals contributes to its successful inhibitory effect on *P. aeruginosa* biofilm formation.[77]

3.7.1.7.2 Utilization of Non-pathogenic Bacteria for Prevention of Colonization. Non-pathogenic bacteria are capable of colonizing on the surface and can resist the attachment of various microbes to the surface.[78] *E. coli* 83972 strain has few virulence factors and UTI-associated adhesins, due to which it is unable to cause inflammation but can compete with virulent bacteria for surface adhesion.[79]

3.7.1.7.3 Antimicrobial Lipids (AMLs). The antimicrobial capability of monoglycerides and unsaturated fats has been constantly tested against different microorganisms.[80,81] AMLs have the capability to improve film penetrability as well as destabilize bacterial cell membranes *via* direct or indirect inhibitory effects. Studies have focused on the surface transduction framework, electron transport chain annihilation, cell lysis, and bacterial enzyme inhibition.[82] A few investigations have reported the adequacy of AMLs to inhibit biofilm at lower doses for various strains.[83]

3.8 Management of Biofilm Infections

As revealed in a few studies, biofilm infections are challenging to manage and often antimicrobial treatment alone is insufficient. For the most part, the procedures can be categorized as including a foreign body or not. If a foreign body is not included, long-term treatment with high doses and often utilizing combination therapy of antimicrobials with various eradication mechanisms can sometimes eliminate the infections. If a foreign body is involved, removal of the material is usually vital for a successful result. In some cases, eradication of the biofilm is conceivable, followed by constant biofilm suppressive treatment.

3.8.1 Removal of Foreign Bodies

According to certain research, a foreign body downregulates the phagocytosis and intracellular bactericidal impacts of polymorphonuclear leukocytes. The foreign body provides an optimal surface to microbes for connection, although polymorphonuclear leukocyte functions are impaired because of the presence of the foreign body. In this way, the presence of a foreign body enhances the chance of biofilm infections. Depending on the antimicrobial resistance characteristics of the biofilm, it is currently difficult to resolve biofilm infections by conventional antimicrobial treatments. Hence the removal of the foreign body is essential for the treatment of such biofilm diseases. It is consequently strongly recommended to remove the devices implanted in patients for medical reasons or to replace the infected device with a new one, to cure the biofilm infections.

3.8.1.1 Change of Infected Central Venous Catheter (CVC) or Dialysis Catheter

At the point when microorganisms form a biofilm on a CVC or dialysis catheter, intermittent bacteremia with an indistinguishable bacterial stain could be expected. To treat such catheter biofilm diseases, a change of the infected catheter is critical, followed by a brief treatment with sensitive antibiotics intravenously to remove the bacteria released from the infected catheter into the bloodstream. In the case that a change of catheter is not possible temporarily, antimicrobials and other treatments might assist in limiting the arrival of planktonic bacterial cells from the catheter biofilm, and that implies the instillation of large groupings of antimicrobials regardless of whether hostile to coagulants or of 70% ethanol or hydrochloric acid into the lumen of the CVC.[84–87]

3.8.1.2 Change of Infected Urinary Catheter (UC)

Catheter-related UTIs are the most widely recognized nosocomial disease, and are related to the formation of microbial biofilms in a UC. Notwithstanding intermittent UTIs with indistinguishable pathogens, they can likewise result in urosepsis. Changing an infected UC is easy; nonetheless, the time needed to change it is significant. It is necessary to change the contaminated UC after 48 h of adequate and sensitive antibiotic treatment in order to limit the bacterial concentration in the bladder and urinary tract; in any case, the new UC would be colonized rapidly by the microbes to form a new biofilm.

3.8.1.3 Change of Infected Joint Prostheses

Prosthesis-related infection is a serious complication in patients with joint replacement, and it is a biofilm-related disease with a poor prognosis. In the event that prosthesis infection is detected, the difference in the infected prosthesis in most cases is the principal factor. Assuming that a prosthetic insert loosens due to biofilm infection, an exchange of the prosthesis in combination with delicate and aggressive antibiotic treatment is recommended.[88,89]

3.8.1.4 Changes of Other Infected Indwelling Devices

Endocarditis patients with prosthetic heart valves or cardiovascular pacemakers are in danger of intermittent sepsis, cardiac insufficiency, and infective embolic complications. In that case, a change of the infected device in combination with aggressive and delicate antimicrobial treatment becomes necessary. For patients with biofilm infections in biliary stents, endotracheal tubes, dead bones (constant osteomyelitis), and biliary and urinary stones (biliary and urinary tract diseases), viable antibiotic treatments and expulsion of the infected foreign bodies are essential in order to treat the infections.[90,91]

3.9 Early and Aggressive Antibiotic Treatments Against Biofilm Infections

In vitro results showed that the early stage of biofilms could be simply cleared by antimicrobial treatment, in contrast to developed biofilms.[92] Hence early and aggressive antibiotic therapies are suggested for biofilm infections. However, early determination of biofilm infection is currently troublesome, and a large proportion of clinical biofilm diseases are really developed biofilms that are generally difficult to destroy with antimicrobial treatment.[93,94] It is consequently significant and urgent to apply currently available antimicrobials in the treatment of biofilm contaminations.

Various accompanying factors ought to be considered when an antimicrobial treatment against biofilm disease is to be chosen, as outlined below.

3.9.1 Selection of Antibiotics

Treatment of biofilm disease requires sensitive and well-penetrating antimicrobials to guarantee an adequate concentration of efficacious antimicrobial at the site of biofilm infection. Generally, macrolides, lincosamides, tetracyclines, rifamycins, quinolones, fusidic acid, nitroimidazole, sulfonamides, and oxazolidinones penetrate better in tissues and cells than β-lactams (including penicillins, cephalosporins, and carbapenems), aminoglycosides, glycopeptides, and polymyxins. It is notable that disease could prompt inflammation, which brings about faster metabolism and a huge utilization of oxygen locally or systemically. In the event that the oxygen supply could not satisfy the need, glycolysis will occur, prompting acidosis, and the impact of antimicrobials could be affected by the pH value. It has been reported that a low pH (pH 5.2) could reduce the impact of β-lactam antimicrobial agents and increase the impact of rifamycin. Consequently, antimicrobial treatment and correction of acid-base balance disorders could be important for the treatment of biofilm infections.

3.9.2 Administration of Antibiotics

Higher doses of antimicrobials considering renal and hepatic safety are recommended depending upon the antimicrobial tolerance and resistance in biofilms.[48] Mixed antimicrobial treatment was subsequently suggested for the treatment of biofilm infections. According to the character of antimicrobial tolerance and resistance in biofilms and the high minimum inhibitory concentration (MIC) and MBC of biofilm cells, high doses of antimicrobials under the protected scope of renal and hepatic capabilities are recommended.[94] Moreover, an appropriate duration of antibiotic treatment is also important.

3.9.3 Pharmacokinetics (PK) and Pharmacodynamics (PD) of Antibiotics in Biofilm Infections

Microbes growing in a biofilm could become 10–1000 times more resistant and tolerant to antimicrobials compared with their planktonic counterparts. Available antimicrobials for the treatment of highly resistant bacterial infections are limited. Accordingly, dose optimization of currently available antimicrobial agents is critical for further developments to improve efficacy. The PK and PD of antimicrobial agents can be utilized to predict the effect of antimicrobial regimens to achieve the greatest bactericidal impact against infections. Recent studies have shown the different PK and PD profiles of antimicrobials among planktonic and biofilm infections. PK and PD data on the effect of antimicrobial agents on biofilm-related microorganisms can be applied to optimize dosage regimens for biofilm infections. The minimum biofilm inhibitory concentration (MBIC) and minimum biofilm eradication concentration are two PD parameters for antimicrobials in biofilm infections.[95]

3.10 Conclusion

The difficulty of eradicating biofilm-associated infections has a huge effect on the healthcare management system. Biofilm development is a multiple-step active process involving a cascade of events. Therefore, a variety of approaches can be utilized to target various sites for the eradication of biofilms. Usually, device-associated infection management involves the removal of the device followed by subsequent treatment with broad-spectrum antibiotics.

References

1. Z. Khatoon, C. McTiernan, E. J. Suuronen, T. F. Mah and E. I. Alarcon, *Heliyon*, 2018, **4**(12), 1–36.
2. D. Machado, J. Castro, A. Palmeira-de-Oliveira, J. Martinez-de-Oliveira and N. Cerca, *Front. Microbiol.*, 2016, **6**, 1–13.
3. E. C. Von Rosenvinge, G. A. O'May, S. Macfarlane, G. T. Macfarlane and M. E. Shirtliff, *Pathog. Dis.*, 2013, **67**(1), 25–38.
4. T. H. T. Elkashab, L. A. Adel, M. S. Nour, M. Mahran and M. Elkaffas, *J. Lab. Physicians*, 2018, **10**(3), 309–315.
5. A. P. Vieira Colombo, C. B. Magalhães, F. A. Hartenbach, R. Martins do Souto and C. Maciel da Silva-Boghossian, *Microb. Pathog.*, 2016, **94**, 27–34.
6. C. Delcaru, I. Alexandru, P. Podgoreanu, M. Grosu, E. Stavropoulos, M. C. Chifiriuc and V. Lazar, *Pathogens*, 2016, **5**(4), 1–12.
7. A. Niedielski, L. P. Chmielik and T. Stankiewicz, *Int. J. Environ. Res. Public Health*, 2021, **18**(7), 3555–3572.
8. P. D. Martino, *AIMS Microbiol.*, 2018, **4**(2), 274–288.
9. R. M. Donlan and J. W. Costerton, *Clin. Microbiol. Rev.*, 2002, **15**(2), 167–193.
10. O. E. Petrova and K. Sauer, *J. Bacteriol.*, 2012, **194**(10), 2413–2425.
11. Y. H. Li and X. Tian, *Sensors*, 2012, **12**(3), 2519–2538.
12. V. Lazar, A. M. Holban, C. Curutiu and M. C. Chifiriuc, *Front. Microbiol.*, 2021, **12**, 1–18.
13. A. Vetrivel, M. Ramasamy, P. Vetrivel, S. Natchimuthu, S. Arunachalam, G. S. Kim and R. Murugesan, *Biologics*, 2021, **1**(3), 312–336.
14. X. Li, L. Sun, P. Zhang and Y. Wang, *Coatings*, 2021, **294**(11), 1–31.
15. M. Jamal, W. Ahmad, A. Saadia, J. Fazal, I. Muhammad, M. I. Nawaz, T. A. Hussain, R. M. Muhammad and M. A. Kamil, *J. Chin. Med. Assoc.*, 2018, **81**(1), 7–11.
16. R. Vasudevan, *J. Microbiol. Exp.*, 2014, **1**(3), 84–98.

17. P. Saxena, Y. Joshi, K. Rawat and R. Bisht, *Indian J. Microbiol.*, 2019, **59**(1), 3–12.
18. J. Wille and T. Coenye, *Biofilm*, 2020, **2**, 1–10.
19. M. Otto, *Annu. Rev. Med.*, 2013, **64**, 175–188.
20. J. F. Gonzalez, M. M. Hahn and J. S. Gunn, *Pathog. Dis.*, 2018, **76**, 1–7.
21. D. Banerjee, P. M. Shivapriya, P. K. Gautam, K. Misra, A. K. Sahoo and S. K. Samanta, *Proc. Natl. Acad. Sci. U. S. A.*, 2019, 1–17.
22. Z. Khatoon, C. McTiernan, E. J. Suuronen, T. F. Mah and E. I. Alarcon, *Heliyon*, 2018, **4**(12), 1–36.
23. B. Naik, T. Dugge, V. Rajinikanth, K. R. Reddy, N. Soldatos, S. S. Shukla, M. Nadagouda and T. M. Aminabhavi, Clay-Loaded Chitosan-Based Biocompatible Composite Films in Dentistry, *Indian J. Appl. Res.*, 2019, **9**, 49–52.
24. L. K. Vestby, T. Grønseth, R. Simm and L. L. Nesse, *Antibiotics*, 2020, **59**(9), 1–29.
25. M. Jamal, W. Ahmad, S. Andleeb, F. Jalil, M. Imran, M. Asif Nawaz, T. Hussain, M. Ali, M. Rafiq and M. Atif Kamil, *J. Chin. Med. Assoc.*, 2017, 1–5.
26. C. Uruén, G. Chopo-Escuin, J. Tommassen, R. C. Mainar-Jaime and J. Arenas, Biofilms as promoters of bacterial antibiotic resistance and tolerance, *Antibiotics*, 2021, **10**(3), 1–36.
27. G. M. Teitzel and M. R. Parsek, *Appl. Environ. Microbiol.*, 2003, **69**(4), 2313–2320.
28. G. Orazi and G. A. Toole, *J. Bacteriol., Res.*, 2019, 1–47.
29. S. Singh, S. K. Singh, I. Chowdhury and R. Singh, *Open Microbiol. J.*, 2017, **11**(2017), 1–10.
30. G. Kapoor, S. Saigal and A. Elongavan, *J. Anaesthesiol., Clin. Pharmacol.*, 2017, **33**(3), 300–305.
31. K. Lewis, *Annu. Rev. Microbiol.*, 2010, **64**, 357–372.
32. R. L. Rodrigues, D. C. Lima, K. X. Sena and M. A. V. Maciel, *Rev. Soc. Bras. Med. Trop.*, 2020, **53**(20200399), 1–7.
33. G. Gebreyohannes, A. Nyerere, C. Bii and D. B. Sbhatu, *Heliyon*, 2019, **5**(8), 1–7.
34. N. P. Shetti, S. J. Malode, S. Roy, P. Chandra, K. R. Reddy and S. Chatterjee, Electroanalytical techniques for investigating biofilms: Applications in biosensing and biomolecular interfacing, in *Nanomaterials in diagnostic tools and devices*, Elsevier, 2020, ch. 11, pp. 293–329.
35. J. M. Rodriguez-Martinez, S. Ballesta and A. Pascual, *Int. J. Antimicrob. Agents*, 2007, **30**(4), 366–368.
36. D. Lebeaux, J. M. Ghigo and C. Beloin, *Microbiol. Mol. Biol. Rev.*, 2014, **78**(3), 510–543.
37. P. Baudoux, N. Bles, S. Lemaire, M. P. Mingeot-Leclercq, P. M. Tulkens and F. Van Bambeke, *J. Antimicrob. Chemother.*, 2007, **59**(2), 246–253.
38. S. J. Pamp, M. Gjermansen, H. K. Johansen and T. Tolker-Nielsen, *Mol. Microbiol.*, 2008, **68**(1), 223–240.

39. M. Kvist, V. Hancock and P. Klemm, *Appl. Environ. Microbiol.*, 2008, **74**(23), 7376–7382.
40. L. Zhang and T. F. Mah, *J. Bacteriol.*, 2008, **190**(13), 4447–4452.
41. N. Kaldalu, V. Hauryliuk, K. J. Turnbull, A. L. Mensa, M. Putrinš and T. Tenson, *Microbiol. Mol. Biol. Rev.*, 2020, **84**(4), 1–24.
42. J. Urbaniec, Y. Xu, Y. Hu, S. H. Wilson and J. McFadden, *FEMS Microbiol. Rev.*, 2022, **46**(1), 1–16.
43. Y. Tanouchi, A. J. Lee, H. Meredith and Y. Lingchong, *Trends Microbiol.*, 2013, **21**(6), 265–270.
44. K. R. Allison, M. P. Brynildsen and J. J. Collins, *Nature*, 2011, **473**(7346), 216–220.
45. A. Chauhan, D. Lebeaux, J. M. Ghigo and C. Beloin, *Antimicrob. Agents Chemother.*, 2012, **56**(12), 6310–6318.
46. B. P. Conlon, E. S. Nakayasu, L. E. Fleck, M. D. LaFleur, V. M. Isabella, K. Coleman, S. N. Leonard, R. D. Smith, J. N. Adkins and K. Lewis, *Nature*, 2013, **503**(7476), 365–370.
47. Y. Dhar and Y. Han, *Eng. Reg.*, 2020, **1**, 64–75.
48. S. Xie, F. Yang, Y. Tao, D. Chen, W. Qu, L. Huang, Z. Liu, Y. Pan and Z. Yuan, *Sci. Rep.*, 2017, **7**(1), 1–9.
49. O. Ciofu, E. Rojo-Molinero, M. D. Macià and A. Oliver, *APMIS*, 2017, **125**(4), 304–319.
50. M. Yasir, M. Willcox and D. Dutta, *Materials*, 2018, **11**(12), 2468–2483.
51. A. A. Bahar and D. Ren, *Pharmaceuticals*, 2013, **26**(12), 1543–1575.
52. B. Bechinger and S. U. Gorr, *J. Dent. Res.*, 2017, **96**(3), 254–260.
53. M. C. Jennings, L. E. A. Ator, T. J. Paniak, P. C. Kevin and W. M. Wuest, *ChemBioChem*, 2014, **15**(15), 2211–2215.
54. A. Vereshchagin, A. M. Gordeeva, N. A. Frolov, P. I. Proshin, K. A. Hansford and M. P. Egorov, *Eur. J. Org. Chem.*, 2019, **26**, 4123–4127.
55. B. Yoon, J. Jackman, V. G. Elba and N. M. Cho, *Int. J. Mol. Sci.*, 2018, **19**(4), 1114–1128.
56. B. Cezairliyan, N. Vinayavekhin, D. Grenfell-Lee, G. J. Yuen, A. Saghatelian, F. M. Ausubel and D. S. Schneider, *PLoS Pathog.*, 2013, **9**(1), 1–10.
57. A. Büyükcam, O. Tuncer, D. Gür, B. Sancak, M. Ceyhan, C. A. Bülent and K. Ateş, *J. Infect. Public Health*, 2017, **2**(4), 1–6.
58. N. Guttenberger, W. Blankenfeldt and R. Breinbauer, *Bioorg. Med. Chem.*, 2017, **25**(22), 6149–6166.
59. L. Yang, E. S. Feura, M. J. R. Ahonen and M. H. Schoenfisch, *Adv. Healthcare Mater.*, 2018, 1–18.
60. R. Smith, J. Russo, J. Fiegel and N. Brogden, *Antibiotics*, 2020, **9**(2), 1–56.
61. G. Hughes and M. A. Webber, *Br. J. Pharmacol.*, 2017, **174**(14), 2237–2246.
62. O. Ciofu, E. Rojo-Molinero, M. D. Macià and A. Oliver, *APMIS*, 2017, **125**(4), 304–319.
63. R. Berndsen, T. Cunningham, L. Kaelin, M. Callender, W. D. Boldog and B. Viering, *ACS Med. Chem. Lett.*, 2022, **13**(2), 483–491.
64. C. Z. Bello, *Bioorg. Med. Chem. Lett.*, 2017, **27**(18), 4221–4228.

65. A. Scarano, M. Piattelli, S. Caputi, G. A. Favero and A. Piattelli, *J. Periodontol.*, 2004, **75**, 292–296.
66. A. L. Hook, C. Y. Chang, J. Yang, J. Luckett, A. Cockayne, S. Atkinson, Y. Mei, R. Bayston, D. J. Irvine, R. Langer, D. G. Anderson, P. Williams, M. C. Davies and M. R. Alexander, *Nat. Biotechnol.*, 2012, **30**(9), 868–875.
67. B. Grossner-Schreiber, J. Teichmann, M. Hannig, C. Dorfer, D. F. Wenderoth and S. J. Ott, *Clin. Oral. Implants Res.*, 2009, **20**(8), 817–826.
68. X. Khoo, P. Hamilton, G. A. O'Toole, B. D. Snyder, D. J. Kenan and M. W. Grinstaff, *J. Am. Chem. Soc.*, 2009, **131**(31), 10992–10997.
69. O. Rendueles and J. M. Ghigo, *Microbiol. Rev.*, 2012, **36**(5), 972–989.
70. A. E. Zeraik and M. Nitschke, *Curr. Microbial.*, 2010, **61**(6), 554–559.
71. D. Lebeaux, J. M. Ghigo and C. Beloin, *Microbiol. Mol. Biol. Rev.*, 2014, **78**(3), 1–34.
72. L. Cegelski, J. S. Pinkner, N. D. Hammer, C. K. Cusumano, C. S. Hung, E. Chorell, V. Aberg, J. N. Walker, P. C. Seed, F. Almqvist, M. R. Chapman and S. J. Hultgren, *Nat. Chem. Biol.*, 2009, **5**(12), 913–919.
73. C. K. Cusumano, J. S. Pinkner, Z. Han, S. E. Greene, B. A. Ford, J. R. Crowley, J. P. Henderson, J. W. Janetka and S. J. Hultgren, *Sci. Transl. Med.*, 2011, **3**(109), 109–115.
74. P. Avery, S. Morton, J. Raitt, H. M. Lossius and D. Lockey, *J. Scand. Trauma, Resusc. Emerg. Med.*, 2021, **29**(1), 64.
75. S. Xuan, H. F. Zhang, C. L. Ma, W. Hua and B. M. Li, *J. Infect. Dis. Med. Microbiol.*, 2021, **30**(2), 1–11.
76. M. Kopel, E. Degtyar and E. Banin, *Biofouling*, 2011, **27**(7), 701–710.
77. A. Kumar, J. O'Bryan and P. Krause, *Pathogens*, 2021, **10**(2), 1447–1461.
78. D. Ribet and P. Cosart, *Microbes Infect.*, 2015, **17**(3), 173–183.
79. H. U. Klemm, D. Steven and C. Johnsen, *Heart Rhythm*, 2007, **4**(2), 587–592.
80. S. C. Karthikeyan, S. Velmurugan, M. B. S. Donio, M. Michaelbabu and T. Citarasu, *Ann. Clin. Microbiol. Antimicrob.*, 2014, **13**(2014), 1–11.
81. W. Wang, R. Wang, G. Zhang, F. Chen and B. Xu, *J. Food Prot.*, 2020, **83**, 331–337.
82. A. D. Verderosa, M. Totsika and K. E. Fairfull-Smith, *Front. Chem.*, 2019, **7**, 824–841.
83. K. T. Yuyama, M. Rohde, G. Molinari, M. Stadler and W. R. Abraham, *Antibiotics*, 2020, **9**(11), 788–798.
84. N. Fernandez-Hidalgo and B. Almirante, *Expert Rev. Anti-Infect. Ther.*, 2014, **12**(1), 117–129.
85. M. A. Vandenhend, J. Buret, F. Camou, P. Morlat and F. Bonnet, *Diagn. Microbiol. Infect. Dis.*, 2014, **78**(4), 497–498.
86. M. Tan, J. Lau and B. J. Guglielmo, *Ann Pharmacother.*, 2014, **48**(5), 607–615.
87. M. Madsen and S. Rosthoj, *Scand. J. Infect. Dis.*, 2013, **45**(1), 38–44.
88. Z. Song, L. Borgwardt, N. Høiby, H. Wu, S. T. Sorensen and A. Borgwardt, *Orthop. Rev.*, 2013, **5**(2), 65–71.

89. W. Zimmerli and C. Moser, *FEMS Immunol. Med. Microbiol.*, 2012, **65**(2), 158–168.
90. R. Mocchegiani and M. Nataloni, *Cardiovasc. Hematol. Disord.: Drug Targets*, 2009, **9**(4), 240–248.
91. M. Nataloni, M. Pergolini, G. Rescigno and R. Mocchegiani, *Eur. J. Cardiovasc. Med.*, 2010, **11**(12), 869–883.
92. W. Hengzhuang, H. Wu, O. Ciofu, Z. Song and N. Høiby, *Antimicrob. Agents Chemother.*, 2010, **11**(12), 869–883.
93. N. Høiby, O. Ciofu, H. K. Johansen, Z. J. Song, C. Moser, O. P. Jensen, S. Molin, M. Givskov, T. T. Nielsen and T. Bjarnsholt, *Int. J. Oral. Sci.*, 2011, **3**(2), 55–65.
94. W. Hengzhuang, H. Wu, O. Ciofu, Z. Song and N. Høiby, *Antimicrob. Agents Chemother.*, 2012, **56**(5), 2683–2690.
95. S. M. Moskowitz, J. M. Foster, J. Emerson and J. L. Burns, *J. Clin. Microbiol.*, 2004, **42**(5), 1915–1922.

CHAPTER 4

Green Nanomaterials Zinc Oxide and Chitosan for Antimicrobial Activity Against Oral Pathogens

A. S. RAJESHKUMAR,* B. D. PAVITHRA, C. M. THARANI, D. G. SULOCHANA AND E. A. JAYASREE

Nanobiomedicine Laboratory, Centre for Global Health Research, Saveetha Medical College and Hospitals, Saveetha Institute of Medical and Technical Sciences, Chennai, Tamil Nadu 602105, India
*Email: ssrajeshkumar@hotmail.com

4.1 Introduction

Nanotechnology involves the design and manipulation of materials at the nanoscale level to create unique properties for various applications.[1] Over the past decade, the potential of nanotechnology has been recognized due to the effectiveness of metal nanoparticles (NPs) as antimicrobial agents against pathogenic microorganisms such as bacteria, fungi, yeasts, algae, and viruses.[2] This is due to the high area-to-volume ratio of nanoscale materials, which can address the issue of multidrug-resistant microorganisms.[3]

Nanoparticles, owing to their unique properties at the nanoscale level, have found extensive applications in various fields, including dentistry. In the dental field, NPs have become increasingly popular owing to their potential to improve oral health, prevent and treat dental diseases, and

Nanoscience & Nanotechnology Series No. 62
Smart Nanomaterials for Infectious Diseases
Edited by Suvardhan Kanchi, Nidhi Sharotri, Rajasekhar Chokkareddy, Deepali Sharma and Falah H. Hussein

Published by the Royal Society of Chemistry, www.rsc.org

enhance the performance and longevity of dental materials and devices. The utilization of NPs in the dental field has been observed in a number of products and applications, such as toothpastes, dental sealants, restorative materials, and more.[4]

One of the most common applications of NPs in the dental field is in toothpastes, where they are used to improve oral hygiene and prevent the formation of plaque and cavities.[5] The main component of these toothpastes is often titanium dioxide, a white pigment that is used to improve the toothpaste's visual appeal and to act as a mild abrasive to remove surface stains. Additionally, silica, another type of NP, is also commonly used in toothpastes to serve as a mild abrasive and improve the toothpaste's consistency.[6]

Another application of NPs in the dental field is in the form of dental sealants, which are thin, protective coatings applied to the surfaces of teeth to prevent the formation of cavities.[7] Dental sealants typically contain NPs of a resin material that is cured to form a durable and long-lasting protective layer. The use of NPs in these materials helps to increase their strength, improve their bonding to the teeth, and reduce the chances of premature failure.[8]

In addition to toothpastes and dental sealants, NPs are also utilized in various restorative materials used in dentistry, such as composite resins, ceramics, and glass ionomer cements.[9] The use of NPs in these materials enhances their physical and mechanical properties, making them more durable, resistant to wear and tear, and longer lasting.[10]

Furthermore, NPs are also being explored for their potential to prevent and treat various dental diseases, such as periodontitis and gingivitis. For instance, NPs of silver have been shown to have strong antibacterial properties, and hence have been used to develop antimicrobial dental materials that can effectively kill harmful bacteria and prevent the growth of plaque and gingivitis.[11]

One example of the use of NPs in dentistry is in the development of scanning sprays for CAD/CAM (computer-assisted design/computer-aided manufacturing) systems. These sprays release NPs to aid in the creation of precise digital models of teeth and gums, which are used in the design and production of dental restorations.[12] In addition, NPs can also be found in occlusion indicator foils, which are used to help dentists identify areas of bite-related problems.[13]

Another source of NPs in the dental field is as by-products from milling processes for fillers. Many dental materials, including resin-based composites, cements, and impression materials, contain fillers, and it is estimated that around 3500 dental materials contain NPs. This chapter specifically highlights the antimicrobial activity of zinc oxide and chitosan NPs against oral pathogens.[14]

Zinc oxide (ZnO) NPs have attracted significant attention in dentistry owing to their unique properties, including biocompatibility, antimicrobial activity, and ability to enhance the mechanical and physical properties of dental materials. In dentistry, ZnONPs have been used in various applications, such as the following:

- Dental fillings: ZnONPs have been incorporated into dental composites and fillings to enhance their physical properties, including improved mechanical strength and increased resistance to wear.
- Dental cements: ZnONPs have been added to dental cements to improve their antimicrobial properties and to enhance the adhesion to tooth structures.[15]
- Toothpastes: ZnONPs have been used in toothpastes as an antimicrobial agent, helping to prevent the growth of bacteria and plaque formation.
- Tissue engineering: ZnONPs have been used in tissue engineering to promote the growth and regeneration of dental tissues, including periodontal and pulp tissues.[16]
- Dental implants: ZnONPs have been added to dental implant surfaces to improve osseointegration and to enhance the antimicrobial properties of the implants.[17]

Chitosan NPs have been actively researched for their potential applications in the field of dentistry. These biocompatible and biodegradable NPs are made from chitin, a naturally occurring polysaccharide. Chitosan NPs have been explored for use in dental treatments and preventions, such as the following:

- Drug delivery: chitosan NPs can be used as a carrier for drugs and antimicrobials in dentistry. The NPs can protect the drugs from degradation and enhance their release, allowing for more effective treatment.[18]
- Caries prevention: chitosan NPs have been shown to have antifungal and antibacterial properties, making them a potential candidate for preventing dental caries and promoting oral health.
- Wound healing: the biocompatibility and wound-healing properties of chitosan NPs make them a promising option for promoting tissue regeneration in the oral cavity.
- Teeth whitening: chitosan NPs have been studied for their ability to absorb pigments and stains, making them a potential candidate for teeth whitening treatments.[19]
- Dental implant surface modification: chitosan NPs can be used to modify the surface of dental implants, promoting osseointegration and improving the success rate of implantation procedures.[20]

In conclusion, nanotechnology has proven to be a valuable tool in the advancement of various fields, including dentistry. The unique properties of NPs such as small size and high surface area-to-volume ratio have allowed the development of new materials and applications with improved performance and functionality. ZnO and chitosan NPs, for example, have shown promise in various applications within dentistry, such as drug delivery, wound healing, and dental tissue engineering. As research and development in this field continue, the potential applications of nanotechnology in dentistry are likely to expand and further enhance the quality of care provided to patients.

4.2 Oral Pathogens

The oral cavity is one of the major sources of all microorganisms in the human body, which is home to trillions of them. The colonies of microorganisms in the human mouth range in number from 700 to 1000 species.[21] Oral microbial communities, also known as "dental plaque," are one of the most complex bacterial floras associated with the human body. More than 700 different bacterial species have been identified in the human oral cavity, most of which are linked to dental plaque.[22] The major source of nutrients for better and healthier associated biofilms is saliva. The oral microbiota's metabolism of host glycoproteins and proteins requires microbial collaboration and the sharing of a diverse repertoire of glycosidases and proteases. The catabolism of these salivary molecules allows oral microorganisms to grow with minimal changes in the pH level of the environment.[23] Microbial species that grow optimally in the pH range 6.5–7.5 result in biofilm homeostasis and the formation of a variety of nutritional interdependencies among the bacteria in the presence of a stable pH.[24]

Oral microbes have made important contributions to the development and treatment of oral illness and oral cancer. Through a damaged oral mucosa, oral bacteria can enter the blood circulation system and increase systemic antibody levels, which increases the risk of developing a number of cardiovascular problems[25,26] (Figure 4.1).

Oral infections may arise either from an endogenous source, one yielding microorganisms normally found in the mouth, such as the main plaque related diseases, *e.g.* dental caries and periodontal disease, or from an exogenous source, yielding microorganisms not normally found as part of the oral microflora.[27] Dental caries and periodontal disease involve the adherence of bacteria to and the development of biofilms on both natural and restored tooth surfaces.[28] Inflammatory periodontal disease and dental caries are the two most common diseases in the world. Bacteria play an important role in dental plaque and disease progression in both of these oral diseases.[29] Dental caries cause tooth decomposition due to the loss of hydroxyapatite (calcium phosphate) crystals from the enamel matrix, as described in ancient writings about tooth worms and demons.[30]

Extensive *in vitro* and animal studies have revealed a plethora of bacterial interspecies interactions.[31] Among these direct interactions are the following:

- Bacterial competition for nutrients.
- Synergistic interactions can help one or more resident microbes grow or survive.

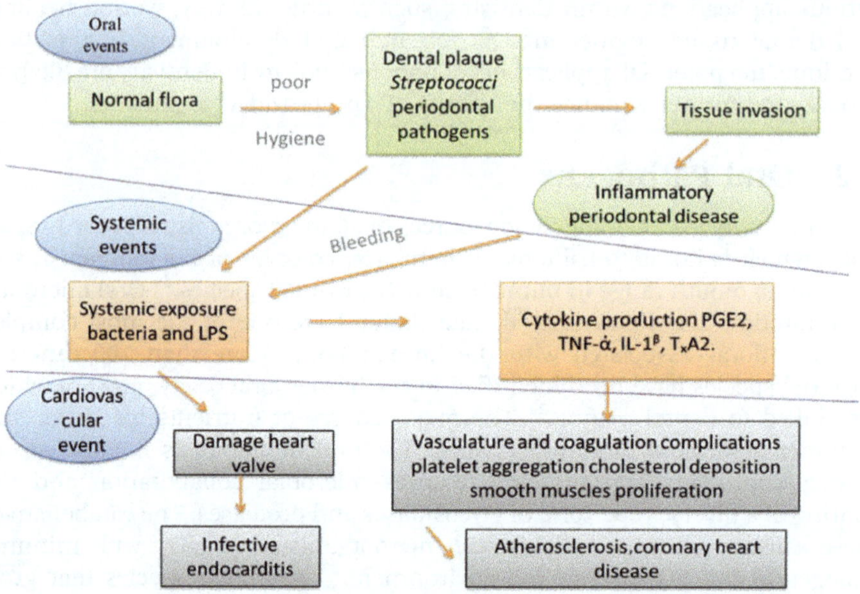

Figure 4.1 A model of a pathway from dental plaque to infective endocarditis and atherosclerosis and coronary heart disease.

- One resident produces an antagonist that inhibits the growth of other organisms.
- Another residential organism neutralizes a virulence factor produced by one of the organisms by interfering with one organism's growth-dependent signalling mechanism with another.[32]

Streptococcus mutans bacteria were observed in 36 of 50 teeth analysed. The acid level increased rapidly and the medium level increased, the ordinary level being pH 7. Reactions occurred within 24 h at pH 4.2. All of the strains were isolated and produced acid by fermentation of glucose, lactose, raffinose, mannitol, inulin, and salicin. On blood agar, there was usually no haemolysis or hyperpigmentation.[33]

MS agar (Murashige and Skoog agar) is the most commonly used medium for isolating *S. mutans* and other oral streptococcal species. Trypan Blue in MS agar has been shown to inhibit the growth of most *S. mutans* strains. Recently, researchers developed MSFA (mannitol–sorbitol–fuchsin agar) for *S. mutans*, a new selective medium for *S. mutans* that contains mannitol, sorbitol, basic fuchsin, and sodium azide. Lectin binds to a component on the surface of microbial cells, causing agglutination.[34]

To demonstrate the presence of chains in Gram-positive or Gram-negative bacteria, microbial colonies were Gram stained and examined under a light microscope.[36] Gram-positive bacteria would appear purple or violet in colour and have the shape of a coccus. Gram-negative bacteria have a pink or red appearance and are shaped like a rod. *S. mutans* colonies were chosen at random and tested for purity by plating onto Columbia blood agar and incubating the inoculated plates anaerobically at 37 °C for 24 h. Growth on the purity plates was observed.[37] If the organism was generally considered pure, it was subjected to rapid carbohydrate fermentation activity.[38]

Lactobacillus is a Gram-positive, rod-shaped, catalase-negative bacterium that can be found in a wide variety of ecosystems.[39] Lactobacillus species are also found in humans, including in the oral cavity, intestine, and vagina.[40] *Lactobacillus* species occurring in healthy resident flora are typically epitopes, and some have been used as probiotics.[41]

4.2.1 Essential Oils and Herbs

One of the main risks to the effective treatment of microbial disease is the distribution of drugs that are pathogen resistant. Those who are interested in treating oral diseases always employ natural products, such as essential oils and other plant extracts.[42] The fragrant oil liquids known as essential oils, also referred to as volatile oils, are derived from plant materials such as flowers, buds, seeds, leaves, twigs, bark, herbs, timber, fruits, and roots.[43]

The small terpenoids and phenolic chemicals that make up essential oils are what give them their antibacterial properties.[44] Many of these are generally regarded as safe.[45] There is a long history of using essential oils such as tea tree oil, lavender oil, thyme oil, peppermint oil, and eugenol oil. The root canal environment changes after the application of a chemical treatment procedure, providing unfavourable conditions for microorganisms. Reduced oxygen tension, a lack of readily available nutrients, and antimicrobial medications all play a role in the survival balance of the bacteria in the root canal system.[46]

Microorganisms can be found in root canal dentinal tubules, and a bacterial biofilm may develop at the apex of the root canal and in the extra-radicular areas.[47] A typical endodontic treatment procedure includes irrigation with a broad range of antimicrobial agents and the placement of intracanal medicaments. The fractions of essential oils that include antibacterial activity are extracted from numerous species and herbs. Eugenol, or clove oil, has been utilized in dentistry for thousands of years, and has been applied topically to treat pulpitis pain and dentinal hypersensitivity. In addition to its analgesic effects, eugenol also has an irritating impact.[48] Therefore, a study was conducted to ascertain the *in vitro* antimicrobial activity of eugenol, thyme oil, tea tree oil, peppermint oil, and lavender oil against four pathogenic oral microorganisms in the quest for plant essential oils containing antimicrobial activity.[49]

In the distant past, only herbs and a few therapeutic plants were used for brushing teeth. Brushes and dentifrices are two products that are now most frequently used to maintain dental health.[50] Different civilizations have utilized the traditional miswak toothbrush or chewing stick for ages. Miswak's antimicrobial properties include the following:

- germs that cause periodontal disease
- carcinogenic microbes
- antifungal outcome
- pathogens in endodontics
- gingivitis and plaque.[51]

Some traditional products obtained from plants are listed in Table 4.1.

4.3 Nanoparticles

Owing to their potential for medical uses such as immunization, diagnostic imaging procedures, cancer therapy, and long-term drug delivery, engineered NPs have grown in prominence in recent years.[52] Nanocontainers, nanofilms, nanomembranes, nanoscaffolds, and composites that are a combination of these are examples of nanomaterials with components smaller than 100 nm in at least one dimension.[53]

There are numerous systems that can promote and improve drug administration by the buccal route and they can be divided into three categories:

- NPs for mucoadhesive film delivery
- mucoadhesive gels containing NPs
- NPs for solid mucoadhesive delivery.

In recent years, there has been a surge of interest in the use of chitosan and chitin in dental medicine.[54] They have control over delivery properties and also an ability to support oral tissue regeneration in applied fields of dentistry such as endodontics, periodontics, regenerative dentistry, invasive dentistry, and implantology.[55]

Table 4.1 Traditional dental products from plants.

Plant	Common name	Useful parts in the plant	Useful properties
Syzygium aromaticum	Clove	Flower	Pulpitis pain and dentinal hypersensitivity
Solvodora persica	Miswak	Fresh young part of stem, branches	Astringent, antibacterial
Azadirachta indica	Neem	Stem, branches	Bitter, astringent, antiseptic, antilactose, analgesic, anti-inflammatory, antiviral, antifungal
Pongamia pinnata	Karanj	Stem, root	Astringent, styptic, antimicrobial, antiseptic
Terminalia arjuna	Arjun	Stem, branches	Astringent, styptic, cooling branch demulcent, antibacterial, antifungal

Recent research on chitosan NPs, which are used in new bone formation therapies, is paving the way for future applications in implantology, periodontology, and dental surgery.[56] The use of chitosan NPs in dental endodontics has been studied and they can be a successfully applied in root canal treatment.[57] *Enterococcus faecalis* is resistant to chitosan NPs. To treat bacterial biofilms in the oral cavity, chitosan NPs integrated into toothpaste or dental prophylactic therapies have been used. The low molecular weight of chitosan is very effective against *S. mutans*. It has a high antimicrobial effect and can eliminate more than 95% of *S. mutans* cells. *S. mutans* is a cariogenic microorganism and eliminating this bacterium is an effective method in preventive dentistry.[58] Metals have been used as antimicrobial agents for centuries. Ag, Cu, Ti, and Zn have attracted considerable attention, and Cu and Zn salts have also been studied and used in the treatment of tinea pedis (athlete's foot) as antiseptic and antifungal agents.[59] Several oral foods and materials, including toothpaste, now contain powdered zinc citrate or acetate to prevent the formation of dental plaque.[60] Powdered titanium oxide is also commonly used in toothpastes as a whitener.[61]

4.3.1 Mechanism of Antimicrobial Activity Using Nanoparticles

Nanoparticles can attach to and penetrate the cell walls of both Gram-positive and Gram-negative bacteria, causing cell functions to be disrupted.[62] Their use is aimed at preventing and treating diseases caused by drug-resistant microorganisms, and also preventing the formation of biofilms.[63] The mechanism of antimicrobial activity of NPs is illustrated in Figure 4.2.

The antibacterial, antifungal, and antiviral properties of sulfide NPs have been extensively studied in comparison with various metals.[64] Silver NPs have been used successfully in a variety of biomedical applications, including use as an antibacterial agent in dental resin composites.[65]

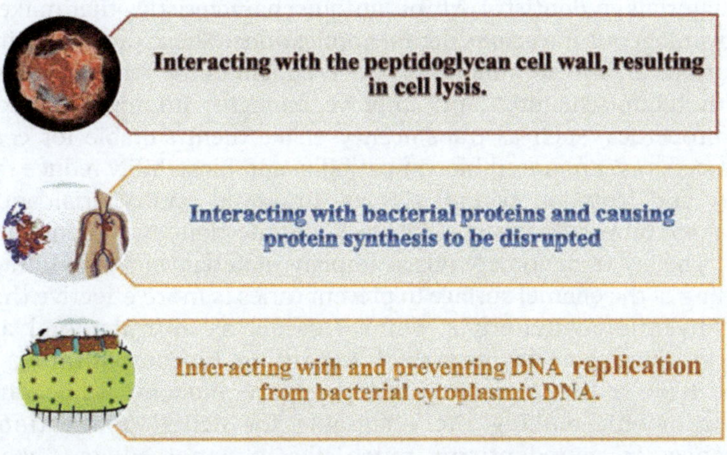

Figure 4.2 Mechanism of antimicrobial activity using nanoparticles.

The prevention of dental caries and periodontal infection is typically aimed at the automatic or non-specific control of the plaque biofilm; biofilms are present in our daily lives, such as when we brush our teeth.[66] Bactericidal agents are an expensive alternative to mechanical plaque control.[67] However, actual periods of exposure to an antimicrobial agent during tooth brushing and mouth rinsing can be extremely brief, lasting only 30 s.[68] Since antiquity, various populations around the world have practised oral hygiene measures,[69] which vary depending on factors such as cultural background, religious norms, educational level, and socioeconomic status.[70]

4.4 General Nanomaterials and Their Applications in Dentistry

Nanotechnology is a cutting-edge field of science that holds immense potential in the field of medicine. Nanotechnology is the discipline that deals with synthesizing and utilizing materials and devices with structural elements in the nanometre range, *i.e.* around one billionth of a metre. It encompasses design, synthesis, analysis, and application.[71] Dentistry has embraced nanotechnology to enhance the quality of dental materials, such as tooth-coloured fillings, adhesives, and coatings.[72] Nanoscale fillers are incorporated to bolster the strength and longevity of composite resins. Additionally, nanosized particles in dental sealants have enhanced the adhesion to tooth structures.[73] Nanotechnology has also played a role in developing antibacterial coatings for dental implants, which helps to prevent infections.[73] When selecting NPs for nanodentistry, consideration is given to the chemical, physical, and biological properties of the nanostructure. Nanoparticles are utilized to create oral disease-preventive medications, prosthetics, and teeth implants. Nanomaterials can deliver oral fluids or drugs, prevent certain oral diseases, such as oral cancer, and significantly improve oral health.[74]

Nanomaterials in dentistry exhibit unique characteristics that make them highly advantageous in various dental applications. These properties include a high surface area-to-volume ratio, which enhances the material's ability to bond with tooth structures and improve strength. Improved optical and physical properties, such as translucency, make them suitable for cosmetic dental procedures. Enhanced biocompatibility and biostability reduce the risk of toxicity and irritation of oral tissues. Improved antibacterial and anti-inflammatory properties reduce the risk of infection and promote faster healing.[72] The use of nanomaterials as implant materials in dental fillings and for polishing of the enamel surface to prevent caries is more effective than the use of conventional materials. Some NPs act as antimicrobial agents, thus preventing bacterial growth.[74] Regarding biocompatibility, nanomaterials have a low toxicity profile and are biocompatible with the oral environment, making them suitable for dental applications.[75,76] The advances in nanodentistry, much like in nanomedicine, can revolutionize oral healthcare through the adoption of innovative nanomaterials and biotechnologies, such as tissue engineering and nanorobots, leading to near-perfect oral health.[77] The objectives of nanotechnology are to examine structures at the nanoscale, understand their physical properties, create nanoscale structures, develop devices with nanoscale accuracy, and bridge the gap between the nanoscale and macroscopic universes by developing appropriate methods.[78]

4.4.1 Types of Nanomaterials in Dentistry

Nanoscale materials are substances whose elements possess at least one dimension in the nanometre range (1–100 nm). These materials display distinct physical, chemical, and biological characteristics due to their small size, high surface area-to-volume ratio, and quantum behaviour. These unique features of nanoscale materials make them suitable for a wide range of applications.[81]

In dentistry, several types of NPs with unique characteristics, properties, and applications are commonly used. Nanomaterials include NPs, nanocomposites, nanocoatings, carbon nanotubes, and dendrimers.[79,80] The role of NPs in dentistry is illustrated in Figure 4.3.

In general, nanomaterials comprise organic nanomaterials, inorganic nanomaterials, and polymeric nanomaterials (Figure 4.4).

4.4.1.1 Organic Nanomaterials

Organic nanomaterials and polymeric nanomaterials encompass many biodegradable and non-toxic particles such as dendrimers, micelles, liposomes,

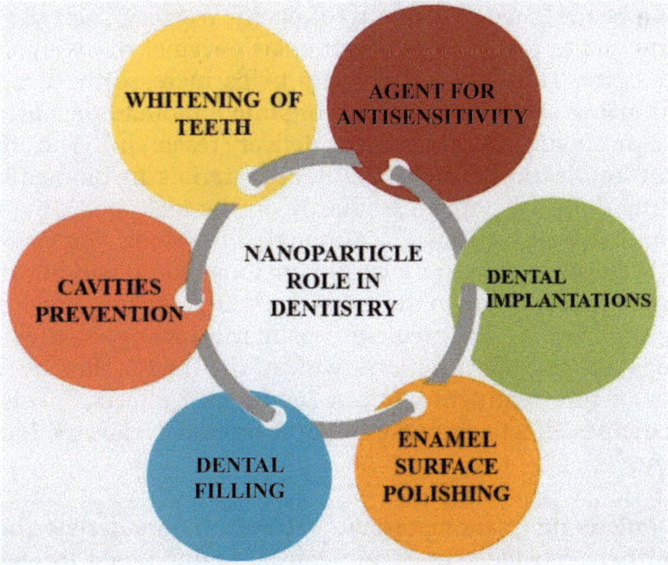

Figure 4.3 Diagram illustrating the role of nanoparticles in dentistry.

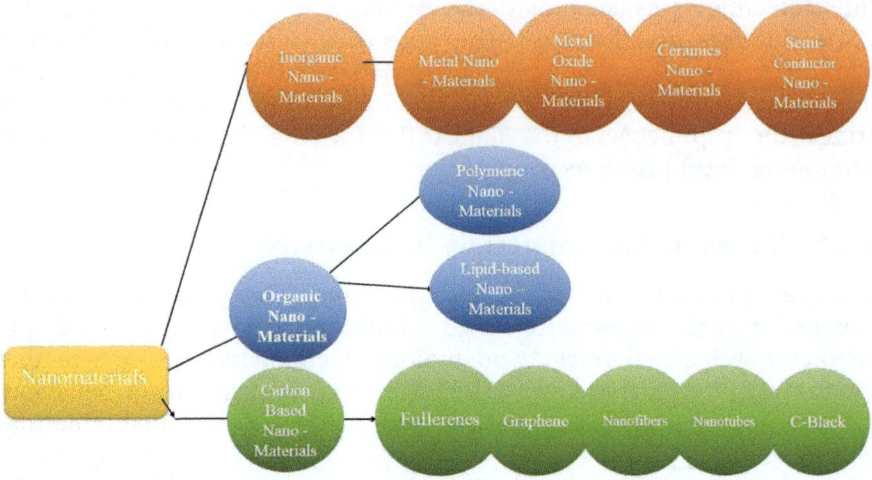

Figure 4.4 Schematic illustration of conventional nanomaterials.

and ferritin. These materials are known for their unique properties, including a hollow core in some cases, such as micelles and liposomes, referred to as nanocapsules. These nanocapsules exhibit sensitivity towards external stimuli, including thermal and electromagnetic radiation such as heat and light. These properties make organic nanomaterials a promising area of research for various applications.[82] The unique properties of organic nanomaterials, including their biodegradability, non-toxicity, and sensitivity to external stimuli, make them ideal for drug delivery applications. For example, their hollow core can be filled with a therapeutic agent. The nanocapsules can be designed to respond to specific triggers, such as a change in temperature or the presence of a particular enzyme, releasing the drug at the targeted site. This targeted delivery helps increase the drug's efficacy while minimizing side effects. Hence organic nanomaterials are highly regarded as a promising avenue for drug delivery research.[83] The effectiveness and type of applications of organic nanomaterials in biomedicine, especially in drug delivery, depend primarily on the absorbed drug's stability, capacity, and delivery systems.[84] Although important, the physical properties, such as the nanomaterials' dimensions, composition, and surface morphology, are secondary to these factors.[85] Organic nanomaterials have a wide range of uses in biomedicine, primarily owing to their ability to target drugs to specific locations within the body, thereby increasing the efficacy of the treatment while reducing side effects.[86] This ability to deliver medicines effectively makes organic nanomaterials a valuable tool in biomedicine.[87]

4.4.1.1.1 Polymeric Nanomaterials. These nanomaterials have nanosphere and nanocapsule shapes and can be manipulated for easy functionalization in the biomedical and medical fields.[82]

4.4.1.1.2 Lipid-based Nanomaterials. These nanomaterials, ranging in diameter between 10 and 1000 nm, include lipid moieties with practical applications in numerous biomedical fields and dentistry.[82]

4.4.1.2 Inorganic Nanomaterials

Metal and metal-based nanomaterials can be classified generally under inorganic nanomaterials.

4.4.1.2.1 Metal and Metal Oxide Nanomaterials. Metal and metal oxide NPs are being studied for various applications in dentistry, including antimicrobial agents, components in dental restorative materials, and tooth whitening.[88] These NPs exhibit unique properties due to their small size and high surface area, making them suitable for various dental applications. In terms of antimicrobial properties, silver and ZnONPs are effective against oral pathogens, which could lead to the development of new endodontic treatments.[89] Furthermore, titanium dioxide NPs are being explored as potential components in dental materials owing to their potential to improve the appearance of teeth.[90] These studies highlight the potential of metal and metal oxide NPs to advance dental care by providing new and improved materials and treatments.[89]

4.4.1.2.2 Ceramic and Semiconductor Nanomaterials. Ceramic nanomaterials, such as zirconia, have been shown to have high strength and durability, making them suitable for use in dental restorations, such as crowns, bridges, and implants.[91] The high surface area of ceramic nanomaterials can also increase their bonding strength to other dental materials, further improving their performance as restorative materials. Additionally, the biocompatibility of ceramic nanomaterials makes them a suitable option for dental applications.[92]

Overall, the unique properties of ceramic and semiconductor nanomaterials, including their small size, high surface area, and biocompatibility, make them suitable for various dental applications[93] (Figure 4.5).

4.4.1.3 Carbon-based Nanomaterials

Carbon-based nanomaterials, such as carbon nanotubes and graphene, are being studied for their potential uses in dentistry. These materials have unique properties, including high mechanical strength, high electrical conductivity, and biocompatibility, making them suitable for various dental applications.[82]

4.4.1.3.1 Fullerenes, Nanotubes, Graphene, Nanofibres, and Carbon Black. Fullerenes, nanotubes, graphene, nanofibres, and carbon black

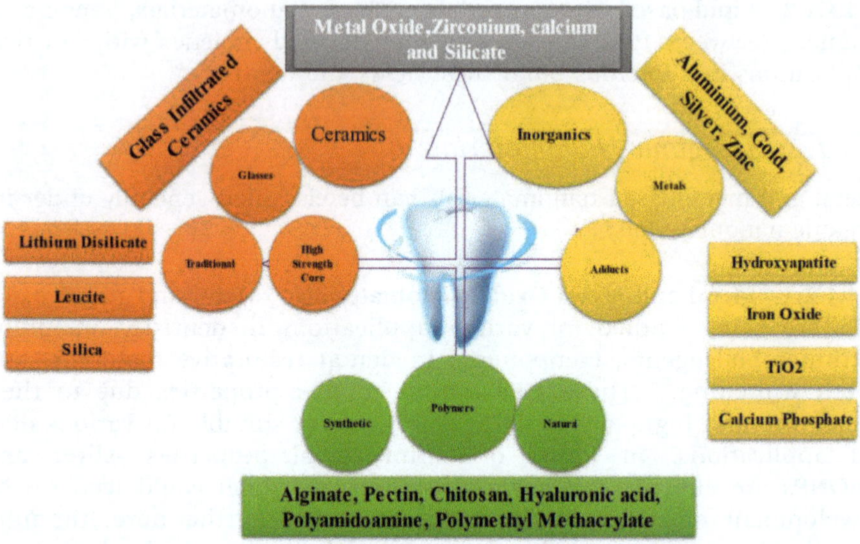

Figure 4.5 Schematic representation of different types of nanomaterials.

are all carbon-based nanomaterials being investigated for their potential uses in dentistry.[94] These materials have unique properties, including high mechanical strength, high electrical conductivity, and biocompatibility, making them suitable for various dental applications.[95]

Carbon nanotubes have been used as a reinforcement material in dental composites to improve their strength and durability.[96] Graphene has also been investigated for use in dental composites, as it has excellent mechanical properties and antimicrobial activity.[98] Nanofibres have been explored for use in tissue engineering, where they can support the growth of cells and the formation of new tissue.[97] Carbon black has been used as a filler material in dental materials to improve their mechanical properties and resistance to wear.[99]

4.5 Nanoparticles

Nanoparticles (NPs) are tiny particles with at least one dimension in the nanoscale range, typically between 1 and 100 nm. They exhibit unique physical, chemical, and biological properties due to their small size, high surface area-to-volume ratio, and quantum effects.[100] NPs can be made from various materials, including metals, ceramics, polymers, and carbon-based materials. These materials can have different shapes and structures, including spherical, rod-like, and layered. The properties of NPs can be tailored by controlling their size, shape, composition, and surface characteristics, making them useful for various applications in fields such as electronics, energy, biomedicine, and materials science.[101,102]

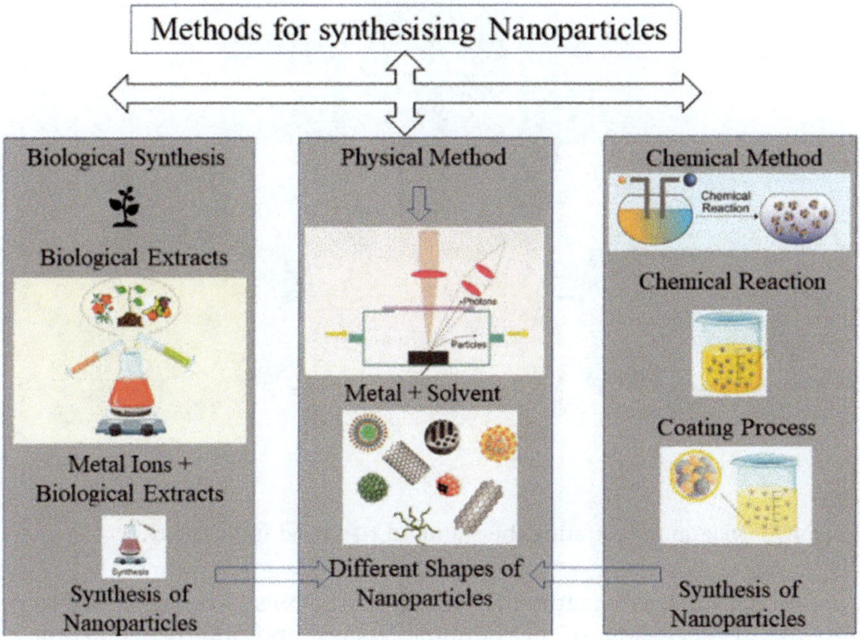

Figure 4.6 Diagram illustrating the different types of synthesis of NPs.

4.5.1 Methods of Synthesizing Nanoparticles

The conventional synthesis of NPs can be divided into three types, chemical, physical, and biological synthesis methods (Figure 4.6).

4.5.2 Types of Nanoparticles Used in Dentistry

Some commonly used NPs in dentistry include metal and metal oxide NPs, such as hydroxyapatite, iron oxide, zirconia, silica, silver, and titania, which are used for tissue regeneration and antimicrobial and anti-inflammatory applications[80] (Figure 4.7).

4.5.2.1 Hydroxyapatite

Hydroxyapatite NPs have become popular in medicine and dentistry owing to their similarity in composition to human teeth and bones.[103] This makes hydroxyapatite a biocompatible substance, which means that it is well tolerated by the human body and does not cause adverse reactions. As a result, hydroxyapatite NPs have been used in various medical and dental applications, such as tissue engineering, drug delivery, and filling materials for teeth restoration.[104] Hydroxyapatite NPs have been incorporated into dental filling materials to improve their physical and mechanical properties.[105] For tooth enamel reconstruction, hydroxyapatite NPs have been used to recreate the natural mineral structure of tooth enamel and strengthen its remineralization.

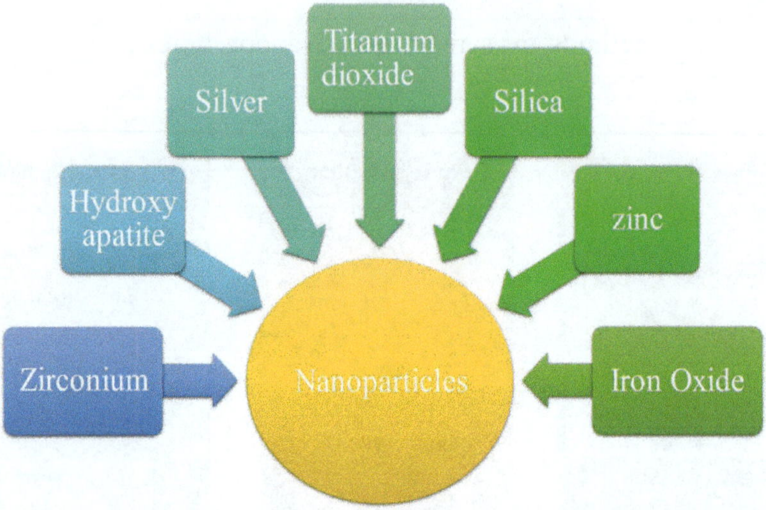

Figure 4.7 Diagram illustrating the standard NPs used in dentistry.

In periodontal disease treatment, hydroxyapatite NPs have been used to promote the regeneration of periodontal tissues and improve outcomes.[106] Hydroxyapatite NPs can fit well in the tiny cavities present in the enamel originating from acidic erosion.[107] Granular hydroxyapatite is employed in dental clinical practices to overcome periodontal shortcomings.[108]

4.5.2.2 Iron Oxide

Bacterial biofilms on dental implants can be challenging to eliminate with antibiotics owing to the protective layer of exopolymers that encase the microorganisms, creating an impenetrable matrix resistant to both immune cells and many antibiotics.[110] To address this issue, nanoparticulate substances are currently being used to control microbial infections. One such substance is iron oxide nanoparticles (IONPs), which have been widely used to eradicate biofilms on dental implants. This demonstrates the potential of nanotechnology in improving oral health and preventing infections associated with dental implants.[109]

IONPs have several properties that make them attractive for use in dental applications, including biocompatibility, magnetic responsiveness, and the ability to encapsulate drugs. IONPs have also been used in the field of tissue engineering, specifically in scaffold design for bone and tooth regeneration.[111] The magnetic properties of IONPs allow them to be easily positioned in the desired location, and their biocompatibility makes them suitable for use near biological tissues.[112]

4.5.2.3 Zirconia

Zirconia is a metal oxide[113] that is insoluble in water, thus reducing adhesion and cytotoxicity.[114] These properties make zirconia a widely used biomaterial for dentistry implants.[115] Zirconia NPs are a bio-inert material, thus easily encapsulating weak animal tissues and the traces remain virtually unnoticeable.[116]

4.5.2.4 *Silica*

Silica NPs have found applications in the field of dentistry as dental fillers. The surface properties of silica NPs, such as surface area and surface chemistry, can also play a role in their biocompatibility and ability to interact with other materials in dental fillings.[117] Understanding the relationships between the dimensions and shape of silica NPs and their properties can lead to the development of more effective dental fillers for use in restorative dentistry.[118] In dentistry, it is important to develop biomaterials that can effectively treat and prevent dental problems, particularly those related to dentine. Dentine is a vital part of the tooth that helps to protect the pulp, the innermost layer of the tooth that contains nerves and blood vessels. When dentine is exposed, it can lead to sensitivity and other issues. Therefore, it is necessary to develop biomaterials that can penetrate deep inside the dentinal tubules without irritating the pulp, provide rapid action, and offer lasting tubule occlusion. Such materials should be biocompatible, non-toxic, and able to block effectively fluid flow and other stimuli that can lead to dental sensitivity.[119] Mesoporous silica has attracted significant attention in research owing to its unique characteristics, including well-defined structure, high surface area, adjustable pore size, and excellent thermal and hydrothermal stability.[120]

4.5.2.5 *Silver*

Silver NPs have attracted attention in dentistry owing to their antimicrobial properties.[121] They have been shown to have potent activity against a range of oral bacteria, including those associated with caries and periodontal disease. As a result, they are being explored as a potential alternative to conventional antimicrobial agents for use in dental applications.[122] Silver NPs can be incorporated into dental materials, such as resins, cement, and coatings, to provide sustained release of the active silver ions and improve the antimicrobial efficacy of the material. Silver NPs can also be used as topical agents for treating oral infections and other dental conditions. Despite their promise, further research is needed to understand fully the potential of silver NPs in dentistry and their safety and toxicity profiles.[123]

4.5.2.6 *Titania*

Titanium dioxide (titania, TiO_2) NPs are widely used in medicine and dentistry. However, the placement of dental implants can trigger allergic reactions by activating complex reactions involving antigens and antibodies of type I and type IV.[124] Additionally, the adhesion of microbes to TiO_2

implants can significantly impact the healing process of teeth and lead to long-term consequences for the implants.[125] Surface modification improves the hardness and strength of dental resin-based composites and affects the linkage and dispersion of TiO_2 present inside a resin matrix.[126]

4.5.3 Role of Other Nanoparticles in Dentistry

Other NPs include zinc oxide, chitosan-based, gold, and calcium phosphate NPs. In addition to the NPs mentioned previously, different types of NPs are utilized in dentistry for various purposes. For instance, zinc oxide NPs, owing to their anti-inflammatory and antibacterial properties, are incorporated into dental cement, coatings, and restorative materials.[127] Chitosan, a biocompatible and biodegradable polysaccharide, is utilized in the form of NPs for drug delivery and wound dressing in dentistry because of its wound-healing and antimicrobial properties.[128] Gold NPs are used in dentistry as dental fillings, alloys, and antimicrobial agents owing to their high biocompatibility and stability.[129] Calcium phosphate NPs, which can mimic the natural mineral structure of bones and teeth, are utilized for bone regeneration, drug delivery, and as components in dental cement because of their biocompatibility.[130]

4.5.3.1 Dental Regeneration and Tissue Engineering

Nanomaterials play a significant role in dental regeneration and tissue engineering. They can be used to design scaffolds for the renewal of bone and teeth and for the fabrication of controlled microscale surface topographies on dental implants.[137] Additionally, nanoscale layers produced through techniques such as attrition can improve the mechanical properties of dental implants, such as hardness and wettability. The small size of nanomaterials also makes them ideal for use in non-invasive imaging techniques, such as magnetic resonance imaging (MRI) and computed tomography (CT) scans, for the diagnosis and imaging of dental conditions. Overall, nanomaterials are an essential factor in advancing dental regeneration and tissue engineering.[138]

Nanomaterials have been extensively studied for their potential applications in dental regeneration and tissue engineering.[139] The unique physical and chemical properties of nanomaterials, such as their high surface area, biocompatibility, and controlled release capabilities, make them promising candidates for tissue regeneration and repair. For instance, nanofibres and nanoparticles have been utilized in scaffold design to enhance mechanical strength and provide a three-dimensional structure for cell attachment and growth.[140] Additionally, nanomaterials have been used as vehicles for drug delivery, allowing localized controlled release of therapeutic agents to improve tissue regeneration outcomes.[141] The use of nanomaterials in dental regeneration and tissue engineering has the potential to provide new and improved treatments for oral and dental conditions[136] (Figure 4.8).

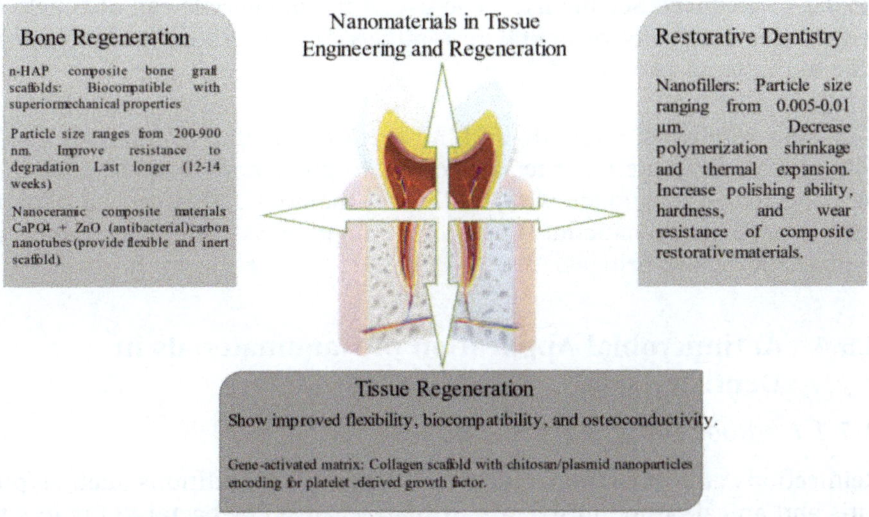

Figure 4.8 Schematic representation of nanomaterials and their role in tissue engineering and regeneration.

4.5.3.2 Dental Restorative Nanomaterials

Nanomaterials play a significant role in dental restoratives such as crowns and bridges. They are used to enhance the physical and mechanical properties of these materials. However, using nanofillers such as nanoceramics and nanohybrids can improve the strength and toughness of dental composites, making them more suitable for use in high-stress load-bearing applications such as dental crowns and bridges.[142] Additionally, nanoscale materials can improve the aesthetics of dental restoratives by creating more natural-looking tooth-coloured materials with improved translucency and colour stability. The use of nanotechnology in dentistry is a rapidly developing field, and more innovative applications are expected to emerge in the near future.

4.5.3.2.1 Improved Biocompatibility. The use of nanoscale materials in dental restoratives can also improve their biocompatibility, which is important for reducing the risk of adverse reactions in the patient.[143]

4.5.3.2.2 Increased Durability. Nanoscale materials can increase the durability of dental restoratives, helping to ensure that they last longer and resist wear and tear over time.[143]

4.5.3.2.3 Better Adhesion. Nanotechnology can also improve the adhesion of dental restoratives to the tooth structure, which is important for maintaining the stability of the restoration and preventing leakage.[144]

4.5.3.2.4 Reduced Sensitivity. The use of nanomaterials can also help to reduce the sensitivity of dental restoratives, making them more comfortable for patients to wear.[145]

4.5.3.2.5 Improved Strength and Aesthetics. Dental restoratives made with nanoscale materials are also often stronger and more aesthetically pleasing than those made with traditional materials. This can be particularly beneficial for patients who require high-stress restorations such as dental crowns and bridges.[146]

4.5.4 Antimicrobial Application of Nanomaterials in Dentistry

4.5.4.1 Root Canal Treatment

Reinfections after root canal treatment, leading to conditions such as pulpitis and apical periodontitis, are primarily caused by bacterial organisms and their by-products. Among these, *E. faecalis* is particularly responsible for long-term secondary dental caries.[147] Biosynthesized silver NPs showed an antimicrobial ability against *E. faecalis*.[148] Nanomaterials have been shown to exhibit antimicrobial activity in root canal treatment. This is due to their small size, high surface area, and ability to penetrate into bacterial cell walls, leading to bacterial death.[149] However, silver and titanium dioxide NPs have been shown to have strong antibacterial properties against common root canal pathogens, such as *S. mutans* and *E. faecalis*.[150] Additionally, nanocomposites containing antimicrobial agents have been developed for use in root canal treatment, providing controlled release of the active agent to inhibit bacterial growth. The use of nanotechnology in root canal treatment is still a developing field, but initial results suggest that nanomaterials may offer a promising solution for improving the outcome of this treatment.[151]

4.5.4.2 Modified Implants With Antimicrobial Nanomaterials

Antimicrobial nanomaterials have been used to modify dental implants to enhance efficacy against bacterial infections.[147] The small size and high surface area of nanomaterials allow for better penetration into bacterial cell walls, leading to greater bacterial death. In dental implantation, nanomaterials with antimicrobial properties, such as silver and titanium dioxide NPs, have been incorporated into implant materials to provide controlled release of the active agent and inhibit bacterial growth.[152] This modification has shown promise in reducing the incidence of implant-related infections and improving the success rate of implantations. The application of nanotechnology in dental implantation is a growing field, with the potential to provide a safer and more effective solution for patients.[153]

4.5.4.2.1 Increased Surface Area. Nanomaterials have a high surface area-to-volume ratio, which allows for a large number of antimicrobial agents to be present on the surface of the implant material. This increased exposure to antimicrobial agents enhances the ability to inhibit bacterial growth and reduce the risk of implant-related infections.[153,154]

4.5.4.2.2 Sustained Release. The use of nanomaterials in implant materials allows the sustained release of antimicrobial agents, providing a long-term barrier against bacterial colonization. This is particularly important in the case of dental implants, where bacterial infections can be difficult to treat owing to the lack of blood flow in the implant area.[155]

4.5.4.2.3 Improved Material Properties. The addition of nanomaterials to implant materials can also improve the mechanical and physical properties of the implant, making it stronger and more durable. This can lead to a longer lifespan for the implant and reduce the risk of implant failure.[156]

4.5.4.2.4 Customization. The use of nanotechnology in dental implantation provides the opportunity for customization, allowing the creation of implant materials specifically designed for each patient's individual needs.[157] This can include tailoring the implant material to specific patient demographics, such as age and oral health, and also adjusting the release of the antimicrobial agents to meet specific treatment goals.[158]

4.5.5 Role of Nanomaterials in Orthodontics

Nanomaterials are playing an increasing role in orthodontics, particularly in the development of new orthodontic devices and materials. The unique properties of nanomaterials, such as their high surface area-to-volume ratio and improved mechanical strength, make them ideal for use in orthodontics.[159] Nanostructured materials can be used to develop braces and other orthodontic devices that are lighter, more comfortable, and more effective in realigning teeth. Additionally, nanotechnology is also being used to develop new types of dental adhesives, coatings, and biocompatible materials for use in orthodontic procedures. However, the use of nanomaterials in orthodontics is still an emerging field and further research is needed to understand fully their potential benefits and risks.[160]

The use of nanotechnology in medicine offers numerous benefits, but also raises concerns about safety for humans and the environment. NPs have the potential to penetrate tissues and alter biological processes, making it crucial to study thoroughly their toxicological and environmental properties before implementing their use. The following subsections summarize the current utilization of NPs with antibacterial properties in orthodontics, including in resins, brackets, and archwires. It is imperative to evaluate the risks involved to ensure the sustainable application of nanomaterials in orthodontics.[161]

4.5.5.1 *Nanocoating in Orthodontic Archwires*

There are different purposes for nanocoating in orthodontic archwires, *e.g.* reducing friction on orthodontic archwires, delivering NPs from an elastomeric ligature, shaping memory polymers in orthodontics, and control of oral biofilms during orthodontic treatment.[161]

4.5.5.2 *Bracket Materials*

Brackets play a crucial role in dentistry, particularly in orthodontic treatment. They serve as anchorage points for the archwires, which help to guide and align the teeth during treatment. The choice of bracket material can impact the effectiveness, comfort, and duration of treatment. There are several types of bracket materials used in dentistry, including metals, ceramics, and plastics.[162] Each material has its own unique properties and advantages, such as strength, biocompatibility, and aesthetics. The selection of the bracket material is determined by factors such as the patient's age, oral hygiene, and treatment goals. To ensure the best results, it is important to consider carefully the properties of each material and choose the one that is best suited for the patient's needs.[163] Other orthodontic factors include nanomaterials in orthodontics, the relationship between the orthodontic arch and bracket materials, microbial colonization associated with different kinds of fixed orthodontic appliances, and introduction of nanofibres to orthodontic adhesives.[162]

4.5.6 Role of Nanomaterials in Endodontics

The branch of dentistry that deals with the morphology and physiology of the root canal is endodontics. It combines various aspects of this field, such as aetiology, pathology, epidemiology, prophylaxis, and, above all, treatment of endodontic and periapical diseases.[164]

Thanks to the expanding variety of NPs, such as bioactive glass, zirconia, chitosan, hydroxyapatite, silver particles, and zinc oxide, the properties of materials used in dentistry, such as durability, tissue regeneration, and bactericidal properties, can be improved. Biofilm is the cause of many failures in root canal treatment, and nanotechnology can provide better bactericidal properties of materials.[165]

Nanomaterials have shown great potential in the field of endodontics. They offer numerous advantages, such as improved biocompatibility, enhanced antimicrobial activity, and extended release of therapeutic agents.[165] NPs can be used in the preparation of endodontic materials, such as sealers, intracanal medicaments, and scaffolds for tissue engineering. Their small size enables them to penetrate deep into the root canal system, helping to eradicate bacteria and promote tissue regeneration.[166] Additionally, the use of nanomaterials in endodontics can increase the mechanical strength of endodontic materials and improve their physical and chemical properties. Despite the promising benefits, it is essential to evaluate thoroughly the safety and toxicity of these materials before widespread implementation. The role of nanomaterials in endodontics is an exciting area of research and development, with great potential to improve the outcomes of endodontic treatment.[167]

4.5.7 Role of Nanomaterials in Periodontics

Nanotechnology has been identified as a promising area of research for improving periodontal treatments. The use of nanomaterials can enhance the biocompatibility and effectiveness of periodontal products such as scaffolds for tissue engineering and drug delivery systems.[168] These materials are able to penetrate into gingival tissues, allowing the targeted delivery of therapeutic agents and improved treatment outcomes. Additionally, nanomaterials can be utilized to create more effective antimicrobial agents for the treatment of periodontal disease caused by bacterial infection.[169]

It is crucial to assess thoroughly the safety and toxicity of these materials before they can be widely used in periodontics. Despite this, the potential benefits of nanomaterials in the field of periodontics are an exciting area of study that has the potential to revolutionize the way in which periodontal treatments are performed.[170]

4.6 Future Perspectives and Challenges in Nanodentistry

Nanodentistry has several advantages over traditional dental systems, such as improved regeneration, enhanced hardness of composite materials, better sealing of fillings, and a powerful antimicrobial effect due to anti-biofilm properties.[171] Despite these benefits, the high cost, precise placement required, potential toxicity, expensive development, and international regulations are obstacles that limit the widespread use of nanodentistry in clinical settings.[172] Researchers are striving to find affordable methods to synthesize NP scaffolds that meet regulatory requirements and are effective in the correct placement of NPs.[172] Newer NP types, such as nanodiamonds, quantum dots, nanoshells, and carbon nanotubes, have shown promising results in research and are being considered for future use in commercial markets. These materials are attractive owing to their superior surface and chemical properties, making them well suited for use as fillers in dental nanocomposites.[171]

In conclusion, nanomaterials have revolutionized the field of dentistry with their unique properties that make them ideal for use in a variety of dental applications. Their small size and high surface area-to-volume ratio make them highly effective in creating new dental materials with improved physical and biological properties. From nanocomposites for dental fillings to nanocoatings for dental implants and orthodontic devices, and to nanosensors for monitoring oral health, nanomaterials have made a significant impact in the field of dentistry. The use of nanotechnology in dentistry is constantly evolving and advancing, making it possible to provide patients with more effective and efficient dental treatments. Despite the potential benefits of nanomaterials in dentistry, it is important to consider their safety and toxicity, and also their long-term effects on oral health. Further research and development are needed to understand fully the potential applications and limitations of nanomaterials in dentistry. However, it is clear that nanomaterials have already made a significant impact on the field of dentistry and have the potential to greatly improve dental treatments and patient outcomes.

4.7 Green Synthesis of Zinc Oxide Nanoparticles

Green synthesis of NPs is the process of synthesizing NPs by using plants and microorganisms with pharmaceutical and biomedical applications. This method is eco-friendly, sustainable, economical, biocompatible, and safe.[173] In comparison with NPs made by chemical procedures, coating biological molecules on the surface of NPs makes them more biocompatible.[174] The use of agricultural waste (agro-waste) or plants and their components has become an effective alternative to chemical synthetic methods because it does not involve any toxic chemicals and the materials are easy to extract and give high production yields.[175] ZnO is a lustrous, bluish white coloured n-type semi-conducting metal oxide. ZnO has a wide range of applications in various fields, including the electronics, food, textile, cosmetics, and pharmaceutical industries.[176] Owing to their low production costs, safety, and ease of preparation, ZnONPs are of great interest. ZnO has been recognized as a GRAS (generally recognized as safe) metal oxide by the US Food and Drug Administration (FDA).[177] ZnO exhibits distinctive optical and chemical properties that are easily controlled by altering its morphology.[178]

Green synthesis of NPs can be achieved using plants, microalgae, macroalgae, bacteria, fungi, and various bio-wastes.[179] Plants have become the most popular source of NPs for synthesis, because of their ability to produce stable NPs on a large scale and they can generate NPs with a variety of shapes and sizes.[180] The synthesis of ZnONPs using microorganisms such as bacteria, fungi, and algae is a simple, environmentally friendly, low toxicity and biocompatible method. NPs synthesized using bacteria have been shown to possess good antimicrobial properties against many pathogenic microorganisms and have many therapeutic properties, including antioxidant, anticancer, anticoagulant, *etc.*[181] Table 4.2 illustrates some extensive investigations of different biological sources used for the synthesis of ZnONPs.

4.7.1 Antimicrobial Applications of Zinc Oxide Nanoparticles

The mechanism of antimicrobial action of ZnONPs is crucial to understanding the use of ZnONPs effectively in food products and to generating non-toxic antimicrobial derivatives.[199] When scaled down to the nanometre range, ZnO particles will begin to exhibit strong antimicrobial properties. When ZnONPs enter the bacterial cell, they interact with the surface and/or the core of the cell and exhibit specific bactericidal mechanisms.[200] Several

Table 4.2 Green synthesis of zinc oxide nanoparticles.

Biosource	Examples	Zinc source	Size/nm[a]	Shape	Ref.
Plants	Dried rhizome of *Coptis chinensis*	Zinc nitrate	2.9–25.2 (TEM)	Spherical or rod	182
	Leaves of *Camelia sinensis*	Zinc nitrate	8 (HRTEM)	Spherical	183
	Leaves of *Plectranthus amboinicus*	Zinc nitrate	50–180 (SEM)	Rod	184
	Peel of *Lycopersicon esculentum*	Zinc nitrate	9 (TEM)	Polyhedron	185
	Seeds of *Peganum harmala*	Zinc nitrate	40 (TEM)	Irregular	186
Bacteria	*Lactobacillus plantarum* (Gram positive)	Zinc sulfate	7–19 (FETEM)	Spherical	187
	Aeromonas hydrophila (Gram negative)	Zinc oxide	57–72 (AFM)	Spherical	188
	Acinetobacter schindleri (Gram negative)	Zinc nitrate	20–100 (HRTEM)	Spherical	189
	Pseudomonas aeruginosa (Gram negative)	Zinc nitrate	35–80	Spherical	190
	Bacillus megaterium	Zinc acetate dihydrate	45–95	Rod and cubic	191
Fungi	*Acremonium potronii*	Zinc acetate hexahydrate	13–15 (TEM)	Spherical	192
	Aspergillus fumigatus	Zinc nitrate	3.8 (DLS)	Spherical	193
	Aspergillus niger	Zinc nitrate	61 (SEM)	Spherical	194
	Phanerochaete chrysosporium	Zinc sulfate	50 (TEM)	hexagonal	195
Algae	*Sargassum murticum*	Zinc acetate	30–57 (FESEM)	Spherical	196
	Chlorella extract	Zinc nitrate	20–50 (SEM)	Spherical	197
	Cystoseira crinita	Zinc sulfate	23–200 (TEM)	Rectangular	198

[a] AFM, atomic force microscopy; DLS, dynamic light scattering; FESEM, field emission scanning electron microscopy; FETEM, field emission transmission electron microscopy, HRTEM, high-resolution transmission electron microscopy; SEM, scanning electron microscopy; TEM, transmission electron microscopy.

mechanisms of action of ZnONPs have been postulated, including the generation of reactive oxygen species (ROS), which increases membrane lipid peroxidation and leads to membrane leakage of reducing sugars, DNA, and proteins and lowers cell viability.[201] The generation of intercellular ROS, such as hydrogen peroxide (H_2O_2), is toxic to bacterial cells. Additionally, it has been observed that UV and visible light can activate ZnO, producing highly reactive oxygen species, such as OH, H_2O_2, and O_2. These radicals and superoxide cannot enter inside the cells and are likely to remain on the cell surface, but H_2O_2 can enter the cells.[202] Green synthesized ZnONPs showed antibacterial activity against many pathogenic bacteria. For example, ZnONPs synthesized using agro-waste showed good antibacterial activity against *Salmonella typhimurium* and *Bacillus subtilis*,[203] and the ZnONPs prepared using plant extract of *Cassia fistula* showed effective antibacterial activity against *Escherichia coli*, *Klebsiella aerogenes*, *Pseudomonas desmolyticum*, and *Staphylococcus aureus*.[204]

Several mechanisms mediated the antiviral effect of ZnONPs, such as the inhibition of viral entry, replication, and organ dissemination – all of which can lead to ROS, oxidative damage, and viral death. Zinc-containing drugs have shown antiviral effectiveness against different viruses through several mechanisms, including physical processes such as virus adherence, inhibition of virus infection, uncoating, and inhibition of viral polymerases and protease enzymes.[205] ZnONPs have shown antiviral activity towards many viruses. For instance, the green synthesized ZnONPs obtained using *Plumbago indica* L. exhibit strong antiviral activity against herpes simplex virus-1.[206] Nidovirus replication was inhibited by Zn^{2+} ions, and increasing intracellular Zn^{2+} concentration also inhibited the replication of several RNA viruses.[207] Zinc oxide tetrapods (ZnOTs) significantly inhibit HSV-2 (herpes simplex virus-2) entry into target cells and also inhibit the spread of the virus. Infected human vaginal epithelium and HeLa cells, which are natural target cells, also show large reductions in infectivity due to the ability of ZnOTs to neutralize HSV-2 virions.[208] Inducing the immune system to combat the HSV-2 virus is one of the therapeutic effects of zinc oxide tetrapod NPs (ZOTEN).[209]

Numerous studies have demonstrated the antifungal mechanism of ZnONPs, which includes altering the cellular structure, inhibiting biological macromolecular activity (protein synthesis inhibition), DNA damage, cell wall disruption, mitochondrial damage, and destabilizing the antioxidant system *via* ROS- and/or Zn^{2+}-mediated processes.[210,211] Through the inhibition of glutathione (GSH)-producing enzymes, treatment with ZnONPs may decrease GSH levels, reducing the antioxidant potential of fungal cells.[212] Green synthesized ZnONPs showed antifungal activity against many pathogenic fungi. ZnONPs prepared using *Cinnamomum camphora* showed antifungal activity against *Alternaria alternata* fungal species[213] (Figure 4.9) and another study showed that *Aspergillus flavus* displayed variable degrees of alterations in the expression of genes involved in oxidative stress, zinc ion binding, transmembrane transport, and oxidative phosphorylation after being exposed to ZnONPs.[210]

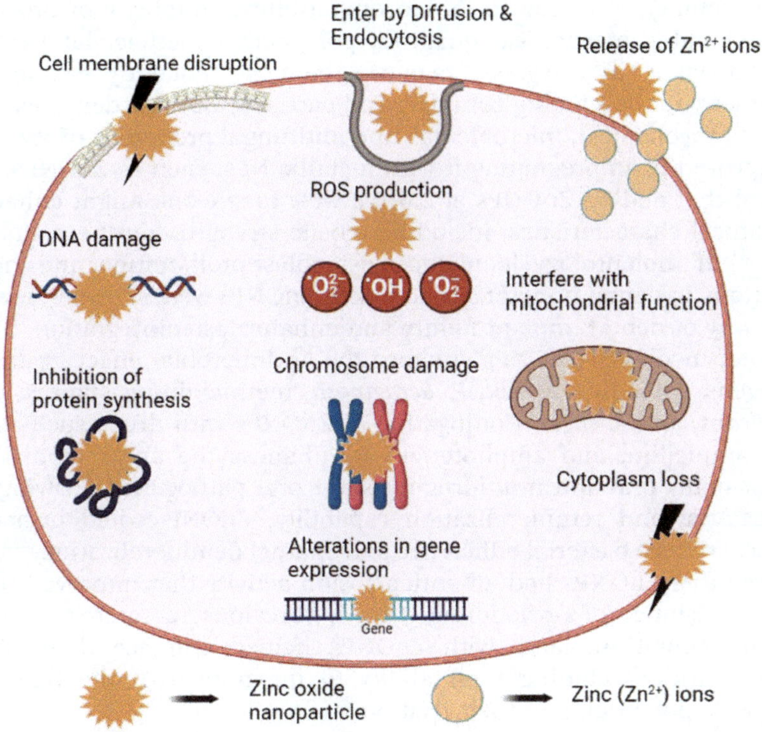

Figure 4.9 Antifungal mechanism of ZnONPs.

4.7.2 Dental Applications of Zinc Oxide Nanoparticles

ZnONPs have a wide variety of applications in various fields of dentistry, including prosthodontics, endodontics, oral medicine, orthodontics, and many other branches. Numerous studies have shown the effectiveness of ZnONPs as reinforcements, to improve their mechanical and physical properties in addition to their antibacterial abilities.[214] In restorative medicine, ZnONPs have been observed to enhance the mechanical and bactericidal qualities of dental restorative materials.[215] It was reported that adding 3% ZnONPs to a flowable resin composite reduced microleakage,[216] and adding it to a commercial resin composite changed the resin composite for improved biofilm inhibition.[217] In endodontics, ZnONPs were shown to improve fracture resistance in roots,[218] and incorporating ZnONPs into an endodontic sealer improved its properties. ZnONPs have an antibacterial effect against many endodontic infections, caused by pathogenic organisms such as *E. faecalis*, *S. aureus*, and *Candida albicans*.[219]

In prosthodontics, ZnONPs prevent the formation of *C. albicans* biofilm on dental tenure without showing a cytotoxic effect on host cells,[220] and it was observed that adding 1.5% ZnONPs to silicone prosthetics improved their colour stability.[221] Owing to its biocompatibility, simplicity of processing, and acceptable mechanical qualities, poly(methyl methacrylate) (PMMA), often known as "acrylic glass," is one of the most frequently used materials in dentistry for developing complete and partial removable dentures.[222] The physical, mechanical, microbicidal, and antifungal properties of PMMA can be improved by impregnating it with metallic NPs, such as ZnONPs. It was claimed that adding ZnONPs at 2.0–2.5 wt% to silicone might enhance its mechanical characteristics, including tensile strength, hardness, and flexibility.[221] In implantology, it enhanced osteoblast proliferation, and chemical alterations in dental implant surfaces with ZnONPs were shown to lower the possibility of dental implant failure and enhance osteointegration.[223]

In oral medicine, ZnONPs enhance the antimicrobial effect against oral pathogens, including *E. coli*, *P. aeruginosa*, methicillin-resistant *S. aureus*, *C. albicans*, and *A. niger.* Conjugation of ZnONPs with drugs such as quercetin, ampicillin, and amphotericin B enhances the antimicrobial effect against many oral and multidrug-resistant oral pathogens.[224] Owing to its antibacterial and remineralization capability, ZnONP-coated orthodontic products reduce bacterial adherence and enamel demineralization.[225] It was reported that ZnONPs had an anticorrosion activity that improved the corrosion resistance of orthodontic wires.[226] Previous research showed that covering dental implants with ZnONPs helped enhance their physical characteristics, including their ability to integrate with the bone, their osteogenic potential, and their fixation.[227]

4.7.3 Antimicrobial Activity of Zinc Oxide Nanoparticles Against Oral Pathogens

More than 25 000 distinct bacterial phylotypes reside in the oral cavity, and some of them also cause systemic infections in addition to periodontal and dental problems. The development of new antimicrobial medicines that are secure, stable, and relatively inexpensive is necessary owing to the emergence of various antibiotic resistances among these bacteria.[228] ZnONPs are well known for their high antibacterial action against microorganisms[229] and they exhibit selective toxicity towards bacterial cells while having a minimal impact on human cells.[230] To prevent the formation of dental plaque, many oral products, such as toothpaste, now include powdered (micron-sized) zinc citrate or acetate.[231]

It was reported that root canal surfaces treated with cationic antibacterial NPs, such as ZnO alone and a combination of ZnO and chitosan NPs, can dramatically inhibit *E. faecalis* adhesion to dentine.[232] Several investigations have been conducted to determine the precise effects of the size scale and aspect ratio of ZnONPs on the cytotoxicity and antibacterial activity. It was reported that the antibacterial effectiveness of poly(ethylene glycol) (PEG)-capped ZnONPs increased with a decrease in particle size.[233] It was observed that ZnONPs with sharp edges exhibit greater antimicrobial activity against *E. coli* and *S. aureus* than the relatively smooth rod- and sphere-shaped NPs.[234]

The release of zinc ions from ZnONPs is one of the potential factors in bacterial growth inhibition.[235] The production of ROS by ZnONPs, such as hydrogen peroxide, contributes to their toxicity to microorganisms.[236] Using scanning electron microscopy (SEM) and transmission electron microscopy (TEM), it has been shown that ZnONPs can cause leakage of intracellular content, cell wall damage, and membrane disruption in *E. coli*.[237]

Microbial biofilms are complex communities adhering to solid surfaces through the secretion of the extracellular matrix that surrounds the bacteria and, compared with planktonic cells, these biofilms display considerably better resistance to toxins and antibiotics.[238] An effective antibiofilm agent should have strong antibacterial action in addition to the capacity to penetrate the biofilm efficiently.[239] It was discovered that *Streptococcus sobrinus* biofilm formation was strongly inhibited by ZnONPs incorporated into various composites at concentrations greater than 10 wt% over a three-day test period.[240] When ZnO and CuO NPs were evaluated for inhibiting biofilm formation activity by a diverse oral bacterial community on artificial dental surfaces and the surface of polystyrene plates, they demonstrated considerable antibiofilm activity at concentrations lower than 100 $\mu g\,mL^{-1}$. ZnONPs decrease dentine demineralization in addition to the targeted antibacterial and antibiofilm-forming capabilities.[241]

4.8 Preparation and Characterization of Chitosan Nanoparticles

Chitosan is a natural biopolymer that is created when acetyl groups are removed from chitin, making it an appealing substance for use in a variety of biomedical applications. Its advantageous biological characteristics, including biocompatibility, biodegradability, non-toxicity, low immunogenicity, and mucoadhesivity, make it one of the most widely used biopolymers in pharmaceutical preparations.[242,243] Additionally, it is the perfect candidate for drug delivery applications owing to its excellent coating properties.[244] Chitosan's many biological attributes, including its antibacterial, antitumor, antiviral, lipid-lowering, and immune-boosting capabilities, have been studied. Chitosan's high positive charge density enables it to penetrate specifically the negatively charged tumour cell membrane, where it exerts its antitumour effect by promoting apoptosis, inhibiting proliferation, and inhibiting angiogenesis.[245,246] Furthermore, chitosan NPs (CNPs) can bind to receptors on the surface of tumour cells and have specific tumour-targeting properties that increase the effectiveness of drug delivery. This is a key consideration in the field of biomedicine, where efficient and focused medication delivery is necessary to reduce adverse effects and boost therapeutic efficacy.[247,248]

To create CNPs, chitosan solution, prepared by dissolving chitosan at room temperature in a small amount of acetic acid, typically at 0.15%, is combined with a certain kind of ionic solution, such as tripolyphosphate (TPP), which is introduced dropwise into the chitosan solution. The ionic interactions between the positive charges of chitosan and the negative charges of TPP during the ionic gelation process between chitosan and TPP result in the formation of NPs. Through these interactions, a gel-like network is created, encasing the chitosan to create CNPs. The colloidal suspension is stirred for a predetermined time after the creation of CNPs to allow the NPs to solidify. To homogenize the suspension and remove any agglomerates, the resultant dispersion is next placed in an ultrasonic bath for 10 min. To obtain a clear suspension of CNPs, the suspension is centrifuged at high speed, typically 10 000 rpm, for 10 min. The unencapsulated chitosan and other contaminants in the suspension can be separated from the CNPs by centrifugation. The resulting CNPs are smooth and spherical in shape, with a clearly defined size distribution. By changing the amount of TPP supplied, the rate of agitation, and the length of ultrasonic treatment, the size of the CNPs can be altered.

Owing to their biocompatibility, biodegradability, and capacity to encapsulate different medications, bioactive compounds, and other therapeutic agents, CNPs are used in a variety of industries. A wide variety of medicinal compounds with different sizes and charges can be encapsulated thanks to the adaptability of the ionic gelation process, which is a quick and efficient way to prepare CNPs.

Calvo *et al.* reported that the ionic gelation process with TPP is a commonly used and successful approach for the synthesis of CNPs. The process entails creating a well-defined suspension of CNPs by dissolving a chitosan solution in an aqueous TPP solution and combining the resulting suspension. The resulting CNPs are a viable tool for numerous biomedical applications because they offer a number of appealing properties, such as biocompatibility, biodegradability, and the capacity to encapsulate different medicinal compounds.[249]

The development of potential biomedical uses of CNPs requires careful consideration of their characterization. The physical and chemical characteristics of the NPs, such as size, shape, surface charge, and drug-loading capability, have been studied using a variety of approaches. The characterization of CNPs offers crucial information for enhancing the formulation and manufacture of these particles and also for assessing their potential for drug administration.[250]

Dynamic light scattering (DLS) is one of the most popular methods for determining the size of NPs. Both the size distribution and mean size of NPs may be estimated quickly and accurately using DLS.[251] Additionally, the morphological characteristics of the NPs, such as their size, shape, and surface structure, may be revealed using TEM.[252] Determining the surface charge of CNPs is a crucial component of their characterization. This can be determined using methods such as zeta potential measurement, which shows how stable the NPs are in aqueous suspension. A high value of the zeta potential denotes high stability, whereas a low value indicates a high propensity for particle aggregation.[253]

Another crucial feature of the characterization of CNPs is their ability to load drugs. This attribute is essential in assessing the potential of CNPs as a drug delivery system since it indicates how much medication can be contained within them. Both the amount of drug loaded into the NPs and the drug's release behaviour from the NPs over time can be assessed using methods such as high-performance liquid chromatography (HPLC).[254] The toxicity and biocompatibility of the NPs and their capacity to encapsulate and release medications are assessed in *in vitro* investigations, which can include cell culture tests. Animal studies that assess the pharmacokinetics and biodistribution of the NPs might be a part of *in vivo* research.[255] A combination of various techniques, including DLS, TEM, zeta potential measurement, HPLC, and *in vitro* and *in vivo* studies, can provide a comprehensive understanding of the properties of CNPs and their potential for drug delivery.

4.8.1 Antimicrobial Applications of Chitosan Nanoparticles

CNPs have been shown in various studies to have antimicrobial activity. This is due to the positive charge and biocompatibility of chitosan, which can interact with the negatively charged bacterial cell wall, leading to bacterial inhibition and death.[256] Additionally, chitosan has been reported to increase the permeability of the bacterial cell membrane, allowing the entry of antimicrobial agents into the cell. The antimicrobial activity of CNPs can be further enhanced by incorporating antimicrobial agents into the particles.[257]

Recently, Sathiyaseelan *et al.* investigated the enhanced antimicrobial and wound-healing applications of bimetallic NPs (AgPt NPs) coated with chitosan-fabricated cotton gauze. The size of the bimetallic NPs was found to be 21.8 ± 6.32 nm and the shape was spherical. The antimicrobial activity results showed a greater zone of inhibition than with standard antibiotics against yeast, filamentous fungi, and bacteria. The bimetallic NPs coated with cotton gauze exhibited an excellent hydrophobic nature with potent antimicrobial and antioxidant activity. This reflects the enhanced antimicrobial properties of chitosan.[258]

Several studies indicated chitosan and CNPs as the best choice in drug delivery systems.[259,260] In many drug delivery systems, CNPs are reported to shine as a promising class of bioactive encapsulators.[259] In another study, CNPs were used as a separate encapsulation or co-encapsulating agent for natural products such as *Origanum syriacum* essential oil and for a synthetic complex, Zn(II)salen. The co-encapsulation of CNPs resulted in the development of a potent antimicrobial nanoplatform.[261]

Ionic gelation was used in a study to incorporate *Citrus aurantium* essential oil into CNPs. The final NPs were spherical in shape and ranged in size from 20 to 60 nm. The loading capacity and encapsulation efficiency of the essential oil ranged from 2.51 to 5.89% and from 4.51 to 15.33%, respectively. Following the Fickian diffusion model, the essential oil was released from the CNPs. During cold storage, white button mushrooms had lower microbial counts, lower peroxidase activity, and higher antioxidant enzyme activity as a result of the use of CNPs as a delivery vehicle for the essential oil. These findings point to a potential function for CNPs in enhancing the antimicrobial safety and antioxidant activity of white button mushrooms.[262]

Kalaivani *et al.* examined the synthesis and characterization of silver NPs (AgNPs) produced by chitosan. Their results demonstrated that the addition of chitosan increased the production efficiency of spherical and crystalline AgNPs while reducing their size. When used against Gram-positive and Gram-negative bacteria, and also fungal pathogens, the synthesized AgNPs demonstrated strong antibacterial and antifungal activities. The antibacterial activity of the AgNPs was observed to be higher than that of the common antibiotics ampicillin and amphotericin B, and it was found that using crustacean waste can produce AgNPs at a reasonable cost and open up new markets for seafood items such as chitin and chitosan.[263]

Similarly, lactoferrin (L), chitosan (C), and gellan (G) NPs were produced through electrostatic complexation to improve their antimicrobial properties. Ternary complexes with an L : C : G ratio of 4.5 : 4.5 : 1.0 showed the highest positive charge density and smallest hydrodynamic diameter, leading to the highest antimicrobial action against *S. aureus*. When applied to fresh strawberries, the NPs effectively preserved their properties and were more effective than lactoferrin and chitosan alone. This once again confirmed the antimicrobial-enhancing properties of CNPs and their potential applications in food preservation.[264]

4.8.2 Dental Applications of Chitosan Nanoparticles

Owing to its biocompatibility, biodegradability, and antibacterial qualities, chitosan, a natural biopolymer derived from chitin, has shown significant promise for a range of dental applications. This potential for dental applications was further increased by the discovery of CNPs. The use of CNPs in several dental applications, including dental implant coatings, dental pulp regeneration, oral medicine administration, and periodontal disease management, has attracted considerable attention in recent years.[265,266]

4.8.2.1 Dental Implant Coatings

Missing teeth are frequently replaced with dental implants. However, the emergence of peri-implantitis, which is brought on by bacterial infection, frequently compromises the effectiveness of these implants. In order to lessen bacterial adherence and minimize peri-implantitis, CNPs have been investigated as a suitable covering material for dental implants. CNP coatings have been shown to significantly reduce bacterial cell adherence and increase antibacterial effectiveness against oral infections.[267]

Calcium phosphate (hydroxyapatite)–chitosan coatings were produced by the pulse electrodeposition method with variation of the deposition cycle and chitosan content. The addition of chitosan, a biocompatible and biodegradable polysaccharide, improved the electrochemical properties of the hydroxyapatite coatings and made them more resistant to corrosion. The composite coatings showed a significant improvement over the bare Ti–6Al–4V substrate, making them a promising candidate for biomedical applications owing to their improved durability and stability. Chitosan played a crucial role in improving the corrosion resistance and electrochemical properties of the coatings.[268]

4.8.2.2 Dental Pulp Regeneration

Hot and cold sensations are brought on by the dental pulp, which is a soft tissue inside the tooth. The dental pulp may become inflamed or necrotic as a result of dental illness or damage, necessitating a pulpectomy, which entails the removal of the entire pulp. CNPs have been investigated for application in dental pulp regeneration. The growth and differentiation of dental pulp cells and also the regeneration of the pulp tissue have both been demonstrated to be enhanced by CNPs.[269,270]

Elgendy and Fayyad reported the role of chitosan in regenerative endodontics. Their study compared the cytotoxicity and apoptotic changes of chitosan and CNPs on dental pulp stem cells. The results showed that CNPs had lower cytotoxicity and DNA fragmentation compared with the original particles. The biocompatibility of CNPs increased over time, making it a promising candidate for use in endodontic regeneration. Overall, the results indicate that chitosan and its NPs are biocompatible and can be used for endodontic regeneration purposes.[271]

4.8.2.3 Oral Drug Delivery

Effective oral drug delivery is essential for the management of various oral diseases, including dental caries, periodontitis, and oral cancer.[272] CNPs have been studied as a potential oral drug delivery system owing to their biocompatibility, biodegradability, and mucoadhesive properties. It has been demonstrated that CNPs can effectively encapsulate various drugs, including antibiotics and anti-inflammatory agents, and can deliver them to the desired site in the oral cavity with improved efficacy.[273] Chitosan is a naturally occurring polymer that has been investigated for its potential as a drug delivery vehicle with oral administration. It has the ability to encapsulate drugs and shield them from degradation in the gastrointestinal tract, allowing a pH-responsive release of the drug in the upper intestinal region. Chitosan has been shown to be a promising delivery system for drugs with poor solubility, such as the anti-tuberculosis drug rifampicin.[274] In another study, the chitosan nanocarrier system was reported to be able to deliver 75% of rifampicin over a 24 h period at simulated intestinal pH. Chitosan is a biodegradable and biocompatible polymer that is being investigated for its potential to improve the bioavailability and efficacy of drugs *via* the oral administration route.[275]

4.8.2.4 Periodontal Disease Management

Periodontal disease is a chronic inflammatory condition that affects the alveolar bone and gums, which support the teeth. The treatment of periodontal disease must include antimicrobial therapy. For the treatment of periodontal disease, CNPs have been investigated as a potential local antibacterial agent. CNPs have been shown to stop periodontal pathogen growth effectively and accelerate wound healing.[276]

Recently, Garg and Tirgar used chitosan as a carrier material for metronidazole, an antibacterial drug, to treat periodontitis. Chitosan was chemically modified to the *N*-trimethyl quaternary ammonium derivative to enhance its solubility in an acidic environment and increase its antibacterial activity. The modified chitosan was used to prepare NPs by ionic gelation methodology. The results showed that the optimized batches of CNPs had a size of 117.01 ± 0.03 nm, a high entrapment efficiency ($81.45 \pm 0.03\%$) of metronidazole, and a smooth surface. The CNPs showed sustained release of the drug and enhanced antibacterial activity against periodontal infections, making it a promising material for the treatment of periodontitis. This study highlights the role of chitosan as an effective carrier material for the delivery of metronidazole in the treatment of periodontitis.[277]

Hence CNPs have demonstrated significant promise for a range of dental applications, such as periodontal disease treatment, oral drug delivery, dental implant coatings, and dental pulp regeneration. To understand completely the mechanisms of action of CNPs and to maximize their utilization in dental applications, more research is required. The outcomes of recent research, however, are positive, and CNPs may offer important advantages in dental applications.

4.8.3 Antimicrobial Activity of Chitosan Nanoparticles Against Oral Pathogens

The growth of oral diseases, such as dental caries, oral cancer, and periodontitis, has greatly affected the quality of life of many people.[278] Biomaterials and nanotechnology have led to the development of various systems for treating oral diseases. Chitosan and its derivatives, which are biocompatible, biodegradable, and bioadhesive polysaccharides, have been widely used to treat oral diseases owing to their antibacterial properties and ability to control drug release. Chitosan has been shown to enhance the effectiveness of therapy for various oral diseases.[279]

Chitin, which is present in the exoskeleton of crustaceans, is a source of the biopolymer chitosan. Its antibacterial effects against various Gram-negative and Gram-positive bacteria have been extensively researched. Gram-negative bacteria are frequently utilized in antibacterial investigations to assess the effectiveness of chitosan, including *E. coli*, *Porphyromonas gingivalis*, *Klebsiella pneumoniae*, *P. aeruginosa*, *Salmonella* spp., and *Serratia marcescens*.[280] Because they are known to be resistant to several common antibiotics, these bacteria present a serious challenge to the medical community.

Similarly, studies on the antibacterial action of chitosan have also been focused on Gram-positive bacteria such as *S. mutans*, *S. aureus*, and *Bacillus* spp.[281] *S. mutans* is an intriguing target for applications in oral healthcare because it plays a significant role in dental caries. Infections of the skin and open wounds are frequently brought on by *S. aureus*, which is also a serious risk factor for pneumonia, sepsis, and meningitis.[282] Owing to its capacity to interact with bacterial cell membranes and compromise their integrity, chitosan has been discovered to possess strong antibacterial effects against both Gram-negative and Gram-positive bacteria. In contrast to conventional antibiotics, which focus on particular enzymes or metabolic processes within the bacterial cell, this method of action is distinct.[283,284]

The antibacterial characteristics of chitosan have been thoroughly investigated, and its prospective uses in a variety of applications, including food preservation, wound healing, and oral healthcare, have been investigated. It is a desirable substitute for traditional antibiotics owing to its biocompatibility, biodegradability, and lack of harmful by-products.[285]

Natural isoflavonoids have been found to have antibacterial activity against oral infections, but their limited practical application is due to their low water solubility and weak absorption. Licoricidin-loaded CNPs (LC-CNPs) were created in a study to increase licoricidin dispersion and boost antibacterial performance. The minimum inhibitory concentration of LC-CNPs against *S. mutans* was lowered twofold to 26 g mL^{-1} compared with free licoricidin, and the particles were capable of destroying 70% of biofilms. The adsorption of LC-CNPs to the surface of bacterial cell membranes was thought to be the cause of the enhanced activity. Additionally, the development of probiotics was not inhibited by LC-CNPs. According to mechanistic studies, LC-CNPs changed the transmembrane potential, which led to the formation of ROS and ultimately apoptosis in bacterial cells. This showed how chitosan-based NPs could potentially improve the dispersion and antibacterial activity of insoluble isoflavonoids, providing a promising method for treating oral infections.[286]

Recently, *Ocimum basilicum* L. (basil) essential oil (BEO) was nanoencapsulated using chitosan as a carrier. Analytical methods such Fourier transform infrared (FTIR) spectroscopy, UV-visible spectrophotometry, X-ray diffraction, and SEM imaging showed that chitosan was effective at loading BEO. The CNPs produced showed an encapsulation effectiveness of 50.39–5.13% and a particle size range of 198.7–373.4 nm. The BEO-loaded CNPs showed potent antibacterial and antibiofilm efficacy against *E. coli* and *S. aureus* with inhibitory diameters of 15.3 and 21.0 mm, respectively. According to this study, CNPs play the role of an antibacterial agent by weakening bacterial cell membranes and causing biological macromolecule leakage.[287]

Maheo described a green approach for creating chitosan–plant extract-mediated copper oxide (CPC) NPs that used chitosan and an aqueous leaf extract of *Eupatorium adenophorum*. Several methods were used to characterize the synthesized CPC NPs, including high-resolution transmission electron microscopy (HRTEM), energy-dispersive X-ray analysis (EDAX), Fourier transform infrared (FTIR) spectroscopy, X-ray diffraction (XRD), field emission scanning electron microscopy (FESEM), UV-visible spectrophotometry, and selected area electron diffraction (SAED). The NPs produced featured flake-like and rod-like architectures, and their diameters ranged from 30 to 80 nm, according to the FESEM and HRTEM images. The EDAX study verified the purity of the elemental constitutions. Testing of the antibacterial effectiveness of the CPC NPs involved four bacterial species, namely two Gram-positive (*E. faecalis* and *S. aureus*) and two Gram-negative (*E. coli* and *Salmonella typhi*). The greatest zones of inhibition were 18, 12, 11, and 12 mm, respectively. The study emphasizes how chitosan and plant extract work together to influence the form and size of the NPs.[288]

Arif *et al.* functionalized chitosan with mannose and produced NPs to serve as an alternative antibacterial treatment for multidrug-resistant *Helicobacter pylori* gastric infections. The NPs were shown to disrupt bacterial adherence and biofilm formation effectively and induce membrane disruption in *H. pylori* cells through hydrogen bond interactions with active site residues in a target protein. This makes mannose-functionalized CNPs an excellent candidate for alternative therapies against resistant gastric infections.[289]

Amoxicillin (AMX) was encapsulated for use in biomedical applications using *N*-2-hydroxypropyltrimethylammonium chloride chitosan (*N*-2-HACC) and *N*,*O*-carboxymethyl chitosan (CMCS) NPs. The encapsulated AMX had increased bioavailability and prolonged, delayed release, and the NPs exhibited broad-spectrum antibacterial activities. Excellent cytocompatibility and antibacterial activity against different microorganisms were also displayed by the NPs. The *N*-2-HACC/CMCS/AMX NPs could be employed as a drug delivery vehicle to produce prolonged release, increase bioavailability, and decrease the dosage of the medication because they were stable towards storage.[290]

4.8.4 Challenges and Future Directions

The use of CNPs for antimicrobial activity has seen significant growth in recent years, but there are still some challenges that need to be addressed. One of the major challenges is the low stability of CNPs in biological environments, as they tend to aggregate and lose their efficacy over time. Another challenge is the lack of specificity in targeting harmful microorganisms, as CNPs often have broad-spectrum activity, which can lead to the development of resistance.[291]

In addition, the biocompatibility of CNPs is still a concern, as they can cause cytotoxicity and trigger immune responses in certain situations. The mechanism of action of CNPs is also not fully understood, and further research is needed to explore the underlying mechanisms and identify new targets.[292,293]

In the future, the development of targeted and sustained-release CNPs, and the exploration of new and improved chitosan-based delivery systems, are expected to overcome these challenges and enhance the efficacy of CNPs for antimicrobial applications. Furthermore, the integration of CNPs with other antimicrobial agents and the development of combination therapies may also lead to more effective and sustainable solutions for controlling microbial infections.

Acknowledgements

We would like to express our sincere thanks to Saveetha Dental College and Hospital, SIMATS, for providing encouragement and constant support.

References

1. S. Bayda, M. Adeel, T. Tuccinardi, M. Cordani and F. Rizzolio, The history of nanoscience and nanotechnology: from chemical–physical applications to nanomedicine, *Molecules*, 2019, **25**(1), 112.
2. S. S. Salem and A. Fouda, Green synthesis of metallic nanoparticles and their prospective biotechnological applications: an overview, *Biol. Trace Elem. Res.*, 2021, **199**, 344–370.
3. M. K. Rai, S. D. Deshmukh, A. P. Ingle and A. K. Gade, Silver nanoparticles: the powerful nanoweapon against multidrug-resistant bacteria, *J. Appl. Microbiol.*, 2012, **112**(5), 841–852.
4. I. X. Yin, J. Zhang, I. S. Zhao, M. L. Mei, Q. Li and C. H. Chu, The antibacterial mechanism of silver nanoparticles and its application in dentistry, *Int. J. Nanomed.*, 2020, 2555–2562.
5. F. Carrouel, S. Viennot, L. Ottolenghi, C. Gaillard and D. Bourgeois, Nanoparticles as anti-microbial, anti-inflammatory, and remineralizing agents in oral care cosmetics: a review of the current situation, *Nanomaterials*, 2020, **10**(1), 140.
6. D. Glavina, *Materials for caries prevention*, Expert Editor, 2022, p. 84.
7. C. Buzea, I. I. Pacheco and K. Robbie, Nanomaterials and nanoparticles: sources and toxicity, *Biointerphases*, 2007, **2**(4), MR17–MR71.
8. I. De la Calle, M. Menta and F. Séby, Current trends and challenges in sample preparation for metallic nanoparticles analysis in daily products and environmental samples: A review, *Spectrochim. Acta, Part B*, 2016, **125**, 66–96.
9. M. Pasha, N. Muhammad, S. Shahnawaz, Y. Najmi, N. Shahroz and S. Liaqat, Ceramic nanomaterials in dental applications, *Nanoeng. Biomater.*, 2022, 123–144.
10. P. Yudaev, V. Chuev, B. Klyukin, A. Kuskov, Y. Mezhuev and E. Chistyakov, Polymeric dental nanomaterials: Antimicrobial action, *Polymers*, 2022, **14**(5), 864.
11. G. M. De Souza, Nanoparticles in restorative materials, in *Nanotechnology in endodontics: current and potential clinical applications*, 2015, pp. 139–171.
12. G. Orsini, V. Tosco, R. Monterubbianesi, G. Orilisi and A. Putignano, A new era in restorative dentistry, in *The First Outstanding 50 Years of*

"Università Politecnica delle Marche", Research Achievements in Life Sciences, 2020, pp. 319–334.
13. L. Thangavelu, A. H. Adil, S. Arshad, E. Devaraj, S. K. Mallineni, R. Sajja, A. Chakradhar and M. I. Karobari, Antimicrobial properties of silver nitrate nanoparticle and its application in endodontics and dentistry: a review of literature, *J. Nanomater.*, 2021, **2021**, 1–2.
14. H. Nagarajan and P. S. Karuppanan, Nanoparticles in dentistry-An updated review, *Int. J. Community Dent.*, 2021, **9**(2), 74.
15. H. Moradpoor, M. Safaei, H. R. Mozaffari, R. Sharifi, M. M. Imani, A. Golshah and N. Bashardoust, An overview of recent progress in dental applications of zinc oxide nanoparticles, *RSC Adv.*, 2021, **11**(34), 21189–21206.
16. M. Kachoei, A. Nourian, B. Divband, Z. Kachoei and S. Shirazi, Zinc-oxide nanocoating for improvement of the antibacterial and frictional behavior of nickel-titanium alloy, *Nanomedicine*, 2016, **11**(19), 2511–2527.
17. P. Makvandi, J. T. Gu, E. N. Zare, B. Ashtari, A. Moeini, F. R. Tay and L. N. Niu, Polymeric and inorganic nanoscopical antimicrobial fillers in dentistry, *Acta Biomater.*, 2020, **101**, 69–101.
18. S. M. Ahsan, M. Thomas, K. K. Reddy, S. G. Sooraparaju, A. Asthana and I. Bhatnagar, Chitosan as biomaterial in drug delivery and tissue engineering, *Int. J. Biol. Macromol.*, 2018, **110**, 97–109.
19. R. P. Allaker, The use of nanoparticles to control oral biofilm formation, *J. Dent. Res.*, 2010, **89**(11), 1175–1186.
20. T. A. Ahmed and B. M. Aljaeid, Preparation, characterization, and potential application of chitosan, chitosan derivatives, and chitosan metal nanoparticles in pharmaceutical drug delivery, *Drug Des., Dev. Ther.*, 2016, 483–507.
21. X. Chen, E. B. Daliri, N. Kim, J. R. Kim, D. Yoo and D. H. Oh, Microbial etiology and prevention of dental caries: exploiting natural products to inhibit cariogenic biofilms, *Pathogens*, 2020, **9**(7), 569.
22. X. S. He and W. Y. Shi, Oral microbiology: past, present and future, *Int. J. Oral Sci.*, 2009, **1**(2), 47–58.
23. M. Sanz, D. Beighton, M. A. Curtis, J. A. Cury, I. Dige, H. Dommisch, R. Ellwood, R. A. Giacaman, D. Herrera, M. C. Herzberg and E. Könönen, Role of microbial biofilms in the maintenance of oral health and in the development of dental caries and periodontal diseases. Consensus report of group 1 of the Joint EFP/ORCA workshop on the boundaries between caries and periodontal disease, *J. Clin. Periodontol.*, 2017, **44**, S5–S11.
24. R. P. Allaker and K. Memarzadeh, Nanoparticles and the control of oral infections, *Int. J. Antimicrob. Agents*, 2014, **43**(2), 95–104.
25. D. H. Meyer and P. M. Fives-Taylor, Oral pathogens: from dental plaque to cardiac disease, *Curr. Opin. Microbiol.*, 1998, **1**(1), 88–95.
26. S. Hamada and H. D. Slade, Biology, immunology, and cariogenicity of Streptococcus mutans, *Microbiol. Rev.*, 1980, **44**(2), 331–384.

27. E. Franco, Dental disease, caries related microflora and salivary of children with severe congenital cardiac disease: an epidemiologlcal and oral microbial survey, *Pediatr. Dent.*, 1996, **18**, 228–235.
28. L. M. Prescott, J. P. Harley and D. A. Klein, *Microbiology*, McGraw-Hill, Boston, MA, 2002.
29. R. Teanpaisan, S. Piwat and G. Dahlen, Inhibitory effect of oral Lactobacillus against oral pathogens, *Lett. Appl. Microbiol.*, 2011, **53**(4), 452–459.
30. S. Prabuseenivasan, M. Jayakumar and S. Ignacimuthu, In vitro antibacterial activity of some plant essential oils, *BMC Complementary Altern. Med.*, 2006, **6**, 39.
31. M. E. Hibbing, C. Fuqua, M. R. Parsek and S. B. Peterson, Bacterial competition: surviving and thriving in the microbial jungle, *Nat. Rev. Microbiol.*, 2010, **8**(1), 15–25.
32. M. Oussalah, S. Caillet, L. Saucier and M. Lacroix, Antimicrobial effects of selected plant essential oils on the growth of a pseudomonas putida strain isolated from meat, *Meat Sci.*, 2006, **73**, 236–244.
33. F. Sharififar, M. H. Moshafi, S. H. Mansouri, M. Khodashenas and M. Khoshnoodi, In vitro evaluation of antibacterial and antioxidant activities of the essential oil and methanol extract of endemic Zataria multiflora Boiss, *Food Control*, 2007, 7, 800–805.
34. S. George, A. Kishen and K. P. Song, The role of environmental changes on monospecies biofilm formation on root canal wall by Enterococcus faecalis, *J. Endod.*, 2005, **31**, 867–872.
35. Y. Noiri, A. Ehara, T. Kawahara, N. Takemura and S. Ebisu, Participation of bacterial biofilms in refractory and chronic periapical periodontitis, *J. Endod.*, 2002, **28**, 679–683.
36. I. B. Sneddon and R. C. Glew, Contact dermatitis due to propanidid in an anaesthetist, *Practitioner*, 1973, **211**, 321–323.
37. N. Thosar, S. Basak, R. N. Bahadure and M. Rajurkar, Antimicrobial efficacy of five essential oils against oral pathogens: An in vitro study, *Eur. J. Dent.*, 2013, **7**(S 01), S071–S077.
38. A. Sukkarwalla, S. M. Ali, P. Lundberg and F. Tanwir, Efficacy of miswak on oral pathogens, *Dent. Res. J.*, 2013, **10**(3), 314.
39. M. J. R. Virlan, D. Miricescu, R. Radulescu, C. M. Sabliov, A. Totan, B. Calenic and M. Greabu, Organic nanomaterials and their applications in the treatment of oral diseases, *Molecules*, 2016, **21**(2), 207.
40. R. P. Allaker and K. Memarzadeh, Nanoparticles and the control of oral infections, *Int. J. Antimicrob. Agents*, 2014, **43**(2), 95–104.
41. E. Giertsen, Effects of mouthrinses with triclosan, zinc ions, copolymer, and sodium lauryl sulphate combined with fluoride on acid formation by dental plaque *in vivo*, *Caries Res*, 2004, **38**(5), 430–435.
42. C. Malarkodi, K. Chitra and S. Rajeshkumar, *et al*, Novel eco-friendly synthesis of titanium oxide nanoparticles by using *Planomicrobium* sp. and its antimicrobial evaluation, *Der Pharm. Sin.*, 2013, **4**(3), 59–66.

43. S. Rajeshkumar, C. Malarkodi, K. Paulkumar, M. Vanaja, G. Gnanajobitha and G. Annadurai, Intracellular and extracellular biosynthesis of silver nanoparticles by using marine bacteria *Vibrio alginolyticus*, *Nanosci. Nanotechnol.*, 2013, **3**(1), 21–25.
44. F. Hans-Curt and G. G. Grill, Biofouling and biocorrosion in industrial water system, in *Proceedings of the International Workshop on Industrial Biofouling and Biocorrosion*, Springer, 1990, p. 200.
45. P. C. Baehni and Y. Takeuchi, Anti-plaque agents in the prevention of biofilm-associated oral diseases, *Oral Dis.*, 2003, **9**(1), 23–29.
46. C. Malarkodi, S. Rajeshkumar, K. Paulkumar, M. Vanaja, G. Gnanajobitha and G. Annadurai, Biosynthesis and antimicrobial activity of semiconductor nanoparticles against oral pathogens, *Bioinorg. Chem. Appl.*, 2014, **2014**.
47. S. G. Asadi and Z. G. Asadi, Chewing sticks and the oral hygiene habits of the adult Pakistani population, *Int. Dent. J.*, 1997, **47**, 275–278.
48. T. Chen, K. Abbey, W. J. Deng and M. C. Cheng, The bioinformatics resource for oral pathogens, *Nucleic Acids Res.*, 2005, **33**(suppl_2), W734–W740.
49. K. A. Boggess, K. Moss, P. Madianos, A. P. Murtha, J. Beck and S. Offenbacher, Fetal immune response to oral pathogens and risk of preterm birth, *Am. J. Obstet. Gynecol.*, 2005, **193**(3), 1121–1126.
50. T. Verspecht, E. Rodriguez Herrero, L. Khodaparast, L. Khodaparast, N. Boon, K. Bernaerts, M. Quirynen and W. Teughels, Development of antiseptic adaptation and cross-adapatation in selected oral pathogens *in vitro*, *Sci. Rep.*, 2019, **9**(1), 1–3.
51. G. Hajishengallis, A. Sharma, M. W. Russell and R. J. Genco, Interactions of oral pathogens with toll-like receptors: possible role in atherosclerosis, *Ann. Periodontol.*, 2002, **7**(1), 72–78.
52. F. A. Al-Bayati and K. D. Sulaiman, In vitro antimicrobial activity of Salvadora persica L. extracts against some isolated oral pathogens in Iraq, *Turk. J. Biol.*, 2008, **32**(1), 57–62.
53. F. Sedighinia, A. S. Afshar, J. Asili and K. Ghazvini, Antibacterial activity of Glycyrrhiza glabra against oral pathogens: an *in vitro* study, *Avicenna J. Phytomed.*, 2012, **2**(3), 118.
54. C. G. Pupe, M. Villardi, C. R. Rodrigues, H. V. Rocha, L. C. Maia, V. P. de Sousa and L. M. Cabral, Preparation and evaluation of antimicrobial activity of nanosystems for the control of oral pathogens Streptococcus mutans and Candida albicans, *Int. J. Nanomed.*, 2011, 2581–2590.
55. S. H. Abdollahzadeh, R. Y. Mashouf, H. Mortazavi, M. H. Moghaddam, N. Roozbahani and M. Vahedi, Antibacterial and antifungal activities of Punica granatum peel extracts against oral pathogens, *J. Dent.*, 2011, **8**(1), 1.
56. M. F. Zarco, T. J. Vess and G. S. Ginsburg, The oral microbiome in health and disease and the potential impact on personalized dental medicine, *Oral Dis.*, 2012, **18**(2), 109–120.
57. R. A. Bapat, T. V. Chaubal, S. Dharmadhikari, A. M. Abdulla, P. Bapat, A. Alexander, S. K. Dubey and P. Kesharwani, Recent advances of gold

nanoparticles as biomaterial in dentistry, *Int. J. Pharm.*, 2020, **586**, 119596.
58. S. Garner and M. E. Barbour, Nanoparticles for controlled delivery and sustained release of chlorhexidine in the oral environment, *Oral Dis.*, 2015, **21**(5), 641–644.
59. B. Zhu, L. C. Macleod, T. Kitten and P. Xu, Streptococcus sanguinis biofilm formation & interaction with oral pathogens, *Future Microbiol.*, 2018, **13**(08), 915–932.
60. P. D. Marsh, Dental plaque as a microbial biofilm, *Caries Res.*, 2004, **38**(3), 204–211.
61. C. Rompelberg, M. B. Heringa, G. van Donkersgoed, J. Drijvers, A. Roos, S. Westenbrink, R. Peters, G. van Bemmel, W. Brand and A. G. Oomen, Oral intake of added titanium dioxide and its nanofraction from food products, food supplements and toothpaste by the Dutch population, *Nanotoxicology*, 2016, **10**(10), 1404–1414.
62. I. Kleinberg, A mixed-bacteria ecological approach to understanding the role of the oral bacteria in dental caries causation: an alternative to Streptococcus mutans and the specific-plaque hypothesis, *Crit. Rev. Oral Biol. Med.*, 2002, **13**(2), 108–125.
63. K. H. Metwalli, S. A. Khan, B. P. Krom and M. A. Jabra-Rizk, Streptococcus mutans, Candida albicans, and the human mouth: a sticky situation, *PLoS Pathog.*, 2013, **9**(10), e1003616.
64. J. A. Lemos, S. R. Palmer, L. Zeng, Z. T. Wen, J. K. Kajfasz, I. A. Freires, J. Abranches and L. J. Brady, The biology of Streptococcus mutans, *Microbiol. Spectrum*, 2019, **7**(1), 10–128.
65. B. Köhler, I. Andréen and B. Jonsson, The effect of caries-preventive measures in mothers on dental caries and the oral presence of the bacteria Streptococcus mutans and lactobacilli in their children, *Arch. Oral Biol.*, 1984, **29**(11), 879–883.
66. L. Montanaro, D. Campoccia, S. Rizzi, M. E. Donati, L. Breschi, C. Prati and C. R. Arciola, Evaluation of bacterial adhesion of Streptococcus mutans on dental restorative materials, *Biomaterials*, 2004, **25**(18), 4457–4463.
67. M. Okada, Y. Soda, F. Hayashi, T. Doi, J. Suzuki, K. Miura and K. Kozai, PCR detection of Streptococcus mutans and S. sobrinus in dental plaque samples from Japanese pre-school children, *J. Med. Microbiol.*, 2002, **51**(5), 443–447.
68. S. S. Ahirwar, M. K. Gupta and S. K. Snehi, Dental caries and lactobacillus: Role and ecology in the oral cavity, *Int. J. Pharm. Sci. Res.*, 2019, **11**, 4818–4829.
69. E. Caglar, B. Kargul and I. Tanboga, Bacteriotherapy and probiotics' role on oral health, *Oral Dis.*, 2005, **11**(3), 131–137.
70. S. M. Michalek, M. Hirasawa, H. Kiyono, K. Ochiai and J. R. McGhee, Oral ecology and virulence of Lactobacillus casei and Streptococcus mutans in gnotobiotic rats, *Infect. Immun.*, 1981, **33**(3), 690–696.
71. R. Saini, S. Saini and S. Sharma, Nanotechnology: the future medicine, *J. Cutaneous Aesthetic Surg.*, 2010, **3**(1), 32.

72. T. Lainović, L. Blažić and M. Potran, Nanotechnology in dentistry: current state and future perspectives, *Stomatol. Glas. Srb.*, 2012, **59**(1), 44–50.
73. N. Gligorijević and M. Kostić, Nanotechnology in dentistry: Current state and future perspectives, *Acta Stomatol. Naissi*, 2015, **31**(72), 1538–1545.
74. S. Priyadarsini, S. Mukherjee and M. Mishra, Nanoparticles used in dentistry: a review, *J. Oral Biol. Craniofacial Res.*, 2018, **8**(1), 58–67.
75. G. Moraes, C. Zambom and W. L. Siqueira, Nanoparticles in dentistry: A comprehensive review, *Pharmaceuticals*, 2021, **14**(8), 752.
76. S. Q. Jamshed, *J. Pharm. BioAllied Sci.*, 2016, **8**(1), 1.
77. R. A. Freitas Jr, Nanodentistry, *J. Am. Dent. Assoc.*, 2000, **131**(11), 1559–1565.
78. S. T. Ozak and P. Ozkan, Nanotechnology and dentistry, *Eur. J. Dent.*, 2013, **7**(01), 145–151.
79. B. Parhi, D. Bharatiya, S. S. Purohit and S. K. Swain, Chitosan-Based Nano Biomaterials and Their Applications in Dentistry, in *Chitosan Nanocomposites: Bionanomechanical Applications*, Springer Nature Singapore, Singapore, 2023, pp. 325–348.
80. V. Harish, D. Tewari, M. Gaur, A. B. Yadav, S. Swaroop, M. Bechelany and A. Barhoum, Review on nanoparticles and nanostructured materials: Bioimaging, biosensing, drug delivery, tissue engineering, antimicrobial, and agro-food applications, *Nanomaterials*, 2022, **12**(3), 457.
81. L. K. Foong, M. M. Foroughi, A. F. Mirhosseini, M. Safaei, S. Jahani, M. Mostafavi, N. Ebrahimpoor, M. Sharifi, R. S. Varma and M. Khatami, Applications of nano-materials in diverse dentistry regimes, *RSC Adv.*, 2020, **10**(26), 15430–15460.
82. M. E. Harb, S. Ebrahim, M. Soliman and M. Shabana, Fabrication of organic transistors using nanomaterials for sensing applications, *J. Electron. Mater.*, 2018, **47**, 353–358.
83. S. Palazzolo, S. Bayda, M. Hadla, I. Caligiuri, G. Corona, G. Toffoli and F. Rizzolio, The clinical translation of organic nanomaterials for cancer therapy: a focus on polymeric nanoparticles, micelles, liposomes and exosomes, *Curr. Med. Chem.*, 2018, **25**(34), 4224–4268.
84. Y. Y. Broza, R. Vishinkin, O. Barash, M. K. Nakhleh and H. Haick, Synergy between nanomaterials and volatile organic compounds for non-invasive medical evaluation, *Chem. Soc. Rev.*, 2018, **47**(13), 4781–4859.
85. Y. Yu, Y. Shi and B. Zhang, Synergetic transformation of solid inorganic–organic hybrids into advanced nanomaterials for catalytic water splitting, *Acc. Chem. Res.*, 2018, **51**(7), 1711–1721.
86. M. Oveisi, M. A. Asli and N. M. Mahmoodi, MIL-Ti metal-organic frameworks (MOFs) nanomaterials as superior adsorbents: Synthesis and ultrasound-aided dye adsorption from multicomponent wastewater systems, *J. Hazard. Mater.*, 2018, **347**, 123–140.
87. L. A. Mercante, R. S. Andre, J. B. Macedo, A. Pavinatto and D. S. Correa, Nanofibras eletrofiadas e suas aplicações: avanços na última década, *Quim. Nova*, 2021, **44**, 717–736.

88. S. Bhandi, D. Mehta, M. Mashyakhy, H. Chohan, L. Testarelli, J. Thomas, H. Dhillon, A. T. Raj, T. Madapusi Balaji, S. Varadarajan and S. Patil, Antimicrobial efficacy of silver nanoparticles as root canal irrigant's: a systematic review, *J. Clin. Med.*, 2021, **10**(6), 1152.
89. N. P. Panpaliya, P. T. Dahake, Y. J. Kale, M. V. Dadpe, S. B. Kendre, A. G. Siddiqi and U. R. Maggavi, In vitro evaluation of antimicrobial property of silver nanoparticles and chlorhexidine against five different oral pathogenic bacteria, *Saudi Dent. J.*, 2019, **31**(1), 76–83.
90. Z. G. Azzawi, T. I. Hamad, S. A. Kadhim and G. A. Naji, Osseointegration evaluation of laser-deposited titanium dioxide nanoparticles on commercially pure titanium dental implants, *J. Mater. Sci.: Mater. Med.*, 2018, **29**, 1–1.
91. S. Gali, Mica glass ceramics for dental restorations, *Mater. Technol.*, 2019, **34**(1), 2–11.
92. Y. Zhou, B. Tan, X. Wang, Z. Li, Z. Liu and Y. Sun, Antimicrobial activity of silicon dioxide nanoparticles against oral microorganisms, *J. Oral Sci.*, 2018, **60**(1), 49–55.
93. A. Kumar and M. Verma, Ceramic-reinforced composites for dental restorations: a review, *J. Dent. Mater. Tech.*, 2017, **6**(2), 53–59.
94. X. Li, S. Li, Y. Li and G. Huang, Application of fullerenes in dentistry, *J. Nanomater.*, 2015.
95. I. A. Aljuffali, B. Neppolian, N. Al-Ansari and M. Al-Saeed, Fullerenes in medicine and biology: synthesis, properties, and biomedical applications, *J. Appl. Pharm. Sci.*, 2017, **7**(11), 102–109.
96. N. A. Kamarudin, *et al.*, Carbon nanotubes and their application in medical field, *J. Med. Syst.*, 2017, **41**(11), 193.
97. H. Toutenburg, *et al.*, Electrospun nanofibers in regenerative dentistry, *J. Dent.*, 2016, **44**(2), 143–151.
98. J. Y. Min, *et al.*, Graphene oxide in dentistry: a review of the current status and future perspectives, *Biomed. Mater.*, 2017, **12**(3), 035004.
99. Z. Sun, *et al.*, Carbon black-reinforced nanocomposites for dental filling materials: preparation and characterization, *Mater. Sci. Eng., C*, 2021, **124**, 111892.
100. X. Chen, J. Fang and Y. Liu, *et al.*, Progress and prospects in nanomedicine: A review, *J. Nanopart. Res.*, 2020, **22**(9), 196.
101. B. S. Kumar and R. K. Avti, Nanoparticle drug delivery systems: A review, *J. Controlled Release*, 2020, **320**, 125–148.
102. M. Jeyaraj, S. Gurunathan, M. Qasim, M. H. Kang and J. H. Kim, A comprehensive review on the synthesis, characterization, and biomedical application of platinum nanoparticles, *Nanomaterials*, 2019, **9**(12), 1719.
103. J. Ferreira, P. T. Pires, C. Almeida, S. Jerónimo and P. R. Melo, Avaliação da Eficácia do nanoXIM CarePaste na Oclusão dos Túbulos Dentinários/Evaluation of the Efficacy of nanoXIM CarePaste in Dentinal Tubule Occlusion, *InInternational Poster Journal of Dentistry and Oral Medicine-XXIII Congresso OMD*, 2014.

104. S. Khetawat and S. Lodha, Nanotechnology (nanohydroxyapatite crystals): recent advancement in treatment of dentinal hypersensitivity, *J. Interdiscip. Med. Dent. Sci.*, 2015, **3**, 181.
105. M. A. Al-Wahadni, *et al*, Hydroxyapatite Nanoparticles in Dentistry: A Review of Literature, *J. Int. Soc. Prev. Community Dent.*, 2020, **10**(4), 405–410.
106. S. Sameni, *et al.*, The Potential of Hydroxyapatite Nanoparticles in Dental Applications: A Review, *J. Nanomater.*, 2020, 903247814 pages, 2020.
107. L. Li, H. Pan, J. Tao, X. Xu, C. Mao, X. Gu and R. Tang, Repair of enamel by using hydroxyapatite nanoparticles as the building blocks, *J. Mater. Chem.*, 2008, **18**(34), 4079–4084.
108. E. Pepla, L. K. Besharat, G. Palaia, G. Tenore and G. Migliau, Nano-hydroxyapatite and its applications in preventive, restorative and regenerative dentistry: a review of literature, *Ann. Stomatol.*, 2014, **5**(3), 108.
109. M. B. Sathyanarayanan, R. Balachandranath, Y. Genji Srinivasulu, S. K. Kannaiyan and G. Subbiahdoss, The effect of gold and iron-oxide nanoparticles on biofilm-forming pathogens, *ISRN Microbiol.*, 2013, **2013**.
110. J. K. Lee, *et al.*, Efficacy of Iron Oxide Nanoparticles in Eradicating Bacterial Biofilms on Dental Implants, *J. Dent. Res.*, 2020, **99**(10), 1230–1236.
111. Y. Wang, *et al.*, Iron Oxide Nanoparticles for Dental Drug Delivery, *J. Dent. Mater.*, 2020, **37**(8), 889–902.
112. H. J. Kim, *et al.*, Iron Oxide Nanoparticle-Based Diagnostic Tools in Dentistry, *J. Dent. Res.*, 2019, **98**(2), 195–201.
113. F. Zarone, S. Russo and R. Sorrentino, From porcelain-fused-to-metal to zirconia: clinical and experimental considerations, *Dent. Mater.*, 2011, **27**(1), 83–96.
114. V. Lughi and V. Sergo, Low temperature degradation-aging-of zirconia: A critical review of the relevant aspects in dentistry, *Dent. Mater.*, 2010, **26**(8), 807–820.
115. T. R. Ramesh, M. Gangaiah, P. V. Harish, U. Krishnakumar and B. Nandakishore, Zirconia Ceramics as a Dental Biomaterial–An Over view, *Trends Biomater. Artif. Organs*, 2012, **26**, 3.
116. Y. Josset, Z. Oum'Hamed, A. Zarrinpour, M. Lorenzato, J. J. Adnet and D. Laurent–Maquin, In vitro reactions of human osteoblasts in culture with zirconia and alumina ceramics, *J. Biomed. Mater. Res.*, 1999, **47**(4), 481–493.
117. M. H. Chen, Update on dental nanocomposites, *J. Dent. Res.*, 2010, **89**(6), 549–560.
118. S. B. Mitra, D. Wu and B. N. Holmes, An application of nanotechnology in advanced dental materials, *J. Am. Dent. Assoc.*, 2003, **134**(10), 1382–1390.
119. S. H. Wu, C. Y. Mou and H. P. Lin, Synthesis of mesoporous silica nanoparticles, *Chem. Soc. Rev.*, 2013, **42**(9), 3862–3875.

120. Y. C. Chiang, H. P. Lin, H. H. Chang, Y. W. Cheng, H. Y. Tang, W. C. Yen, P. Y. Lin, K. W. Chang and C. P. Lin, A mesoporous silica biomaterial for dental biomimetic crystallization, *ACS Nano*, 2014, **8**(12), 12502–12513.
121. A. M. Al-Wahadni, H. M. Al-Zahrani and F. Alqahtani, Silver nanoparticles: a review on antimicrobial activity and toxicity mechanisms, *Microb. Pathog.*, 2017, **108**, 6–17.
122. Y. Chen and X. Xiong, Applications of silver nanoparticles in dentistry, *J. Dent.*, 2018, **68**, 33–41.
123. N. Naghsh and H. Tavakoli, Silver nanoparticles: applications and toxicities, *J. Appl. Toxicol.*, 2015, **35**(9), 951–961.
124. H. Zhang, H. Zhang, Y. Chen, H. Liu and D. Wang, The cytotoxicity of titanium dioxide nanoparticles and its mechanism, *Front. Physiol.*, 2018, **9**, 877.
125. B. Baysal, E. Bayramli, M. Koc, Y. Cicek and N. Ertugrul, Allergic Reactions Induced by Dental Implants, *J. Clin. Exp. Dent.*, 2018, **10**(4), e423–e429.
126. A. Santini, D. Paolino, B. J. Givens, M. Del Fabbro and J. Tolman, The Adhesion of Microorganisms on Dental Implants, *J. Oral Implantol.*, 2017, **43**(3), 227–234.
127. F. Esmaeilpour, H. Zare and S. Shirani, Zinc Oxide Nanoparticles in Dentistry: A Comprehensive Review, *J. Appl. Biomater. Funct. Mater.*, 2018, **16**(1), 1–7.
128. S. J. Yeon, Y. S. Kim and H. C. Jang, *et al.*, Chitosan Nanoparticles for Drug Delivery and Wound Healing in Dentistry, *J. Biomed. Mater. Res., Part A*, 2017, **105**(9), 2312–2321.
129. S. R. Deo, A. K. Panda and S. N. Bhadange, Gold Nanoparticles in Dentistry: A Review, *J. Clin. Exp. Dent.*, 2018, **10**(3), e295–e300.
130. J. K. Song, H. C. Jang and Y. S. Kim, *et al.*, Calcium Phosphate Nanoparticles in Dentistry: A Review, *J. Biomed. Nanotechnol.*, 2017, **13**(5), 7.
131. L. Zhang and G. Huang, Nanocomposites in dental restorative materials, *J. Dent.*, 2016, **44**, 4–10.
132. V. Shetty, K. Bhat and A. Kumar, Nanocomposites in dentistry: A review, *J. Orofacial Sci.*, 2019, **11**(1), 1–7.
133. X. Yang, H. Yang, H. Yang, J. Yang and L. Sun, Recent advances in nanocomposites for dental implant applications, *J. Mater. Sci.: Mater. Med.*, 2016, **27**(2), 39.
134. S. Mukherjee and S. Sivaraman, The role of nanotechnology in dental implants: a review, *J. Appl. Biomater. Biomech.*, 2015, **13**(1), 1–12.
135. Y. Zhang, K. Gulati, Z. Li, P. Di and Y. Liu, Dental Implant Nano-Engineering: Advances, Limitations and Future Directions, *Nanomaterials*, 2021, **11**(10), 2489.
136. R. N. AlKahtani, The implications and applications of nanotechnology in dentistry: A review, *Saudi Dent. J.*, 2018, **30**(2), 107–116.
137. V. P. Singh, D. G. Nayak, A. S. Uppoor and D. Shah, Clinical and radiographic evaluation of nano-crystalline hydroxyapatite bone graft

(Sybograf) in combination with bioresorbable collagen membrane (Periocol) in periodontal intrabony defects, *Dent. Res. J.*, 2012, **9**(1), 60–67.
138. B. T. Reves, J. A. Jennings, J. D. Bumgardner and W. O. Haggard, Preparation and functional assessment of composite chitosan-nano-hydroxyapatite scaffolds for bone regeneration, *J. Funct. Biomater.*, 2012, **3**, 114–130.
139. C. Kailasanathan, N. Selvakumar and V. Naidu, Structure and properties of titania reinforced nano-hydroxyapatite/gelatin bio-composites for bone graft materials, *Ceram. Int.*, 2012, **38**(1), 571–579.
140. S. Liao, W. Wang, M. Uo, S. Ohkawa, T. Akasaka, K. Tamura, F. Cui and F. Watari, A three-layered nano-carbonated hydroxyapatite/collagen/PLGA composite membrane for guided tissue regeneration, *Biomaterials*, 2007, **26**(36), 7564–7571.
141. L. Peng, X. Cheng and R. Zhuo, *et al*, Novel gene-activated matrix with embedded chitosan/plasmid DNA nanoparticles encoding PDGF for periodontal tissue engineering, *J. Biomed. Mater. Res.*, 2009, **90**(2), 564–576.
142. N. Jain, P. Jain and V. Jain, Nanomaterials in Dentistry: A Review, *J. Clin. Exp. Dent.*, 2014, **6**(5), e451–e457.
143. N. Jain, P. Jain and V. Jain, Nanotechnology in Dentistry, *J. Conservative Dent.*, 2012, **15**(3), 252–258.
144. X. Gao, Y. Lu and J. Feng, Dental Nanomaterials, *Int. J. Nanomed.*, 2012, **7**, 6233–6244.
145. R. Bhandari, A. Arolkar, S. Mishra and R. Jain, Nanotechnology in Dentistry: A Review, *J. Oral Biol. Craniofacial Res.*, 2016, **6**(3), 157–163.
146. M. Nejatian, M. R. Bouschlicher and D. H. Pashley, Nanotechnology in Restorative Dentistry, *J. Adhes. Dent.*, 2011, **13**(1), 1–14.
147. W. Song and S. Ge, Application of Antimicrobial Nanoparticles in Dentistry, *Molecules*, 2019, **24**(6), 1033.
148. K. R. Halkai and J. A. Mudda, Evaluation of Antibacterial Efficacy of Fungal-Derived Silver Nanoparticles against Enterococcus faecalis, *Contemp. Clin. Dent.*, 2018, **9**, 45–48.
149. K. R. Halkai and J. A. Mudda, Evaluation of antibacterial efficacy of biosynthesized silver nanoparticles derived from fungi against endo-perio pathogens Porphyromonas gingivalis, Bacillus pumilus, and Enterococcus faecalis, *J. Conservative Dent.*, 2017, **20**, 398–404.
150. M. Samiei, A. Farjami, S. M. Dizaj and F. Lotfipour, Nanoparticles for antimicrobial purposes in Endodontics: A systematic review of in vitro studies, *Mater. Sci. Eng., C*, 2016, **58**, 1269–1278.
151. K. P. Rakesh, S. Hemalatha and A. Prabhakar, Antimicrobial activity of nanoparticle-based systems for the management of endodontic infections, *J. Endod.*, 2011, **37**(10), 1301–1307.
152. S. E. Bishara, N. Roohi and S. Koshy, *et al.*, Antimicrobial efficacy of nanotechnology in endodontics: A systematic review, *J. Endod.*, 2014, **40**(11), 1641–1649.

153. A. Kishen, *Nanotechnology in endodontics*, Springer, Cham, Switzerland, 2016.
154. A. R. De Souza, A. H. Farias and A. M. Lopes, *et al.*, In vitro antimicrobial activity of silver nanoparticles against oral microorganisms, *J. Appl. Microbiol.*, 2009, **107**(4), 1154–1159.
155. A. H. Faria, A. R. De Souza and C. R. Cardoso, *et al.*, The antimicrobial activity of titanium dioxide nanoparticles against oral microorganisms, *Microbiol. Immunol.*, 2008, **52**(3), 249–253.
156. K. P. Rakesh, S. Hemalatha and A. Prabhakar, Antimicrobial activity of nanoparticle-based systems for the management of endodontic infections, *J. Endod.*, 2011, **37**(10), 1301–1307.
157. M. Hulsmann and M. Zehnder, Dental implant-associated infections: diagnosis and treatment, *Swiss Dent. J.*, 2008, **118**(10), 935–943.
158. N. Roohi, S. E. Bishara and S. Koshy, *et al.*, Nanotechnology in endodontics: a new frontier in restorative dentistry, *J. Endod.*, 2012, **38**(9), 1071–1077.
159. J. Sharan, S. Singh, S. V. Lale, M. Mishra, V. Koul and O. P. Kharbanda, Applications of nanomaterials in dental science: A review, *J. Nanosci. Nanotechnol.*, 2017, **17**, 2235–2255.
160. M. Behnaz, K. Dalaie, H. Mirmohammadsadeghi, H. Salehi, V. Rakhshan and F. Aslani, Shear bond strength and adhesive remnant index of orthodontic brackets bonded to enamel using adhesive systems mixed with TiO_2 nanoparticles, *Dent. Press J. Orthod.*, 2018, **23**, 43.e1–43.e7.
161. W. Zakrzewski, M. Dobrzynski, W. Dobrzynski, A. Zawadzka-Knefel, M. Janecki and K. Kurek, *et al*, Nanomaterials Application in Orthodontics, *Nanomaterials*, 2021, **11**, 337.
162. T. Eliades, S. Zinelis, C. Bourauel and G. Eliades, Manufacturing of Orthodontic Brackets: A Review of Metallurgical Perspectives and Applications, *Recent Pat. Mater. Sci.*, 2010, **1**, 135–139.
163. T. Ogiński, B. Kawala, M. Mikulewicz and J. Antoszewska-Smith, A Clinical Comparison of Failure Rates of Metallic and Ceramic Brackets: A Twelve-Month Study, *BioMed Res. Int.*, 2020, **2020**, 9725101.
164. W. Zakrzewski, M. Dobrzyński, A. Zawadzka-Knefel, A. Lubojański, W. Dobrzyński, M. Janecki, K. Kurek, M. Szymonowicz, R. J. Wiglusz and Z. Rybak, Nanomaterials Application in Endodontics, *Materials*, 2021, **14**(18), 5296.
165. K. Jhajharia, A. Parolia, K. V. Shetty and L. K. Mehta, Biofilm in endodontics: a review, *J. Int. Soc. Prev. Community Dent.*, 2015, **5**(1), 1.
166. A. Lubojanski, M. Dobrzynski, N. Nowak, J. Rewak-Soroczynska, K. Sztyler, W. Zakrzewski, W. Dobrzynski, M. Szymonowicz, Z. Rybak and K. Wiglusz, *et al*, Application of selected nanomaterials and ozone in modern clinical dentistry, *Nanomaterials*, 2021, **11**, 259.
167. M. J. R. Virlan, D. Miricescu, R. Radulescu, C. M. Sabliov, A. Totan, B. Calenic and M. Greabu, Organic nanomaterials and their applications in the treatment of oral diseases, *Molecules*, 2016, **21**, 207.

168. T. Parvathi, R. Vijayalakshmi, R. Theyagarajan, C. Burnicenalinakumari and J. Mahendra, Nanotechnology in Periodontics: An Overview, *Medico-Legal Update*, 2020, **20**, 2272–2277.
169. V. Noronha, M. Venkatesh and N. De Souza, *et al.*, Nanotechnology in Dentistry-A Review of Current Status, Future Perspectives and Challenges, *J. Clin. Exp. Dent.*, 2020, **12**(9), e1654–e1661.
170. N. Roohi, S. E. Bishara and S. Koshy, *et al.*, Antimicrobial efficacy of nanotechnology in endodontics: A systematic review, *J. Endod.*, 2014, **40**(11), 1641–1649.
171. D. Tziafas, Nanodentistry: current status and future prospects, *Front. Bioeng. Biotechnol.*, 2021, **8**, 624770.
172. J. Datta and S. Ghosh, Nanotechnology in dentistry: An overview of current scenario and future prospects, *J. Indian Soc. Periodontol.*, 2019, **23**(2), 139–147.
173. H. A. Salam, R. Sivaraj and R. Venkatesh, Green synthesis and characterization of zinc oxide nanoparticles from Ocimum basilicum L. var. purpurascens Benth.-Lamiaceae leaf extract, *Mater. Lett.*, 2014, **131**, 16–18.
174. M. M. Chikkanna, S. E. Neelagund and K. K. Rajashekarappa, Green synthesis of zinc oxide nanoparticles (ZnONPs) and their biological activity, *SN Appl. Sci.*, 2019, **1**, 117.
175. N. Saifuddin, C. W. Wong and A. A. Yasumira, Rapid biosynthesis of silver nanoparticles using culture supernatant of bacteria with microwave irradiation, *J. Chem.*, 2009, **6**(1), 61–70.
176. H. Agarwal, S. V. Kumar and S. Rajeshkumar, A review on green synthesis of zinc oxide nanoparticles–An eco-friendly approach, *Resour.-Effic. Technol.*, 2017, **3**(4), 406–413.
177. M. Wang, G. Tan, H. Ren, A. Xia and Y. Liu, Direct double Z-scheme O-g-C_3N_4/Zn_2SnO_4N/ZnO ternary heterojunction photocatalyst with enhanced visible photocatalytic activity, *Appl. Surf. Sci.*, 2019, **492**, 690–702.
178. D. Sharma, J. Rajput, B. S. Kaith, M. Kaur and S. Sharma, Synthesis of ZnO nanoparticles and study of their antibacterial and antifungal properties, *Thin Solid Films*, 2010, **519**(3), 1224–1229.
179. A. Mukherjee, D. Sarkar and S. Sasmal, A review of green synthesis of metal nanoparticles using algae, *Front. Microbiol.*, 2021, **12**, 693899.
180. J. Qu, X. Yuan, X. Wang and P. Shao, Zinc accumulation and synthesis of ZnO nanoparticles using Physalis alkekengi L, *Environ. Pollut.*, 2011, **159**(7), 1783–1788.
181. G. Grasso, D. Zane and R. Dragone, Microbial nanotechnology: challenges and prospects for green biocatalytic synthesis of nanoscale materials for sensoristic and biomedical applications, *Nanomaterials*, 2019, **10**(1), 11.
182. P. C. Nagajyothi, T. V. Sreekanth, C. O. Tettey, Y. I. Jun and S. H. Mook, Characterization, antibacterial, antioxidant, and cytotoxic activities of ZnO nanoparticles using Coptidis Rhizoma, *Bioorg. Med. Chem. Lett.*, 2014, **24**(17), 4298–4303.

183. O. J. Nava, P. A. Luque, C. M. Gómez-Gutiérrez, A. R. Vilchis-Nestor, A. Castro-Beltrán, M. L. Mota-González and A. Olivas, Influence of Camellia sinensis extract on Zinc Oxide nanoparticle green synthesis, *J. Mol. Struct.*, 2017, **1134**, 121–125.
184. L. Fu and Z. Fu, Plectranthus amboinicus leaf extract–assisted biosynthesis of ZnO nanoparticles and their photocatalytic activity, *Ceram. Int.*, 2015, **41**(2), 2492–2496.
185. O. J. Nava, C. A. Soto-Robles, C. M. Gómez-Gutiérrez, A. R. Vilchis-Nestor, A. Castro-Beltrán, A. Olivas and P. A. Luque, Fruit peel extract mediated green synthesis of zinc oxide nanoparticles, *J. Mol. Struct.*, 2017, **1147**, 1–6.
186. M. Fazlzadeh, R. Khosravi and A. Zarei, Green synthesis of zinc oxide nanoparticles using Peganum harmala seed extract, and loaded on Peganum harmala seed powdered activated carbon as new adsorbent for removal of Cr(VI) from aqueous solution, *Ecol. Eng.*, 2017, **103**, 180–190.
187. E. Selvarajan and V. J. Mohanasrinivasan, Biosynthesis and characterization of ZnO nanoparticles using Lactobacillus plantarum VITES07, *Mater. Lett.*, 2013, **112**, 180–182.
188. C. Jayaseelan, A. A. Rahuman, A. V. Kirthi, S. Marimuthu, T. Santhoshkumar, A. Bagavan, K. Gaurav, L. Karthik and K. B. Rao, Novel microbial route to synthesize ZnO nanoparticles using Aeromonas hydrophila and their activity against pathogenic bacteria and fungi, *Spectrochim. Acta, Part A*, 2012, **90**, 78–84.
189. S. Busi, J. Rajkumari, S. Pattnaik, P. Parasuraman and S. Hnamte, Extracellular synthesis of zinc oxide nanoparticles using Acinetobacter schindleri SIZ7 and its antimicrobial property against foodborne pathogens, *J. Microbiol., Biotechnol. Food Sci.*, 2021, 407–411.
190. B. N. Singh, A. K. Rawat, W. Khan, A. H. Naqvi and B. R. Singh, Biosynthesis of stable antioxidant ZnO nanoparticles by Pseudomonas aeruginosa rhamnolipids, *PLoS One*, 2014, **9**(9), e106937.
191. M. Saravanan, V. Gopinath, M. K. Chaurasia, A. Syed, F. Ameen and N. Purushothaman, Green synthesis of anisotropic zinc oxide nanoparticles with antibacterial and cytofriendly properties, *Microb. Pathog.*, 2018, **115**, 57–63.
192. F. Ameen, T. Dawoud and S. AlNadhari, Ecofriendly and low-cost synthesis of ZnO nanoparticles from Acremonium potronii for the photocatalytic degradation of azo dyes, *Environ. Res.*, 2021, **202**, 111700.
193. R. Raliya and J. C. Tarafdar, ZnO nanoparticle biosynthesis and its effect on phosphorous-mobilizing enzyme secretion and gum contents in Clusterbean (Cyamopsis tetragonoloba L.), *Agric. Res.*, 2013, **2**, 48–57.
194. V. N. Kalpana, B. A. Kataru, N. Sravani, T. Vigneshwari, A. Panneerselvam and V. D. Rajeswari, Biosynthesis of zinc oxide nanoparticles using culture filtrates of *Aspergillus niger*: Antimicrobial textiles and dye degradation studies, *OpenNano*, 2018, **3**, 48–55.

195. J. L. Sharma, V. Dhayal and R. K. Sharma, White-rot fungus mediated green synthesis of zinc oxide nanoparticles and their impregnation on cellulose to develop environmental friendly antimicrobial fibers, *3 Biotech*, 2021, **11**(6), 269.
196. S. Azizi, M. B. Ahmad, F. Namvar and R. Mohamad, Green biosynthesis and characterization of zinc oxide nanoparticles using brown marine macroalga Sargassum muticum aqueous extract, *Mater. Lett.*, 2014, **116**, 275–277.
197. T. Khalafi, F. Buazar and K. Ghanemi, Phycosynthesis and enhanced photocatalytic activity of zinc oxide nanoparticles toward organosulfur pollutants, *Sci. Rep.*, 2019, **9**(1), 6866.
198. A. A. Elrefaey, A. D. El-Gamal, S. M. Hamed and E. F. El-belely, Algae-mediated biosynthesis of zinc oxide nanoparticles from Cystoseira crinite (Fucales; Sargassaceae) and it's antimicrobial and antioxidant activities, *Egypt. J. Chem.*, 2022, **65**(4), 231–240.
199. S. Dwivedi, R. Wahab, F. Khan, Y. K. Mishra, J. Musarrat and A. A. Al-Khedhairy, Reactive oxygen species mediated bacterial biofilm inhibition *via* zinc oxide nanoparticles and their statistical determination, *PLoS One*, 2014, **9**(11), e111289.
200. A. Sirelkhatim, S. Mahmud, A. Seeni, N. H. Kaus, L. C. Ann, S. K. Bakhori, H. Hasan and D. Mohamad, Review on zinc oxide nanoparticles: antibacterial activity and toxicity mechanism, *Nano-Micro Lett.*, 2015, **7**, 219–242.
201. V. Tiwari, N. Mishra, K. Gadani, P. S. Solanki, N. A. Shah and M. Tiwari, Mechanism of anti-bacterial activity of zinc oxide nanoparticle against carbapenem-resistant Acinetobacter baumannii, *Front. Microbiol.*, 2018, **9**, 1218.
202. G. Parthasarathy, M. Saroja, M. Venkatachalam, S. Shankar and V. K. Evanjelene, Green synthesis of zinc oxide nanoparticles-review paper, *World J. Pharm. Pharm. Sci.*, 2016, **5**(4), 922–931.
203. M. M. Chikkanna, S. E. Neelagund and K. K. Rajashekarappa, Green synthesis of zinc oxide nanoparticles (ZnONPs) and their biological activity, *SN Appl. Sci.*, 2019, **1**, 117.
204. D. Suresh, P. C. Nethravathi, H. Rajanaika, H. Nagabhushana and S. C. Sharma, Green synthesis of multifunctional zinc oxide (ZnO) nanoparticles using Cassia fistula plant extract and their photodegradative, antioxidant and antibacterial activities, *Mater. Sci. Semicond. Process.*, 2015, **31**, 446–454.
205. I. N. Alrabayah, S. S. Elhawary, Z. A. Kandil, E. M. El-Kadder, Y. S. Moemen, A. M. Saleh and M. A. El Raey, Green Synthesized Zinc Oxide Nanoparticles Based on Cestrum diurnum L. of Potential Antiviral Activity against Human Corona 229-E Virus, *Molecules*, 2022, **28**(1), 266.
206. K. Jannat, A. K. Paul, T. A. Bondhon, A. Hasan, M. Nawaz, R. Jahan, T. Mahboob, V. Nissapatorn, P. Wilairatana, M. D. Pereira and

M. Rahmatullah, Nanotechnology applications of flavonoids for viral diseases, *Pharmaceutics*, 2021, **13**(11), 1895.

207. T. Ishida, Review on the role of Zn^{2+} ions in viral pathogenesis and the effect of Zn^{2+} ions for host cell-virus growth inhibition, *SN Appl. Sci.*, 2019, **2**(1), 28–37.
208. T. E. Antoine, Y. K. Mishra, J. Trigilio, V. Tiwari, R. Adelung and D. Shukla, Prophylactic, therapeutic and neutralizing effects of zinc oxide tetrapod structures against herpes simplex virus type-2 infection, *Antiviral Res.*, 2012, **96**(3), 363–375.
209. T. E. Antoine, S. R. Hadigal, A. M. Yakoub, Y. K. Mishra, P. Bhattacharya, C. Haddad, T. Valyi-Nagy, R. Adelung, B. S. Prabhakar and D. Shukla, Intravaginal zinc oxide tetrapod nanoparticles as novel immunoprotective agents against genital herpes, *J. Immunol.*, 2016, **196**(11), 4566–4575.
210. Q. Sun, J. Li and T. Le, Zinc oxide nanoparticle as a novel class of antifungal agents: current advances and future perspectives, *J. Agric. Food Chem.*, 2018, **66**(43), 11209–11220.
211. T. R. Lakshmeesha, M. Murali, M. A. Ansari, A. C. Udayashankar, M. A. Alzohairy, A. Almatroudi, M. N. Alomary, S. M. Asiri, B. S. Ashwini, N. K. Kalagatur and C. S. Nayak, Biofabrication of zinc oxide nanoparticles from Melia azedarach and its potential in controlling soybean seed-borne phytopathogenic fungi, *Saudi J. Biol. Sci.*, 2020, **27**(8), 1923–1930.
212. J. Li, H. Sang, H. Guo, J. T. Popko, L. He, J. C. White, O. P. Dhankher, G. Jung and B. Xing, Antifungal mechanisms of ZnO and Ag nanoparticles to Sclerotinia homoeocarpa, *Nanotechnology*, 2017, **28**(15), 155101.
213. W. Zhu, C. Hu, Y. Ren, Y. Lu, Y. Song, Y. Ji, C. Han and J. He, Green synthesis of zinc oxide nanoparticles using Cinnamomum camphora (L.) Presl leaf extracts and its antifungal activity, *J. Environ. Chem. Eng.*, 2021, **9**(6), 106659.
214. E. A. Eid, A. N. Fouda and E. S. Duraia, Effect of adding 0.5 wt% ZnO nanoparticles, temperature and strain rate on tensile properties of Sn–5.0 wt% Sb–0.5 wt% Cu (SSC505) lead free solder alloy, *Mater. Sci. Eng., A*, 2016, **657**, 104–114.
215. C. Pushpalatha, J. Suresh, V. S. Gayathri, S. V. Sowmya, D. Augustine, A. Alamoudi, B. Zidane, N. H. Mohammad Albar and S. Patil, Zinc oxide nanoparticles: a review on its applications in dentistry, *Front. Bioeng. Biotechnol.*, 2022, **10**, 917990.
216. K. Teymoornezhad, H. Alaghehmand, G. Daryakenari, S. Khafri and M. Tabari, Evaluating the microshear bond strength and microleakage of flowable composites containing zinc oxide nano-particles, *Electron. Physician*, 2016, **8**(11), 3289.
217. A. A. Balhaddad, A. A. Kansara, D. Hidan, M. D. Weir, H. H. Xu and M. A. Melo, Toward dental caries: Exploring nanoparticle-based

platforms and calcium phosphate compounds for dental restorative materials, *Bioact. Mater.*, 2019, **4**, 43–55.
218. Z. Jowkar, S. A. Hamidi, F. Shafiei and Y. Ghahramani, The effect of silver, zinc oxide, and titanium dioxide nanoparticles used as final irrigation solutions on the fracture resistance of root-filled teeth, *Clin., Cosmet. Invest. Dent.*, 2020, 141–148.
219. M. J. Alves, L. Grenho, C. Lopes, J. Borges, F. Vaz, I. P. Vaz and M. H. Fernandes, Antibacterial effect and biocompatibility of a novel nanostructured ZnO-coated gutta-percha cone for improved endodontic treatment, *Mater. Sci. Eng., C*, 2018, **92**, 840–848.
220. M. Cierech, J. Wojnarowicz, A. Kolenda, A. Krawczyk-Balska, E. Prochwicz, B. Woźniak, W. Łojkowski and E. Mierzwińska-Nastalska, Zinc oxide nanoparticles cytotoxicity and release from newly formed PMMA–ZnO nanocomposites designed for denture bases, *Nanomaterials*, 2019, **9**(9), 1318.
221. D. Charoenkijkajorn and S. Sanohkan, The effect of nano zinc oxide particles on color stability of MDX4-4210 silicone prostheses, *Eur. J. Dent.*, 2020, **14**(04), 525–532.
222. M. Cierech, A. Kolenda, A. M. Grudniak, J. Wojnarowicz, B. Woźniak, M. Gołaś, E. Swoboda-Kopeć, W. Łojkowski and E. Mierzwińska-Nastalska, Significance of polymethylmethacrylate (PMMA) modification by zinc oxide nanoparticles for fungal biofilm formation, *Int. J. Pharm.*, 2016, **510**(1), 323–335.
223. N. Gomez and C. E. Schmidt, Nerve growth factor-immobilized polypyrrole: Bioactive electrically conducting polymer for enhanced neurite extension, *J. Biomed. Mater. Res., Part A*, 2007, **81**(1), 135–149.
224. N. Akbar, Z. Aslam, R. Siddiqui, M. R. Shah and N. A. Khan, Zinc oxide nanoparticles conjugated with clinically-approved medicines as potential antibacterial molecules, *AMB Express*, 2021, **11**(1), 6.
225. S. Tahmasbi, F. Mohamadian, S. Hosseini and L. Eftekhar, A review on the applications of nanotechnology in orthodontics, *Nanomed. J.*, 2019, **6**, 1.
226. M. Kachoei, A. Nourian, B. Divband, Z. Kachoei and S. Shirazi, Zinc-oxide nanocoating for improvement of the antibacterial and frictional behavior of nickel-titanium alloy, *Nanomedicine*, 2016, **11**(19), 2511–2527.
227. F. Parnia, J. Yazdani, V. Javaherzadeh and S. M. Dizaj, Overview of nanoparticle coating of dental implants for enhanced osseointegration and antimicrobial purposes, *J. Pharm. Pharm. Sci.*, 2017, **20**, 148–160.
228. S. T. Khan, A. A. Al-Khedhairy and J. Musarrat, ZnO, and TiO_2 nanoparticles as novel antimicrobial agents for oral hygiene: a review, *J. Nanopart. Res.*, 2015, **17**, 1–6.
229. R. P. Allaker and K. Memarzadeh, Nanoparticles and the control of oral infections, *Int. J. Antimicrob. Agents*, 2014, **43**(2), 95–104.
230. R. Brayner, R. Ferrari-Iliou, N. Brivois, S. Djediat, M. F. Benedetti and F. Fiévet, Toxicological impact studies based on *Escherichia coli* bacteria

in ultrafine ZnO nanoparticles colloidal medium, *Nano Lett.*, 2006, **6**(4), 866–870.

231. E. Giertsen, Effects of mouthrinses with triclosan, zinc ions, copolymer, and sodium lauryl sulphate combined with fluoride on acid formation by dental plaque *in vivo*, *Caries Res.*, 2004, **38**(5), 430–435.
232. A. Kishen, Z. Shi, A. Shrestha and K. G. Neoh, An investigation on the antibacterial and antibiofilm efficacy of cationic nanoparticulates for root canal disinfection, *J. Endod.*, 2008, **34**(12), 1515–1520.
233. S. Nair, A. Sasidharan, V. V. Divya Rani, D. Menon, S. Nair, K. Manzoor and S. Raina, Role of size scale of ZnO nanoparticles and microparticles on toxicity toward bacteria and osteoblast cancer cells, *J. Mater. Sci.: Mater. Med.*, 2009, **20**, 235–241.
234. N. Talebian, S. M. Amininezhad and M. Doudi, Controllable synthesis of ZnO nanoparticles and their morphology-dependent antibacterial and optical properties, *J. Photochem. Photobiol., B*, 2013, **120**, 66–73.
235. H. Fukui, M. Horie, S. Endoh, H. Kato, K. Fujita, K. Nishio, L. K. Komaba, J. Maru, A. Miyauhi, A. Nakamura and S. Kinugasa, Association of zinc ion release and oxidative stress induced by intratracheal instillation of ZnO nanoparticles to rat lung, *Chem.-Biol. Interact.*, 2012, **198**(1–3), 29–37.
236. M. Eshed, J. Lellouche, S. Matalon, A. Gedanken and E. Banin, Sonochemical coatings of ZnO and CuO nanoparticles inhibit Streptococcus mutans biofilm formation on teeth model, *Langmuir*, 2012, **28**(33), 12288–12295.
237. Y. J. Liu, L. L. He, A. Mustapha, H. Li, Z. Q. Hu and M. S. Lin, Antibacterial activities of zinc oxide nanoparticles against *Escherichia coli* O157: H7, *J. Appl. Microbiol.*, 2009, **107**(4), 1193–201.
238. M. E. Davey and G. A. O'Toole, Microbial biofilms: from ecology to molecular genetics, *Microbiol. Mol. Biol. Rev.*, 2000, **64**(4), 847–867.
239. J. Fabrega, R. Zhang, J. C. Renshaw, W. T. Liu and J. R. Lead, Impact of silver nanoparticles on natural marine biofilm bacteria, *Chemosphere*, 2011, **85**(6), 961–966.
240. A. Almaguer-Flores, L. A. Ximénez-Fyvie and S. E. Rodil, Oral bacterial adhesion on amorphous carbon and titanium films: effect of surface roughness and culture media, *J. Biomed. Mater. Res., Part B*, 2010, **92**(1), 196–204.
241. T. Takatsuka, K. Tanaka and Y. Iijima, Inhibition of dentine demineralization by zinc oxide: *in vitro* and *in situ* studies, *Dent. Mater.*, 2005, **21**(12), 1170–1177.
242. K. Yu, J. Xu, L. Zhou, L. Zou and W. Liu, Effect of chitosan coatings with cinnamon essential oil on postharvest quality of mangoes, *Foods.*, 2021, **10**(12), 3003.
243. L. Vlaia, I. Olariu, A. M. Muţ, G. Coneac, V. Vlaia, D. F. Anghel, M. E. Maxim, G. Stângă, A. Dobrescu, M. Suciu and Z. Szabadai, New, biocompatible, chitosan-gelled microemulsions based on essential oils

and sucrose esters as nanocarriers for topical delivery of fluconazole, *Pharmaceutics*, 2022, **14**(1), 75.
244. S. A. de Lapena, M. Terra-Garcia, R. A. da Costa Ward, R. D. Rossoni, V. M. Melo and J. C. Junqueira, Enhancing effect of chitosan on methylene blue-mediated photodynamic therapy against C. albicans: A study in planktonic growth, biofilms, and persister cells, *Photodiagn. Photodyn. Ther.*, 2022, **38**, 102837.
245. H. S. Adhikari and P. N. Yadav, Anticancer activity of chitosan, chitosan derivatives, and their mechanism of action, *Int. J. Biomater.*, 2018, **2018**.
246. M. M. Silva, R. Calado, J. Marto, A. Bettencourt, A. J. Almeida and L. M. Gonçalves, Chitosan nanoparticles as a mucoadhesive drug delivery system for ocular administration, *Mar. Drugs*, 2017, **15**(12), 370.
247. A. Kefayat, M. Hosseini, F. Ghahremani, N. A. Jolfaie and M. Rafienia, Biodegradable and biocompatible subcutaneous implants consisted of pH-sensitive mebendazole-loaded/folic acid-targeted chitosan nanoparticles for murine triple-negative breast cancer treatment, *J. Nanobiotechnol.*, 2022, **20**(1), 169.
248. N. Ghobadi-Oghaz, A. Asoodeh and M. Mohammadi, Fabrication, characterization and *in vitro* cell exposure study of zein-chitosan nanoparticles for co-delivery of curcumin and berberine, *Int. J. Biol. Macromol.*, 2022, **204**, 576–586.
249. P. Calvo, C. Remunan-Lopez, J. L. Vila-Jato and M. J. Alonso, Novel hydrophilic chitosan-polyethylene oxide nanoparticles as protein carriers, *J. Appl. Polym. Sci.*, 1997, **63**(1), 125–132.
250. M. H. Sultan, S. S. Moni, O. A. Madkhali, M. A. Bakkari, S. Alshahrani, S. S. Alqahtani, N. A. Alhakamy, S. Mohan, M. Ghazwani, H. A. Bukhary and Y. Almoshari, Characterization of cisplatin-loaded chitosan nanoparticles and rituximab-linked surfaces as target-specific injectable nano-formulations for combating cancer, *Sci. Rep.*, 2022, **12**(1), 468.
251. M. Elsabahy and K. L. Wooley, Design of polymeric nanoparticles for biomedical delivery applications, *Chem. Soc. Rev.*, 2012, **41**(7), 2545–2561.
252. J. Stetefeld, S. A. McKenna and T. R. Patel, Dynamic light scattering: a practical guide and applications in biomedical sciences, *Biophys. Rev.*, 2016, **8**, 409–427.
253. D. S. Rao, K. Muraleedharan and C. J. Humphreys, TEM specimen preparation techniques, in *Microscopy: Science, Technology, Applications and Education*, 2010, vol. 2, p. 1232.
254. A. H. Shah and M. A. Rather, Effect of calcination temperature on the crystallite size, particle size and zeta potential of TiO_2 nanoparticles synthesized *via* polyol-mediated method, *Mater. Today: Proc.*, 2021, **44**, 482–488.
255. L. Mu and S. S. Feng, A novel controlled release formulation for the anticancer drug paclitaxel (Taxol®): PLGA nanoparticles containing vitamin E TPGS, *J. Controlled Release*, 2003, **86**(1), 33–48.

256. S. Rana, J. Singh, A. Wadhawan, A. Khanna, G. Singh and M. Chatterjee, Evaluation of *in vivo* toxicity of novel biosurfactant from *Candida parapsilosis* loaded in PLA-PEG polymeric nanoparticles, *J. Pharm. Sci.*, 2021, **110**(4), 1727–1738.
257. S. Rashki, K. Asgarpour, H. Tarrahimofrad, M. Hashemipour, M. S. Ebrahimi, H. Fathizadeh, A. Khorshidi, H. Khan, Z. Marzhoseyni, M. Salavati-Niasari and H. Mirzaei, Chitosan-based nanoparticles against bacterial infections, *Carbohydr. Polym.*, 2021, **251**, 117108.
258. M. Potara, E. Jakab, A. Damert, O. Popescu, V. Canpean and S. Astilean, Synergistic antibacterial activity of chitosan–silver nanocomposites on *Staphylococcus aureus*, *Nanotechnology*, 2011, **22**(13), 135101.
259. A. Sathiyaseelan, K. Saravanakumar and M. H. Wang, Bimetallic silver-platinum (AgPt) nanoparticles and chitosan fabricated cotton gauze for enhanced antimicrobial and wound healing applications, *Int. J. Biol. Macromol.*, 2022, **220**, 1556–1569.
260. K. Divya and M. S. Jisha, Chitosan nanoparticles preparation and applications, *Environ. Chem. Lett.*, 2018, **16**, 101–112.
261. S. E. Elbehairi, L. A. Ismail, M. Y. Alfaifi, R. F. Elshaarawy and H. S. Hafez, Chitosan nano-vehicles as biocompatible delivering tools for a new Ag(I) curcuminoid-Gboxin analog complex in cancer and inflammation therapy, *Int. J. Biol. Macromol.*, 2020, **165**, 2750–2764.
262. Y. A. Hassan, A. I. Khedr, J. Alkabli, R. F. Elshaarawy and A. M. Nasr, Co-delivery of imidazolium Zn(II) salen and Origanum Syriacum essential oil by shrimp chitosan nanoparticles for antimicrobial applications, *Carbohydr. Polym.*, 2021, **260**, 117834.
263. R. Karimirad, M. Behnamian and S. Dezhsetan, Bitter orange oil incorporated into chitosan nanoparticles: Preparation, characterization and their potential application on antioxidant and antimicrobial characteristics of white button mushroom, *Food Hydrocolloids*, 2020, **100**, 105387.
264. R. Kalaivani, M. Maruthupandy, T. Muneeswaran, A. H. Beevi, M. Anand, C. M. Ramakritinan and A. K. Kumaraguru, Synthesis of chitosan mediated silver nanoparticles (Ag NPs) for potential antimicrobial applications, *Front. Lab. Med.*, 2018, **2**(1), 30–35.
265. L. G. Duarte and C. S. Picone, Antimicrobial activity of lactoferrin-chitosan-gellan nanoparticles and their influence on strawberry preservation, *Food Res. Int.*, 2022, **159**, 111586.
266. E. Fakhri, H. Eslami, P. Maroufi, F. Pakdel, S. Taghizadeh, K. Ganbarov, M. Yousefi, A. Tanomand, B. Yousefi, S. Mahmoudi and H. S. Kafil, Chitosan biomaterials application in dentistry, *Int. J. Biol. Macromol.*, 2020, **162**, 956–974.
267. C. P. Jiménez-Gómez and J. A. Cecilia, Chitosan: a natural biopolymer with a wide and varied range of applications, *Molecules*, 2020, **25**(17), 3981.
268. A. D. Pye, D. E. Lockhart, M. P. Dawson, C. A. Murray and A. J. Smith, A review of dental implants and infection, *J. Hosp. Infect.*, 2009, **72**(2), 104–110.

269. K. H. Park, S. J. Kim, M. J. Hwang, H. J. Song and Y. J. Park, Pulse electrodeposition of hydroxyapatite/chitosan coatings on titanium substrate for dental implant, *Colloid Polym. Sci.*, 2017, **295**, 1843–1849.
270. P. Sanap, V. Hegde, D. Ghunawat, M. Patil, N. Nagaonkar and V. Jagtap, Current applications of chitosan nanoparticles in dentistry: A review, *Int. J. Appl. Dent. Sci.*, 2020, **6**(4), 81–84.
271. C. Zhang, D. Hui, C. Du, H. Sun, W. Peng, X. Pu, Z. Li, J. Sun and C. Zhou, Preparation and application of chitosan biomaterials in dentistry, *Int. J. Biol. Macromol.*, 2021, **167**, 1198–1210.
272. A. A. Elgendy and D. M. Fayyad, Cell viability and apoptotic changes of dental pulp stem cells treated with propolis, chitosan, and their nano counterparts, *Tanta Dent. J.*, 2017, **14**(4), 198.
273. S. Nguyen and M. Hiorth, Advanced drug delivery systems for local treatment of the oral cavity, *Ther. Delivery*, 2015, **6**(5), 595–608.
274. S. J. Cao, S. Xu, H. M. Wang, Y. Ling, J. Dong, R. D. Xia and X. H. Sun, Nanoparticles: oral delivery for protein and peptide drugs, *AAPS PharmSciTech*, 2019, **20**, 1.
275. K. S. Putri, L. S. Ramadhani, T. Rachel, G. Suhariyono and S. Surini, Promising chitosan-alginate combination for rifampicin dry powder inhaler to target active and latent tuberculosis, *J. Appl. Pharm. Sci.*, 2022, **12**(5), 098–103.
276. R. Ghosh, S. Mondal, D. Mukherjee, A. Adhikari, S. A. Ahmed, R. I. Alsantali, A. S. Khder, H. M. Altass, Z. Moussa, R. Das and M. Bhattacharyya, Oral drug delivery using a polymeric nanocarrier: Chitosan nanoparticles in the delivery of rifampicin, *Mater. Adv.*, 2022, **3**(11), 4622–4628.
277. H. Gao, N. Wu, N. Wang, J. Li, J. Sun and Q. Peng, Chitosan-based therapeutic systems and their potentials in treatment of oral diseases, *Int. J. Biol. Macromol.*, 2022, **222**, 3178–3194.
278. K. Garg and P. Tirgar, Local Delivery of Metronidazole-loaded N-Trimethyl Quaternary Ammonium Chitosan Nanoparticles for Periodontitis Treatment, *Curr. Drug Ther.*, 2022, **17**(4), 233–265.
279. D. Kandelman, P. E. Petersen and H. Ueda, Oral health, general health, and quality of life in older people, *Spec. Care Dent.*, 2008, **28**(6), 224–236.
280. J. Gupta, Nanotechnology applications in medicine and dentistry, *J. Invest. Clin. Dent.*, 2011, **2**(2), 81–88.
281. J. C. Vilar Junior, D. R. Ribeaux, C. A. Alves da Silva, D. Campos-Takaki and G. Maria, Physicochemical and antibacterial properties of chitosan extracted from waste shrimp shells, *Int. J. Microbiol.*, 2016, **58**, 33–41.
282. E. Pazos-Ortiz, J. H. Roque-Ruiz, E. A. Hinojos-Márquez, J. López-Esparza, A. Donohué-Cornejo, J. C. Cuevas-González, L. F. Espinosa-Cristóbal and S. Y. Reyes-López, Dose-dependent antimicrobial activity of silver nanoparticles on polycaprolactone fibers against Gram-positive and Gram-negative bacteria, *J. Nanomater.*, 2017, **2017**, 4752314.

283. G. Batoni, G. Maisetta and S. Esin, Therapeutic potential of antimicrobial peptides in polymicrobial biofilm-associated infections, *Int. J. Mol. Sci.*, 2021, **22**(2), 482.
284. R. Guiglia, A. Musciotto, D. Compilato, M. Procaccini, L. L. Russo, D. Ciavarella, L. L. Muzio, V. Cannone, I. Pepe, M. D'Angelo and G. Campisi, Aging and oral health: effects in hard and soft tissues, *Curr. Pharm. Des.*, 2010, **16**(6), 619–630.
285. M. A. Ramos, P. B. Da Silva, L. Spósito, L. G. De Toledo, B. V. Bonifacio, C. F. Rodero, K. C. Dos Santos, M. Chorilli and T. M. Bauab, Nanotechnology-based drug delivery systems for control of microbial biofilms: a review, *Int. J. Nanomed.*, 2018, **13**, 1179.
286. S. S. Al-Zahrani, R. S. Bora and S. M. Al-Garni, Antimicrobial activity of chitosan nanoparticles, *Biotechnol. Biotechnol. Equip.*, 2021, **35**(1), 1874–1880.
287. Y. Wu, H. Gao, J. Liu and H. Liang, Chitosan nanoparticles efficiently enhance the dispersibility, stability and selective antibacterial activity of insoluble isoflavonoids, *Int. J. Biol. Macromol.*, 2023, 123420.
288. M. Cai, Y. Wang, R. Wang, M. Li, W. Zhang, J. Yu and R. Hua, Antibacterial and antibiofilm activities of chitosan nanoparticles loaded with Ocimum basilicum L. essential oil, *Int. J. Biol. Macromol.*, 2022, **202**, 122–129.
289. A. R. Maheo, Biosynthesis and characterization of Eupatorium adenophorum and chitosan mediated Copper oxide nanoparticles and their antibacterial activity, *Results Surf. Interfaces*, 2022, **6**, 100048.
290. M. Arif, R. Ahmad, M. Sharaf, J. Muhammad, M. Abdalla, W. A. Eltayb and C. G. Liu, Antibacterial and antibiofilm activity of mannose-modified chitosan/PMLA nanoparticles against multidrug-resistant Helicobacter pylori, *Int. J. Biol. Macromol.*, 2022, **223**, 418–432.
291. Z. Zhao, J. Han, S. Xu, Z. Jin, T. H. Yin and K. Zhao, Amoxicillin encapsulated in the N-2-hydroxypropyl trimethyl ammonium chloride chitosan and N,O-carboxymethyl chitosan nanoparticles: Preparation, characterization, and antibacterial activity, *Int. J. Biol. Macromol.*, 2022, **221**, 613–622.
292. M. Chandrasekaran, K. D. Kim and S. C. Chun, Antibacterial activity of chitosan nanoparticles: A review, *Processes*, 2020, **8**(9), 1173.
293. S. Y. Ong, J. Wu, S. M. Moochhala, M. H. Tan and J. Lu, Development of a chitosan-based wound dressing with improved hemostatic and antimicrobial properties, *Biomaterials*, 2008, **29**(32), 4323–4332.

CHAPTER 5

Biopolymer-based Drug Delivery Systems: Synthesis Strategies, Functionalization, and Applications

MANU SHARMA,*[a] EASHA PANDEY,[a] RANJU SONI,[a] NAMITA GUPTA,[a] TAPASVI GUPTA,[a] SWATI SAINI,[a] VEERA SADHU,[b] SHEELU SHARMA[c] AND RAGHAVA REDDY KAKARLA*[d]

[a] Department of Pharmacy, Banasthali Vidyapith, Banasthali, Rajasthan 304022, India; [b] School of Physical Sciences, Kakatiya Institute of Technology and Science (KITS), Warangal, Telangana 506015, India; [c] Pt. Deendayal Upadhayay National Institute for Persons with Physical Disabilities, Delhi University, New Delhi 11002, India; [d] School of Chemical and Biomolecular Engineering, University of Sydney, Sydney, NSW 2006, Australia

*Emails: aasmanu2018@gmail.com; reddy.chem@gmail.com

5.1 Introduction

Microencapsulation technology has become progressively important in the advancement of pharmaceutical formulations and oral delivery systems. It includes the preparation of microemulsions, microparticles, microspheres, and microcapsules.[1] In general, a microsphere is a matrix of polymer with an active ingredient dispersed in it, whereas a microcapsule is a reservoir-type

Nanoscience & Nanotechnology Series No. 62
Smart Nanomaterials for Infectious Diseases
Edited by Suvardhan Kanchi, Nidhi Sharotri, Rajasekhar Chokkareddy, Deepali Sharma and Falah H. Hussein

Published by the Royal Society of Chemistry, www.rsc.org

structure with a definite core and coat/envelope.[2] Microencapsulation is a technique of enclosing small discrete particles or droplets using a coating material to bestow various useful properties on microparticles. The inner material in the microcapsule comprises the core, internal phase or filler, whereas the exterior of the microcapsule is known as the coating, shell or membrane, which encloses one or more active ingredients.[3,4] Basically, this technique can provide a substantial barrier between the core material and other materials of the product. It can encapsulate all types of substances (solids, liquids, proteins, drugs, stem cells, bacterial cells, *etc.*) into a film of a microencapsulating agent.[5] The distinctiveness of the microcapsules depends on their composition, size, and function. Mostly, microcapsules vary in size from 1 to 100 μm.[6]

Microencapsulation is a well-developed technique utilized in drug delivery to overcome the problems associated with conventional drug delivery, resulting in increased patient compliance by delivering the drug at the affected/targeted site.[7] It is generally used to protect active pharmaceutical ingredients from UV radiation, humidity, and oxidation either to avoid undesirable effects or to prolong their storage lifetime. It can also be aimed at masking bitter tastes and odors, preserving volatility, and resolving incompatibilities.[8] Other applications include the development of controlled-release therapeutic formulations to reduce dosing frequency, improve therapeutic effect, and minimize toxicity.[9] The fabrication of microcapsules involves a mixture of several materials such as lipids and polymers to encapsulate the active pharmaceutical ingredient (API). They protect the drug from environmental conditions and shield the body from the irritative and mucosal-damaging effects of the active ingredient. The same principle can be used in agricultural and paint applications, where the release of pesticides for a prolonged duration might be essential.[10]

This chapter provides an up-to-date overview of microencapsulation techniques and classification and comprehensively discusses their application in drug delivery. First, advantages and disadvantages are listed.

5.2 Advantages

- Masking of bitter-tasting drugs, such as an orodispersible tablet of ofloxacin microspheres prepared using Eudragit E 100[11].
- Protection against environmental conditions such as light, heat, humidity, oxidation, *etc.*, *e.g.* vitamin A palmitate.
- Reduce the hygroscopicity of bioactive food proteins and hydrolyzates and peptides.[12]
- Conversion of liquid to pseudo-solids by complex coacervation technique.
- Decrease in vaporization of volatile drugs, *e.g.* methyl salicylate microcapsules were prepared by complex coacervation using gum arabic and gelatine.[13]
- Prevention of incompatibilities between drugs, *e.g.* microencapsulation of aspirin (acidic drug) combined with chlorpheniramine maleate (basic drug).

- Used for preparing prolonged-action dosage forms, *e.g.* clarithromycin microcapsules were prepared by a solvent diffusion method using ethylcellulose and PEG 6000.[14]
- Reduce gastric and other gastrointestinal tract irritations, *e.g.* indomethacin microcapsules were prepared for oral administration using polymers (hydroxypropylmethylcellulose and ethylcellulose).[15]
- Preparation of enteric dosage forms, *e.g.* astaxanthin oleoresin microcapsules were prepared by a coacervation technique to improve their stability.[16]
- Delivery of vaccines, *e.g.* type A botulism-loaded microcapsules were prepared using poly-DL-lactide-*co*-glycolide to provide timed-release microspheres.[17]
- Preparation of multilayered microcapsules for controlled release of a drug, *e.g.* layer-by-layer deposition of chitosan and sodium alginate was used to prepare multilayered microcapsules of isoniazid utilized for controlled delivery.[18]

5.3 Disadvantages

- Possibility of cross-reaction between core and coating material.
- Difficult to achieve a homogeneous and continuous film.
- High production cost.
- Decrease in the shelf life of hygroscopic drugs.
- Require advanced skills and knowledge.

5.4 Fundamental Considerations

Microencapsulation techniques are useful to prepare colloidal size ranged particles ranging in size from 0.1 to 5000 μm. However, microspheres usually prepared in size range of 1 to 100 μm are used in drug delivery and for various distinctive applications. The fundamental components for microencapsulation are core materials, coating materials, and vehicles. The core material may be physically solid or liquid. The composition of a liquid core comprises either a dispersion and/or dissolved material (drug), whereas a solid core comprises APIs, excipients, diluents, stabilizers, and accelerators or release retardants. The vehicle used to prepare the coating solution could be an aqueous or non-aqueous solvent. In order to improve the characteristic features of core material, a coating material alone or in combination is used. Polymeric materials belonging to various categories such as natural, synthetic, and semisynthetic are widely used in the pharmaceutical industry for the microencapsulation of drugs. An ideal coating material should exhibit the following properties:

- It should stabilize the core material.
- It should be compatible and chemically inert towards APIs.

- It should show release rate-controlling behavior.
- It should be tasteless, a film former, flexible, and stable.
- The substance used for coating should be of low viscosity, non-hygroscopic, and cost-effective.
- It should exhibit diverse solubility (aqueous medium/solvent/melting).
- The material of the coating can be hard, thin, brittle, flexible, *etc.*

5.5 Types of Polymers Used for Coating

5.5.1 Natural Polymers

- Proteins: *e.g.* albumin, gelatine, collagen;
- carbohydrates: *e.g.* starch, agarose, chitosan;
- waxes: *e.g.* beeswax, carnauba wax, paraffin wax.

5.5.2 Synthetic Polymers

- Biodegradable polymers: *e.g.* polyphosphazenes, polyamines (nylon), poly(glycolic acid), poly(lactic acid), polybutyrate;
- non-biodegradable polymers: *e.g.* poly(methyl methacrylate) (PMMA), epoxy polymers, acrolein.

5.6 Classification of Microparticles

Depending on the arrangement of the core material and coating material in microparticles, they are classified broadly as microcapsules and microspheres (see Figure 5.1). Generally, in microcapsules, coating material forms an inert diffusional barrier around the core material, whereas core material is dissolved or dispersed in the inert polymer in the case of microspheres.

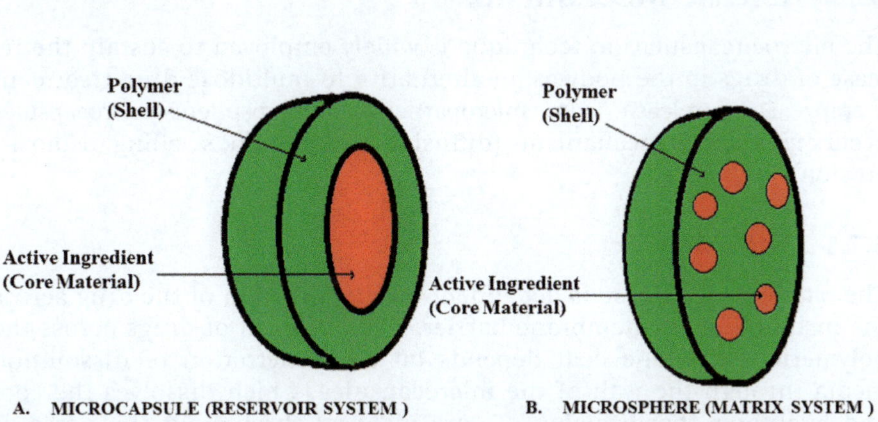

Figure 5.1 Pictorial representation of (A) microcapsules and (B) microspheres.

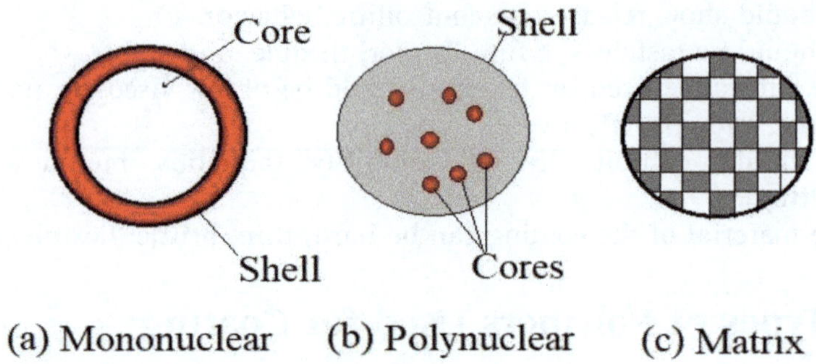

Figure 5.2 Different microcapsules with respect to the core material distribution in the coating composition.

Microcapsules are further categorized as follows (see Figure 5.2):

(1) Mononuclear or continuous core–shell microcapsules exhibit a shell around the core.
(2) Polynuclear or multinuclear microcapsules have many cores enclosed inside the shell. Polynuclear capsules include an outer capsule enclosing various inner capsules, wherein each of the inner capsules includes a shell and a core.
(3) Matrix-type microcapsules contain core material dissolved or dispersed homogeneously in the coating material. This is the ideal process to change liquids or very fine powders into free-flowing granulates or powders. The preparation of matrix-type microcapsules also provides protection against oxidation by preventing direct contact with other components of a formulation where incompatibility is troublesome.

5.7 Release Mechanisms

The microencapsulation technique is widely employed to sustain the release of drugs in the body as an alternative to multidose drug treatment therapy. Drug release from microparticulates/microspheres/microcapsules occurs *via* various mechanisms (diffusion, dissolution, swelling, osmosis, erosion).

5.7.1 Diffusion

The rate of drug release is dependent on the diffusion of the drug across the insoluble inert membrane barrier. The diffusion of drugs across the polymeric membrane coat depends on the penetration of dissolution media through the wall of the microcapsules, which dissolves the core and facilitates the diffusion of core material throughout the pores or interstitial channels formed.[19] However, intact release depends on the following:

- rate of fluid penetration into the wall of the microcapsule;
- rate of drug solubilization in the dissolution fluid;
- rate at which the dissolved drug reaches the surface of the system and concentration of drug in the saturated aqueous layer.

5.7.2 Dissolution

In a dissolution-controlled system, the drug is enclosed/coated with a slowly dissolving or erodible matrix. In reservoir systems, slowly dissolving materials [cellulose, polymethacrylates, poly(ethylene glycol)s, *etc.*] are used to coat or encapsulate the drug particles by microencapsulation techniques to control the drug release. The rates of penetration and dissolution of the coating by the dissolution fluid regulate the release of the drug from the microcapsule. The rate of dissolution of the coating material depends on the thickness and solubility of the coating in the dissolution fluid.[20]

In a matrix system, the drug is uniformly dispersed all over in a rate-controlling medium. Waxes such as hydrogenated castor oil, carnauba wax, beeswax, *etc.*, are employed to control the drug release from microcapsules and also to control the rate of infiltration of the dissolution fluid in the matrix.

5.7.3 Swelling

In this system, the drug release rate is controlled by the degree of swelling of the polymeric coat depending on the interfacial tension between the polymer coat and dissolution fluid. Microcapsules in the dissolution fluid absorb water and swell. Swelling increases the water content inside the microcapsule and also the polymer mesh size, resulting in the formation of new pores, facilitating drug diffusion through the swollen net into the exterior environment.[21]

5.7.4 Osmosis

Microcapsules fabricated by the use of a semipermeable membrane facilitate the development of an osmotic gradient that controls the drug release rate in osmotically controlled drug delivery systems. The osmotic gradient developed due to the difference in osmotic pressure between the outside and inside of the microcapsules forces the drug solution out of the microcapsules by small pores present in the coating, independent of pH.[22]

5.7.5 Erosion

In an erodible system, the drug release mechanism is facilitated *via* erosion. It may be surface erosion or bulk erosion. It is a slow process and is used for prolonged therapies for days or months. Erosion of the polymeric layer is governed by oxidation, hydrolysis, and/or enzymatic degradation. Polymers utilized for preparing erodible systems include polyorthoesters, polyanhydrides, stearyl alcohol, glyceryl monostearate and beeswax.[23]

5.8 Microencapsulation Techniques

Various physical, chemical, and physicochemical methods are utilized commercially for the encapsulation of core material (see Figure 5.3). Commonly utilized physical methods include spray drying, fluidized bed/pan coating, centrifugal extrusion, use of a vibrating nozzle, and spinning disk microencapsulation. Chemical methods of microencapsulation involve interfacial and *in situ* polymerization processes. Physicochemical processes combine physical and chemical methods to encapsulate the core material. In other words, the coat-forming materials are allowed to precipitate from the solution by varying the temperature, pH, or electrolyte concentration to facilitate their gradual deposition on the surface of the core material. Ionotropic gelation, polyelectrolyte complexation, coacervation, and supercritical fluid technology are common examples of physicochemical encapsulation techniques.

5.8.1 Coacervation Phase Separation Techniques

Coacervation is a method developed many years ago that is widely utilized for microencapsulation. It involves the formation of a colloid-rich viscous

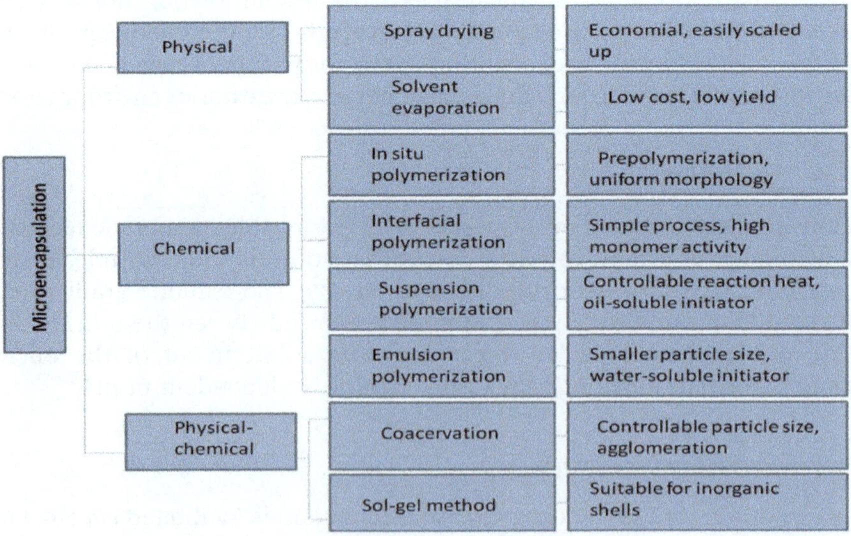

Figure 5.3 Microencapsulation techniques utilized for coating core material.

liquid phase that may be separated from a colloidal solution with the addition of another component. It is a process of desolvation or differentiation of colloids into a colloid-rich phase (coacervate) and a non-colloidal region (equilibrium phase), before coagulation. A high encapsulation efficiency and productive regulation over particle size, along with a narrow size distribution, contribute to its promising potential in the microencapsulation of therapeutic agents.[24] According to Bungenberg de Jong, coacervation is classified into two types, simple and complex coacervation processes.[28]

5.8.1.1 *Simple Coacervation Process*

This involves the utilization of a single polymer or polyelectrolyte, or a mixture of polyelectrolytes. The phase separation of dissolved matrix materials and their deposition on an insoluble core material lead to the formation of a simple coacervate. Basically, the forces accountable for the formation of simple coacervates are hydrophobic forces.

5.8.1.2 *Complex Coacervation Process*

This is a three-phase system comprising solvent (vehicle phase), active material (core material phase), and coating material phase. According to Voorn–Overbeek theory, coacervation proceeds by an electrostatic interaction between two or more hydrophilic polymers such as proteins and polysaccharides under continuous agitation.[95] The coacervate phase is in either suspended or emulsified form around active ingredients in the reaction medium. This technique involves the fabrication of nanoparticles by utilizing a lipid core in a colloidal system.[25] Electrostatic forces mainly manage the formation of complex coacervates and are responsible for higher encapsulation efficiency, integrity, and controlled-release behavior of complex coacervates.[26]

The coacervation process basically involves three steps:[27,28]

(1) *Fabrication of three immiscible phases:* in order to prepare three immiscible phases, the core material is dispersed in a coating material solution where the solvent of coating material is in liquid state. An immiscible coating material phase is formed by choosing a suitable phase separation technique, *i.e.* by altering the temperature of the aqueous phase, adding salt, adding non-solvent, adding incompatible polymers, and polymer–polymer interaction.

(2) *Deposition of coating material around the core:* this step involves the deposition of a liquid polymer coat over the core material by reducing the interfacial tension at their interface. The separated phase has polymeric coating material in a dispersed state deposited over the core material. Viscosity, speed, and temperature control the mixing conditions along with the rate of deposition.

(3) *Rigidizing the coating:* this step involves rigidization of the coated layer by utilizing suitable thermal, crosslinking, or desolvation techniques to form microcapsules.

5.8.2 Rapid Expansion of Supercritical Fluids

In this technique, the core material and coating material are dissolved in a supercritical fluid (SCF). An SCF can be defined as a fluid (homogeneous phase) whose thermodynamic condition is above the critical temperature and pressure, where the gas/liquid boundary disappears within the SCF zone. An SCF exhibits properties such as zero surface tension, high liquid density, and gas–liquid transport ability. The viscosity and diffusivity remain the same between the gas and liquid, while a low viscosity induces a kinetic reaction in an SCF medium. SCFs can facilitate the formation of micron-sized particles by rapid depressurization or by exceeding the saturation point of a solute by dilution or by a combination of the two methods. Solubility of solutes (coating material) in an SCF medium can be achieved by maintaining high pressure. Subsequently, an SCF solution containing dissolved coating material maintained at high pressure is passed through a nozzle with a small orifice in order to expand. This leads to desolvation of the coating material and facilitates its deposition over the core material to form microcapsules (see Figure 5.4). It is a cost-effective and non-toxic technique for microcapsule production. This process can eliminate the use of organic solvents and avoid rigid operating conditions in comparison with other techniques.[29] The most extensively used SCF is supercritical carbon dioxide. It is widely used because of its low critical temperature of 31.4 °C, which protects thermolabile compounds from thermal degradation. It is widely employed as it is non-toxic, cost-effective, inert, and non-flammable.[30,31]

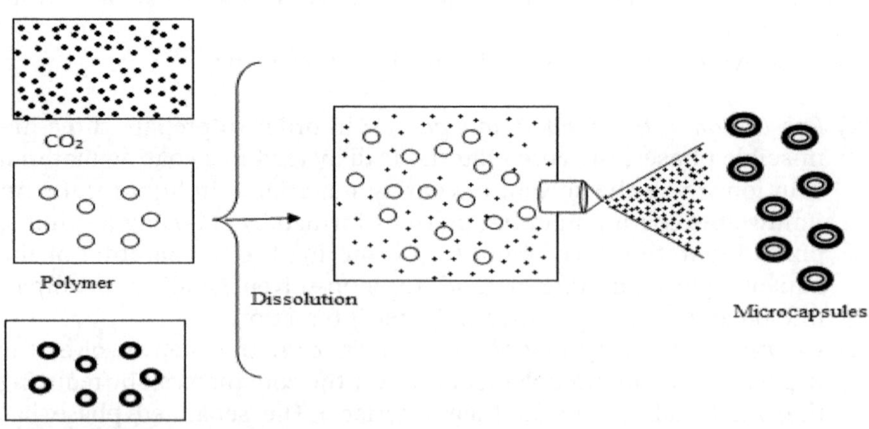

Figure 5.4 Microcapsule preparation method using rapid expansion of a supercritical fluid.

5.8.3 Layer-by-layer (LBL) Assembly Technique

This is a simple multifaceted technique also known as planar layer assembly. It facilitates the deposition of alternate layers of oppositely charged polymeric materials with a washing step in between. A wide range of natural, semisynthetic, and synthetic macromolecules, including biological materials such as proteins and DNA, are extensively used to prepare microcapsules by the LBL method.[32] The fabrication of a multilayer capsule utilizes a variety of driving forces such as hydrogen bonding, covalent bonding, base-pair interaction, guest–host interaction, and van der Waals interactions to crosslink the oppositely charged materials deposited as thin films to assemble into capsule walls. Crosslinking between different layers increases stability and boosts multi-layer properties of the capsule. This process controls the shape, size, wall thickness, permeability, and mechanical strength of the capsules, and makes them strong enough to survive in harsh conditions of high ionic strength, high pH, and strongly polar organic solvents.[33,34] Crosslinking of *N*-methyl-2-nitrodiphenylamine-4-diazoresin (NDR) and *m*-methylphenol–formaldehyde resin (MPR) in menthol led to the formation of covalently linked LBL microcapsules.[35] In the case of bone implants and skin patches, LBL assembly released around 1–5% of the total core content over a period of 1–100 h.[32] Antifog and UV-protective coatings, cell adhesion, and tissue engineering are some of the functional outcomes of the LBL assembly technique.

5.8.4 Sol–Gel Encapsulation

A model of the solution–gelation (sol–gel) technique was proposed by Graham in 1864 through his work on silica rods.[36] The sol–gel technique provides certain benefits with marked utility in medical and pharmaceutical formulations (silica gels, aerogels, organo-modified silicates, enzyme immobilization, coatings, *etc.*), separation chromatography, catalysis, diagnostics, biotechnology, waterproofing of buildings, optical lenses, controlled release, and restoration. The method permits powderless processing of glass, ceramics, and thin films/fibers immediately from solution. During the encapsulation of active materials, initially the selection of precursors (solvent and catalyst) should be carried out on the basis of their chemical properties and the environmental conditions. Essentially, the sol–gel method involves the use of tetraethyl orthosilicate (TEOS) as a network-forming agent. TEOS is insoluble in water and allows easy inclusion of organic molecules, polymers, and biomolecules at low temperatures and pH. The sol–gel technique can be used in the synthesis of combined organic–inorganic materials with numerous biomedical applications.[36,37] Superparamagnetic nanoparticles prepared *via* the sol–gel technique are very important in separation processes and diagnosis, *e.g.* for performing magnetic resonance imaging (MRI) and the development of targeted drug delivery systems.[38]

5.8.5 Spray Drying and Spray Congealing

Spray drying is the most extensively used method of encapsulation as it bestows faster evaporation of the solvent and maintains a low temperature in the particles. It involves uniform dispersion of core material in polymer solution by intensive homogenization. Subsequently, the prepared slurry is atomized as a mist at a high rate of around 300 mPa, by pumping the slurry through a rotating disk into the hot compartment of a spray drier. The latent heat of vaporization of hot air leads to evaporation of the solvent and transforms the fine mist of slurry into a powder (see Figure 5.5). The properties of the final product depend on the operating conditions and properties of the slurry. It is the most economical and flexible method of encapsulation with less chance of contamination. However, exposure to hot air led to the formation of highly porous particulate structures, which are prone to oxidation and are highly hygroscopic.[39,40]

Spray congealing is a solvent-free technique of microencapsulation. The spray congealing process is similar to spray drying. In spray congealing, the core material is dispersed in the molten coating material. The coating material is solidified by spraying the hot slurry into a cool stream of air to solidify the coating material over the core material *via* the thermal congealing process. Solidification of dissolved coating material can also be achieved by incorporating the coating–core mixture into a non-solvent. Subsequently, the non-solvent is removed from the products *via* sorption, extraction, and evaporation techniques. The particle size of the spray congealed product is controlled by the feed rate, atomizing velocity, feed material viscosity, and other processing variables.[40]

5.8.6 Fluidized Bed Coating

Fluidized-bed coating is also called air suspension coating. In this method, solid particles are suspended in a jet of air followed by spraying of coating material onto the moving particles of the core material. During each pass of

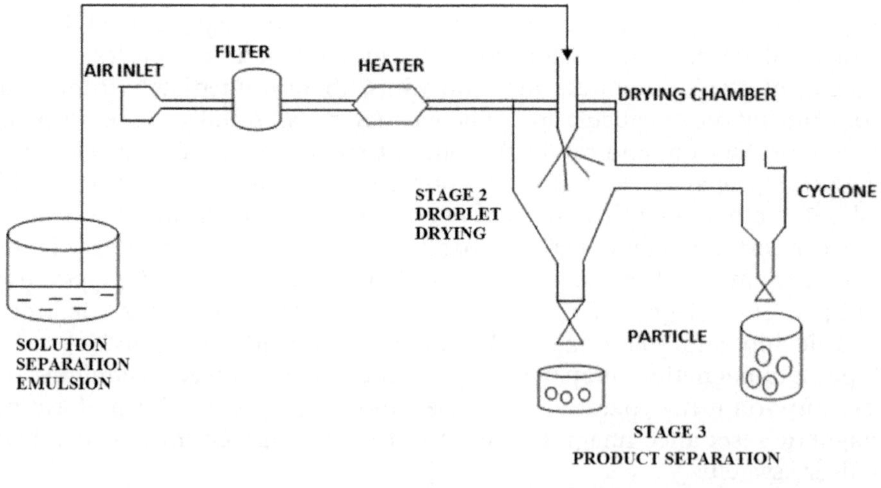

Figure 5.5 Schematic representation of microcapsule fabrication *via* spray drying.

core material through the coating zone, it gains coating material. This cyclic process is repeated several times until the desired thickness of coat is achieved for the purpose of microencapsulation. The jet of air serves a dual function, *i.e.* fluidization of core material and drying of the coated surface of the core material, which facilitates faster drying. The quality of the finished product is governed by the nature of the core material, the concentration of the coating material, the rate of application of the coating material, the volume of air needed to fluidize the bed and inlet, and the outlet temperature of air. During fluidized-bed coating, a faster rate of heat and mass transfer can be achieved with uniform mixing and spreading rate of the coating material. However, this method causes relatively higher dust generation and there is a higher probability of metallic contamination of the finished product.

The air suspension method can be operated in different kinds of fluidized-bed coaters having design differences in the spraying pattern of the coating solution:

(1) *Top spray:* Nozzle position directed downwards countercurrent to the particle flow.
(2) *Wurster method (bottom spray):* Nozzle position directed upwards countercurrent to the low-density particle flow.
(3) *Tangential (rotatory spray):* Nozzle position directed tangentially with high-density particle flow.
(4) *Tangential (static spray):* Nozzle position directed diagonally concurrent with high-density particle flow.[41]

5.8.7 Centrifugal Extrusion

A volatile and unstable compound in a glassy carbohydrate matrix is encapsulated using the extrusion method (see Figure 5.6). This technique consists of an aligned feed tube through which coating and core material are forced through multiple orifices mounted on the surface of the device. The core material circulates *via* the center of the tube while the coating material proceeds through the remaining tube. The whole device is connected to a rotating shaft, where its front head revolves around its vertical axis. When the shaft revolves, core and coating material are co-extruded *via* multiple orifices as "rot", thus giving a wrapped effect around the coating material. Finally, the breakdown of particles takes place under centrifugal force applied outwards to the "rot". Surface tension further works to enfold the core material by using coating material to achieve encapsulation.[42–44] This is a useful technique to coat heat-labile active materials with a high production rate of up to 22.5 $kg\,h^{-1}$ of microcapsules.

5.8.8 Inclusion Complexation (β-Cyclodextrin Inclusion)

At the molecular level, encapsulation can be achieved using cyclodextrin in a polymeric matrix. The cyclodextrin's inner core acts as a non-polar molecule

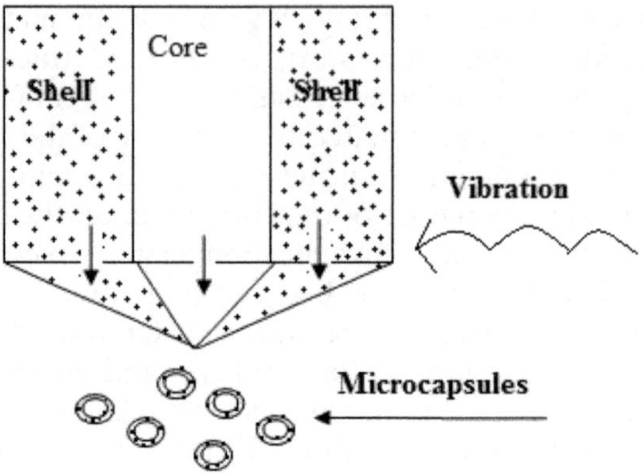

Figure 5.6 Process of microcapsule formation *via* centrifugal extrusion.

whereas the outer core has a polar nature. An inclusion complex is formed with addition to many organic compounds. In this process of micro-encapsulation, the outer core of the cyclodextrin molecule is hydrophilic whereas the inner core is hydrophobic. Non-polar bioactive molecules fit dimensionally into the non-polar inner matrix of the cyclodextrin *via* hydrophobic bonding. Additionally, other interactive forces such as dipole–dipole interactions and van der Waals forces are also involved in the formation of complexes. β-Cyclodextrin-based encapsulation can be achieved *via* coprecipitation, lyophilization, and spray drying.[27,39] An inclusion complex provides protection, increases the solubility of hydrophobic drugs, and controls their release rate from the encapsulated formulation. However, the route of administration of the cyclodextrin complex is limited.

5.8.9 Solvent Evaporation

This is a widely used technique for the encapsulation of drugs, foods, flavors, and many more materials, and shows a high encapsulation efficiency with low residual solvent content. Particles ranging in size from nanometers to micrometers can be easily prepared. Various formulation and process variables such as solvent type, drug-to-polymer ratio, solvent volume, internal aqueous phase volume, external phase volume, and ultrasonic and other mechanical forces are applied to tailor the physicochemical properties of the microspheres formed.[45] There are four basic steps that are involved in the encapsulation of core material by the solvent evaporation technique:

(1) Hydrophobic drug dissolution in the organic phase along with the polymer.
(2) Emulsification of the organic phase in the aqueous phase, *i.e.* continuous phase.

(3) Solvent extraction from the dispersed phase by continuous phase accompanying solvent evaporation and forming solid particles from droplets of the dispersed phase.
(4) Elimination of residual solvent by drying microspheres.

However, this technique is not suitable for highly hydrophilic drugs owing to the poor solubility of the drug in the organic phase, and second, diffusion of the drug into the continuous phase may occur during the formation of the emulsion, which may lead to drug loss.[46]

5.8.10 Interfacial Polymerization

This involves the dispersion of reactive monomer constituting one phase into a second immiscible phase containing a second monomer. Both monomers are soluble in their respective solvents constituting the aqueous and organic phase, respectively, and approach close to each other at the interface and form a polymeric film. The basic criterion for the formation of microcapsules by interfacial polymerization is high reactivity of the monomers under mild conditions to facilitate the rapid formation of polymeric membranes. Other parameters such as membrane strength, thickness, composition, degradability, and permeability also influence the selection of suitable monomers. Various conformations of materials have been appraised for the preparation of microcapsules by interfacial polymerization, but only four kinds of polymers have been developed, *i.e.* polyamides, polyurethanes, polyureas, and polyesters. This technique assists in the formation of more stable microspheres than any other microencapsulation method owing to the irreversibility of the reaction.[47] It is widely used in the pesticide industry and for the production of cosmetic and pharmaceutical products. It allows a high load of active ingredients with an easy process and helps in achieving high yields.[48] The initial state of these formulations has high pH and contains toxic monomers, solvents, and reactive products. Further, to improve the conditions, interfacial polymerization is processed, *e.g.* encapsulating enzymes and their cofactors in semipermeable nylon leads to the formation of an artificial cell.[47]

5.8.11 *In Situ* Polymerization

This is a chemical method of microencapsulation in which two or more monomers are added to a continuous phase containing uniformly dispersed core material. Monomers react to form oligomers or polymers in the system. Subsequently formed polymer deposition and precipitation at the interface of the core and coating polymer are regulated by the addition of precipitants and changing the pH, temperature, or solvent. Typically, coating materials used for *in situ* polymerization are aminoplast resins, such as melamine–formaldehyde, urea–formaldehyde, urea–melamine–formaldehyde, or resorcinol-modified melamine–formaldehyde polymers. Based on the solubility of the monomer and polymer formed, *in situ* polymerization is categorized into three types: suspension polymerization, precipitation condensation, and dispersion polycondensation.[49]

Suspension polymerization occurs when, in a dispersion medium, the monomer is insoluble, leading to the formation of suspended monomer droplets polymerizing in solution to form polymer microparticles. To maintain a uniform size distribution, the polymerization reactor and stirring rate are the main parameters to be maintained.[50,51] On the other hand, precipitation condensation occurs when only the monomer is soluble in the dispersion medium and not the polymer. When the reaction is carried out, flocculation and aggregation of polymers yield particles leading to the formation of a broad size distribution and irregular shape. Dispersion polycondensation occurs when the dispersion medium shows good solubility for the monomer but poor solubility for the polymer. This method led to formation of small-sized microparticles as particle growth is regulated by the sustained addition of monomers.

5.9 Recent Advances in Microencapsulation Processes

Topics considered here are the use of an ultrasonic atomizer based on interfacial solvent exchange, melt prilling in a fluidized bed, and deagglomerating jet spray coating.

5.9.1 Use of an Ultrasonic Atomizer Based on Interfacial Solvent Exchange

This is a process for generating reservoir-type microcapsules, using a double microdispenser system consisting of two inkjet nozzles. One of the nozzles creates aqueous drops and the other produces a polymer solution, which combine so that the drops collide to form microdrops. After the collision, two phases become differentiated, namely core and membrane, due to the resulting interfacial tension between the phases. This process is followed by interfacial solvent exchange, the end result being a polymeric membrane encasing aqueous drops. This technique uses interfacial mass transfer among two mutually soluble liquids to harvest reservoir-type microcapsules. As soon as the water-insoluble polymer encounters an aqueous solution, a solid polymer film is established at the interface between the two liquids as a consequence of mutual transfer of solvents (*i.e.* solvent exchange).[40,52,53]

5.9.2 Melt Prilling in a Fluidized Bed

Melt prilling provides an additional coating of molten lipophilic material (*e.g.* wax) on a solid particle, which becomes further solidified into micron-sized droplets, after its dispersal into a cold gas stream. During the progression, the heated core material is atomized from a nozzle to produce droplets of the desired size.

These droplets fall into the fluidized bed with simultaneous cooling and coating with the wall material. Here, the heat released from the core droplets

Table 5.1 Applications of microencapsulation to improve the properties of the core material.

No.	Core material	Characteristic property	Purpose of encapsulation
1	Prednisolone	Insoluble in water	Taste masking[55]
2	Cetirizine	Poorly water soluble	Taste masking[56]
3	α-Amylase	Water-soluble solid	Increase gastrointestinal stability, improve substrate specificity, enhance its usability under *in vitro* conditions[49,57–59]
4	Papain	Enzyme in water-soluble solid state	Improve gastrointestinal stability[60,61]
5	Curcumin	Water-insoluble solid	Taste masking, improve stability and bioavailability[62]
6	Bromelain	Water-soluble enzyme	Stability, prolonged drug release[63,64]
7	Menthol	Volatile solution	Reduction of volatility, sustained release[65]
8	Potassium chloride	Water-soluble enzyme	Reduction of gastric irritation[66]
9	Clarithromycin	Insoluble in water	Enhance solubility and bioavailability[67]
10	Scutellarin	Insoluble in water	Improve solubility and bioavailability[68]
11	Aspirin	Slightly water-soluble solid	Taste masking, extended drug release, reduction of gastric irritation[69]
12	Vitamin A palmitate	Non-volatile liquid	Improve stability to oxidation[70]
13	β-Carotene	Insoluble in water	Enhance solubility[71]
14	*Stevia* leaf	Thermosensitive	Enhance thermal and physical stability[72]
15	β-Carotene	Light sensitive	Photostability[73]
16	Anchovy oil	Prone to oxidation	Oxidative stability[74]
17	Soybean oil and curcumin	Prone to oxidation	Controlled release, oxidative stability[75]
18	Rosuvastatin calcium	Poor solubility	Sustained release, enhance solubility[76–78]
19	L-Casein	Improvement in lactose intolerance	Controlled release, protection from gastric degradation[79,80]
20	Allyl methyl disulfide	Volatile in nature	Controlled release, reduction of volatility[81]
21	Piroxicam	Poor water solubility	Controlled release[82]
22	Ibuprofen	Bitter taste, poor water solubility	Controlled release[14]
23	Roxithromycin	Bitter taste, poor water solubility	Controlled release, improve pH sensitivity[83]
24	Fenofibrate	Poor solubility	Increase solubility and flowability[84,85]
25	Metoprolol succinate	Plasma concentration fluctuations, adverse effects	Sustained release, reduce adverse effects[86]
26	Chia essential oil	Instability, volatility	Stability and reduction of volatility[87]

Table 5.1 (*Continued*)

No.	Core material	Characteristic property	Purpose of encapsulation
27	Peppermint oil	Volatility, instability	Stability and reduction of volatility[88]
28	Citronella oil	Volatility, instability	Controlled release, volatility, instability[89]
29	Poly-L-lysine	Instability	Enhance stability[90]
30	Flavorzyme	Instability	Controlled release, enhance stability[91]
31	Neodymium	Poor solubility	Selective absorption and separation[92]
32	Clinosorbent-5	Poor solubility and unstable	Sustained release, selective separation[93]
33	Mussel protein hydrolyzate	Bitter taste	Taste masking[94]
34	Diclofenac sodium	Bitter taste	Taste masking[92]
35	Enalapril maleate	Bitter taste	Taste masking[93]
36	Levofloxacin	Bitter taste	Taste masking, controlled release[94]

is transmitted to the wall material, which leads to the wall material sticking to the core surface. The product obtained is a coherent capsule wall structure, which is then screened separately from the fluidizing column and capsule. The remaining bed material is returned to the system, and the process is carried out on a continuous basis. Applications of this procedure have provided slow-release glycerine capsules and biologically active encapsulated products.[40,54]

5.9.3 Deagglomerating Jet Spray Coating

A conventional fluidized-bed dryer is associated with the disadvantage of agglomeration of core particles during the microencapsulation process. One of several modifications of the conventional fluidized-bed dryer is a deagglomeration unit. A high-speed gas jet spray and a conical conduit are used as the deagglomeration unit, which partially coats the core particles before additional coating material is applied from the coating spray nozzle. The partial coating helps to overcome the problem of agglomeration of core particles. This process is used to encapsulate solid or liquid particles (*e.g.* pharmaceuticals, resins, catalysts, inorganic salts, *etc.*) smaller than about 300 g.[40]

5.10 Applications

Microcapsules are well known for their exclusive properties as a carrier of various pharmaceuticals, and these can be customized to stick to targeted tissue systems. The process of microencapsulation offers diverse prospects such as stability, taste masking, targeting tissues, controlled release, *etc.* (Table 5.1). Hence this approach has been used extensively in various applications, *e.g.* for targeted delivery of an active constituent to a specific tissue, for controlled release, and to improve the stability of active pharmaceutical agents.

5.11 Conclusion

Microparticles can be prepared satisfactorily by selecting a suitable coating material and suitable techniques of microencapsulation to achieve the desired therapeutic potential. Microparticles offer potential promise as unique carrier systems for several pharmaceuticals. They can be customized for targeted tissue systems and also controlled release at a specific site in the body. Hence microencapsulation offers prospects for the development of safe and efficient therapies for the management of various ailments.

References

1. M. Lengyel, N. K. Szabó, V. Antal, A. J. Laki and I. Antal, *Sci. Pharm.*, 2019, **87**(3), 20.
2. M. Singh, J. S. Dua, M. Menra, M. Soni and D. N. Prasad, *Int. J. Adv. Res.*, 2016, **4**(6), 2094.
3. N. Agnihotri, R. Mishra, C. Goda and M. Arora, *Indo. Glob. J. Pharm. Sci.*, 2012, **2**(1), 1.
4. M. Sharma, A. Rathore, S. Sharma, V. Sadhu, K. R. Reddy and R. V. Kulkarni, Recent progress in mucoadhesive polymers for buccal drug delivery applications, in *Nanomaterials in Diagnostic Tools and Devices*, 2020, ch. 8, pp. 213–240.
5. S. Bale, A. Khurana, A. S. S. Reddy, M. Singh and C. Godugu, *Crit. Rev. Ther. Drug Carrier Syst.*, 2016, **33**(4), 309.
6. M. N. Singh, K. S. Y. Hemant, M. Ram and H. G. Shivakumar, *Res. Pharm. Sci.*, 2010, **5**(2), 65.
7. N. V. Patil, N. V. Wadd, S. S. Thorat and S. S. Upadhye, *Am. J. Pharm. Tech Res.*, 2020, **10**(2), 286.
8. M. Peanparkdee, S. Iwamoto and R. Yamauchi, *Rev. Agric. Sci.*, 2016, **4**, 56.
9. M. Sharma, A. Deohra, K. R. Reddy, V. Sadhu, Biocompatible *in situ* gelling polymer hydrogels for treating ocular infection, in *Methods in Microbiology*, Academic Press, 2019, vol. 46, ch. 5, pp. 93–114.
10. A. Latnikova and M. Jobmann, *Top. Curr. Chem.*, 2017, **375**(3), 64.
11. K. Malik, G. Arora and I. Singh, *Sci. Pharm.*, 2011, **79**(3), 653.
12. A. Mohan, S. R. C. K. Rajendran, Q. S. He, L. Bazinet and C. C. Udenigwe, *RSC Adv.*, 2015, **5**(97), 79270.
13. S. Mendes, A. Catarino, A. Zille, N. Fernandes and F. M. Bezerra, *Materials*, 2021, **14**(5), 1087.
14. L. Song, S. He and Q. Ping, *Exp. Ther. Med.*, 2017, **13**(5), 2435.
15. L. Usha Ayalasomayajula, S. D. N. Siva Giri, T. Prasanthi, G. Sneha and S. C. S. Kumar, *Int. J. Pharm. Phytopharmacol. Res.*, 2015, **5**(3), 48.

16. R. Li, R. Chen, W. Liu, C. Qin and J. Han, *Pharm. Dev. Technol.*, 2018, **23**(7), 674.
17. R. L. Whalen, D. J. Dempsey, L. M. Thompson, K. Bucknell, R. Kunitomo, Y. Okazaki and H. Harasaki, *ASAIO J.*, 1996, **42**(5), M649.
18. S. Tiwari and B. Mishra, *Daru*, 2011, **19**(1), 41.
19. C. Berkland, M. J. Kipper, B. Narasimhan, K. Y. Kim and D. W. Pack, *J. Control. Rel.*, 2004, **94**, 129.
20. C. Wang, Q. Ge, D. Ting, D. Nguyen, H. R. Shen, J. Chen, H. N. Eisen, J. Heller, R. Lan and D. Tutnam, *Nat. Mater.*, 2004, **3**, 190.
21. C. S. Brazel and N. A. Peppas, *Eur. J. Pharm. Biopharm.*, 2000, **49**(1), 47.
22. F. M. Veronese, F. Marsilio, S. Lora, P. Caliceti, P. Passi and P. Orsolini, *Biomaterials*, 1999, **20**(1), 91.
23. S. Lakshmi, D. S. Katti and C. T. Laurencin, *Adv. Drug Delivery Rev.*, 2003, **55**(4), 467.
24. X. Li, W. Huo, Y. Wang, N. Han, J. Wang, N. Wang and X. Zhang, *Carbohydr. Polym.*, 2018, **200**(15), 602.
25. A. Melnyk, J. Namieśnik and L. Wolska, *Trends Anal. Chem.*, 2015, **71**, 282.
26. Y. L. Du, H. Q. Huang, H. O. Wang and J. X. Xiao, *J. Food Eng.*, 2018, **234**(4), 91.
27. Y. Timilsena, M. A. Haque and B. Adhikari, *Food Sci. Nutr.*, 2020, **11**(6), 481.
28. P. L. Madan, *Drug Dev. Ind. Pharm.*, 1978, **4**(1), 95.
29. W. C. Tsai and Y. Wang, *Prog. Polym. Sci.*, 2019, **98**, 101161.
30. R. Couto, V. Alvarez and F. Temel, *J. Supercrit. Fluids*, 2017, **120**(2), 432.
31. N. Esfandiari, *J. Supercrit. Fluids*, 2015, **100**, 129.
32. M. N. Antipina, M. V. Kiryukhin and A. G. Skirtach, *Int. Mater. Rev.*, 2014, **59**(4), 224.
33. W. Tong, X. Song and C. Gao, *Chem. Soc. Rev.*, 2012, **41**(18), 6103.
34. E. V. Lengert, S. I. Koltsov, J. Li, A. V. Ermakov, B. V. Parakhonskiy, E. V. Skorb and A. G. Skirtach, *Coatings*, 2020, **10**(11), 1131.
35. Y. Zhang, S. Yang, Y. Guan, W. Cao and J. Xu, *Macromolecules*, 2003, **36**(11), 4238.
36. G. J. Owens, R. K. Singh, F. Foroutan, M. Alqaysi, C. M. Han, C. Mahapatra, J. C. Knowles and H. W. Kim, *Prog. Mater. Sci.*, 2016, **77**, 1.
37. F. Salaün, C. Butstraen and E. Devaux, in *Science and Technology Behind Nanoemulsions*, ed. S. Karakus, InTechopen, 2018, DOI: 10.5772/intechopen.74299.
38. D. E. Lee, H. Koo, I. C. Sun, J. H. Ryu, K. Kim and I. C. Kwon, *Chem. Soc. Rev.*, 2012, **41**(7), 2656.
39. N. Choudhury, M. Meghwal and K. Das, *Food Front.*, 2021, 1.
40. T. Farheen, A. Shaikh and S. Shahi, *Int. J. Pharma Res. Health Sci.*, 2017, **5**(5), 1823.
41. R. Zhang, T. Hoffmann and E. Tsotsas, *Processes*, 2020, **8**(12), 1525.
42. Y. Srivastava, A. D. Semwal and G. K. Sharma, *Int. J. of Food. Ferment. Technol.*, 2013, **3**(1), 1.
43. V. Suganya and V. Anuradha, *Int. J. Pharm.*, 2017, **9**(3), 233.

44. A. Poshadri and A. Kuna, *J. Res. ANGRAU.*, 2010, **38**(1), 86.
45. T. Ansari, M. Farheen, M. N. Hoda and A. K. Nayak, *Elixir Pharm.*, 2012, **47**, 8821.
46. M. Li, O. Rouaud and D. Poncelet, *Int. J. Pharm.*, 2008, **363**(1–2), 26.
47. C. Perignon, G. Ongmayeb, R. Neufeld, Y. Frere and D. Poncelet, *J. Microencapsulation*, 2015, **32**(1), 1.
48. F. Ricardo, D. Pradilla, R. Luiz and O. A. A. Solano, *Polymers*, 2021, **13**(4), 644.
49. Y. Zhou, B. Qiu, X. Yin, H. Liu and L. Zhu, *Int. J. Polym. Mater.*, 2019, **69**(11), 1.
50. D. Sun, J. Li, Q. Pan, C. Hao and G. Lai, *J. Spectrosc.*, 2013, 164275.
51. K. K. Palodkar, N. M. Rao, S. Iyer, R. T. Puttalingaiah, V. Sadhu, T. M. Aminabhavi, K. R. Reddy and A. V. S. Sainath, Maltose-based methacrylated polymer architectures and their biocompatibility, *Mater. Today Chem.*, 2022, **23**, 100669.
52. Y. Yeo and K. Park, *J. Controlled Release*, 2004, **100**(3), 379.
53. Y. Yeo and K. Park, *Polym. Eng. Sci.*, 2003, **89**, 143.
54. M. Nickerson, C. Yan, S. Cloutier and W. Zhang, in *Microencapsulation in the Food Industry*, ed. A. K. G. Gaonkar, N. Vasisht, A. R. Khare and R. Sobel, Academic Press, 2014, pp. 485–500.
55. W. Brniak, E. Maślak and R. Jachowicz, *Eur. J. Pharm. Sci.*, 2015, **75**, 81.
56. M. Stojanov, R. Wimmer and K. L. Larsen, *J. Pharm. Sci.*, 2011, **100**, 3177.
57. M. Sharma, V. Sharma and D. K. Majumdar, *Adv. Sci. Eng. Med.*, 2013, **5**, 1120.
58. M. Sharma, V. Sharma, A. K. Panda and D. K. Majumdar, *Pharm. Dev. Technol.*, 2013, **18**(3), 560.
59. M. Sharma, V. Sharma and D. K. Majumdar, *Int. Scholarly Res. Not.*, 2014, **2014**, 936129.
60. M. Sharma, V. Sharma, A. K. Panda and D. K. Majumdar, *Int. J. Nanomed.*, 2011, **6**, 2097.
61. M. Sharma, V. Sharma, A. K. Panda and D. K. Majumdar, *Yakugaku Zasshi*, 2011, **131**(5), 697.
62. M. Sharma, S. Sharma and J. Wadhawa, *Artif. Cells, Nanomed., Biotechnol*, 2019, **47**(1), 45.
63. M. Sharma and R. Sharma, *RSC Adv.*, 2018, **8**, 2541.
64. M. Sharma and D. Chaudhary, *Int. J. Pharm.*, 2021, **594**, 120176.
65. R. Kaur, D. Kukkar, S. K. Bhardwaj, K. H. Kim and A. Deep, *J. Controlled Release*, 2018, **285**, 81.
66. M. Kulkarni, B. Vishwakarma, S. Sen, S. Anupuram and A. A. Date, *AAPS Open*, 2019, **5**(1), 1.
67. M. Sharma, N. Gupta and S. Gupta, *RSC Adv.*, 2016, **6**, 76621.
68. J. Wang, J. Tan, J. Luo, P. Huang, W. Zhou, L. Chen, L. Long, L. M. Zhang, B. Zhu and L. Yang, *J. Nanobiotechnol.*, 2017, **15**, 1.
69. S. A. Chime, P. A. Akpa, C. C. Ugwuanyi and A. A. Attama, *Recent Pat. Inflammation Allergy Drug Discovery*, 2020, **14**(1), 78.

70. B. Albertina, M. Di-Sabatino, G. Calogera, N. Passerini and L. Rodriguez, *J. Microencapsulation*, 2010, **27**(2), 150.
71. E. De Paz, A. Martín, Á. Bartolomé, M. Largo and M. J. Cocero, *Food Hydrocolloids*, 2014, **37**, 14.
72. M. Y. Mutmainah, Y. D. Franyoto and L. Kusmita, *Library J.*, 2019, **6**, E5925.
73. M. J. Sáiz-Abajo, C. González-Ferrero, A. Moreno-Ruiz, A. Romo-Hualde and C. J. González-Navarro, *Food Chem.*, 2013, **138**(2–3), 1581.
74. Q. Xia, T. O. Akanbi, B. Wang, R. Li, W. Yang and C. J. Barrow, *Mar. Drugs*, 2019, **17**(143), 1.
75. M. L. N. Le, H. N. L. Thi and V. T. Nguyen, *J. Food Qual.*, 2021, 5593065.
76. H. Zhao, X. Fei, L. Cao, B. Zhang and X. Liu, *Materials*, 2019, **12**(3), 393.
77. T. Gupta, M. Sharma and R. Gilhotra, *Indian J. Pharm. Educ. Res.*, 2017, **2**(2), 245.
78. S. Butt, S. M. F. Hasan, M. M. Hassan, K. M. Alkharfy and S. H. Neau, *Saudi Pharm. J.*, 2019, **27**, 619.
79. S. Basu, D. Banerjee, R. Chowdhury and P. Bhattacharya, *J. Food Eng.*, 2018, **238**, 61.
80. J. Hadzieva, K. Mladenovska, M. S. Crcarevska, M. G. Dodov, S. Dimchevska, N. Geškovski, A. Grozdanov, E. Popovski, G. Petruševski, M. Chachorovska, T. P. Ivanovska, L. Petruševska-Tozi, S. Ugarkovic and K. Goracinova, *Food Technol. Biotechnol.*, 2017, **55**(2), 173.
81. M. Wang, T. Doi and D. J. McClements, *Food Res. Int.*, 2019, **119**, 6.
82. W. L. Lee, J. W. M. Tan, C. N. Tan and S. C. J. Loo, *PLoS One*, 2014, **9**(12), 1.
83. D. A. R. González, J. C. Olivares, A. R. Guerrero, M. E. R. Huezo, E. J. V. Carter and C. P. Alonso, *J. Food Eng.*, 2012, **111**(1), 102.
84. Z. Dong, Y. Ma, K. Hayat, C. Jia, S. Xia and X. Zhang, *J. Food Eng.*, 2011, **104**(33), 455.
85. A. Yousaf, D. W. Kim, J. K. Kim, J. O. Kim, C. Yong and H. G. Ch, *Powder Technol.*, 2015, 275.
86. W. C. Hsieh, C. P. Chang and Y. L. Gao, *Colloids Surf., B*, 2006, **53**(2), 209.
87. T. Haque, H. Chen, W. Ouyang, C. Martoni, B. Lawuyi, A. M. Urbanska and S. Prakash, *Mol. Pharm.*, 2005, **2**(1), 29.
88. K. Anjani, K. Kailasapathy and M. Phillips, *Int. Dairy J.*, 2007, **17**(1), 79.
89. L. Zhang, D. Wu, B. Zhu, Y. Yang and L. Wang, *J. Chem. Eng. Data*, 2011, **56**(5), 2280.
90. Z. Yaneva, D. Ivanova and N. Popov, *Molecules*, 2021, **26**(6), 1655.
91. N. R. Breternitz, H. M. A. Bolini and M. D. Hubinger, *Sci. Technol.*, 2017, **83**, 141.
92. M. F. Al-Omran, S. A. Al-Suwayeh, A. M. El-Helw and S. I. Saleh, *J. Microencapsulation*, 2002, **19**(1), 45.
93. Y. Georgieva, M. Kassarova, V. Kokova, E. Apostolova and B. Pilicheva, *Pharmazi*, 2020, **75**(2), 61.
94. N. Hassan, S. Latif, H. Afzal, N. Abbas, S. Naheed, R. Khokar and A. Shah, *Indian. J. Pharm. Sci.*, 2019, **81**(5), 843.
95. A. Veis, *Adv. Colloid Interface Sci.*, 2011, **167**(1–2), 2–11.

CHAPTER 6

Mechanism of Silver and Gold Nanoparticles Against Infectious Disease-causing Pathogens

E. A. JAYASREE AND A. S. RAJESHKUMAR*

Nanobiomedicine Laboratory, Centre for Global Health Research, Saveetha Medical College and Hospitals, Saveetha Institute of Medical and Technical Sciences, Chennai, Tamil Nadu 602105, India
*Email: ssrajeshkumar@hotmail.com

6.1 Introduction

The pathogenic microorganisms that cause infectious diseases, such as bacteria, viruses, parasites, and fungi, can spread a disease from one person to another either directly or indirectly.[1] This is the second leading cause of mortality worldwide.[2] Infectious diseases can be divided into two categories: newly emerging and re-emerging. Emerging infectious diseases are those that have been discovered recently whereas re-emerging infectious diseases are those that have already been discovered but have developed drug resistance when they reoccur, making them challenging to treat or control.[3]

The most common harmful infectious agents (pathogens) that cause infections include the following:

1. *Bacteria* are small unicellular organisms belonging to the prokaryotes group (*e.g. Mycobacterium tuberculosis*, *Streptococcus* spp., *Clostridium tetani*, *Klebsiella pneumoniae*).

Nanoscience & Nanotechnology Series No. 62
Smart Nanomaterials for Infectious Diseases
Edited by Suvardhan Kanchi, Nidhi Sharotri, Rajasekhar Chokkareddy, Deepali Sharma and Falah H. Hussein

Published by the Royal Society of Chemistry, www.rsc.org

2. *Viruses* are living things that can only reproduce in their host [*e.g.* human immunodeficiency virus (HIV), Ebola virus, dengue virus, coronavirus].
3. *Fungi* are eukaryotic filamentous organisms (*e.g. Candida* spp., *Aspergillus* spp., *Cryptococcus* spp., *Histoplasma* spp.).
4. *Parasites* are organisms that reside in or on a host and derive sustenance from the host. Parasites are divided into two main groups:
   (a) Unicellular organisms (*e.g. Plasmodium* spp., *Babesia* spp., *Toxoplasma gondii*, *Cryptosporidium* spp.).
   (b) Multicellular organisms (*e.g. Ascaris* spp., *Wuchereria bancrofti*, *Trichinella* spp.).[4,5]

Antibiotic resistance is currently increasing to extremely high levels and the ability to cure common infectious diseases is under threat as new resistance mechanisms emerge and spread globally. As antibiotics lose their effectiveness, it has become more and more difficult, and in some cases impossible, to treat infections.[6] High doses of antibiotics will be used to treat these multidrug-resistant organisms (MDROs) that cause infections, but they may have very toxic and undesirable side effects, necessitating the development of alternative methods.[7]

Nanoparticles (NPs), with sizes between 1 and 100 nm, are of particular interest in nanoscience and nanotechnology because of both their unique features and their potential as building blocks for more sophisticated nanostructures.[8] NPs are categorized according to their size, morphology, physical characteristics, and chemical composition, and include carbon-based NPs, ceramic NPs, metal NPs, semiconductor NPs, polymeric NPs, and lipid NPs.[9] The use of NPs provides an effective method to control infections caused by MDROs[10] and other pathogens. When NPs enter the cell membrane of pathogenic microbes, they disrupt important molecular pathways and produce unique antimicrobial mechanisms.[11] A variety of NP types are considered to be non-toxic, and some of them even provide non-toxic benefits to health.[12] One of these beneficial health strategies is the use of NPs, which have antibacterial properties, to combat spoilage and harmful microbes.[13]

The precise mechanisms associated with the antibacterial properties of NPs are still under investigation. The fundamental mechanisms that inhibit bacteria, fungi, and viruses include microbial cell wall penetration, the production of reactive oxygen species (ROS), damaged DNA and proteins, and loss of cellular integrity. Before other mechanisms come into play in various microbial cell inhibition processes, cell penetration is often the first stage. The primary penetration method involves NPs adsorbing or diffusing on the surface of the cell. The negatively charged functional groups of proteins can be bound by NPs to achieve adsorption, which leads to the degradation of proteins and the death of cells.[14] Also, there is evidence that the diffusion mechanism can result in ROS entering pathogenic cells. In addition, microbial inactivation and destruction caused by interactions between several surface-exposed groups of bacteria and NPs appear to be a potential mechanism.[15] Additionally, it has been shown that the morphological and physicochemical properties of NPs have an impact on the antibacterial actions of metals. It has been demonstrated that small NPs have more potent bactericidal effects. The ability of metal NPs with positive surface charges to bind with bacteria having negative surface charges may improve their bactericidal activities.[16] A possible mechanism of the antiviral activity of a few NPs is to inhibit viral replication and stop viral particles from entering host cells. Some NPs prevent the attachment of the virus to host cells, stop the cleavage process, interfere with the release of progeny viruses from the host cells, and subsequently stop the spread of infection.[17]

6.1.1 General Introduction to Infectious Disease

An infection is an invasion of tissues by pathogens, their growth, and the reaction of host tissues to the infectious agent and the toxins they release.[18] Infectious diseases are transferred from one person to another by direct contact, by insects or different animals, through airborne routes, and by consuming contaminated food.[19] Infections are the primary cause of disease worldwide. Lower respiratory tract infections, diarrhoea, HIV infection, malaria, tuberculosis (TB), and other conditions are associated with high death rates.[20] When a potential host is exposed to an infectious agent, the result of that exposure depends on the dynamic interaction between the intrinsic host determinants of susceptibility to infection and the agent determinants of infectivity, pathogenicity, virulence, and disease. Physical and social behavioural and environmental factors are extrinsic determinants of the susceptibility of a host to exposure.[21]

Numerous chronic conditions that were not previously believed to be linked to microbial infections were subsequently found to be either directly or indirectly caused by infectious microorganisms (*e.g. Helicobacter pylori* bacteria might cause peptic ulcers and gastric carcinoma). The development of microbial drug resistance has been a challenge in medicine owing to the inappropriate use of antimicrobial medicines,[2] and drug access is hindered by the presence of several infections inside cells, either active or latent. Pathogens can develop resistance through a variety of processes, including DNA mutations, changes in membrane permeability, the development of multidrug efflux pumps, activation of drug-degrading enzymes, or an expansion of their intracellular life cycle.[22] Most medicines remain in the extracellular environment, but the few substances that can enter host cells quickly break down there, or their concentration does not reach therapeutic levels once inside the cell.[23] The major problem is that intracellular pathogenic microorganisms such as *Mycoplasma pneumoniae* and *Chlamydophila pneumoniae* can also persist and proliferate inside photogenic cells. As a result, these cells are the first to become "infected", and they serve as a reservoir for drugs that cannot reach them. They enhance the intracellular replication of infections and the spread of infection by protecting them from treatment.[24]

6.1.2 General Introduction to Nanoparticles

Nanotechnology is the area of science and engineering that focuses on substances with dimensions of less than 100 nm.[25] Nanoparticles and nanomaterials are used for applications in medicine as drug delivery systems, therapeutic agents, diagnostic imaging systems, and many other purposes.[26] The use of NPs for infections is attractive owing to the lack of novel medicines under development and the treatment failures caused by antibiotic resistance. NPs provide a solution for resolving these issues and can be used in applications such as vaccines, theranostic systems, methods for drug delivery to infected bones, lungs, and the gastrointestinal tract, and antimicrobial agents against various infectious pathogens, multidrug-resistant strains, and biofilms.[11]

The purpose of drug targeting, as stated by the Paul Erlich "magic bullet" concept, is to deliver drugs to the correct location at the right concentration for the right amount of time.[27] Owing to their prolonged release based on environmental factors, NPs permit drugs to be specifically targeted and to reach therapeutic intracellular levels. They prevent drug degradation and unintended interactions before the drug reaches the target tissue or cell. The physicochemical properties of pharmaceuticals are optimized at the same time, allowing clinical use or administration in more patient-friendly ways. Additionally, they can contain various drugs in a single formulation, which is extremely beneficial when treating diseases such as HIV and *Helicobacter pylori* infection that require combinations of drugs.[28]

Metal NPs have long been used to combat microorganisms. Owing to their numerous potential uses, metal NPs have attracted considerable attention as the need for nanotechnology has increased. Because of their high efficiency, low production costs, and eco-friendliness, chemical and biological methods of NP synthesis are preferred. In the chemical synthesis of NPs, precursors, reducing agents, and stabilizers can all be utilized. Biological systems utilize plant parts and plant extracts of non-pathogenic fungi, bacteria, and yeasts. This method is mostly preferred because of its fast production, low toxicity, environmental friendliness, and cost-effectiveness.[29]

6.2 Antimicrobial Activity of Metal Nanoparticles

NPs have been increasingly used as an alternative to antibiotics to target bacteria in recent years. Metal and metal oxide NPs are the most promising nanomaterials for biological applications because of their intrinsic antibacterial activity.[30] Several metal NPs have found numerous biomedical uses and show antimicrobial activity against infectious and non-infectious diseases. Metal NPs are submicron-sized particles formed from pure metals (such as silver, titanium, zinc, gold, platinum, cerium, iron, and thallium) or their compounds (*e.g.* oxides, phosphates, hydroxides, sulfides, fluorides, and chlorides).[31] The electrical, magnetic, catalytic, and optical properties of metal NPs are distinct from those of the bulk metals.[32]

The behaviour of bacteria when in contact with metal NPs is explained by particular traits. It is important to understand the differences between the cell walls of Gram-positive and Gram-negative bacteria, since the primary toxicological effect of antimicrobial agents on bacteria is caused by direct contact with the cell surface.[29] The surfaces of bacteria, both Gram-positive and Gram-negative, are negatively charged. Owing to electrostatic interactions, negatively charged bacterial cell walls attract positively charged NPs to their surface. Positively charged metal-based NPs form a solid bond with membranes, disrupting cell walls and increasing permeability as a result. Additionally, metal ions released by NPs from the extracellular space have the ability to penetrate cells and interfere with biological processes.[33]

The size, shape, and metal ions of metal NPs determine their effectiveness and antiviral activity.[34] The recognized modes of action for metal NPs can take place inside or outside the host cells. The NP can interact with the gp120 proteins, outbid the virus for binding sites on host cells, interfere with viral attachment, and prevent virus binding or penetration. Other possible modes of action include interacting with the viral genome, attaching to the viral particles, and inactivating the virus particles prior to cellular entry. The virally encoded and host cellular components necessary for viral replication and the generation of progeny virions are also in abundance in the intracellular compartment of an infected cell. Hence another antiviral mechanism of action involves the interaction of metal NPs with these replication factors.[35]

Metal NPs have also been shown to have a variety of impacts on the growth of fungi, including the destruction of the cell wall and membrane, inhibition of sporulation and germination, alteration of the colour of colonies of filamentous fungi, and inhibition of yeast budding.[36] Some metal NPs bind with biological functional groups of membrane proteins, allowing them to pass through cell walls and fungal membranes, causing cell death. The possibility of these NPs producing metal ions, which result in ROS and cause DNA damage, protein denaturation, and cell material leakage, has also been reported.[37] Another possible mechanism of action is the interruption of electron transport, interference with nutrient absorption and DNA synthesis, and even the simultaneous occurrence of many mechanisms.[38]

Metal NPs interfere with the cell membrane of parasites, damage electron transport, and cause disruption of DNA. Some metal NPs accumulate inside the cell and produce free radicals. This causes cytotoxicity in parasites by producing reactive oxygen. Some NPs cause strong electrostatic attractions, accumulate at the cell surface, and interact and disrupt the cell membrane.[39] Figure 6.1 shows some of the applications of AgNPs and AuNPs against harmful infectious diseases.

6.3 General Introduction to Silver Nanoparticles

AgNPs are small silver particles that range in size from 1 to 100 nm in diameter.[40] Since silver is known to be poisonous to a wide range of small

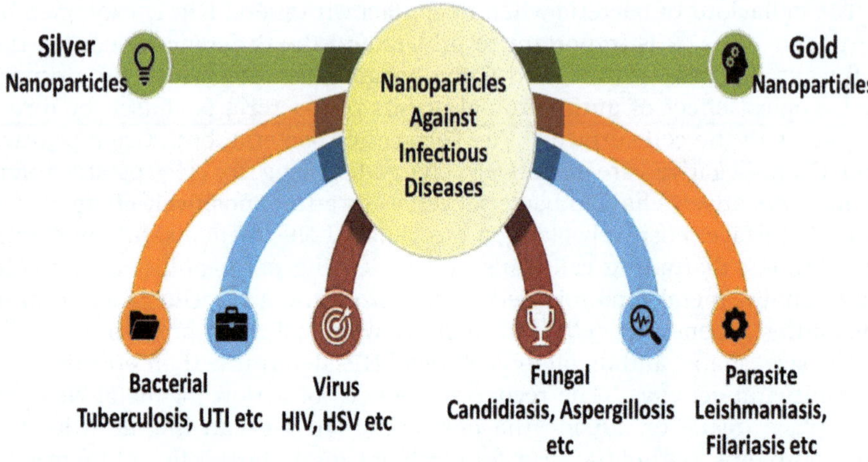

Figure 6.1 Applications of AgNPs and AuNPs against infectious diseases.

organisms, silver-based combinations have been extensively used in antibacterial applications. AgNPs have been used as a material medium for the transportation of antibiotics, as a blending agent to form composites used in purification processes, and as casing material.[41] Nanostructured silver materials are widely utilized as biomedical probes in the treatment of burns and wounds and in the prevention of many infectious diseases.[42] It was observed that silver ions disrupt enzyme activity and membrane function to operate as an antiseptic with a broad biocidal impact against various microbial species.[43] Based on these findings, stabilizing agents and AgNP size and concentration are all important determinants of the biological activity of AgNPs. Generally, the antibacterial mechanism works by impairing bacterial cell walls and inhibiting the actions of bacterial proteins, which prevents bacterial growth. AgNPs showed the most significant antifungal activity against various pathogenic fungal species including *Candida* spp. The growth of many fungal species was successfully inhibited by functionalized AgNPs. Similarly, functionalized AgNPs have demonstrated a high affinity to bind with particular viral species, inhibiting viral RNA synthesis. Functional AgNPs are generally considered as potential antimicrobial agents for inhibiting a broad range of microbial species in diverse clinical applications.[44]

6.3.1 Antimicrobial Activity of Silver Nanoparticles

6.3.1.1 Antibacterial Activity of Silver Nanoparticles

AgNPs have been widely used as a potential tool for bactericidal drugs to combat a variety of microbial infections.[45] When AgNPs interact with bacteria, they release silver ions, which hinder membrane permeability, deactivate

Table 6.1 Antibacterial potential of silver nanoparticles.

AgNPs	Therapeutic efficacy	Ref.
Spherical-shaped AgNPs with *Cestrum nocturnum* as a reducing agent	Strong antibacterial activity against *Vibrio cholerae*, *Citrobacter*, *Salmonella typhi*, *Escherichia coli*, and *Proteus vulgaris*	47
Spherical-shaped AgNPs with serine as a reducing agent	Effective against *Bacillus cereus*, *Micrococcus luteus*, *Pseudomonas aeruginosa*, *S. typhi*, and *E. coli*	48
Spherical-shaped AgNPs with *Emblica officinalis* as a reducing agent and AgNPs with *Clitoria ternatea* as a reducing agent	Good antibacterial activity against enteropathogenic bacteria such as *E. coli*, *K. pneumoniae*, *Staphylococcus aureus*, and *Bacillus subtilis*	49 and 50
AgNPs	Effective against *M. tuberculosis*, *B. cereus*, *E. coli*, and *Mycobacterium smegmatis*	51
Spherical-shaped AgNPs in a urinary catheter	Effective against urinary tract infection-causing organisms	52
Triangular and ellipsoidal-shaped AgNPs with *Spatoglossum asperum* as a reducing agent	Potent antimicrobial activity against oral pathogens	53
AgNPs with *Planomicrobium* spp.	Effective against food-borne disease-causing pathogens such as *E. coli*, *Klebsiella planticola*, *K. pneumoniae*, *B. subtilis*, and *Serratia nematodiphila*	54
AgNPs combined with doxycycline, tetracycline, erythromycin, polymixin B, and rifampicin	Effective synergistic effects in the treatment of *Acinetobacter baumannii* infection	55 and 56
Spherical-shaped AgNPs with *Spirulina platensis* as a reducing agent	Good antibacterial activity against opportunistic nosocomial pathogens	57

cellular enzymes, and cause cell death.[46] Several groups have reported the antibacterial activity of AgNPs (see Table 6.1).

6.3.1.2 *Antiviral Activity of Silver Nanoparticles*

Numerous researchers have been inspired to search for novel antiviral treatments as a result of recent epidemics of infectious diseases brought on by pathogenic viruses that are still in the process of developing and have become resistant to current antiviral drugs. The most important approach for developing new antiviral drugs is to disrupt the interactions between the virus ligand and cell membrane, preventing the virus from adhering to and entering the cells. The most promising choice for an antiviral drug is AgNPs[58] and the antiviral activities of AgNPs against several types of pathogenic viruses are shown in Table 6.2.

Table 6.2 Antiviral potential of silver nanoparticles.

AgNPs	Therapeutic efficacy	Ref.
Polyvinylpyrrolidone (PVP)-coated AgNPs	Interact with gp120 of HIV (*Retroviridae* family)	34
Tannic acid-modified AgNPs	Interact with glycoproteins and interfere with cell attachment of herpes simplex virus type 2 (HSV-2) (*Herpesviridae* family)	59
Mercaptoethanesulfonate-coated AgNPs and AuNPs	Competition for the binding of the herpes simplex virus type 1 (HSV-1) (*Herpesviridae* family) to the cell	60
AgNPs	Inhibits the synthesis of hepatitis B (HBV) viral RNA and extracellular virions as a result of interaction with the HBV viral particles	61
PVP-coated AgNPs	Interfere with respiratory syncytial virus (*Paramyxoviridae* family) attachment to the cell	62
AgNPs	Effective against influenza viruses (*Orthomyxoviridae* family), causing morphological damage to them and preventing viral binding to host receptors	63
Magnetic hybrid colloid AgNPs	Damage proteins of the bacteriophage MS2 (*Leviviridae* family) and murine norovirus (*Caliciviridae* family) viral coat	64
Polysaccharide-coated AgNPs and uncoated AgNPs	Inactivation of Tacaribe virus (TCRV) (*Arenaviridae* family) particles prior to entry	65
AgNPs	Damage adenovirus type 3 (Ad3) (*Adenoviridae* family) viral particles and binding to the viral DNA	66
Polysaccharide-coated AgNPs and uncoated AgNPs	Block the virus–host cell binding and penetration in monkeypox virus (*Poxviridae* family)	67

6.3.1.3 *Antifungal Activity of Silver Nanoparticles*

The silver ions released by AgNPs into the fungal cell have an antifungal effect on the cell when they attach to the cell membrane and penetrate the fungus. Then, AgNPs attach to the respiratory sequence, inhibiting cell division to cause cell death.[68] Several groups have reported the antifungal activity of AgNPs (see Table 6.3).

6.3.1.4 *Antiparasitic Activity of Silver Nanoparticles*

AgNPs have the ability to produce ROS, which are capable of eliminating infectious parasites,[74] and the antiparasitic abilities of AgNPs against several types of pathogenic parasites are shown in Table 6.4.

6.4 General Introduction to Gold Nanoparticles

AuNPs are small gold particles that range in size from 1 to 100 nm in diameter. Because of their small size and high surface area-to-volume ratio,

Table 6.3 Antifungal potential of silver nanoparticles.

AgNPs	Therapeutic efficacy	Ref.
Spherical-shaped AgNPs with *Ocimum sanctum* as a reducing agent	Good antifungal activity against *Candida albicans*, *Candida glabrata*, and *Candida tropicalis*	69
AgNPs	Effective antifungal activity against ocular pathogenic fungi such as *Fusarium*, *Aspergillus*, and *Alternaria alternata*	70
AgNPs synthesized by the fungus *Fusarium oxysporum*	Antifungal activity against food contamination caused by *Aspergillus flavus*, *Aspergillus nomius*, *Aspergillus parasiticus*, *Aspergillus melleus*, and *Aspergillus ochraceus*	71
Spherical-shaped AgNPs with sodium citrate as a reducing agent and in combination with nystatin and chlorhexidine digluconate	Good antifungal activity against denture-related stomatitis caused by *C. albicans* and *C. glabrata*	72
AgNPs with alternan as a stabilizing agent	Good efficiency against food-borne disease-causing pathogens such as *Botrytis cinerea*, *Fusarium oxysporum*, *A. alternata*, *Aspergillus niger*, *Penicillium chrysogenum*, and *A. parasiticus*	73

Table 6.4 Antiparasitic potential of silver nanoparticles.

AgNPs	Therapeutic efficacy	Ref.
Round-shaped AgNPs	Effective antileishmanial effects on *Leishmania tropica* parasites	74
Spherical-shaped AgNPs with *Euphorbia prostrata* as a reducing agent	Larvicidal activity against the filariasis-causing vector *Culex quinquefasciatus*	75
Circular-shaped AgNPs with *Nelumbo nucifera* as a reducing agent	Larvicidal activity against larvae of dengue fever-causing *Anopheles subpictus* Grassi and filariasis-causing *C. quinquefasciatus*	76
Spherical-shaped AgNPs in combination with the bioactive compound fucan	Antiparasitic activity against Chagas disease (also called American trypanosomiasis) caused by *Trypanosoma cruzi*	77
Irregular- and spherical-shaped AgNPs with *Phoenix dactylifera* and *Ziziphus spina-christi* extracts as a reducing agent	Antitoxoplasmic activity against *T. gondii*	78
AgNPs with *Andrographis paniculata* as a reducing and capping agent	Antiplasmodial activity against malaria-causing *Plasmodium falciparum*	79

AuNPs interact effectively with microbes and can therefore exert a variety of antimicrobial effects.[80] AuNPs are of low toxicity, chemically inert, and simple to modify and functionalize and can be used to treat various infectious diseases. The biocompatibility and biodistribution characteristics of AuNPs make them ideal for clinical use. They have good immune-modulatory and adjuvant properties.[81] Owing to their high surface area-to-volume ratio, AuNPs offer strong binding affinity, unique electronic structure, plasmon excitation, and huge surface energies. Additionally, AuNPs can engage with many functional groups or ligands with great affinity. Colloidal gold has already been used to treat a wide range of diseases with a low level of cytotoxicity because of its inherent optical and magnetic properties.[82]

6.4.1 Antimicrobial Activity of Gold Nanoparticles

6.4.1.1 Antibacterial Activity of Gold Nanoparticles

AuNPs are very stable and have the capability to interact with bacterial cells, making them potential antibacterial agents.[83] Their mechanism of action against bacteria is due to the reduction of adenosine triphosphate (ATP) synthase activity, thereby disrupting metabolic processes, reducing ribosomal subunits for tRNA binding, interfering with bacterial proteins and

Table 6.5 Antibacterial potential of gold nanoparticles.

AuNPs	Therapeutic efficacy	Ref.
Spherical-, decahedral-, and triangular-shaped AuNPs with *Ananas comosus* as a reducing agent	Antibacterial activity against *S. aureus* and *P. aeruginosa*	85
Spherical-, triangular-, and hexagonal-shaped AuNPs with *Galaxaura elongata* as a reducing agent	Antibacterial activity against *E. coli*, *K. pneumoniae*, *S. aureus*, *P. aeruginosa*, and methicillin-resistant *S. aureus* (MRSA)	86
AuNPs	Effective against urinary tract infection caused by Gram-negative, Gram-positive, and multidrug-resistant pathogens	87
AuNPs	Effective against enteric bacterial pathogens such as *E. coli*, *K. pneumoniae*, *S. aureus*, and *B. subtilis*	88
AuNPs combined with ampicillin, streptomycin, and kanamycin	Antibacterial activity against *E. coli*, *M. luteus*, and *S. aureus*	89
AuNPs with cefaclor as a capping agent	Effective against *E. coli* and *S. aureus*	90
AuNPs	Antibacterial activity against *E. coli*, *S. typhimurium*, and *S. aureus*	91
Spherical-shaped AuNPs synthesized by the Actinobacteria *Streptomyces* spp.	Antibacterial activity against *E. coli*, *K. pneumoniae*, *Proteus mirabilis*, *Salmonella infantis*, *P. aeruginosa*, and *B. subtilis*	92
Spherical-shaped AuNPs with *Crescentia cujete* as a reducing agent	Antibacterial activity against *E. coli*, *P. aeruginosa*, *V. cholerae*, *S. typhi*, *Shigella flexneri*, and *B. subtilis*	93

cytoplasm, and targeting mechanisms that allow them to target specific bacteria.[84] Several reports have shown that AuNPs are effective antibacterial agents (see Table 6.5).

6.4.1.2 *Antiviral Activity of Gold Nanoparticles*

AuNPs inhibit viral entry and also act as a viral-neutralizing agent. AuNPs are employed as delivering agents to enhance the efficacy of antiviral drugs. They hinder viral replication at early stages and enhance the inhibition of some viral infections.[17] Several groups have reported the antiviral activity of AuNPs (see Table 6.6).

6.4.1.3 *Antifungal Activity of Gold Nanoparticles*

The antifungal activity of AuNPs has been studied by several groups, and it seems to depend on the size, shape, and concentration. Smaller particles had the greatest impact since their surface area increased. The fungicidal effect of AuNPs is rapid and irreversible. AuNPs have the ability to burst the cell wall of fungi[100] and the antifungal activities of AuNPs against several types of pathogenic fungi are shown in Table 6.7.

6.4.1.4 *Antiparasitic Activity of Gold Nanoparticles*

AuNPs increase the generation of ROS in parasites, which eventually leads to damage to the cell membrane of parasites. AuNPs show strong

Table 6.6 Antiviral potential of gold nanoparticles.

AuNPs	Therapeutic efficacy	Ref.
Cuboidal-shaped AuNPs combined with azidothymidine	Inhibit viral replication at early stages in HIV (*Retroviridae* family)	94
AuNPs functionalized with sialic acid	Inhibit the viral infection with the influenza virus (*Orthomyxoviridae* family)	95
AuNPs	Prevent viral attachment and cell penetration and eliminate infections of herpes simplex virus-1 (*Herpesviridae* family)	60
Spherical-shaped gallic acid-induced AuNPs	Inhibit proliferation of viral infection and viral attachment of herpes simplex virus (*Herpesviridae* family)	96
Spherical-shaped AuNPs loaded with ribavirin	Highest antiviral activity against measles virus (*Paramyxoviridae* family)	97
Spherical-shaped AuNPs coated with poly(ethylene glycol) (PEG)	Inhibit viral entry and also act as a neutralizing agent against HIV (*Retroviridae* family)	98
AuNPs conjugated with peptide triazoles	Enhanced antiviral functions against HIV-1 (*Retroviridae* family)	99

Table 6.7 Antifungal potential of gold nanoparticles.

AuNPs	Therapeutic efficacy	Ref.
Spherical-shaped AuNPs with tin(II) chloride ($SnCl_2$) and sodium borohydride ($NaBH_4$) as a reducing agent	Effective against clinical strains of *Candida* spp. such as *C. albicans*, *C. glabrata*, and *C. tropicalis*	101
Spherical-shaped AuNPs with *Abelmoschus esculentus* seed extract as a reducing agent	Effective against *Puccinia graminis tritci*, *A. niger*, *A. flavus*, and *C. albicans*	102
Spherical-shaped AuNPs	Effective against onychomycosis (fungal nail infection) caused by *A. flavus*, *A. niger*, *C. albicans*, *C. tropicalis*, *Epidermophyton floccosum*, and *Trichophyton mentagrophytes*	103
Spherical-shaped AuNPs synthesized using cell-free *Phaffia rhodozyma* yeast	Effective against cutaneous mycosis caused by *Microsporum*, *Trichophyton dermatophytes*, *Trichophyton mentagrophytes*, and *Cryptococcus neoformans* and five different *Candida* spp.	104
Round-shaped AuNPs synthesized using *Mentha piperita* plant essential oil	Antifungal activity against *C. albicans*, *Candida kefyr*, *C. tropicalis*, *A. niger*, and *A. flavus*	105
Spherical-shaped AuNPs with *Annona muricata* leaf extract as a reducing agent	Antifungal activity against *Fusarium oxysperium*, *A. flavus*, *Penicillium camemeri*, and *C. albicans*	106

Table 6.8 Antiparasitic potential of gold nanoparticles.

AuNPs	Therapeutic efficacy	Ref.
Polygonal-shaped *Streptomyces* spp.-mediated AuNPs	Effective against malaria-causing *Plasmodium berghei* parasite	107
Spherical-shaped AuNPs synthesized using *Nigrospora oryzae* phytopathogenic fungus	Effective against helminth infections caused by *Raillietina* spp.	108
Spherical-shaped AuNPs conjugated with quercetin	Effective against the leishmanial parasite *Leishmania donovani*	109
AuNPs conjugated with anti-*T. gondii* antibody	Antiparasitic activity against tachyzoites stage of *T. gondii*	110
Spherical-shaped AuNPs synthesized using *Terminalia chebula*	Antifilarial activity against lymphatic filariasis caused by *W. bancrofti*	111

electrostatic attraction that accumulates at cell surfaces and interacts with the cell membranes of the parasites.[39] Several reports have shown AuNPs to be effective antiparasitic agents (see Table 6.8)

6.5 Mechanism of Silver and Gold Nanoparticles Against Infectious Disease-causing Pathogens

6.5.1 Mechanism of Silver Nanoparticles Against Microbial Infections

6.5.1.1 Mechanism of Silver Nanoparticles Against Bacterial Infections

Compared with other metallic NPs, the antibacterial activity of AgNPs is more complex and they exhibit lower cell toxicity.[112] AgNPs have four known antibacterial effects: (1) adherence to the bacterial surface membrane, (2) entry of AgNPs into the cells causes disruption of biomolecules and intracellular damage, (3) they induce cellular toxicity by the generation of ROS, which puts the cells in an oxidative stress state, and (4) disruption of their signal transduction pathways.[68]

When bacteria are exposed to AgNPs, the NPs tend to adhere to the bacterial cell wall owing to the electrostatic attraction between the positively charged AgNPs and the negatively charged bacterial cell membrane.[113] The sulfur-containing proteins in the microbial cell wall also have a high level of affinity for AgNPs. The morphological changes in the structure of the cell membrane are irreversibly altered by the attachment of AgNPs to the membrane of bacterial cells and it can also cause a loss of the permeability of the cell membrane and integrity of the lipid bilayer.[114] The ability of the cell to regulate activity consistently can be impacted by changes in the cell structure that cause an increase in membrane permeability. An increased cell membrane permeability may also cause the loss or leakage of cellular components such as cytoplasm, proteins, ions, and ATP, the energy reservoir of the cell, which can allow the bacteria to act like ghost cells. When the contents of a cell or bacteria are expelled, it causes the ghost cell effect in bacteria, which results in a hollow bacterial envelope.[115] Figure 6.2 shows some of the antibacterial mechanisms of AgNPs.

6.5.1.2 Mechanism of Silver Nanoparticles Against Viral Infections

The exact mechanism through which AgNPs destroy viruses remains uncertain. AgNPs can be used to prevent a range of viral infections, such as those caused by the herpes simplex virus (HSV), respiratory syncytial virus, and adenovirus type 3 (ADV3), either by preventing viral infection in cells or by directly inactivating the viruses.[116] AgNPs have a high antiviral activity due to their large surface area, which makes interactions with virus particles easier. Viral nanoparticle interactions are critically influenced by the size of the AgNPs.[117] The viral genomic material is eventually destroyed or prevented from penetrating the cell membrane as a result of initial interactions between AgNPs and the

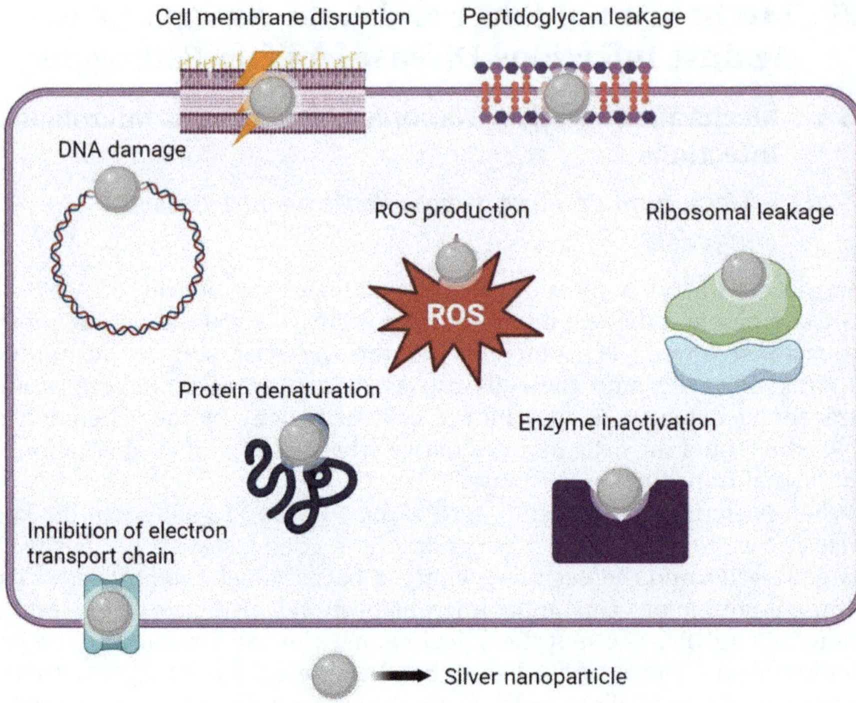

Figure 6.2 Antibacterial mechanisms of silver nanoparticles.

viral surface. To prevent the virus from interacting with the cell membrane, AgNPs adhere to the viral entities. AgNPs also inhibit the nucleocapsids of the viral entity inside the cell. AgNPs bind with viral genomic material, preventing host cell genome replication. Finally, they stop cellular processes such as protein synthesis to prevent the viral entity from replicating. Although other metals have antiviral capabilities, AgNPs are the most potent; regardless of the virus family, they have demonstrated strong antiviral activity.[118]

6.5.1.3 Mechanism of Silver Nanoparticles Against Fungal Infections

In recent years, the prevalence of fungal infections has increased, and AgNPs have attracted attention as possible antifungal agents. Patients with impaired immune systems from cancer chemotherapy or viral infections are more susceptible to fungal infections.[119] By rupturing cell membranes and preventing the normal process of cell division, AgNPs have remarkable antifungal efficacy against fungal infections.[120] AgNPs have an antifungal effect on fungi by disrupting membrane-bound lipids and enzymes that cause cell lysis and forming insoluble compounds with sulfhydryl groups in the cell walls of fungi.[121] Increased membrane permeability and the release of potassium ions (K^+) are the results of cell wall and membrane damage.[122]

AgNPs also prevent the cellular processes necessary for yeast budding, most likely by compromising the integrity of the membrane.[120]

6.5.1.4 Mechanism of Silver Nanoparticles Against Parasitic Infections

AgNPs function in a variety of ways, including free radical generation to kill the parasites, disruption of DNA, protein synthesis suppression, disruption of the cell membrane, and affecting electron transport.[39] When exposed to AgNPs, *Leishmania tropica* promastigotes were observed to have impaired metabolism, lower survival rates, and inhibited proliferation.[123] AgNPs impaired the mitochondrial membrane of the *T. gondii* parasite, greatly inhibited the infection, and there was no impact on the host cells.[124] The capacity of AgNPs to bind sulfur- and phosphorus-containing groups, and also their large surface areas and small sizes, were expected to enhance their antileishmanial effects. It was reported that AgNPs reduced parasite metabolic activity and proliferation rate.[125]

6.5.2 Mechanism of Gold Nanoparticles Against Microbial Infections

6.5.2.1 Mechanism of Gold Nanoparticles Against Bacterial Infections

AuNPs are believed to be effective in the production of antibacterial drugs because of their biocompatibility, high ability to functionalize, polyvalent effects, simplicity of detection, and photothermal activities. The antibacterial activity of AuNPs is based on their ability to penetrate cell walls.[126] The integrity of bacterial membranes could be damaged and bacterial cells could become toxic when exposed to cationic hydrophobic AuNPs. They showed a mild toxic effect on mammalian cells.[87] Because of the differences in the structure of the bacterial membrane, the antibacterial efficacy of AuNPs differs when applied to Gram-positive and Gram-negative bacteria, indicating the necessity for a higher dose of NPs. The antibacterial properties of AuNPs have been reported to be caused by decreased ATP synthase activity, which disrupts the metabolic process, and a decline in the ribosomal component responsible for tRNA binding, which causes the biological mechanism to collapse.[127] The increased surface area improved the interaction of the NPs with the bacteria. The AuNPs interfered with the bacterial cytoplasm and protein, leading to cell death.[128] Figure 6.3 shows some of the antibacterial mechanisms of AuNPs.

6.5.2.2 Mechanism of Gold Nanoparticles Against Viral Infections

AuNPs have been reported to exhibit antiviral activity against several viruses. Because of their small size, the AuNPs can enter the cell through the

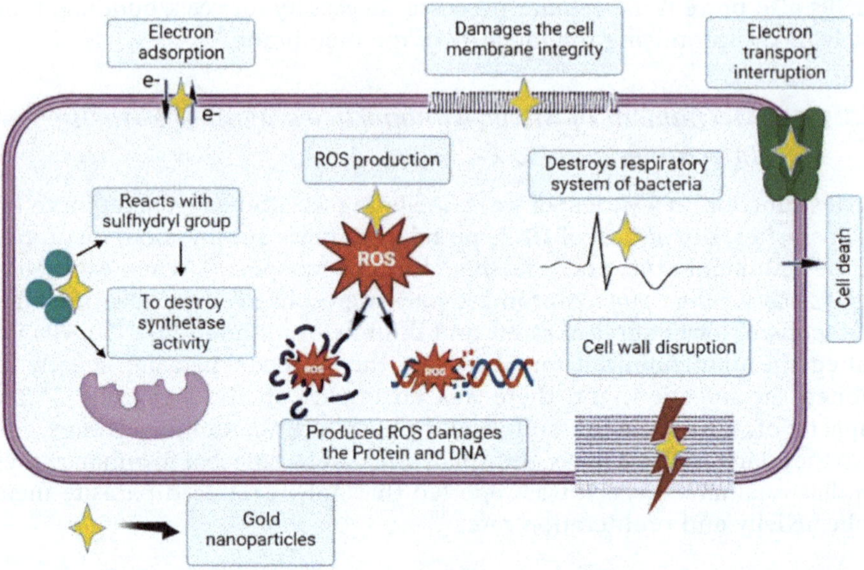

Figure 6.3 Antibacterial mechanisms of gold nanoparticles.

endosome vesicle and may even interfere with the fusion process. Numerous investigations revealed low or no cytotoxicity for AuNPs, and any toxicity was related to the size, shape, and surface charge of the NPs.[129] AuNPs interact with surface proteins of the influenza A virus and cleave their disulfide bonds. This causes inactivation of the proteins, which leads to inhibition of infection. Anionic AuNPs showed inhibition of the attachment of the influenza virus to the host cell surface. It was demonstrated that the antiviral activity of the NPs is significantly influenced by charge density and functional groups.[130] AuNPs coated with multivalent ligands can be used as an anti-HIV agent to target the process of fusion/adsorption in the virus infection.[13] At the early stages of infection and low toxic concentrations, AuNPs inhibited attachment and penetration of the foot and mouth disease virus (FMDV), and also hindered the replication of the virus at later stages.[131]

6.5.2.3 *Mechanism of Gold Nanoparticles Against Fungal Infections*

It has been reported that the antifungal effects of NPs are correlated with their particle size: the smaller the particle, the more potent is the antifungal activity. This feature of NPs can explain their higher/increased interaction with target sites on plasma membranes due to their larger surface area, and the cell membrane can also be easily penetrated by NPs with smaller particle sizes. Sulfur- and phosphorus-containing soft bases react strongly with gold (soft acid). In consequence, AuNPs may interact with DNA bases that contain phosphorus or membrane proteins that contain sulfur. By disrupting normal DNA functions such as repair mechanisms, replication, and synthesis of new strands, these interactions could ultimately result in cell death. In order to understand fully the mechanism underlying the antifungal activity of NPs, the process must be further elaborated.[132] When AuNPs interact with fungal enzymes (ATPases), they may alter their typical conformations, causing them to lose their function and finally cell death occurs.[101]

6.5.2.4 Mechanism of Gold Nanoparticles Against Parasitic Infections

The mechanism of AuNPs against parasites involves increasing the cell wall permeability, which results in leakage of the cell contents and death of the cells. Furthermore, NPs alter the expression of genes associated with parasite cell division.[133] In helminth parasitic infections, AuNPs affect the physiological functioning of the helminth worms, resulting in paralysis and subsequent death.[134] In the most probable mechanism of action of AuNPs against *L. donovani* promastigotes and amastigotes, the surface area of AuNPs in interaction with promastigotes and amastigotes of *L. donovani* may have some effect on the membrane or bind to proteins or DNA, rendering them ineffective.[80] Owing to the hydrophobic and unsaturated nature of membrane lipids, which makes them particularly sensitive to peroxidation, AuNPs promote lipid peroxidation. As a result, lipid free radicals are produced, which ultimately cause damage to the cell membrane of filarial parasites. ROS often induce lipid peroxidation and, consequently, ROS produced during AuNP treatment may be the primary inducer of lipid peroxidation.[135]

6.6 Conclusion

Owing to newly emerging and re-emerging diseases, multidrug resistance, and chronic conditions that appear due to infectious pathogens, it is necessary to discover new antimicrobial drugs or modify existing drugs. NPs target the precise and difficult-to-reach sites where pathogens are harboured, offering promising treatments for infectious diseases. Additionally, even in the case of resistant bacterial strains, they improve the effectiveness of drugs. More studies are still required to understand fully the mechanism of action of NPs as antimicrobial agents, the toxicity of NPs in humans, and the improvement of drug delivery employing nanodrug carriers.

Acknowledgements

We would like to express our sincere thanks to Saveetha Dental College and Hospital, SIMATS, for encouragement and constant support.

References

1. S. Agrebi and A. Larbi, Use of artificial intelligence in infectious diseases, in *Artificial intelligence in precision health*, Academic Press, 2020, pp. 415–438.
2. A. S. Fauci, Infectious diseases: considerations for the 21st century, *Clin. Infect. Dis.*, 2001, **32**(5), 675–685.
3. B. A. Aderibigbe, Metal-based nanoparticles for the treatment of infectious diseases, *Molecules*, 2017, **22**(8), 1370.
4. J. Fierer, D. Looney, M. Kok and J. C. Pechère, Nature and pathogenicity of micro-organisms, in *Infectious Diseases*, 2010, p. 3.
5. R. A. Harvey, *Microbiology*, Lippincott Williams & Wilkins, 2007.
6. D. Chinemerem Nwobodo, M. C. Ugwu, C. Oliseloke Anie, M. T. Al-Ouqaili, J. Chinedu Ikem, U. Victor Chigozie and M. Saki, Antibiotic resistance: The challenges and some emerging strategies for tackling a global menace, *J. Clin. Lab. Anal.*, 2022, **36**(9), e24655.
7. R. Giurazza, M. C. Mazza, R. Andini, P. Sansone, M. C. Pace and E. Durante-Mangoni, Emerging treatment options for multi-drug-resistant bacterial infections, *Life*, 2021, **11**(6), 519.
8. M. Benelmekki, Introduction to nanoparticles and nanotechnology, in *Designing Hybrid Nanoparticles*, IOP Publishing, 2nd edn, 2021.
9. P. Wadhwa, S. Sharma, S. Sahu, A. Sharma and D. Kumar, A Review of Nanoparticles Characterization Techniques, *Curr. Nanomater.*, 2022, **7**(3), 202–214.
10. N. Y. Lee, W. C. Ko and P. R. Hsueh, Nanoparticles in the treatment of infections caused by multidrug-resistant organisms, *Front. Pharmacol.*, 2019, **10**, 1153.
11. H. Zazo, C. I. Colino and J. M. Lanao, Current applications of nanoparticles in infectious diseases, *J. Controlled Release*, 2016, **224**, 86–102.
12. A. Elsaesser and C. V. Howard, Toxicology of nanoparticles, *Adv. Drug Delivery Rev.*, 2012, **64**(2), 129–137.
13. A. Khezerlou, M. Alizadeh-Sani, M. Azizi-Lalabadi and A. Ehsani, Nanoparticles and their antimicrobial properties against pathogens including bacteria, fungi, parasites and viruses, *Microb. Pathog.*, 2018, **123**, 505–526.
14. N. Padmavathy and R. Vijayaraghavan, Interaction of ZnO nanoparticles with microbes—a physio and biochemical assay, *J. Biomed. Nanotechnol.*, 2011, 7(6), 813–822.
15. S. Sharmin, M. M. Rahaman, C. Sarkar, O. Atolani, M. T. Islam and O. S. Adeyemi, Nanoparticles as antimicrobial and antiviral agents: A literature-based perspective study, *Heliyon*, 2021, **7**(3), e06456.
16. J. T. Seil and T. J. Webster, Antimicrobial applications of nanotechnology: methods and literature, *Int. J. Nanomed.*, 2012, **7**, 2767.

17. S. Gurunathan, M. Qasim, Y. Choi, J. T. Do, C. Park, K. Hong, J. H. Kim and H. Song, Antiviral potential of nanoparticles—can nanoparticles fight against coronaviruses?, *Nanomaterials*, 2020, **10**(9), 1645.
18. J. W. Peterson, Bacterial pathogenesis, in *Medical Microbiology*, 4th edn, 1996.
19. J. Chin, *Control of communicable diseases manual*, 2000.
20. A. R. Kirtane, M. Verma, P. Karandikar, J. Furin, R. Langer and G. Traverso, Nanotechnology approaches for global infectious diseases, *Nat. Nanotechnol.*, 2021, **16**(4), 369–384.
21. J. M. Van Seventer and N. S. Hochberg, Principles of infectious diseases: transmission, diagnosis, prevention, and control, in *International Encyclopedia of Public Health*, 2017, p. 22.
22. M. N. Alekshun and S. B. Levy, Molecular mechanisms of antibacterial multidrug resistance, *Cell*, 2007, **128**(6), 1037–1050.
23. E. Briones, C. I. Colino and J. M. Lanao, Delivery systems to increase the selectivity of antibiotics in phagocytic cells, *J. Controlled Release*, 2008, **125**(3), 210–227.
24. A. Mukhopadhyay and S. K. Basu, Intracellular delivery of drugs to macrophages, in *Biotechnology in India I*, 2003, pp. 183–209.
25. O. V. Salata, Applications of nanoparticles in biology and medicine, *J. Nanobiotechnol.*, 2004, **2**(1), 1–6.
26. K. James, J. Highsmith and P. Evers, Nanotechnology market–nanotechnology markets in healthcare & medicine, *Drug Dev. Delivery*, 2014, **11**(12), 43–45.
27. A. Tewabe, A. Abate, M. Tamrie, A. Seyfu and E. A. Siraj, Targeted drug delivery—from magic bullet to nanomedicine: principles, challenges, and future perspectives, *J. Multidiscip. Healthcare*, 2021, **14**, 1711.
28. H. Hillaireau and P. Couvreur, Nanocarriers' entry into the cell: relevance to drug delivery, *Cell. Mol. Life Sci.*, 2009, **66**(17), 2873–2896.
29. L. Wang, C. Hu and L. Shao, The antimicrobial activity of nanoparticles: present situation and prospects for the future, *Int. J. Nanomed.*, 2017, **12**, 1227.
30. D. Franco, G. Calabrese, S. P. Guglielmino and S. Conoci, Metal-based nanoparticles: Antibacterial mechanisms and biomedical application, *Microorganisms.*, 2022, **10**(9), 1778.
31. J. T. Huerta, J. M. Bermudez, D. A. Quinteros, D. A. Allemandi and S. D. Palma, New trends, challenges, and opportunities in the use of nanotechnology in restorative dentistry, in *Nanobiomaterials in Dentistry*, 2016, pp. 133–160.
32. H. Hariharan, N. Al-Harbi, P. Karuppiah and S. Rajaram, Microbial synthesis of selenium nanocomposite using Saccharomyces cerevisiae and its antimicrobial activity against pathogens causing nosocomial infection, *Chalcogenide Lett.*, 2012, **9**(12), 509–515.
33. M. C. Stensberg, Q. Wei, E. S. McLamore, D. M. Porterfield, A. Wei and M. S. Sepúlveda, Toxicological studies on silver nanoparticles:

challenges and opportunities in assessment, monitoring and imaging, *Nanomedicine*, 2011, **6**(5), 879–898.

34. S. Galdiero, A. Falanga, M. Vitiello, M. Cantisani, V. Marra and M. Galdiero, Silver nanoparticles as potential antiviral agents, *Molecules*, 2011, **16**(10), 8894–8918.
35. K. Maduray and R. Parboosing, Metal nanoparticles: a promising treatment for viral and arboviral infections, *Biol. Trace Elem. Res.*, 2021, **199**(8), 3159–3176.
36. A. Ogar, G. Tylko and K. Turnau, Antifungal properties of silver nanoparticles against indoor mould growth, *Sci. Total Environ.*, 2015, **521**, 305–314.
37. A. Brandelli, A. C. Ritter and F. F. Veras, Antimicrobial activities of metal nanoparticles, in *Metal nanoparticles in pharma*, Springer, Cham, 2017, pp. 337–363.
38. M. A. Alghuthaymi, H. Almoammar, M. Rai, E. Said-Galiev and K. A. Abd-Elsalam, Myconanoparticles: synthesis and their role in phytopathogens management, *Biotechnol. Biotechnol. Equip.*, 2015, **29**(2), 221–236.
39. H. U. Bajwa, M. K. Khan, Z. Abbas, R. Riaz, T. U. Rehman, R. Z. Abbas, M. T. Aleem, A. Abbas, M. M. Almutairi, F. A. Alshammari and Y. Alraey, Nanoparticles: Synthesis and Their Role as Potential Drug Candidates for the Treatment of Parasitic Diseases, *Life*, 2022, **12**(5), 750.
40. C. Graf, D. L. Vossen, A. Imhof and A. van Blaaderen, A general method to coat colloidal particles with silica, *Langmuir*, 2003, **19**(17), 6693–6700.
41. S. A. Ahmad, S. S. Das, A. Khatoon, M. T. Ansari, M. Afzal, M. S. Hasnain and A. K. Nayak, Bactericidal activity of silver nanoparticles: A mechanistic review, *Mater. Sci. Energy Technol.*, 2020, **3**, 756–769.
42. J. Elliott, A. Mackie, C. Dibden, A. Blumsohn and F. Creagh, Oral quinine-induced hypoglycaemic seizures, *Pract. Diabetes Int.*, 2010, **27**(1), 32–33.
43. I. Sondi and B. Salopek-Sondi, Silver nanoparticles as antimicrobial agent: a case study on E. coli as a model for Gram-negative bacteria, *J. Colloid Interface Sci.*, 2004, **275**(1), 177–182.
44. S. K. Kailasa, T. J. Park, J. V. Rohit and J. R. Koduru, Antimicrobial activity of silver nanoparticles, in *Nanoparticles in pharmacotherapy*, William Andrew Publishing, 2019, pp. 461–484.
45. R. Holladay, W. Moeller, D. Mehta, J. H. Brooks, R. Roy and M. Mortenson, Inventors; American Silver Llc, assignee, Silver/water, silver gels and silver based compositions, and methods for making and using the same, 2008.
46. I. X. Yin, J. Zhang, I. S. Zhao, M. L. Mei, Q. Li and C. H. Chu, The antibacterial mechanism of silver nanoparticles and its application in dentistry, *Int. J. Nanomed.*, 2020, 2555–2562.
47. A. K. Keshari, R. Srivastava, P. Singh, V. B. Yadav and G. Nath, Antioxidant and antibacterial activity of silver nanoparticles synthesized by Cestrum nocturnum, *J. Ayurveda Integr. Med.*, 2020, **11**(1), 37–44.

48. N. Jayaprakash, J. J. Vijaya, L. J. Kennedy, K. Priadharsini and P. Palani, Antibacterial activity of silver nanoparticles synthesized from serine, *Mater. Sci. Eng., C*, 2015, **49**, 316–322.
49. P. S. Ramesh, T. Kokila and D. Geetha, Plant mediated green synthesis and antibacterial activity of silver nanoparticles using Emblica officinalis fruit extract, *Spectrochim. Acta, Part A*, 2015, **142**, 339–343.
50. R. B. Malabadi, G. S. Mulgund, N. T. Meti, K. Nataraja and S. V. Kumar, Antibacterial activity of silver nanoparticles synthesized by using whole plant extracts of Clitoria ternatea, *Res. Pharm.*, 2015, **2**(4), 11–21.
51. V. L. Praba, M. Kathirvel, K. Vallayyachari, K. Surendar, M. Muthuraj, P. J. Jesuraj, S. Govindarajan and K. V. Raman, Bactericidal effect of silver nanoparticles against Mycobacterium tuberculosis, *J. Bionanosci.*, 2013, **7**(3), 282–287.
52. S. Ravikumar, R. Gokulakrishnan and P. Boomi, In vitro antibacterial activity of the metal oxide nanoparticles against urinary tract infectious bacterial pathogens, *Asian Pac. J. Trop. Dis.*, 2012, **2**(2), 85–89.
53. M. Jeevitha and S. Rajeshkumar, Antimicrobial activity of silver nanoparticles synthesized using marine brown seaweed Spatoglossum asperum against oral pathogens, *Indian J. Public Health*, 2019, **10**(11), 3569.
54. S. Rajeshkumar and C. Malarkodi, In vitro antibacterial activity and mechanism of silver nanoparticles against foodborne pathogens, *Bioinorg. Chem. Appl.*, 2014, **2014**, 01555.
55. G. Wan, L. Ruan, Y. Yin, T. Yang, M. Ge and X. Cheng, Effects of silver nanoparticles in combination with antibiotics on the resistant bacteria Acinetobacter baumannii, *Int. J. Nanomed.*, 2016, **11**, 3789.
56. R. Singh, J. Vora, S. B. Nadhe, S. A. Wadhwani, U. U. Shedbalkar and B. A. Chopade, Antibacterial activities of bacteriagenic silver nanoparticles against nosocomial Acinetobacter baumannii, *J. Nanosci. Nanotechnol.*, 2018, **18**(6), 3806–3815.
57. F. Ameen, M. M. Abdullah, A. A. Al-Homaidan, H. A. Al-Lohedan, A. A. Al-Ghanayem and A. Almansob, Fabrication of silver nanoparticles employing the cyanobacterium Spirulina platensis and its bactericidal effect against opportunistic nosocomial pathogens of the respiratory tract, *J. Mol. Struct.*, 2020, **1217**, 128392.
58. A. Salleh, R. Naomi, N. D. Utami, A. W. Mohammad, E. Mahmoudi, N. Mustafa and M. B. Fauzi, The potential of silver nanoparticles for antiviral and antibacterial applications: A mechanism of action, *Nanomaterials*, 2020, **10**(8), 1566.
59. P. Orlowski, E. Tomaszewska, M. Gniadek, P. Baska, J. Nowakowska, J. Sokolowska, Z. Nowak, M. Donten, G. Celichowski, J. Grobelny and M. Krzyzowska, Tannic acid modified silver nanoparticles show antiviral activity in herpes simplex virus type 2 infection, *PLoS One*, 2014, **9**(8), e104113.
60. D. Baram-Pinto, S. Shukla, A. Gedanken and R. Sarid, Inhibition of HSV-1 attachment, entry, and cell-to-cell spread by functionalized multivalent AuNPs, *Small*, 2010, **6**(9), 1044–1050.

61. L. Lu, R. W. Sun, R. Chen, C. K. Hui, C. M. Ho, J. M. Luk, G. K. Lau and C. M. Che, Silver nanoparticles inhibit hepatitis B virus replication, *Antiviral Ther.*, 2008, **13**(2), 253–262.
62. L. Sun, A. K. Singh, K. Vig, S. R. Pillai and S. R. Singh, Silver nanoparticles inhibit replication of respiratory syncytial virus, *J. Biomed. Nanotechnol.*, 2008, **4**(2), 149–158.
63. D. Xiang, Y. Zheng, W. Duan, X. Li, J. Yin, S. Shigdar, M. L. O'Connor, M. Marappan, X. Zhao, Y. Miao and B. Xiang, Inhibition of A/Human/Hubei/3/2005 (H3N2) influenza virus infection by silver nanoparticles in vitro and in vivo, *Int. J. Nanomed.*, 2013, **8**, 4103.
64. S. Park, H. H. Park, S. Y. Kim, S. J. Kim, K. Woo and G. Ko, Antiviral properties of silver nanoparticles on a magnetic hybrid colloid, *Appl. Environ. Microbiol.*, 2014, **80**(8), 2343–2350.
65. J. L. Speshock, R. C. Murdock, L. K. Braydich-Stolle, A. M. Schrand and S. M. Hussain, Interaction of silver nanoparticles with Tacaribe virus, *J. Nanobiotechnol.*, 2010, **8**(1), 1–9.
66. N. Chen, Y. Zheng, J. Yin, X. Li and C. Zheng, Inhibitory effects of silver nanoparticles against adenovirus type 3 in vitro, *J. Virol. Methods*, 2013, **193**(2), 470–477.
67. J. V. Rogers, C. V. Parkinson, Y. W. Choi, J. L. Speshock and S. M. Hussain, A preliminary assessment of silver nanoparticle inhibition of monkeypox virus plaque formation, *Nanoscale Res. Lett.*, 2008, **3**(4), 129–133.
68. T. C. Dakal, A. Kumar, R. S. Majumdar and V. Yadav, Mechanistic basis of antimicrobial actions of silver nanoparticles, *Front. Microbiol.*, 2016, **7**, 1831.
69. N. Khatoon, A. Mishra, H. Alam, N. Manzoor and M. Sardar, Biosynthesis, characterization, and antifungal activity of the silver nanoparticles against pathogenic Candida species, *Bionanoscience*, 2015, **5**(2), 65–74.
70. Y. Xu, C. Gao, X. Li, Y. He, L. Zhou, G. Pang and S. Sun, In vitro antifungal activity of silver nanoparticles against ocular pathogenic filamentous fungi, *J. Ocul. Pharmacol. Ther.*, 2013, **29**(2), 270–274.
71. K. P. Bocate, G. F. Reis, P. C. de Souza, A. G. Junior, N. Durán, G. Nakazato, M. C. Furlaneto, R. S. de Almeida and L. A. Panagio, Antifungal activity of silver nanoparticles and simvastatin against toxigenic species of Aspergillus, *Int. J. Food Microbiol.*, 2019, **291**, 79–86.
72. D. R. Monteiro, S. Silva, M. Negri, L. F. Gorup, E. R. de Camargo, R. Oliveira, D. B. Barbosa and M. Henriques, Antifungal activity of silver nanoparticles in combination with nystatin and chlorhexidine digluconate against C andida albicans and C andida glabrata biofilms, *Mycoses*, 2013, **56**(6), 672–680.
73. M. T. Yilmaz, H. İspirli, O. Taylan and E. Dertli, Synthesis and characterisation of alternan-stabilised silver nanoparticles and determination of their antibacterial and antifungal activities against foodborne pathogens and fungi, *LWT*, 2020, **128**, 109497.

74. A. M. Allahverdiyev, E. S. Abamor, M. Bagirova, C. B. Ustundag, C. Kaya, F. Kaya and M. Rafailovich, Antileishmanial effect of silver nanoparticles and their enhanced antiparasitic activity under ultraviolet light, *Int. J. Nanomed.*, 2011, **6**, 2705.
75. G. Rajakumar and A. A. Rahuman, Larvicidal activity of synthesized silver nanoparticles using Eclipta prostrata leaf extract against filariasis and malaria vectors, *Acta Trop.*, 2011, **118**(3), 196–203.
76. T. Santhoshkumar, A. A. Rahuman, G. Rajakumar, S. Marimuthu, A. Bagavan, C. Jayaseelan, A. A. Zahir, G. Elango and C. Kamaraj, Synthesis of silver nanoparticles using Nelumbo nucifera leaf extract and its larvicidal activity against malaria and filariasis vectors, *Parasitol. Res.*, 2011, **108**(3), 693–702.
77. A. O. Souza, J. W. Oliveira, C. J. Moreno, M. J. de Medeiros, M. M. Fernandes-Negreiros, F. R. Souza, D. L. Pontes, M. S. Silva and H. A. Rocha, Silver nanoparticles containing fucoidan synthesized by green method have anti-Trypanosoma cruzi activity, *Nanomaterials*, 2022, **12**(12), 2059.
78. R. A. Alajmi, W. A. Al-Megrin, D. Metwally, H. Al-Subaie, N. Altamrah, A. M. Barakat, A. E. Abdel Moneim, T. T. Al-Otaibi and M. El-Khadragy, Anti-Toxoplasma activity of silver nanoparticles green synthesized with Phoenix dactylifera and Ziziphus spina-christi extracts which inhibits inflammation through liver regulation of cytokines in Balb/c mice, *Biosci. Rep.*, 2019, **39**(5), BSR20190379.
79. C. Panneerselvam, S. Ponarulselvam and K. Murugan, Potential antiplasmodial activity of synthesized silver nanoparticle using Andrographis paniculata Nees (Acanthaceae), *Arch. Appl. Sci. Res.*, 2011, **3**(6), 208–217.
80. M. Y. Want, P. Yadav, R. Khan, G. Chouhan, M. Islamuddin, S. Y. Aloyouni, A. P. Chattopadhyay, S. Y. AlOmar and F. Afrin, Critical antileishmanial in vitro effects of highly examined gold nanoparticles, *Int. J. Nanomed.*, 2021, **16**, 7285.
81. A. Sengupta, M. Azharuddin, N. Al-Otaibi and J. Hinkula, Efficacy and Immune Response Elicited by Gold Nanoparticle-Based Nanovaccines against Infectious Diseases, *Vaccines*, 2022, **10**(4), 505.
82. M. L. Personick, M. R. Langille, J. Zhang and C. A. Mirkin, Shape control of gold nanoparticles by silver underpotential deposition, *Nano Lett.*, 2011, **11**(8), 3394–3398.
83. C. Su, K. Huang, H. H. Li, Y. G. Lu and D. L. Zheng, Antibacterial properties of functionalized gold nanoparticles and their application in oral biology, *J. Nanomater.*, 2020, **2020**, 1–3.
84. E. O. Mikhailova, Gold nanoparticles: biosynthesis and potential of biomedical application, *J. Funct. Biomater.*, 2021, **12**(4), 70.
85. M. R. Bindhu and M. Umadevi, Antibacterial activities of green synthesized gold nanoparticles, *Mater. Lett.*, 2014, **120**, 122–125.
86. N. Abdel-Raouf, N. M. Al-Enazi and I. B. Ibraheem, Green biosynthesis of gold nanoparticles using Galaxaura elongata and characterization of their antibacterial activity, *Arabian J. Chem.*, 2017, **10**, S3029–S3039.

87. X. Li, S. M. Robinson, A. Gupta, K. Saha, Z. Jiang, D. F. Moyano, A. Sahar, M. A. Riley and V. M. Rotello, Functional gold nanoparticles as potent antimicrobial agents against multi-drug-resistant bacteria, *ACS Nano*, 2014, **8**(10), 10682–10686.
88. S. Shamaila, N. Zafar, S. Riaz, R. Sharif, J. Nazir and S. Naseem, Gold nanoparticles: an efficient antimicrobial agent against enteric bacterial human pathogen, *Nanomaterials*, 2016, **6**(4), 71.
89. B. Saha, J. Bhattacharya, A. Mukherjee, A. Ghosh, C. Santra, A. K. Dasgupta and P. Karmakar, In vitro structural and functional evaluation of gold nanoparticles conjugated antibiotics, *Nanoscale Res. Lett.*, 2007, **2**(12), 614–622.
90. A. Rai, A. Prabhune and C. C. Perry, Antibiotic mediated synthesis of gold nanoparticles with potent antimicrobial activity and their application in antimicrobial coatings, *J. Mater. Chem.*, 2010, **20**(32), 6789–6798.
91. T. S. Dasari, Y. Zhang and H. Yu, Antibacterial activity and cytotoxicity of gold (I) and (III) ions and gold nanoparticles, *Biochem. Pharmacol.: Open Access*, 2015, **4**(6), 199.
92. M. Składanowski, M. Wypij, D. Laskowski, P. Golińska, H. Dahm and M. Rai, Silver and gold nanoparticles synthesized from Streptomyces sp. isolated from acid forest soil with special reference to its antibacterial activity against pathogens, *J. Cluster Sci.*, 2017, **28**, 59–79.
93. P. Seetharaman, R. Chandrasekaran, S. Gnanasekar, I. Mani and S. Sivaperumal, Biogenic gold nanoparticles synthesized using Crescentia cujete L. and evaluation of their different biological activities, *Biocatal. Agric. Biotechnol.*, 2017, **11**, 75–82.
94. S. Vijayakumar and S. Ganesan, Gold nanoparticles as an HIV entry inhibitor, *Curr. HIV Res.*, 2012, **10**(8), 643–646.
95. I. Papp, C. Sieben, K. Ludwig, M. Roskamp, C. Böttcher, S. Schlecht, A. Herrmann and R. Haag, Inhibition of influenza virus infection by multivalent sialic-acid-functionalized gold nanoparticles, *Small*, 2010, **6**(24), 2900–2906.
96. A. Halder, S. Das, D. Ojha, D. Chattopadhyay and A. Mukherjee, Highly monodispersed gold nanoparticles synthesis and inhibition of herpes simplex virus infections, *Mater. Sci. Eng., C*, 2018, **89**, 413–421.
97. E. M. Ahmed, S. M. Solyman, N. Mohamed, A. A. Boseila and A. Hanora, Antiviral activity of Ribavirin nano-particles against measles virus, *Cell. Mol. Biol.*, 2018, **64**(9), 24–32.
98. R. Kesarkar, G. Oza, S. Pandey, R. Dahake, S. Mukherjee, A. Chowdhary and M. Sharon, Gold nanoparticles: effective as both entry inhibitors and virus neutralizing agents against HIV, *J. Microbiol. Biotechnol. Res.*, 2012, **2**(2), 276–283.
99. A. R. Bastian, A. Nangarlia, L. D. Bailey, A. Holmes, R. V. Sundaram, C. Ang, D. R. Moreira, K. Freedman, C. Duffy, M. Contarino and C. Abrams, Mechanism of multivalent nanoparticle encounter with HIV-1 for potency enhancement of peptide triazole virus inactivation, *J. Biol. Chem.*, 2015, **290**(1), 529–543.

100. I. A. Wani and T. Ahmad, Size and shape dependant antifungal activity of gold nanoparticles: a case study of Candida, *Colloids Surf., B*, 2013, **101**, 162–170.
101. T. Ahmad, I. A. Wani, I. H. Lone, A. Ganguly, N. Manzoor, A. Ahmad, J. Ahmed and A. S. Al-Shihri, Antifungal activity of gold nanoparticles prepared by solvothermal method, *Mater. Res. Bull.*, 2013, **48**(1), 12–20.
102. C. Jayaseelan, R. Ramkumar, A. A. Rahuman and P. Perumal, Green synthesis of gold nanoparticles using seed aqueous extract of Abelmoschus esculentus and its antifungal activity, *Ind. Crops Prod.*, 2013, **45**, 423–429.
103. A. Sami, Antifungal effect of gold nanoparticles on fungi isolated from onychomycosis patients, *Al-Azhar J. Pharm. Sci.*, 2019, **60**(2), 26–42.
104. A. Rónavári, N. Igaz, M. K. Gopisetty, B. Szerencsés, D. Kovács, C. Papp, C. Vágvölgyi, I. M. Boros, Z. Kónya, M. Kiricsi and I. Pfeiffer, Biosynthesized silver and gold nanoparticles are potent antimycotics against opportunistic pathogenic yeasts and dermatophytes, *Int. J. Nanomed.*, 2018, **13**, 695.
105. R. R. Thanighaiarassu, P. Sivamai, R. Devika and B. Nambikkairaj, Green synthesis of gold nanoparticles characterization by using plant essential oil menthapiperita and their antifungal activity against human pathogenic fungi,*J. Nanomed. Nanotechnol.*, 2014, **5**(5), 229–234.
106. A. Folorunso, S. Akintelu, A. K. Oyebamiji, S. Ajayi, B. Abiola, I. Abdusalam and A. Morakinyo, Biosynthesis, characterization and antimicrobial activity of gold nanoparticles from leaf extracts of Annona muricata, *J. Nanostruct. Chem.*, 2019, **9**(2), 111–117.
107. L. Karthik, G. Kumar, T. Keswani, A. Bhattacharyya, B. P. Reddy and K. B. Rao, Marine actinobacterial mediated gold nanoparticles synthesis and their antimalarial activity, *Nanomedicine*, 2013, **9**(7), 951–960.
108. P. K. Kar, S. Murmu, S. Saha, V. Tandon and K. Acharya, Anthelmintic efficacy of gold nanoparticles derived from a phytopathogenic fungus, Nigrospora oryzae, *PLoS One*, 2014, **9**(1), e84693.
109. S. Das, P. Roy, S. Mondal, T. Bera and A. Mukherjee, One pot synthesis of gold nanoparticles and application in chemotherapy of wild and resistant type visceral leishmaniasis, *Colloids Surf., B*, 2013, **107**, 27–34.
110. D. Pissuwan, S. M. Valenzuela, C. M. Miller, M. C. Killingsworth and M. B. Cortie, Destruction and control of Toxoplasma gondii tachyzoites using gold nanosphere/antibody conjugates, *Small*, 2009, **5**(9), 1030–1034.
111. P. Roy, S. K. Saha, P. Gayen, P. Chowdhury and S. P. Babu, Exploration of antifilarial activity of gold nanoparticle against human and bovine filarial parasites: A nanomedicinal mechanistic approach, *Colloids Surf., B*, 2018, **161**, 236–243.
112. O. Akhavan and E. Ghaderi, Cu and CuO nanoparticles immobilized by silica thin films as antibacterial materials and photocatalysts, *Surf. Coat. Technol.*, 2010, **205**(1), 219–223.

113. O. Choi, C. P. Yu, G. E. Fernández and Z. Hu, Interactions of nanosilver with Escherichia coli cells in planktonic and biofilm cultures, *Water Res.*, 2010, **44**(20), 6095–6103.
114. S. S. Abdalla, H. Katas, J. Y. Chan, P. Ganasan, F. Azmi and M. F. Busra, Antimicrobial activity of multifaceted lactoferrin or graphene oxide functionalized silver nanocomposites biosynthesized using mushroom waste and chitosan, *RSC Adv.*, 2020, **10**(9), 4969–4983.
115. R. B. Wakshlak, R. Pedahzur and D. Avnir, Antibacterial activity of silver-killed bacteria: the "zombies" effect, *Sci. Rep.*, 2015, **5**(1), 1–5.
116. D. Morris, M. Ansar, J. Speshock, T. Ivanciuc, Y. Qu, A. Casola and R. P. Garofalo, Antiviral and immunomodulatory activity of silver nanoparticles in experimental RSV infection, *Viruses*, 2019, **11**(8), 732.
117. N. Khandelwal, G. Kaur, N. Kumar and A. Tiwari, Application of silver nanoparticles in viral inhibition: a new hope for antivirals, *Dig. J. Nanomater. Nanomater.*, 2014, **9**(1), 117.
118. Z. A. Ratan, F. R. Mashrur, A. P. Chhoan, S. M. Shahriar, M. F. Haidere, N. J. Runa, S. Kim, D. H. Kweon, H. Hosseinzadeh and J. Y. Cho, Silver nanoparticles as potential antiviral agents, *Pharmaceutics*, 2021, **13**(12), 2034.
119. A. Ahmad, Y. Wei, F. Syed, K. Tahir, R. Taj, A. U. Khan, M. U. Hameed and Q. Yuan, Amphotericin B-conjugated biogenic silver nanoparticles as an innovative strategy for fungal infections, *Microb. Pathog.*, 2016, **99**, 271–281.
120. K. J. Kim, W. S. Sung, B. K. Suh, S. K. Moon, J. S. Choi, J. G. Kim and D. G. Lee, Antifungal activity and mode of action of silver nanoparticles on Candida albicans, *Biometals*, 2009, **22**(2), 235–242.
121. B. Dorau, R. Arango and F. Green, An investigation into the potential of ionic silver as a wood preservative, in *Proceedings of the 2nd wood-frame housing durability and disaster issues conference*, Forest Products Society, Madison, WI, USA, 2004, pp. 133–145.
122. M. Yamanaka, K. Hara and J. Kudo, Bactericidal actions of a silver ion solution on Escherichia coli, studied by energy-filtering transmission electron microscopy and proteomic analysis, *Appl. Environ. Microbiol.*, 2005, **71**(11), 7589–7593.
123. L. Abdellahi, F. Iraji, A. Mahmoudabadi and S. H. Hejazi, Vaccination in Leishmaniasis: A Review Article, *Iran. Biomed. J.*, 2022, **26**(1), 1.
124. E. Bertin, A. Münzer, S. Reichenberger, R. Streubel, T. Vinnay, H. Wiggers, C. Schulz, S. Barcikowski and G. Marzun, Durability study of platinum nanoparticles supported on gas-phase synthesized graphene in oxygen reduction reaction conditions, *Appl. Surf. Sci.*, 2019, **467**, 1181–1186.
125. A. M. Allahverdiyev, E. S. Abamor, M. Bagirova, C. B. Ustundag, C. Kaya, F. Kaya and M. Rafailovich, Antileishmanial effect of silver nanoparticles and their enhanced antiparasitic activity under ultraviolet light, *Int. J. Nanomed.*, 2011, **6**, 2705.
126. M. M. Mohamed, S. A. Fouad, H. A. Elshoky, G. M. Mohammed and T. A. Salaheldin, Antibacterial effect of gold nanoparticles against

Corynebacterium pseudotuberculosis, *Int. J. Vet. Sci. Med.*, 2017, **5**(1), 23–29.

127. E. Lima, R. Guerra, V. Lara and A. Guzmán, Gold nanoparticles as efficient antimicrobial agents for Escherichia coli and Salmonella typhi, *Chem. Cent. J.*, 2013, **7**(1), 1–7.
128. D. MubarakAli, N. Thajuddin, K. Jeganathan and M. Gunasekaran, Plant extract mediated synthesis of silver and gold nanoparticles and its antibacterial activity against clinically isolated pathogens, *Colloids Surf., B*, 2011, **85**(2), 360–365.
129. S. B. Yaqoob, R. Adnan, R. M. Rameez Khan and M. Rashid, Gold, silver, and palladium nanoparticles: a chemical tool for biomedical applications, *Front. Chem.*, 2020, **8**, 376.
130. J. Kim, M. Yeom, T. Lee, H. O. Kim, W. Na, A. Kang, J. W. Lim, G. Park, C. Park, D. Song and S. Haam, Porous gold nanoparticles for attenuating infectivity of influenza A virus, *J. Nanobiotechnol.*, 2020, **18**(1), 1–1.
131. S. Rafiei, S. E. Rezatofighi, M. R. Ardakani and S. Rastegarzadeh, Gold nanoparticles impair foot-and-mouth disease virus replication, *IEEE Trans. NanoBiosci.*, 2015, **15**(1), 34–40.
132. S. M. Dizaj, F. Lotfipour, M. Barzegar-Jalali, M. H. Zarrintan and K. Adibkia, Antimicrobial activity of the metals and metal oxide nanoparticles, *Mater. Sci. Eng., C*, 2014, **44**, 278–284.
133. M. Saleh, G. Kumar, A. A. Abdel-Baki, S. Al-Quraishy and M. El-Matbouli, In vitro antimicrosporidial activity of gold nanoparticles against Heterosporis saurida, *BMC Vet. Res.*, 2016, **12**(1), 1–6.
134. D. Pissuwan, C. Gazzana, S. Mongkolsuk and M. B. Cortie, Single and multiple detections of foodborne pathogens by gold nanoparticle assays, *Wiley Interdiscip. Rev.: Nanomed. Nanobiotechnol.*, 2020, **12**(1), e1584.
135. P. Roy, S. K. Saha, P. Gayen, P. Chowdhury and S. P. Babu, Exploration of antifilarial activity of gold nanoparticle against human and bovine filarial parasites: a nanomedicinal mechanistic approach, *Colloids Surf., B*, 2018, **161**, 236–243.

CHAPTER 7

Nanotoxicology and Its Remediation

KADAMBINI PRADHAN, LOKANATH MISHRA AND MONALISA MISHRA*

Neural Developmental Biology Laboratory, Department of Life Science, National Institute of Technology Rourkela, Rourkela, Odisha 769008, India
*Email: mishramo@nitrkl.ac.in

7.1 Introduction

"Nano" comes from the Greek word for "dwarf." With the advancement of nanotechnology, nanomaterials are increasingly being used in various fields of science. The properties of a nanomaterial often vary depending on the original size of the nanoparticles (NPs), and sometimes good properties are achieved but occasionally the worst properties are obtained with the same material. For example, titanium dioxide loses its color when converted to its nanoform. Large surface areas are achieved with a small size, which helps with the catalytic properties. The tiny size often results in unwanted reactions resulting in toxic chemicals or gases; as a consequence, nanomaterials are checked *in vitro*, *in vivo*, and *ex vivo* to detect their toxicity. The toxicity of nanomaterials leads to cytotoxicity, genotoxicity, and behavioral abnormalities, which has given rise to a new term, "nanotoxicity." Coating nanomaterials with natural organic matter may decrease toxicity. The hydrophobic nature of the organic matter may be responsible for the toxicity of a material.

Many nanomaterials are constantly released into the environment knowingly or unknowingly. These nanomaterials enter the body *via* air, water, and

Nanoscience & Nanotechnology Series No. 62
Smart Nanomaterials for Infectious Diseases
Edited by Suvardhan Kanchi, Nidhi Sharotri, Rajasekhar Chokkareddy, Deepali Sharma and Falah H. Hussein

Published by the Royal Society of Chemistry, www.rsc.org

the skin and cause cytotoxicity by inducing stress and inflammation. Once internalized, they mix with the blood and can cause many defects within the body.[1] Large NPs are easily eliminated from the body and thus may be less harmful than smaller particles.[2] Metallic NPs such as those containing Ag, Au, and Cu interact with proteins and enzymes and cause toxicity within the body. Often the NPs cause an oxygen imbalance and toxicity within the body.[3] Usually, the mitochondria are found to be damaged during this process. NPs used in sunscreen lotion accumulate in the stratum corneum of the epidermis over time. Later, these NPs enter the body and produce undesirable effects. The small NPs are not only chemically active but are also biologically active. The small size also helps to produce large amounts of reactive oxygen species (ROS) within the body. ROS are produced from almost all NPs, including polymeric NPs, as summarized by Mishra and Panda.[3] NP-induced ROS cause damage to cells, macromolecules, and DNA. The damage to macromolecules may affect several biochemical pathways and cause metabolic disorders in humans.

7.1.1 Factors Affecting Nanotoxicity

- The toxicity of a nanomaterial depends on the dose, route of administration, and location of accumulation.
- Properly synthesized NPs are capable of releasing a drug without affecting healthy cells.
- The biocompatibility of a drug further depends on size, structure, and composition.
- Metallic nanocarriers are toxic for some cell lines but may not be toxic to others.
- The half-life of an NP depends on the time of drug delivery, distance traveled, and carriers of cytotoxic properties.
- Biocompatibility depends on the stability in biological media.
- Nanotoxicity depends on the environment in which the NP will be dissolved and the pH of that environment.
- Although the activity of NPs is unknown, ROS are believed to be the key contributor to nanotoxicity.
- The effects of ROS are seen in a three-stage model: (a) enhanced antioxidant enzyme to defend the cell, (b) enhanced proinflammatory cytokine, and (c) mitochondrial perturbation leading to cell death.

7.2 Mechanism of Nanotoxicity

7.2.1 Toxicity by Oxidative Stress

With regard to cell contact, NPs show oxidative stress that may lead to toxicity (see Figure 7.1). Asymmetric oxidants and antioxidants cause these effects. Biochemical reactions such as respiration, phagocytosis, and

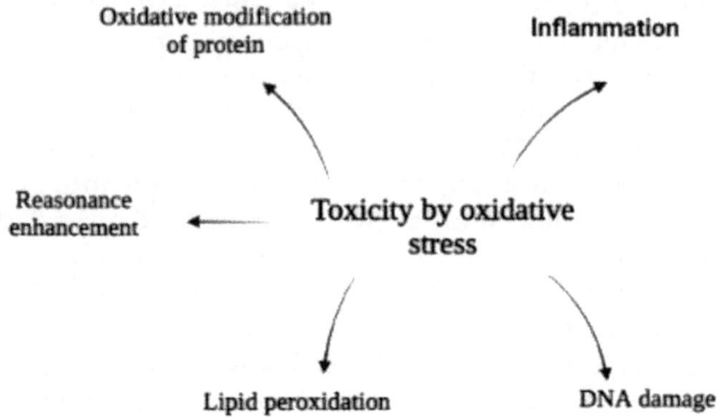

Figure 7.1 Different types of toxicity due to oxidative stress. Created with BioRender.com.

metabolism form ROS,[4] and the generation of ROS contributes to oxidative stress.[5] ROS adversely affect many important reactions, including electron transport chain impairment;[6] these ROS damage genetic material by cross-linking and breaking DNA. Inflammation also causes the formation of ROS.[7] When adenosine triphosphate (ATP) synthesis occurs in mitochondria, some oxygen remains unreduced and forms free radicals. These ROS are formed when NPs are deposited in the cells, leading to oxidative stress. Oxidative stress causes improper physiological function.[8] The chemical characteristics of the NPs determine the production level of ROS.[9] ROS-mediated cytotoxicity is caused by nano-CuO that releases lactate dehydrogenase and generates oxidative stress.[10]

Similarly, nano-ZnO, nano-TiO, and silica NPs also cause ROS-mediated cytotoxicity.[11–16] ROS-mediated apoptosis and genotoxicity are induced by nano-Ag through oxidative stress, and also a mutation in lymphoma cells found in mice. A comparison of the cytotoxicities of NPs was made in catfish and HepG2 cells of humans, and ROS-mediated oxidative stress that induces mitochondrial dysfunction, cancer, cytotoxicity, lipid peroxidation, and DNA damage was examined.[17]

The generation of ROS or oxidative stress was examined by treating with antioxidants and observing whether it reduces toxicity. Dröge *et al.*[18] did this with Bi_2O_3 NPs in MCF-7 cells by administering *N*-acetylcysteine as an antioxidant. Most of the NPs are dose dependent, and some also are time dependent. NPs that are both time and dose dependent, such as cilia, were observed to cause oxidative stress and endothelium dysfunction.[19] NPs used for immunoassays, drug delivery for cancer, and distinction in magnetic resonance imaging (MRI) were observed to cause oxidative stress and many more neurological disorders due to imbalance in NP-related ions.[20,21] Some antimicrobial drugs contain silver NPs (AgNPs) that cause cytotoxicity, leading to oxidative stress. Experiments with *Candida albicans* showed the change in morphology of cell environment and plasma membrane through

AgNP-induced oxidative stress.[22] Severe DNA damage in mice caused by polyvinylpyrrolidone-coated AgNPs was observed. Even the oxidation of silver in AgNPs induces cytotoxicity.[23] Some cancer treatment drugs also hold such characteristics.[24,25]

With regard to ROS-mediated toxicity, there are several types of mechanisms, as outlined below.

7.2.1.1 *Resonance Enhancement by Surface Plasmons*

Metal NPs have free electrons that respond to electromagnetic waves. Thus, when resonance occurs at a frequency of incident photons equal to the NP's free electrons, surface plasmon resonance occurs. When an electromagnetic field is generated, it causes ROS production. For example, under UV irradiation, free radicals are generated by AgNPs.[26] Generally, the generation of ROS-mediated toxicity depends on all the characteristics of NPs such as their dissociation, shape, size, agglomeration, *etc.*[27–29]

7.2.1.2 *Oxidative Modification of Proteins*

The production of excess ROS leads to the oxidation of amino acid residues, and the peptide bond is broken by the free radicals that form covalent crosslinks between proteins.[30]

7.2.1.3 *Lipid Peroxidation*

In lipid peroxidation, free radicals take electrons from the lipid. The free radicals are unstable and readily react with oxygen to form peroxy fatty acids, which induces other fatty acids to produce free radicals.[31]

7.2.1.4 *DNA Damage*

The most critical effect of ROS production is DNA damage and apoptosis. Li *et al.* showed that starting from mutation to fragmentation of chromosomes, NPs induce various adverse effects. At a particular concentration, AuNPs decrease the repair mechanism of DNA. When DNA damage occurs, p53 responds rapidly, and the cell cycle is arrested. Cadmium telluride was observed to enhance the level of p53, which is generally used to treat breast cancer.[32]

7.2.1.5 *Inflammation*

ROS-mediated oxidative stress (see Figure 7.2) is also directly linked with the release of mediators for inflammation. There are many inflammatory cascades, such as NF-κB, PI3, and MAPK pathways.[33] The family of NF-κB plays a vital role in immunity, *i.e.* inflammatory responses. It is a DNA-binding protein that exhibits a dimeric form in its inactivated state. The protein IκB

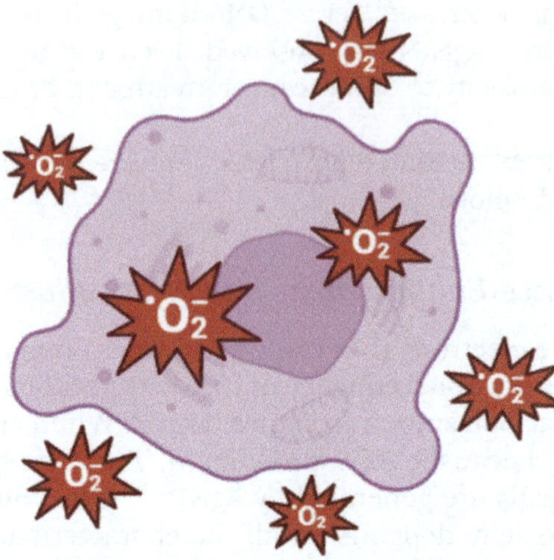

Figure 7.2 ROS-mediated toxicity. Created with BioRender.com.

inhibits its activation. In the activation of NFκB, IκB degrades and the nuclear localization signal (NLS) sequence is revealed. After entering the nucleus, NFκB binds to proinflammatory DNA genes and recruits other proteins and polymerase for transcription. Most metal oxides induce NFκB inflammation. Oxidation of cysteine in P50 is caused by excess ROS inhibiting the proinflammatory gene transcription. Some inflammation also occurs through the MAPK pathway. Administration of titanium dioxide NPs induced bronchial epithelial toxicity through the P38 MAPK pathway that produces IL-8. Immune cells such as macrophages and T cells produce TNF-α, IL-6, IL-1, and IL-8, which also cause cell death.[34,35]

7.2.1.6 Two More Types of Toxicity

Under ROS-mediated cytotoxicity, there are two more types of toxicity, as follows.

7.2.1.6.1 Genotoxicity. Genotoxicity (see Figure 7.3) occurs due to the excess production of ROS and causes damage to genetic materials. NPs generate target cell inflammation and produce oxidative stress. Genotoxicity can be primary or secondary due to the deterioration of genetic material. Primary NPs interact directly with DNA and secondary DNA-mediated damage occurs through the production of NP-mediated reactive nitrogen species (RNS).[36] Mei *et al.* observed DNA lesions by carbon nanotubes in Chinese hamster fibroblasts.[37] In *Saccharomyces cerevisiae*, zebrafish, silkworms, and plants, DNA damage occurs by AgNPs at various concentrations.[15,17,38–40]

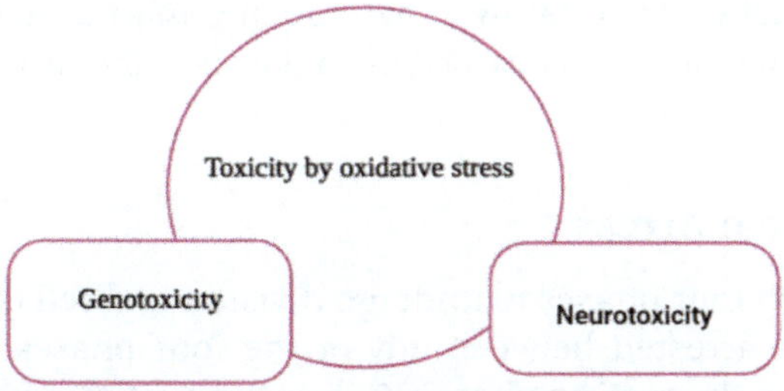

Figure 7.3 Oxidative stress is of two types, genotoxicity and neurotoxicity. Created with BioRender.com.

7.2.1.6.2 Neurotoxicity. According to several reports, the neurotoxicity of various NPs occurs through oxidative stress.[41,42] NP-mediated toxicity affects the physiology and morphology of neurons, hence studies are aimed at producing green NPs that target the brain without harm.

7.2.2 Physiochemical Mechanism

A broad range of investigations must be carried out to create green NPs.[9,12,43–45] Shape and size are very important in cytotoxicity. NPs having smaller sizes are most potent in causing cytotoxicity.[46] Similarly, rod-shaped Fe_2O_3 NPs are more potent cytotoxins than spherical Fe_2O_3 NPs.[47] Surface charge is another factor that decreases the uptake and interaction of NPs with cellular components. NPs with a greater surface charge have a higher potential for cellular uptake and hence higher toxicity.[48] Thus, glycosaminoglycan with a positive charge in mammalian cells induces a greater level of internalization and can interact with DNA and damage it. Some commercial drugs such as dendrimers showed greater cytotoxicity in cationic form than in anionic form owing to their permeability through lipid bilayers.[49,50]

7.2.3 Molecular Mechanism

An increase in intracellular calcium plays a vital role in NP-induced cytotoxicity by cellular imbalance.[51] Calcium is involved in cell signaling. Hence any imbalance in the calcium level causes dysregulation of cellular mechanisms. Excess calcium causes the release of cytochrome *c* from mitochondria and activates the apoptotic pathway. Here ROS are also produced and lead to the death of cells.[52] Some NPs, such as ZnONPs, open the calcium channels to the influx of calcium from outside, causing cytotoxicity and imbalance.[53] The mitochondrial membrane potential is reduced in adenocarcinoma cells of alveoli when it comes in contact with ZnO.[54] It also caused the mitochondria to be squeezed into HeLa cell lines.[55] TiO_2 and Fe_3O_4 are also potent NPs that disrupt the mitochondria membrane potential in neural and hepatoma cells.[10,16,48,56] Bi_2O_3 also induces apoptosis through the mitochondrial pathway by reducing matrix metalloproteinases (MMPs) with genes such as *Bax*/*Bcl-2*.[18] Likewise, metal oxides cause the denaturation of proteins.[57]

7.2.4 Cell Cycle Arrest

Cells pass through four phases to undergo division and cell division is halted if a cell becomes arrested between any of the four phases. Such a halting phase is known as the resting phase G0.[10] Accumulation of NPs may lead to the inhibition of cell proliferation by arresting any one of the cell cycle phases. Like NiONPs, exposure decreases G0/G1 in the A549 cell line but increases it in the BEA-23 cell line. Similarly, the G2/M phase increases in the A549 cell line and decreases in the BEAS-20 cell line.[48]

7.3 Impact of NPs on Different Subcellular Organs

7.3.1 Cell Membrane and Cytoskeleton

Wang *et al.* observed the effect of NPs on the cell membrane.[58,59] Based on the charge of engineered nanomaterial (ENM), NPs slightly disrupt the phospholipid bilayer of the cell membrane (see Figure 7.4), *i.e.* cause pinholes by decreasing lipid or protein. This phenomenon leads to cytotoxicity. Size-based carbon ENM was observed to block the ion channel on the membrane.[60] In some studies it was observed that ENM affects the cytoskeleton after entering into cells. Higher amounts of iron oxide ENMs hinder the activation of actin, leading to detrimental effects on cell migration.[61]

Small-sized NPs can interact with proteins and disturb protein folding, and changes in folding may affect the activity of an enzyme. The activation and catalytic site may become hindered or buried inside. Likewise, proteins become functionally inactive. This represents an important area of study on drug delivery within such nanomaterials. Studies have also shown protein aggregation through ENM–protein interactions that may give rise to neurodegenerative disorders.[62]

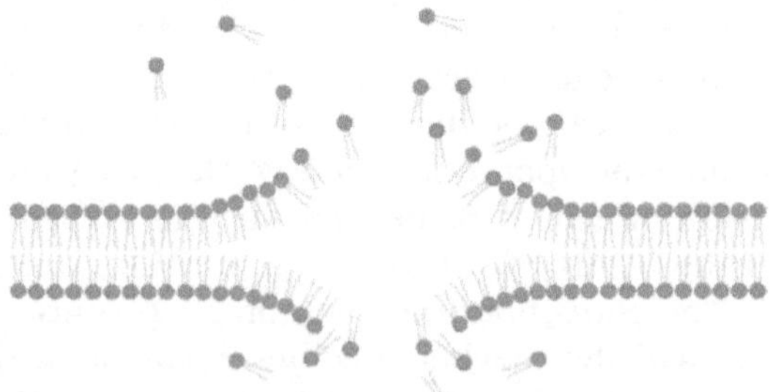

Figure 7.4 Lipid bilayer disruption. Created with BioRender.com.

7.3.2 Lysosomes

The charge is a vital factor determining that cells are negatively charged, and lysosomes are positively charged for their various activities. Hence negatively charged NPs prefer lysosomes to be present.[63] AgNPs are suspected of accumulating in lysosomes and harming the "suicidal bag" activity of lysosomes, which indirectly leads to cytotoxicity.[64] Like protein aggregation in the lysosomes, autophagosome accumulation leads to serious nervous degeneration.[65]

7.3.3 Mitochondria

Mitochondria play a vital role in ATP generation. There are many pathways for apoptosis (see Figure 7.5), among which the mitochondrial pathway can be caused by contact with the ENM.[66] As mitochondria are related to the electron transport chain (ETC), the ENM targets mitochondria and causes oxidative stress. A hypothetical result was given by King-Heiden *et al.* on zebrafish embryos on exposure to quantum dots. They assumed that the malfunction of the embryonic heart might be related to engineered nanoparticle (ENP) administration during a specific phase of development that leads to some extent of apoptosis.[67] Exposure of zebrafish embryos to AuNPs also showed abnormal pigmentation in the eyes. A higher level of apoptosis in the eyes through p53 signaling was suspected.[68]

7.3.4 Nucleus

Studies of ENPs inside the nucleus were carried out by fluorescence labeling, and showed no ENM.[69–71] However, in amphibian oocytes it was found that

Figure 7.5 Apoptosis in a cell. Created with BioRender.com.

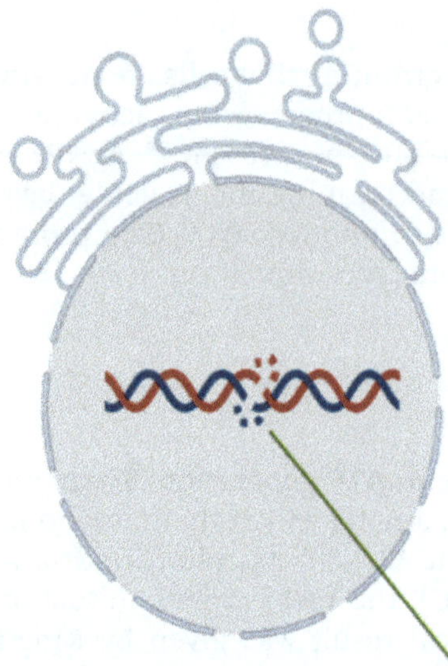

Figure 7.6 DNA damage due to nanomaterial. Created with BioRender.com.

nuclear pore molecules with a diameter of *ca.* 9 nm could be diffused. Hence quantum dots, like small NPs, can enter the nucleus *via* nuclear pore complexes and damage the nucleus (see Figure 7.6).[72] Molecules larger than these NPs can be tagged with the NLS sequence and only enter the nucleus. For example, in the amphibian oocyte, AuNPs having an NLS were observed to be recognized by nuclear pore receptors and able to enter the nucleus.[73] All this information on transportation into the nucleus is applicable only during the non-dividing phase of cells. Exposure to NPs causes considerable degranulation in the nucleus, such as protein aggregation and DNA damage, ultimately leading to many pathological conditions. Genotoxicity has been observed many times upon administration of NPs. The disturbance of cell division by ENMs and the increase in ROS and oxidative stress create the scenario for genotoxicity.[9] Exposure to quantum dots causes dysregulation of the cell cycle by specifically targeting histone proteins.[72]

7.4 Impact on Development and Behavior

Many behavioral studies have been carried out on the impact of ENMs. The effect of titanium dioxide on earthworms was studied by McShane *et al.* and it was found that earthworms avoid NPs by sensing them.[74] Behavioral changes in feeding were observed on administering ZnNPs and AgNPs to snails.[75] *Caenorhabditis elegans* lost its ability to move on exposure to aluminum oxide. Different types of quantum dots produced many harmful effects, such as vascular blockage and malformations in zebrafish.[67] For example, behavioral changes occurred upon exposure to charged AuNPs in the adult stage.[76–78]

In addition, cartilage deformation was observed on administration of SiNPs [79] and altered behavior was caused by titanium dioxide.[80] Mostly in fishes, all NPs changed their swimming behavior.

7.5 Impact of Nanoparticles on the Environment

NPs pose a conditional risk to both the environment and humans. The emerging risks associated with NPs open up a range of challenges. Although many tests to examine toxicity are available, a broad range of investigations must be carried out to create green NPs.[81–85]

Along with protection of human health, we must also take care of our ecological system. NPs from industry, vehicles, and animal waste pose significant environmental risks. Detailed studies on the environmental issues caused by NPs have not yet been carried out, which is a matter of concern. Generally, we observe zebrafish to determine the effect of NP toxicity on our ecosystem.[86–88] The administration of AgNPs to zebrafish changed their lysosomal function, leading to apoptosis in developing embryos. The effect of AgNP toxicity is size dependent.[86] Adult zebrafish also showed toxicity-induced behavioral changes. Administration of polystyrene NPs caused swimming disturbances, neurological disorders, accumulation of NPs in the gonads and other organs, *etc.*[87] A different testing strategy was developed to examine the chemical and NP risks. Mammalian and non-mammalian model organisms have been studied for toxicity checks.[83] Nowadays, stem cells are considered ideal for toxicity tests for their ability to regenerate.[89]

A detailed study is required on perturbances to biological systems caused by NPs to find ways of preventing all these toxic phenomena. A better route needs to be devised to assess NP hazards that would allow tests to be carried out on non-rodents. The modern era is based on nanotechnology, and billions of products, from cosmetics to medicines, are now based on nanotechnology, which represents the future. However, with the race to discover even better NPs, some potentially dire consequences also come into play that may affect the quality of life. Being so small can cause many complexities, which must be of concern.

7.6 Methods to Prevent Nanotoxicity

Several methods have been developed to reduce the toxicity of metal and metal oxide NPs. By altering the shape and size of NPs, non-toxic nanomaterials with desirable properties can be obtained.[90] NPs can be used in drugs that contain various salts and proteins, both of which influence the stability of the nanomaterial. With an increase in salt concentration, the electrostatic repulsion between the NPs decreases, resulting in NP aggregation. When a protein enters the surface of an NP, it causes alterations in the size and charge of the particle.[91] Understanding the behavior of NPs in biological systems is a crucial step in diagnosis and therapy.

7.6.1 Synthesis of Non-toxic Nanomaterials

7.6.1.1 Synthesis of Metal and Metal Oxide Nanoparticles

The toxicity of nanomaterials can be reduced by altering the temperature, pH, energy supply, reagents used, and reaction environment. Chemical, biological, and physical methods can be used to synthesize NPs. With regard to chemical techniques, oxide NPs can be obtained by simple precipitation reactions.[92] Similarly, nanometals can be obtained by chemical reduction of metal ions using inorganic and organic reducing agents.[93] This leads to the production of metallic particles resulting in the NPs sticking together to form agglomerates. These agglomerates cause the formation of colloidal metallic particles. To prevent agglomeration, the pH and the temperature of the system are controlled. The metallic NPs act as a carrier and have two functions: (1) to stabilize the NPs and prevent them from aggregation and (2) to modify the surface so that it can adapt to the environment easily.[92] Among physical methods, standard techniques include evaporation–condensation and laser ablation. Algae, fungi, bacteria, and plants are used to synthesize biological NPs. Kitching *et al.*[94] used fungi to synthesize AuNPs. *Enteromorpha compressa*, a seaweed, synthesizes AgNPs.[95] In biological materials, sulfur, nitrogen, and oxygen are present and react with metal ions, resulting in reduction.[96] Nanogold-modified *p*-aminobenzoic acid–Quat 188–pullulan was used as a carrier of doxorubicin, and due to surface modification a higher drug loading than normal was possible.[95]

7.6.1.2 Morphology of Metal and Metal Oxide Nanoparticles

The cytotoxic properties of metal and metal oxide NPs increase with the size of the NPs. Their small size allows the particles to pass easily through the intracellular membrane and directly release the ions into the cells. NPs are synthesized with circular, cubic, triangular, and elongated shapes, and the shape plays a dynamic role in immunogenic and toxic properties. Variously shaped AgNPs showed different toxicities to cell lines.[97] Platelet AgNPs demonstrated greater toxic effects than spherical and wire-shaped AgNPs. AuNPs with various shapes and concentrations were investigated in cancer cell lines. Low cytotoxicity was observed with spherical NPs at high concentrations. Different shapes of ZnONPs were tested by Dlugosz *et al.* and nanowire-shaped NPs showed higher toxicity than cubic and spherical NPs. Small-sized NPs at high concentration increased the cytotoxic concentration. Increasing the concentration of ZnONPs reduced lysosomal and mitochondrial activity. In zebrafish embryos, silver nanoplates were found to be more harmful than silver nanospheres.[98] Spherical NPs can absorb better than NPs with any other shape.[99] Gold nanorods can absorb better and accumulate less than gold nanospheres.[100] Gold NPs with star, rod, and spherical shapes were investigated by Xie *et al.*[101,102] It was found that star-shaped NPs are more toxic than any other shapes of NPs, as evidenced in the human cell line. This suggests that the cytotoxic and anticancer properties depend on the shape of the cell line. Mesoporous SiNPs have also been used for oral drug delivery.

Needle-shaped NPs showed greater toxicity than spherical NPs owing to their endocytic mechanism, rate of internalization, and greater adhesiveness towards the surface of target cells [103–105]

7.6.2 Coating to Prevent Nanotoxicity

By coating the particle surface, the toxicity of NPs can be reduced. On covering the surface, the NPs change completely, providing new NPs for drug carrier systems. The coating improves stability by preventing oxidation of the NPs and restricting aggregation and agglomeration. Coating AgNPs with SiO_2 minimizes toxicity by preventing the release of ions and interactions with bacteria and/or cells. Iron oxide NPs coated with polyethylenimine (PEI) increased uptake more than PEGylated iron oxide NPs in both cancerous and macrophage cell lines. Furthermore, they reduced the cytotoxicity.[106]

Natural products such as saccharides, hydrocolloids, and polyphenols are used to reduce toxicity. Polysaccharides such as chitosan and sodium alginate improve the penetration load of proteins and peptides in the intestine owing to their mucoadhesive and adsorption-enhancing properties.[107] Hydrocolloid materials protect foods and separate different parts of the food. The hydrocolloid coating prevents the molecules from deteriorating. Pectins (PECs) are polysaccharides composed mainly of galacturonic acid units and have different structures and molecular weights. Strawberry-dipped PEC retains the physicochemical properties of the original PEC. Grass pea flour (GPF) acts as a transglutaminase substrate and is like a bioplastic.[108] Polyphenols are known to have anticancer, antioxidant, and antibacterial properties. They are secondary plant metabolites that positively affect metabolic processes. The polyphenols quercetin, curcumin, eugenol, and epigallocatechin are widely used for encapsulation, especially epigallocatechin. Polyphenol-loaded NPs enhance the physicochemical properties of foods.[109]

7.6.3 Aspect Ratio

The width-to-height ratio of NPs is the aspect ratio. For a sphere, the aspect ratio is 1. The aspect ratio of NPs is directly proportional to toxicity. Major pulmonary toxicity is based on this property of NPs, as observed by Lin *et al.* by injecting high aspect ratio carbon nanotubes into the trachea. Further, aspect ratio-based toxicity checking was also carried out by comparative studies between the lung and gastrointestinal (GI) tract for CeO_2 nanorods.[110] This type of nanotoxicity can be controlled by using toxic aspect ratio windows for high aspect ratio nanomaterials such as anodic alumina nanotubes (AANTs). The aspect ratio of AANTs varies from 7.8 to 63.3, but 7.8 is the toxic aspect ratio window.[111]

7.6.4 Crystallinity

Solids that have identical chemical compositions but different forms and physical properties are called polymorphs. Crystalline compounds have many polymorphs, and they all have different toxicity levels. Many studies have been carried out on the cytotoxicity of various polymorphs of different particles. The cytotoxicity of three polymorphs of 10-hydroxycamptothecin was studied by Liu *et al.*[112] Among the pancake, prismatic, and needle forms, the needle form was the most harmful owing to its shape and size. Further, seven different crystalline structures of polycaprolactone were studied for their different effects on the enzymatic action of particles.[111] Interparticle space and size determine the crystallinity.[113] Novel nanomaterials can be developed by controlling the crystallinity of the polymer.

7.6.5 Dissolution

The dissolution of an NP determines its risk of toxicity and gives an idea of NP diffusion and dispersion. These characteristics of NPs can be used to design novel NPs with a less toxic effect. Many non-specific phenomena affect the diffusion rate of NPs; smaller-sized NPs have a high dissolution rate.[114] Decreasing the dissolution rate by incubating ZnO nanowires with Curosurf, a natural surfactant, was investigated by Theodorou *et al.*[115,116]

7.6.6 Agglomeration

The aggregation of NPs is due to the adhesion of particles to each other to reduce the surface free energy and surface area, and this somehow determines the toxicity of NPs. The impact of agglomerated NPs on lung inflammation was examined by Bantz *et al.*[117] A comparison of hemolytic toxicity by large AgNPs and small agglomerates was made by Zook *et al.*[118] Agglomerate-engineered NPs (ENPs) decrease the chances of translocation through the primary barrier, as in the GI tract, which restricts the ENPs from contacting the secondary organs.[119] When positively charged NPs agglomerated with *Escherichia coli* were injected into *C. elegans*, no harm was caused, but in assays that did not contain *E. coli*, reduced toxicity of ENPs with hindrance of growth and reproduction of *C. elegans* was observed.[120]

7.7 Toxicity Testing

The dose factor plays a vital role in measuring toxicity. Mostly *in vivo* experiments are performed in toxicity tests, such as *in vivo* sedimentation diffusion. This technique clearly shows the accumulation of NPs inside the cells. The timing of the dose administration helps to determine the dose-dependent response.

Many techniques can be used to measure the physicochemical structure of cells, from microscopy to spectroscopy. By observing the chemical nature of a cell by X-ray spectroscopy and examining the physiological nature of the cell by transmission electron microscopy or atomic force microscopy, one can easily evaluate the toxicity.[121]

There are also some methods, such as organs-on-a-chip testing, that evaluate toxicity at the organ level. Toxicity in organs such as the lungs, brain, placenta, *etc.*, is determined by this test.[122]

7.8 Nanotoxicity Remediation

Although plenty of evidence of nanotoxicity can be found, little information is available for its remediation. Research Team for Nano-Associated Safety Assessment (RT-NASA) examined the impact of toxicity in three phases. First, they demonstrated coordination at the international level based on nanosafety. Second, they focused on the toxicity level of NPs on different organs. Third, they observed the efficiency of other techniques for measuring cytotoxicity. They evaluated different responses of other NPs from the hierarchies of people in society.[123] Experts from various agencies and academia of other countries have conducted evaluation procedures for screening NP-related cytotoxicity.[44] Mutual interaction between chemists, biologists, and ecologists is needed to exchange information and ideas in order to solve toxicity-related problems. Combined ideas from studies of NPs in different fields would help scientists find a possible way to create green-worthy NPs that are non-toxic.[124]

Several types of remediation and doping of NPs represent one such approach. Iron-doped NPs such as ZnONPs resulted in decreased dissolution, leading to a decrease in toxicity in zebrafish and rodents. The reduced level of lethality in response to iron doping was studied in various organisms. Accordingly, pH-responsive biodegradable NPs were formed.[125] As NP characterization depends on size, spherical homogeneous NPs were synthesized by studying the codelivery of different chemicals with pH-responsive NPs and the percentage uptake increased. The aggregation of NPs on the surface was reduced, so the toxicity was decreased.[126]

The synthesis of new NPs by modifying their chemical properties would ameliorate the increase of nanotoxicity. Structural modification, surface coating, and many more approaches are available through which we can create novel NPs. Among all the approaches, surface coating is the easiest and most effective means.[127,128] For example, Pluronics (poloxamers) can decrease the toxicity of carbon nanotubes.[129]

7.9 Phytoremediation

"Phyto" means plants, hence phytoremediation is the use of chemicals extracted from different parts of plants, such as fruits and vegetables, to reduce the concentrations or toxic effects of contaminants in the environment. Phytochemicals can be used to decrease nanotoxicity, and instead of using chemical reagents, plant derivatives are used to convert metal ions and stabilize them in aquatic conditions. For NPs that are intended to be used for medical purposes, it is generally preferred that they are produced in this non-toxic and green way. Secondary products of plants such as flavonoids and other compounds have been used to generate AgNPs with a size of *ca.* 8–10 nm with some antibacterial properties. Likewise, antimicrobial properties were also found in AgNPs and AuNPs synthesized from phenolic compounds such as phytosterol.[130] Many phytochemicals, such as *trans*-cinnamic acid, gallic acid,[131] and gellan gum,[132] were used to form antioxidizing AuNPs. These AuNPs are nowadays also used in X-ray imaging and green drug delivery.

Similarly, NPs can be applied as carriers of flavonoids and used to form anticancer SeNPs with a size of *ca.* 50–150 nm.[133] Gum arabic has also been used to formulate iron oxide NPs, which can be used in MRI. NPs usually act as a contrasting agent in MRI and X-ray imaging, *etc.* The efficiency of some NPs, mostly AuPs, in cancer diagnosis has also been measured. Different types of evaluations from cytotoxicity checks to determination of uptake efficiency have been performed by different groups using different phytoremediation NPs.[134]

7.10 Advances in Nanotoxicology

To ensure the safety of nanomedicine and the health of humans, researchers are focusing on developing green NPs and studies of such development broaden the available information on the effects of NPs. Some reviews have covered the clinical safety of NPs. Detection of methicillin-resistant *Staphylococcus aureus* (MRSA) infection and biofilm eradication with the application of nanotheranostics were reviewed by Mosselhy *et al.*[135] Demasco *et al.*[136] studied the development of safe, rationally designed inorganic NPs used in cancer treatment.

7.11 Challenges for Safer Nanoparticle Design

Data gaps may be the greatest challenge for nanotoxicologists attempting to find safer nanomaterials. Lack of knowledge about instrumentation, dose quantities, and analytics is a barrier to designing safer NPs.[137] Scientists studying nanotoxicology have been making great efforts to establish new protocols for the aggregation state of NPs.[138] However, after much research, the community has been unable to find a perfect way to create completely non-toxic NPs.[139] Nevertheless, there are some generalized characteristics for producing safer NPs.[140] The size of the NPs has the most significant effect on toxicity, and the size of NPs must not be too small.[141] The impact of surface charge must be observed before creating NPs, as biological cells have negative charges inside them, so NPs that have positive charges are more toxic.[142] All toxicity indices, such as dissolution, shape, and agglomeration, need to be checked.

It is too difficult to identify the exact cause of the toxicity of a complex NP. The effect of toxicity predicted by changing the parameters *in vivo* and *in vitro* may have different effects in their actual platform.[143]

7.11.1 Plasmonic Properties

"Plasmonics" involves optically active compounds, and light affects plasmonic NPs by inducing electromagnetic fields that influence their characteristics in their biological environment,[144] and this may have positive or negative effects on the toxicity. Hence it is a challenge to make green NPs.

7.11.2 Size

As discussed previously, the size of the NPs principally affects the toxicity index. Similarly, the transport of NPs also depends on their size. NPs with the size of a virus have an exact biological mechanism of uptake, *i.e.* endocytosis. Therefore, it is not easy to reduce their toxicity.[145]

7.11.3 Charged Particles

Positively charged NPs such as silver and ammonium NPs have a more toxic effect on the biological environment as they are highly attracted by bacteria and eukaryotic cells. Owing to the high uptake capacity, these NPs are more harmful to biological entities.[146]

7.11.4 Instability and Defects

Artificially produced synthetic NPs show crystal defects and density defects,[147] and these impurities of the NPs lead to the instability of nanomaterials.

7.12 Conclusion

Nanotechnology has emerged as a revolutionary technology in developing drug delivery systems and various major fields of medical science. However, owing to overuse and misuse, the advantages of nanotechnology are continuously degrading, and nanotoxicity cases are increasingly in the spotlight. Although the mechanism, impact, and toxicity of NPs are partially known, this information is insufficient to establish the proper remediation of nanotoxicity. Some recent studies have determined partial solutions to nanotoxicity by discovering and modifying NPs in different ways. However, the problems remain unsolved, and there is now an urgent need to carry out research in this area to protect our biological and ecological environment.

Acknowledgements

We acknowledge BioRender for providing the platform BioRender.com to create all our images. L. M. is grateful to the UGC for financial assistance for his PhD studies.

References

1. M. Sahoo, S. Vishwakarma, C. Panigrahi and J. Kumar, *Food Front.*, 2021, 2, 3–22.
2. B. Pegaz, E. Debefve, J.-P. Ballini, Y. N. Konan-Kouakou and H. van den Bergh, *J. Photochem. Photobiol., B*, 2006, **85**, 216–222.

3. M. Mishra and M. Panda, *Free Radical Res.*, 2021, **55**, 919–935.
4. C. Egbuna, S. Kumar, J. Ifemeje and J. Kurhekar, *Phytochemistry*, 2019, 131–146.
5. J. Ifemeje, S. Udedi, A. Okechukwu, A. Nwaka, C. Lukong, I. Anene, C. Egbuna and I. Ezeude, *J. Sci. Res. Rep.*, 2015, **6**, 394–403.
6. X. Sun, J.-F. Wang, M. Tseng and L. T. Young, *J. Neuropsychiatry Clin. Neurosci.*, 2006, **31**, 189–196.
7. D. A. M. Kheirallah, L. M. El-Samad and A. M. Abdel-Moneim, *Sci. Total Environ.*, 2021, **753**, 141743.
8. J. Jiang, G. Oberdörster, A. Elder, R. Gelein, P. Mercer and P. Biswas, *Nanotoxicology*, 2008, **2**, 33–42.
9. L. Gonzalez, D. Lison and M. Kirsch-Volders, *Nanotoxicology*, 2008, **2**, 252–273.
10. M. J. Akhtar, M. Ahamed, M. Fareed, S. A. Alrokayan and S. Kumar, *J. Toxicol. Sci.*, 2012, **37**, 139–148.
11. T. Xia, M. Kovochich, M. Liong, L. Madler, B. Gilbert, H. Shi, J. I. Yeh, J. I. Zink and A. E. Nel, *ACS Nano*, 2008, **2**, 2121–2134.
12. H.-m. Chiang, Q. Xia, X. Zou, C. Wang, S. Wang, B. J. Miller, P. C. Howard, J. J. Yin, F. A. Beland and H. Yu, *J. Nanosci. Nanotechnol.*, 2012, **12**, 2126–2135.
13. Z. Fan and J. G. Lu, *J. Nanosci. Nanotechnol.*, 2005, **5**, 1561–1573.
14. J.-J. Yin, J. Liu, M. Ehrenshaft, J. E. Roberts, P. P. Fu, R. P. Mason and B. Zhao, *Toxicol. Appl. Pharmacol.*, 2012, **263**, 81–88.
15. S. Kim and D. Y. Ryu, *J. Appl. Toxicol.*, 2013, **33**, 78–89.
16. M. J. Akhtar, M. Ahamed, S. Kumar, H. Siddiqui, G. Patil, M. Ashquin and I. Ahmad, *Toxicology*, 2010, **276**, 95–102.
17. S. Shaligram and A. Campbell, *Toxicol. In Vitro*, 2013, **27**, 844–851.
18. W. Dröge, *Physiol. Rev.*, 2002, **82**, 47–95.
19. D. A. Butterfield, in *Reactive Oxygen Species in Biological Systems*, Springer, 2002, pp. 609–638.
20. H. Shi, L. G. Hudson and K. J. Liu, *Free Radical Biol. Med.*, 2004, **37**, 582–593.
21. T. Bodamyali, C. R. Stevens, D. R. Blake and P. G. Winyard, in *Free radicals and inflammation*, Springer, 2000, pp. 11–16.
22. P. P. Fu, Q. Xia, X. Sun and H. Yu, *J. Environ. Sci. Health, Part C: Environ. Carcinog. Ecotoxicol. Rev.*, 2012, **30**, 1–41.
23. Q. Xia, H.-M. Chiang, Y.-T. Zhou, J.-J. Yin, F. Liu, C. Wang, L. Guo and P. P. Fu, *Am. J. Chin. Med.*, 2012, **40**, 1271–1288.
24. Q. Xia, J. J. Yin, S.-H. Cherng, W. G. Wamer, M. Boudreau, P. C. Howard and P. P. Fu, *Toxicol. Lett.*, 2006, **163**, 30–43.
25. Q. Xia, J. J. Yin, P. P. Fu and M. D. Boudreau, *Toxicol. Lett.*, 2007, **168**, 165–175.
26. W. Zhang, Y. Li, J. Niu and Y. Chen, *Langmuir*, 2013, **29**, 4647–4651.
27. S. A. Agnihotri, N. N. Mallikarjuna and T. M. Aminabhavi, *J. Controlled Release*, 2004, **100**, 5–28.
28. W. Lu, D. Senapati, S. Wang, O. Tovmachenko, A. K. Singh, H. Yu and P. C. Ray, *Chem. Phys. Lett.*, 2010, **487**, 92–96.

29. A. Nel, T. Xia, L. Madler and N. Li, *Science*, 2006, **311**, 622–627.
30. E. C. Català, J. T. Sumalla and J. R. Salvador, *International Microbiology*, 2000, vol. 3, núm. 1, pp. 3–8.
31. G. Spiteller and M. Afzal, *Chem. - Eur. J.*, 2014, **20**, 14928–14945.
32. J. J. Li, L. Zou, D. Hartono, C. N. Ong, B. H. Bay and L. Y. Lanry Yung, *Adv. Mater.*, 2008, **20**, 138–142.
33. Y.-W. Huang, C.-H. Wu and R. S. Aronstam, *Materials*, 2010, **3**, 4842–4859.
34. M. Serasanambati and S. R. Chilakapati, *South Indian J. Biol. Sci.*, 2016, **2**, 368–387.
35. J. I. Lee and G. J. Burckart, *J. Clin. Pharmacol.*, 1998, **38**, 981–993.
36. Y.-H. Hsin, C.-F. Chen, S. Huang, T.-S. Shih, P.-S. Lai and P. J. Chueh, *Toxicol. Lett.*, 2008, **179**, 130–139.
37. N. Mei, Y. Zhang, Y. Chen, X. Guo, W. Ding, S. F. Ali, A. S. Biris, P. Rice, M. M. Moore and T. Chen, *Environ. Mol. Mutagen.*, 2012, **53**, 409–419.
38. A. Shvedova, V. Castranova, E. Kisin, D. Schwegler-Berry, A. Murray, V. Gandelsman, A. Maynard and P. Baron, *J. Toxicol. Environ. Health, Part A*, 2003, **66**, 1909–1926.
39. A. Schvedova, *J. Toxicol. Environ. Health, Part A*, 2003, **21**, 1166–1170.
40. C. He, Y. Hu, L. Yin, C. Tang and C. Yin, *Biomaterials*, 2010, **31**, 3657–3666.
41. M. Ghosh, M. Bandyopadhyay and A. Mukherjee, *Chemosphere*, 2010, **81**, 1253–1262.
42. K. Hirakawa and T. Hirano, *Chem. Lett.*, 2006, **35**, 832–833.
43. S. K. Brar, M. Verma, R. Tyagi and R. Surampalli, *Waste Manage.*, 2010, **30**, 504–520.
44. G. Oberdörster, A. Maynard, K. Donaldson, V. Castranova, J. Fitzpatrick, K. Ausman, J. Carter, B. Karn, W. Kreyling and D. Lai, *Part. Fibre Toxicol.*, 2005, **2**, 1–35.
45. C. H. Lin, L. W. Chang, Y. H. Wei, S. B. Wu, C. S. Yang, W. H. Chang, Y. C. Chen and P. P. Lin, *Kaohsiung J. Med. Sci.*, 2012, **28**, S53–S62.
46. S. Kawanishi, Y. Hiraku, M. Murata and S. Oikawa, *Free Radicals Biol. Med.*, 2002, **32**, 822–832.
47. D. M. Niedowicz, *Cell Biochem. Biophys.*, 2005, **43**, 289–330.
48. A. Valavanidis, T. Vlachogianni and C. Fiotakis, *J. Environ. Sci. Health, Part C: Environ. Carcinog. Ecotoxicol. Rev.*, 2009, **27**, 120–139.
49. E. A. Hileman, G. Achanta and P. Huang, *Expert Opin. Ther. Targets*, 2001, **5**, 697–710.
50. H. Pelicano, D. Carney and P. Huang, *Drug Resistance Updates*, 2004, **7**, 97–110.
51. T. Xia, M. Kovochich, J. Brant, M. Hotze, J. Sempf, T. Oberley, C. Sioutas, J. I. Yeh, M. R. Wiesner and A. E. Nel, *Nano Lett.*, 2006, **6**, 1794–1807.
52. J. Yin, Q. Xia and P. Fu, *Toxicol. Ind. Health*, 2007, **23**, 625–631.
53. H. Yu, Q. Xia, J. Yan, D. Herreno-Saenz, Y.-S. Wu, I.-W. Tang and P. P. Fu, *Int. J. Environ. Res. Public Health*, 2006, **3**, 348–354.

54. Y. Zhao, Q. Xia, J. J. Yin, G. Lin and P. P. Fu, *Toxicol. Lett.*, 2011, **205**, 302–309.
55. Y. Li, S. Yu, Q. Wu, M. Tang, Y. Pu and D. Wang, *J. Hazard. Mater.*, 2012, **219**, 221–230.
56. F. M. Winnik and D. Maysinger, *Acc. Chem. Res.*, 2013, **46**, 672–680.
57. S. J. Kang, B. M. Kim, Y. J. Lee and H. W. Chung, *Environ. Mol. Mutagen.*, 2008, **49**, 399–405.
58. Z. Wang, N. Li, J. Zhao, J. C. White, P. Qu and B. Xing, *Chem. Res. Toxicol.*, 2012, **25**, 1512–1521.
59. X. Wang, S. P. Reece and J. M. Brown, *Toxicol. Mech. Methods*, 2013, **23**, 168–177.
60. K. H. Park, M. Chhowalla, Z. Iqbal and F. Sesti, *J. Biol. Chem.*, 2003, **278**, 50212–50216.
61. S. J. Soenen, N. Nuytten, S. F. De Meyer, S. C. De Smedt and M. De Cuyper, *Small*, 2010, **6**, 832–842.
62. F. Chiti and C. M. Dobson, *Annu. Rev. Biochem.*, 2006, **75**, 333–366.
63. O. Harush-Frenkel, E. Rozentur, S. Benita and Y. Altschuler, *Biomacromolecules*, 2008, **9**, 435–443.
64. X. Ma, Y. Wu, S. Jin, Y. Tian, X. Zhang, Y. Zhao, L. Yu and X.-J. Liang, *ACS Nano*, 2011, **5**, 8629–8639.
65. S. T. Stern, P. P. Adiseshaiah and R. M. Crist, *Part. Fibre Toxicol.*, 2012, **9**, 1–17.
66. E. Frohlich, *Curr. Drug Metab.*, 2013, **14**, 976–988.
67. T. C. King-Heiden, P. N. Wiecinski, A. N. Mangham, K. M. Metz, D. Nesbit, J. A. Pedersen, R. J. Hamers, W. Heideman and R. E. Peterson, *Environ. Sci. Technol.*, 2009, **43**, 1605–1611.
68. K.-T. Kim, T. Zaikova, J. E. Hutchison and R. L. Tanguay, *Toxicol. Sci*, 2013, **133**, 275–288.
69. A. S. Idil and N. Donaldson, *J. Neural Eng.*, 2018, **15**(2), 021006.
70. W. Busch, S. Bastian, U. Trahorsch, M. Iwe, D. Kühnel, T. Meißner, A. Springer, M. Gelinsky, V. Richter and C. Ikonomidou, *J. Nanopart. Res.*, 2011, **13**, 293–310.
71. D. Kühnel, K. Scheffler, P. Wellner, T. Meißner, A. Potthoff, W. Busch, A. Springer and K. Schirmer, *J. Hazard. Mater.*, 2012, **227**, 418–426.
72. I. Nabiev, S. Mitchell, A. Davies, Y. Williams, D. Kelleher, R. Moore, Y. K. Gun'ko, S. Byrne, Y. P. Rakovich and J. F. Donegan, *Nano Lett.*, 2007, **7**, 3452–3461.
73. N. Panté and M. Kann, *Mol. Biol. Cell*, 2002, **13**, 425–434.
74. H. McShane, M. Sarrazin, J. K. Whalen, W. H. Hendershot and G. I. Sunahara, *Environ. Toxicol. Chem.*, 2012, **31**, 184–193.
75. M.-N. Croteau, S. K. Misra, S. N. Luoma and E. Valsami-Jones, *Environ. Sci. Technol.*, 2011, **45**, 6600–6607.
76. W. Zhang, K. Lin, X. Sun, Q. Dong, C. Huang, H. Wang, M. Guo and X. Cui, *Chemosphere*, 2012, **89**, 52–59.
77. W. Zhang, K. Lin, Y. Miao, Q. Dong, C. Huang, H. Wang, M. Guo and X. Cui, *J. Hazard. Mater.*, 2012, **213**, 413–420.

78. L. Truong, K. S. Saili, J. M. Miller, J. E. Hutchison and R. L. Tanguay, *Comp. Biochem. Physiol., Part C: Toxicol. Pharmacol.*, 2012, **155**, 269–274.
79. J. Y. Xue, X. Li, M. Z. Sun, Y. P. Wang, M. Wu, C. Y. Zhang, Y. N. Wang, B. Liu, Y. S. Zhang and X. Zhao, *Small*, 2013, **9**, 3161–3168.
80. T.-H. Chen, C.-Y. Lin and M.-C. Tseng, *Mar. Pollut. Bull.*, 2011, **63**, 303–308.
81. Y.-H. Lee, F.-Y. Cheng, H.-W. Chiu, J.-C. Tsai, C.-Y. Fang, C.-W. Chen and Y.-J. Wang, *Biomaterials*, 2014, **35**, 4706–4715.
82. H.-W. Chiu, T. Xia, Y.-H. Lee, C.-W. Chen, J.-C. Tsai and Y.-J. Wang, *Nanoscale*, 2015, **7**, 736–746.
83. R.-J. Chen, Y.-Y. Chen, M.-Y. Liao, Y.-H. Lee, Z.-Y. Chen, S.-J. Yan, Y.-L. Yeh, L.-X. Yang, Y.-L. Lee and Y.-H. Wu, *Int. J. Mol. Sci.*, 2020, **21**, 2387.
84. B.-H. Mao, J.-C. Tsai, C.-W. Chen, S.-J. Yan and Y.-J. Wang, *Nanotoxicology*, 2016, **10**, 1021–1040.
85. A. Jain, S. Ranjan, N. Dasgupta and C. Ramalingam, *Crit. Rev. Food Sci. Nutr.*, 2018, **58**, 297–317.
86. Z.-Y. Chen, N.-J. Li, F.-Y. Cheng, J.-F. Hsueh, C.-C. Huang, F.-I. Lu, T.-F. Fu, S.-J. Yan, Y.-H. Lee and Y.-J. Wang, *Int. J. Mol. Sci.*, 2020, **21**, 2864.
87. S. Sarasamma, G. Audira, P. Siregar, N. Malhotra, Y.-H. Lai, S.-T. Liang, J.-R. Chen, K. H.-C. Chen and C.-D. Hsiao, *Int. J. Mol. Sci.*, 2020, **21**, 1410.
88. S. Sarasamma, G. Audira, P. Samikannu, S. Juniardi, P. Siregar, E. Hao, J.-R. Chen and C.-D. Hsiao, *Int. J. Mol. Sci.*, 2019, **20**, 5795.
89. S. Lynch, C. S. Pridgeon, C. A. Duckworth, P. Sharma, B. K. Park and C. E. Goldring, *Biochem. J.*, 2019, **476**, 1149–1158.
90. D. Ungor, I. Dékány and E. Csapó, *Nanomaterials*, 2019, **9**, 1229.
91. F. Sawosz, L. Pineda, A. Hotowy, S. Jaworski, M. Prasek, E. Sawosz and A. Chwalibog, *Arch. Anim. Nutr.*, 2013, **67**, 347–355.
92. A.-H. Lu, E. L. Salabas and F. Schüth, *Angew. Chem., Int. Ed.*, 2007, **46**, 1222–1244.
93. M. A. Malik, M. Y. Wani and M. A. Hashim, *Arabian J. Chem.*, 2012, **5**, 397–417.
94. M. Kitching, M. Ramani and E. Marsili, *Microb. Biotechnol.*, 2015, **8**(6), 904–917.
95. O. Długosz, K. Szostak, A. Staroń, J. Pulit-Prociak and M. Banach, *Materials*, 2020, **13**, 279.
96. R. B. Thurman, C. P. Gerba and G. Bitton, *Crit. Rev. Environ. Sci. Technol.*, 1989, **18**, 295–315.
97. S. George, S. Lin, Z. Ji, C. R. Thomas, L. Li, M. Mecklenburg, H. Meng, X. Wang, H. Zhang and T. Xia, *ACS Nano*, 2012, **6**, 3745–3759.
98. N. B. Abramenko, T. B. Demidova, V. Abkhalimov, B. G. Ershov, E. Y. Krysanov and L. M. Kustov, *J. Hazard. Mater.*, 2018, **347**, 89–94.
99. I. Khan, K. Saeed and I. Khan, *Arabian J. Chem.*, 2019, **12**(7), 908–931.
100. Y. Wang, X. Zhou, C. Xu, Y. Jin and B. Li, *Sci. Rep.*, 2018, **8**, 1–9.
101. X. Xie, J. Liao, X. Shao, Q. Li and Y. Lin, *Sci. Rep.*, 2017, **7**(1), 3827.

102. E. Morgan, D. Wupperfeld, D. Morales and N. Reich, *Bioconjugate Chem.*, 2019, **30**, 853–860.
103. S. Gratton, *Proc. Natl. Acad. Sci. U. S. A.*, 2008, **105**, 11613.
104. H.-H. Chen, C.-C. Chien, C. Petibois, C.-L. Wang, Y. S. Chu, S.-F. Lai, T.-E. Hua, Y.-Y. Chen, X. Cai and I. M. Kempson, *J. Nanobiotechnol.*, 2011, **9**, 1–15.
105. J. Liao, W. Li, J. Peng, Q. Yang, H. Li, Y. Wei, X. Zhang and Z. Qian, *Theranostics*, 2015, **5**, 345.
106. Q. Feng, Y. Liu, J. Huang, K. Chen, J. Huang and K. Xiao, *Sci. Rep.*, 2018, **8**, 1–13.
107. Y. Pan, Y.-J. Li, H.-Y. Zhao, J.-M. Zheng, H. Xu, G. Wei and J.-S. Hao, *Int. J. Pharm.*, 2002, **249**, 139–147.
108. B. Johnston, I. Radecka, D. Hill, E. Chiellini, V. I. Ilieva, W. Sikorska, M. Musioł, M. Zięba, A. A. Marek and D. Keddie, *Polymers*, 2018, **10**, 957.
109. S. Di Bucchianico, M. R. Fabbrizi, S. Cirillo, C. Uboldi, D. Gilliland, E. Valsami-Jones and L. Migliore, *Int. J. Nanomed.*, 2014, **9**, 2191.
110. S. Lin, X. Wang, Z. Ji, C. H. Chang, Y. Dong, H. Meng, Y.-P. Liao, M. Wang, T.-B. Song and S. Kohan, *ACS Nano*, 2014, **8**, 4450–4464.
111. A. Vollrath, C. Kretzer, B. Beringer-Siemers, B. Shkodra, J. A. Czaplewska, D. Bandelli, S. Stumpf, S. Hoeppener, C. Weber and O. Werz, *Polymers*, 2021, **13**, 2557.
112. G. Liu, J. Pang, Y. Huang, Q. Xie, G. Guan and Y. Jiang, *Ind. Eng. Chem. Res.*, 2017, **56**(30), 8517–8527.
113. A. Jabbarzadeh, *Nanomaterials*, 2019, **9**, 1472.
114. J. Hedberg, E. Blomberg and I. Odnevall Wallinder, *Environ. Sci. Technol.*, 2019, **53**, 4030–4044.
115. I. G. Theodorou, P. Ruenraroengsak, A. Gow, S. Schwander, J. Zhang, K. F. Chung, T. D. Tetley, M. P. Ryan and A. E. Porter, *Nanotoxicology*, 2016, **10**(9), 1351–1362.
116. R. Mihalache, J. Verbeek, H. Graczyk, V. Murashov and P. van Broekhuizen, *Nanotoxicology*, 2017, **11**, 7–19.
117. C. Bantz, O. Koshkina, T. Lang, H.-J. Galla, C. J. Kirkpatrick, R. H. Stauber and M. Maskos, *Beilstein J. Nanotechnol.*, 2014, **5** 1774–1786.
118. J. M. Zook, R. I. MacCuspie, L. E. Locascio, M. D. Halter and J. T. Elliott, *Nanotoxicology*, 2011, **5**, 517–530.
119. A. Bruinink, J. Wang and P. Wick, *Arch. Toxicol.*, 2015, **89**, 659–675.
120. S. K. Hanna, A. R. Montoro Bustos, A. W. Peterson, V. Reipa, L. D. Scanlan, S. Hosbas Coskun, T. J. Cho, M. E. Johnson, V. A. Hackley and B. C. Nelson, *Environ. Sci. Technol.*, 2018, **52**, 5968–5978.
121. S. Behrens, A. Kappler and M. Obst, *Environ. Microbiol.*, 2012, **14**, 2851–2869.
122. N. Alepee, A. Bahinski, M. Daneshian, B. De Wever, E. Fritsche, A. Goldberg, J. Hansmann, T. Hartung, J. Haycock and H. T. Hogberg, *Altex*, 2014, **31**, 441.

123. Y.-R. Kim, S. H. Park, J.-K. Lee, J. Jeong, J. H. Kim, E.-H. Meang, T. H. Yoon, S. T. Lim, J.-M. Oh and S. S. A. An, *Int. J. Nanomed.*, 2014, **9**, 3.
124. K. Donaldson, V. Stone, C. Tran, W. Kreyling and P. Borm, *Nanotoxicology*, 2004, **61**, 727–728.
125. E. Ahmadi, N. Zarghami, M. A. Jafarabadi, L. Alizadeh, M. Khojastehfard, M. R. Yamchi and R. Salehi, *J. Drug Delivery Sci. Technol.*, 2019, **51**, 721–735.
126. F. Tang, L. Li and D. Chen, *Adv. Mater.*, 2012, **24**, 1504–1534.
127. M. Chanana, P. Rivera_Gil, M. A. Correa-Duarte, L. M. Liz-Marzán and W. J. Parak, *Angew. Chem., Int. Ed.*, 2013, **52**, 4179–4183.
128. M. J. Osmond-McLeod, R. I. Osmond, Y. Oytam, M. J. McCall, B. Feltis, A. Mackay-Sim, S. A. Wood and A. L. Cook, *Part. Fibre Toxicol.*, 2013, **10**, 1–18.
129. X. Wang, T. Xia, M. C. Duch, Z. Ji, H. Zhang, R. Li, B. Sun, S. Lin, H. Meng and Y.-P. Liao, *Nano Lett.*, 2012, **12**, 3050–3061.
130. K. D. Arunachalam, S. K. Annamalai and S. Hari, *Int. J. Nanomed.*, 2013, **8**, 1307.
131. M. J. Kim, A. R. Seong, J. Y. Yoo, C. H. Jin, Y. H. Lee, Y. J. Kim, J. Lee, W. J. Jun and H. G. Yoon, *Mol. Nutr. Food Res.*, 2011, **55**, 1798–1808.
132. C. Park, H. Youn, H. Kim, T. Noh, Y. H. Kook, E. T. Oh, H. J. Park and C. Kim, *J. Mater. Chem.*, 2009, **19**, 2310–2315.
133. F. Martínez-Esquivias, M. Gutiérrez-Angulo, A. Pérez-Larios, J. A. Sánchez-Burgos, J. S. Becerra-Ruiz and J. M. Guzmán-Flores, *Anti-Cancer Agents Med. Chem.*, 2022, **22**(9), 1658–1673.
134. K. Hola, Z. Markova, G. Zoppellaro, J. Tucek and R. Zboril, *Biotechnol. Adv.*, 2015, **33**, 1162–1176.
135. D. A. Mosselhy, M. Assad, T. Sironen and M. Elbahri, *Nanomaterials*, 2021, **11**, 82.
136. J. A. Damasco, S. Ravi, J. D. Perez, D. E. Hagaman and M. P. Melancon, *Nanomaterials*, 2020, **10**(11), 2186.
137. R. D. Handy, R. Owen and E. Valsami-Jones, *Ecotoxicology*, 2008, **17**, 315–325.
138. K. Donaldson, A. Schinwald, F. Murphy, W.-S. Cho, R. Duffin, L. Tran and C. Poland, *Acc. Chem. Res.*, 2013, **46**, 723–732.
139. S. D. Richardson, M. J. Plewa, E. D. Wagner, R. Schoeny and D. M. DeMarini, *Mutat. Res., Rev. Mutat. Res.*, 2007, **636**, 178–242.
140. S. M. Hussain, D. B. Warheit, S. P. Ng, K. K. Comfort, C. M. Grabinski and L. K. Braydich-Stolle, *Toxicol. Sci*, 2015, **147**, 5–16.
141. O. Bar-Ilan, R. M. Albrecht, V. E. Fako and D. Y. Furgeson, *Small*, 2009, **5**, 1897–1910.
142. A. Albanese, P. S. Tang and W. C. Chan, *Annu. Rev. Biomed. Eng.*, 2012, **14**, 1–16.
143. M. Andrione, B. F. Timberlake, G. Vallortigara, R. Antolini and A. Haase, *Learn. Mem.*, 2017, **24**, 622–629.
144. Q. Zeng, D. Shao, W. Ji, J. Li, L. Chen and J. Song, *Toxicol. Rep.*, 2014, **1**, 137–144.

145. R. R. Arvizo, S. Bhattacharyya, R. A. Kudgus, K. Giri, R. Bhattacharya and P. Mukherjee, *Chem. Soc. Rev.*, 2012, **41**, 2943–2970.
146. B. Reidy, A. Haase, A. Luch, K. A. Dawson and I. Lynch, *Materials*, 2013, **6**, 2295–2350.
147. S. Wang, S. S. Park, C. T. Buru, H. Lin, P.-C. Chen, E. W. Roth, O. K. Farha and C. A. Mirkin, *Nat. Commun.*, 2020, **11**, 1–8.

CHAPTER 8

Biocompatibility and Functionalization of Sustainable Nanomaterials

JYOTHY G. VIJAYAN AND T. NIRANJANA PRABHU*

Department of Chemistry, Faculty of Mathematical and Physical Sciences, M. S. Ramaiah University of Applied Sciences, IV Phase, Peenya Industrial Area, Bengaluru, Karnataka 560058, India
*Email: niranjan1975@gmail.com

8.1 Introduction

Nanomaterials (NMs) have attracted considerable interest from researchers owing to their wide range of applications. NMs exhibit different properties that help in applying them in various fields such as pharmaceutical, biomedical, environmental, *etc.*[1–3] The dimensions of NMs are *ca.* 1–100 nm which possess specific characteristics. They exhibit biocompatibility, biodegradability, and various mechanical and optical properties. NMs are used in clinical areas, especially in the fields of targeted drug delivery, tissue engineering, and disease diagnostics.[4–8] Nanoparticles (NPs) are considered as promising future materials for improving the capacity of major disease diagnostics. Their properties such as high stability, high dispersity, facile production, and biocompatibility help in their use as contrast agents for clinical applications. Tuning NMs with suitable morphology and specific size, suitable biodegradability and biocompatibility, and high safety for specific disease diagnosis are specific recommendations[9] for the efficient diagnosis of cancer. Various metal oxides, such as ZnO, TiO_2, AuO, AgO, *etc.*, and

Nanoscience & Nanotechnology Series No. 62
Smart Nanomaterials for Infectious Diseases
Edited by Suvardhan Kanchi, Nidhi Sharotri, Rajasekhar Chokkareddy, Deepali Sharma and Falah H. Hussein

Published by the Royal Society of Chemistry, www.rsc.org

non-metal oxides such as graphene oxide (GO), reduced graphene oxide, *etc.*, have been modified and used for different applications.[10] These metal oxides are used because of their high surface area, permeability, selectivity, chemical stability, and other properties.

Many NMs have been examined for their applications and usefulness in clinical areas. Owing to the properties of metal oxides, such as a rich valence state, electronic structure, and surface area, they are applied for environmental monitoring, pollution control, and remediation purposes.[11] They have applications from catalysis to clinical diagnoses. Surface-functionalized NMs have useful applications as antimicrobial agents.[12] Fungi are a major cause of infections.[13] Fungi are different kinds of eukaryotic organisms, and among all fungi, around 300 species are dangerous and cause diseases in humans. Human fungal pathogens are the reason for a wide variety of infectious diseases that infect parts of the human body such as the skin, nose, and other tissues.[14–16]

NPs have the potential to be applied in different fields such as in tissue engineering, drug delivery, viral infections, and other disease diagnostics. NPs have the ability to control different mechanisms, from pharmaceutical to clinical. The different forms of NPs, such as nanorods, nanocapsules, nanotubes, nanofibres, and nanoliposomes, have different applications.[17–19] This chapter summarizes the functionalization, biocompatibility, and applications of NMs in the biomedical field.

8.2 Synthesis of Nanomaterials

Owing to their remarkable mechanical, electrical, optical, and magnetic capabilities, NMs have attracted considerable attention recently. The creation of nanostructures or nanoarrays with unique qualities for unique classes is one of the objectives of the emerging field of nanotechnology. The high density and small corner sizes of metal oxide NPs result in their distinctive chemical and physical characteristics. The growing number of surface and interface atoms causes stress or strain and consequent structural disturbances as the size of the NPs decreases.

The production of metal oxide NPs using various techniques, including chemical, sol–gel, and coprecipitation procedures, has been described.[20] They can be produced using a variety of preparation techniques that provide possible means of achieving greater chemical homogeneity. These techniques were employed because they have better qualities. Metal oxide NPs are used in experiments to produce good to exceptional yields in organic processes. Therefore, the main goal of the effort is to establish a low-cost, straightforward, and effective method for synthesizing different heterocyclic skeletons utilizing metal oxide NPs that have been created. There is no bandgap in metals because the valance band and conduction band are combined. Therefore, depending on the band position, either reduction or oxidation occurs. However, because of their extremely wide bandgap, insulators require a tremendous amount of energy for the excitation process.

As a result, owing to their medium-sized bandgap, semiconducting materials perform far better as photocatalysts than metals and insulators. Numerous semiconducting substances are available for use as photocatalysts, and numerous techniques are available for creating metal oxides with different levels of photocatalytic activity. These include sol–gel, sol, micelle and inverse micelle, hydrothermal, solvothermal, direct oxidation, chemical vapour deposition (CVD), physical vapour deposition, electrodeposition, sonochemical, and microwave methods, and others.[21–25]

Conducting polymers are combined with inorganic oxide NPs to create nanocomposites, and the results are useful for various forms of energy production and environmental cleanup. The development of metal oxides and their modified materials with a wide range of applications is the primary focus of studies in the field of photocatalysis. The major goal of characterization is to establish a relationship between the structure, physiochemical makeup, and photocatalytic activity.

8.2.1 Synthesis of Metal Oxide-based Nanomaterials

General chemical reduction methods are used for the synthesis of metal oxide NPs because of their simple and easy laboratory procedures and the non-requirement for special instruments. Techniques such as sol–gel, CVD, hydrothermal, and solvothermal methods are used to prepare metal oxide-based NPs. The properties of metal oxide-based NPs are the following:

1. *Chemical stability*: The particle size of NPs is in the range 1–100 nm, hence they possess a high surface-to-volume ratio. The size of NPs has a high association with nanotoxicity, because NPs with smaller size can enter the cells and cause cell disruption. Nanotoxicity is dependent on the size of NPs.
2. *Oxidative dissolution of metal oxide NPs*: Oxidation of metal oxide NPs causes elimination of metals and metal ions in the solute, which leads to toxicity of the solution. The chemical and physical properties of NPs induce toxicity towards macro- and microorganisms.

Metal oxide NMs, especially in the form of NPs, have dimensions of 1–200 nm. Metal oxide NPs are used in imaging as contrast agents for the detection of cancer or disease therapy, owing to their optical properties. As the size of metal NPs decreases, their toxicity increases.

8.2.1.1 Titanium Dioxide (TiO_2) Nanoparticles

Among different metal oxides and sulfides, titanium dioxide has been extensively used because of its reasonable price, photostability, chemical stability, and photocatalytic activity. Titanium dioxide NPs are good antimicrobial agents that are efficient at killing fungi, algae, protozoa, and viruses. Titanium dioxide is used for the degradation of organic compounds under solar light irradiation.[26] When titanium dioxide NPs are irradiated with UV-visible light they exhibit antibacterial properties.

8.2.1.2 Zinc Oxide Nanoparticles

Apart from other metal oxides, ZnONPs also proved to be an efficient material for antimicrobial applications.[27] They have a wide bandgap and strong oxidation ability, which help in their use in wastewater treatment. ZnONPs can be prepared *via* green methods and are more economical to use than titanium dioxide NPs. Dopants such as anionic, cationic, and rare earth dopants are used to enhance the efficiency of ZnONPs.

8.2.1.3 Iron Oxide Nanoparticles

Owing to the small size of iron oxide NPs, they exhibit good adsorption efficiency. Iron oxide NPs are functionalized to synchronize with different ligands such as ethylenediaminetetraacetic acid (EDTA), thiol-containing poly(ethylene glycol) (PEG-SH), mercaptobenzoic acid (MBA), *etc.* Among iron oxides, haematite has the advantage of being a cheap material. Iron oxide NPs have applications in magnetic resonance imaging (MRI) and computed tomography (CT).[28] They have also been applied in magnetic fluid synthesis, micelle synthesis, and hydrothermal synthesis.

8.2.1.4 Dimension and Morphology of Nanoparticles

NMs are considered to be the backbone of nanoscience and nanotechnology. NMs are used in different areas, including electronics, biomaterials, medical, environmental, and energy production. Norio Taguchi was the first person to use the term "nanotechnology" in 1974.[29] NMs are classified according to the dimensions and morphology.

All three dimensions are narrowed down to the nanometre scale in the case of zero-dimensional nanomaterials. Quantum dots of semiconductor metals and non-metals show novel electronic, optical, magnetic, and catalytic properties owing to the particle size being reduced in all three dimensions to the nanometre scale. An anisotropic shape is why one-dimensional nanostructures of metal oxides show unique optical properties. Nanowires have a diameter of less than 20 nm and are used as interconnectors for electron mobility in nanoelectronic devices. Nanowires can be fabricated with different metals such as Co, Au, Cu, and Si. Nanowire metal oxides are used as anodes in dye-sensitized solar cells.

Nanorods are prepared using different techniques such as microwave-assisted, hydrothermal, and thermal decomposition methods. Titania nanotubes (TNTs) are among potential nanostructured metal oxides with tubular morphology. The first nanotubes were carbon nanotubes (CNTs), which were discovered by Sumio Iijima and have been used in various device applications.[30] Metal oxide nanostructures are very effective as catalysts, sensors, solar cells, and electrodes in lithium batteries. Owing to their small dimensions, metal oxide nanotubes show advanced physical and chemical properties. In 2D NMs, one dimension is reduced to the nanometre range and the other dimensions remain large. Graphene is a 2D NM in which monolayers of carbon atoms are tightly packed into a honeycomb lattice. Compared with pristine titania, graphene-decorated titania shows high phocatalytic activity. Among 3D NMs, the major part includes dispersions of NPs, nanowires, and multilayers of nanotubes.

8.2.1.5 Mixed Metal Oxide Nanocomposites

Oxides that contain two or more kinds of metals are called mixed metal oxides. They are divided into binary, ternary, and quaternary based on the presence of metal cations. Mixed metal oxides can enhance the overall performance that is not available in single metal oxides and can also boost the synergetic properties.

8.3 Characteristics of Metal Oxide Nanoparticles

The properties of metal oxide NPs include the following:

1. optical properties
2. electrical properties
3. mechanical properties
4. catalytic properties.

8.4 Methods of Synthesis of Metal Oxide Nanoparticles

Top-down and bottom-up are the two approaches used to prepare NMs. In the top-down approach, breakdown of larger to smaller materials takes place, and in the bottom-up approach, fusion of the smaller particles in the presence of a stabilizer takes place.[31] The various methods of preparation are outlined below.

8.4.1 Coprecipitation Method

Precipitation is an ancient method in which sparingly soluble products from aqueous solutions are decomposed to yield oxides. In coprecipitation, dissolution of salts in solvents takes place.[32]

8.4.2 Sol–Gel Method

In this method, two materials are used, a sol and a gel, *i.e.* it includes a colloidal phase and a gel phase. Usually, NPs prepared using the sol–gel method are amorphous in nature, and next-level advances are needed to produce a crystal structure. The main merits of the sol–gel method are that it has high homogeneity and purity, a low processing temperature, and a uniform phase distribution.[33]

8.4.3 Microemulsion Technique

Surfactants are used for stabilization in microemulsion synthesis. The main advantages of this method are biocompatibility and biodegradability of synthesized materials.[34]

8.4.4 Solvothermal Methods

These are mainly used for the synthesis of metal oxides. Most solvothermal processes need supercritical solvents, which are used to increase the solubility and reactivity of metal salts and complexes at high temperature and pressure without bringing the solvent to its critical point.[35]

8.4.5 Chemical Vapour Deposition (CVD)

Of bottom-up approaches, CVD is generally used. It is used to make nanotubes, nanowires, and other NPs. The types involved are classical, metallo-organic and plasma-assisted.[36]

8.4.6 Microwave-assisted Synthesis

This is a recent method for the synthesis of metal oxide NMs. It initiates rapid heating of reaction mixtures.[37] The main merits are penetration of radiation, controlled electric field distributions, and self-limiting reactions.

8.4.7 Mechanochemical Method

This is a part of solid-state chemistry where intermolecular bonds are broken mechanically. This involves external mechanical action, which is followed by further chemical reactions.

8.4.8 Combustion Method

Using the combustion method, the preparation of metal oxides and their composites in homogeneous form can be achieved.[38] In the combustion method, reaction between an oxidizer, precursor metal nitrates, and fuels (glycine, urea, and hydrazines) in water takes place. It is classified into two types, namely solution combustion synthesis and gel combustion synthesis.

8.4.9 Sonochemical Precipitation Method

This method involves ultrasound irradiation and produces small-sized particles with a large surface area.[39] In this method, precipitation of metal salts occurs when exposed to high-intensity ultrasound radiation under ambient air for a given duration of time.

8.4.10 Photocatalysis

Metal oxide NPs modified by a simple route can be used as photocatalysts, thereby reducing the cost of production. In 1980, Salomon[40] categorized photocatalysis into two classes:

- photon-generated catalysis;
- catalysed photolysis that does not involve photon catalysis.

The visible light-driven photochemical reaction is the basis of the metal oxide semiconductor photocatalytic process. The catalyst is improved by UV or visible light radiation by creating a redox environment in the aqueous solution.[40] Semiconductors serve as sensitizers for the aquatic environment's light-induced redox environment.

A perfect photocatalyst is very effective, stable, cheap, and non-toxic. TiO_2, ZnO, WO_3, ZnS, Fe_2O_3, and other metal oxides have bandgap energies that are suitable for stimulating a variety of chemical processes. The components of heterogeneous photocatalysts are classified into four categories:

1. to form energy structure and crystal structure;
2. to form energy structure but not crystal structure;
3. to construct impurity levels as dopants; and
4. to be exploited as a cocatalyst.

The following are attributes in the field of photocatalysis:

1. small visible light response;
2. doping with foreign species that act as recombination centres;
3. difficulty in supporting some materials.

8.5 Surface Functionalization of Nanomaterials

Monitoring and controlling the interactions between the NPs and the biosystem is essential for the effective utilization of NPs in biomedicine. A specifically structured NP monolayer structure can enhance cellular internalization, non-cytotoxicity, *etc.* Surface functionalization of NPs helps to tune their specificity. NMs are functionalized with different kinds of ligands such as surfactants, dendrimers, polymers, and other biomolecules.

8.5.1 Functionalization of Nanoparticles by Ligands

For different approaches, a large number of ligands are incorporated in the surface of NPs. Functionalization helps to apply NPs in clinical, pharmaceutical, and environmental scenarios. When NPs are functionalized with ligands, they start to exhibit an affinity towards proteins and cell surface molecules. When NPs are functionalized with positively charged ligands, they exhibit higher internalization into the cells than with negatively charged particles.

8.5.2 Polymer-functionalized Nanoparticles

Polymeric NPs contain C, H, N, and O and are synthesized *via* synthetic and biological methods. Biodegradable polymeric NPs are used for the controlled delivery of drugs, selected targeting, and long circulation times. PEG conjugation is used for drug delivery owing to its hydrophilicity.[41] Polymeric NMs are synthesized by different methods. Polymerization first starts with a monomer, and other methods of synthesis include interfacial polymerization and radical polymerization. Polymeric NPs are divided into natural and synthetic types, and synthetic types are further divided into biodegradable and non-biodegradable, *e.g.* synthetic polymeric NPs include poly(acrylic acid) (PAA), polystyrene (PS), poly(methyl methacrylate) (PMMA), *etc.*, which are non-degradable. Examples of biodegradable polymers are poly(lactic acid) (PLA), poly(lactic-*co*-glycolic acid) (PLGA), and PEG. Natural biopolymers such as cellulose, chitin, chitosan, starch, *etc.*, are used in nanomedicine.

Polymer coatings also provide an alternative to other molecules that confers the characteristics of a macromolecular system to the particle surface. PEG is a biodegradable polymer and coating of NPs with PEG helps to increase the passive targeting of tumour tissue. Another advantage of PEG is that it inhibits the adsorption of blood serum proteins, increases the circulation time, and increases the probability of permeation of the particles into tumour tissues. PEG-coated AuNPs exhibit a good targeting mechanism. Polymer-functionalized NPs can enhance the controlled release of drugs. Polymers that exhibit antimicrobial characteristics have a cidal effect on fungi and yeasts owing to their cationic charge, which gives a high binding affinity to the negatively charged microbial membrane surface. Advantages of using polymeric NPs are that they are inexpensive and are not time consuming to synthesize.

8.5.3 Functionalization of Nanoparticles with Biomolecules

Biomolecules of functionalized NPs have specific characteristics that are difficult to achieve *via* synthetic materials. AuNPs functionalized with oligonucleotides help in molecular diagnostics.[42] For targeted drug delivery, NPs are functionalized with proteins and antibodies. Functionalization also helps in imaging and photothermal therapy.

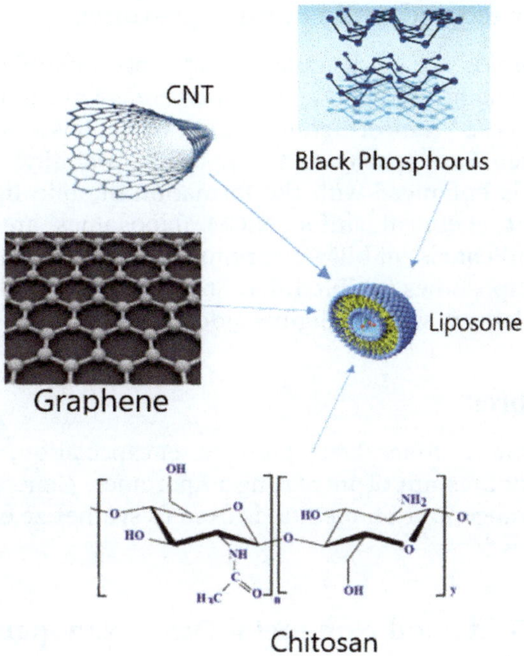

Figure 8.1 Nanomaterials against infections.

An advantage of functionalization of NPs in the biomedical and pharmaceutical fields is the development of a versatile scaffold. Surface functionalization helps to increase biomimetic properties such as non-cytotoxicity, specificity, and stability in biological media. Functionalization of NPs helps to improve theranostic skills by gaining knowledge of cellular uptake. The use of different NMs against infection is illustrated in Figure 8.1.

8.6 Clinical Applications of Nanomaterials Against Infectious Fungi

8.6.1 Peptides

Antimicrobial peptides (AMPs) or host defence peptides are present in animals, plants, or microbes and help to defend the host from invasive diseases. These peptides are amphiphilic and have sequences with fewer than 100 amino acids in length. Antimicrobial peptides possess cationic charge and are rich in histidine, which exhibits high antifungal activity. AMPs are deposited inside the macrophages and lysosomes and are the basis for controlling and preventing fungal infections.[43] Anionic AMPs help to enhance the membrane permeability of microbial membranes. The main limitation of AMPs is their limited half-life. In total, 1211 AMPs with antifungal activities are present throughout the world.

8.6.2 Nanostructured Lipids and Liposomes

Nanostructured lipids are biodegradable and biocompatible and are made up of a blend of a solid lipid and a small percentage of liquid lipid derived from natural sources.[44] Nanostructured lipid carriers possess characteristics such as drug retention capacity and long-term durability. The antifungal activity of lipids is enhanced with the formation of solid lipid NPs or nucleus-independent chemical shifts (NICS). Liposomes are composed of phospholipids and consist of bilayers or multilayers with sizes varying from 30 nm to larger. Liposomes exhibit different properties that allow them to be applied in drug delivery with negligible side effects.[45]

8.6.3 Nanofibres

Owing to their high surface area, porosity, encapsulation, flexibility, and consistency, nanofibres are of increasing importance. Generally, a dynamic high-pressure homogenization method is used to synthesize environmentally friendly nanofibres.[46]

8.6.4 Metal Oxide and Non-metal Oxide Nanoparticles

NPs can be effective antifungal agents with high suppressive effects on the growth of fungi, *e.g.* when AuNPs are combined with photosensitizers they show enhanced photodynamic characteristics. AuNPs possess higher antimicrobial activity along with antifungal activity. Mesoporous SiNPs are effective nanocarriers for the transport of drugs to targeted cells in the body. It is known that all metal oxide and metal NPs induce an inflammatory response.[47] This is completely dependent on size, composition, and shape. Other properties that influence the toxicity of metal NPs include crystallinity, charge, composition, and stability of the NPs. Gold and silver are two noble metals used in nanomedicine. AgNPs possess antimicrobial properties. AuNPs possess show toxicity and are used in diagnostics owing to their surface plasma resonance effect. CuNPs are inexpensive and have useful physical characteristics AlNPs are applied clinically as vaccine adjuvants.[48]

8.6.5 Carbon-based Nanomaterial Composites

Carbon nanomaterials are being used in various forms with varied size. Carbon NMs are known as graphitic NMs,[49] because the carbon atoms are arranged as hexagonal rings as in graphite. Carbon quantum dots are known as a zero-dimensional material,[50] and they are among the smallest NMs. They are prepared by breaking up graphite or CNTs to less than 10 nm in size. Carbon-based NMs are used to activate the inflammasome complex, leading to inflammation strategies to investigate chronic effects.

Membranes made of graphene are typically utilized in the water filtration process. The size exclusion or interaction with graphene-based membranes affects the removal of pollutants. Porous graphene NMs that are atomically thin offer high selectivity and high water flow. This approach offers an alternative to traditional polymer-based nanofiltration and reverse osmosis membranes.[51] Graphene membranes can function at a high pressure of 1750 psi and are quite effective in comparison with reverse osmosis membranes, which operate at 870–1160 psi. Specific regions become wrinkled as a result of membrane breakdown, and these wrinkled areas represent the locations of defects. Cleaning methods for nanofiltration membranes include sonication, air blowing, and flushing. Membranes made of porous graphene show good resilience in high and medium acidity environments.[52] At various pH levels, graphene membranes can also function when sonication waves are present. Membrane wrinkles demonstrate antibacterial action against *Staphylococcus aureus* and *Escherichia coli*. Porous graphene with a large lateral width exhibits greater antibacterial activity. The formation of complexes during the filtration of divalent ions is the main drawback of porous graphene-based membranes.[53] These ions interact with the graphene's functional groups, increasing the interlayer gap and effectively reducing membrane separation.

8.6.6 Nanoceramics

Described as inorganic parts of metals, metalloids, and non-metals with ionic or covalent interactions, ceramics are a diverse class of materials. Microemulsion, precipitation, and hydrothermal synthesis methods are used to create nanoceramics.[54] Applications of ceramic NMs range from tissue engineering and medication delivery to diagnostics and biosensors. The primary benefit of ceramic NPs is that they have an inorganic core that can enclose and safeguard a medication. Ceramics are materials that have a large surface area, high porosity, and a long, regulated drug breakdown time.

8.6.7 Nanochitosan

For the effective removal of contaminants, a variety of biopolymers are combined with chitosan to create blended adsorbents. Alginate and cellulose are added to the biopolymer with chitosan to help in treating contamination.[55] For the purpose of treating water, chitosan and NPs have been combined. A chitosan–ZnO composite was used to remove Acid Black dye from wastewater. Chitosan NPs are very effective at removing pollutants owing to their ability to store water effectively. Because they have functional groups that are ionic in nature and appropriate for separation, pH-sensitive hydrogels made of chitosan NP composites operate well as conductors.[56] Because chitosan beads include hydroxyl groups, they display excellent qualities. Additionally, chitosan demonstrates traits such as acid stability, increased surface area, smaller pores, chemical stability, *etc.* Chitosan is altered using both chemical and physical processes. In order to develop a new, innovative material with improved qualities, two polymers are blended or physically mixed. Blending takes a long time and is very inexpensive to perform. Chitosan membranes work well because of their strong mechanical and physical characteristics. In order to remove heavy metal ions from contaminated water, chitosan adsorptive membranes are utilized. Different functional groups are introduced into the chitosan structure to bring different functionalities.[57] The most promising chemical alteration, which has so far received little attention, is photosensitizer modification. A chemical substance called a photosensitizer is used to absorb light energy. The optimum method for enhancing pollutant degradation is the use of chitosan as a photosensitizer, which is pertinent for synergistic operation with light-derived catalysis. Heavy metal ions and dyes are removed using chitosan and its composites. They contain amino and hydroxyl functional groups, which make them suitable biosorbents. They are also employed as antimicrobial agents against infections.[58]

8.7 Methods of Biocompatibility Measurement

Methods of biocompatibility measurement are illustrated in Figure 8.2 and can be summarized as follows:

1. *Cytotoxicity*: includes qualitative and quantitative cytotoxicity.
2. *Genotoxicity*: *in vitro* and *in vivo* evaluation to find traces of mutagens.
3. *Subchronic toxicity*: to analyse the side effects from longer term influences.
4. *Systemic toxicity*: helps to evaluate any systematic toxicity.
5. *Haemocompatibility*: helps to evaluate the blood compatibility to find safety issues.
6. *Carcinogenic bioassay*: helps to analyse tumorigenic issues.

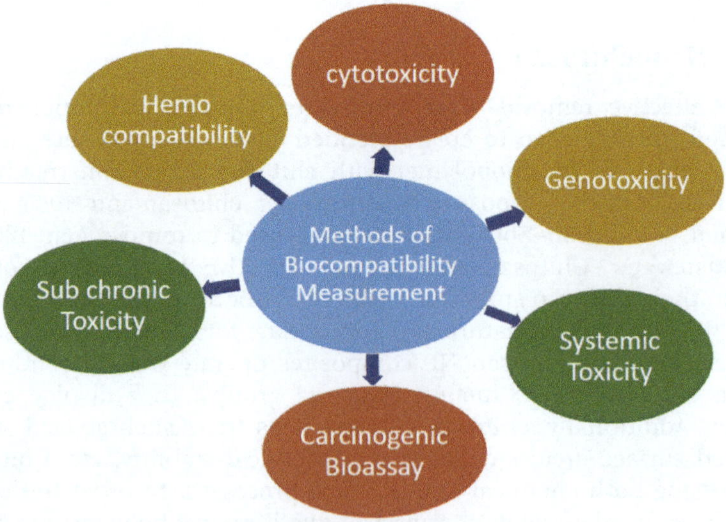

Figure 8.2 Biocompatibility of nanomaterials.

8.8 Biocompatibility and Toxicity of Nanomaterials

NMs that are used in clinical or pharmacological applications must be evaluated in terms of cytotoxicity and viability.[59] The biocompatibility of NMs is evaluated *via in vitro* and *in vivo* methods.[60] It is important to understand the physicochemical features of NMs and their effects on the human system, which are due to different types of NMs varying in chemical composition, shape, size, and surface coating. A lack of standard protocols and increases in the number of *in vivo* tests and their complications are also challenges to the application of NMs in clinical biological processes. Generally, cytotoxicity experiments are based on analysing damaged cell membranes, synthesis of cell death markers, and impaired metabolism. NMs can reduce cell viability through damage to DNA. They also impair metabolic activity based on their composition or geometry. Direct or indirect immunomodulation of NMs has applications in drug delivery. NPs serve as vehicles for vaccines or drugs. In immunomodulation, the main focus is on immune reactions by innate and adaptive immune reflexes. In innate immunity, interactions between NMs and immune cells are considered. In adaptive immunity, T cells and B cells are activated that elicit antigen-specific effects.

Immunomodulation caused by NMs is affected by chemical composition, molecular structure, surface chemistry, size, *etc.* In immunomodulation, the size of the NPs is important. Compared with microsized particles, nanosized particles exhibit a stronger immunoresponse. This leads to the conclusion that a larger size may elicit a stronger immunoglobulin response.

Due to specific feature of NMs, different biological membranes have been developed. AuNPs have been used to eliminate genome damage.[61]

8.9 Cellular Interactions of Nanomaterials

Characterization of NMs can provide information regarding their immunological identity. Interactions between cells and NMs are divided into clathrin-dependent endocytosis, caveolae-dependent endocytosis, phagocytosis, and micropinocytosis.[62] When the size of the NM is larger than 200 nm, phagocytosis and micropinocytosis are the dominant uptake methods. When the size of the NM is between 50 and 80 nm, uptake will depend on either clathrin or caveolae. Both are proteins.

8.9.1 Advantages of Nanomaterials Against Bacterial Infections

NMs can be functionalized to adopt antimicrobial activity without any side effects. Administration of antimicrobial agents with NMs enhances the therapeutic effects. It helps to extend the drug circulation time or extend the half-life. Suspension of drugs and synergetic antibacterial activity are advantages of NMs. The following are the advantages of NMs as antimicrobial drug delivery agents:

- The size of the NMs helps in the design of targeted antibiotics.

- Depending on the nanosize, NMs can cross the cell membrane by phagocytosis and enter the host cell by endocytosis.
- An NM-based drug delivery system can elevate the drug retention time in the blood.
- NMs can protect antibiotics from chemical reaction and show resistance to the targeted bacteria.

8.10 Limitations

Globalization and its effects on social life enhance the prevalence of multidrug-resistant microorganisms. The majority of studies on NMs have solely examined their antibacterial activity *in vitro*, according to a literature analysis. However, only *in vivo* investigations can reveal characteristics such as targeted drug delivery, cytotoxicity, and stability of NMs inside the body. No technique has been developed to pinpoint the aforementioned variables that affect the antibacterial activity of NMs. The antibacterial mechanism hidden behind each particular mechanism must therefore be discovered. Antibacterial resistance can be resolved if we can overcome this obstacle.

8.11 Future Prospects and Conclusions

Surface functionalization of NPs helps to prevent opsonization. Different synthetic and natural polymers have been applied for covering NPs. The main limitations of the former are the absence of robustness and therapeutic and diagnostic value. Different NMs are used in clinical and biomedical applications. To use NMs clinically we need to overcome the following issues:

1. *in vitro* cell uptake
2. *in vitro* photothermal performance
3. *in vivo* MRI
4. *in vivo* photothermal therapy
5. *in vitro* toxicity
6. *in vivo* cytotoxicity
7. *in vivo* biodistribution
8. *in vivo* blood circulation.

Applications of NMs in the biomedical field depend on their toxicity, biocompatibility, sensitivity, and biodegradability. NPs synthesized *via* a green route are generally considered for *in vivo* experiments. NMs and their green synthesis help to reduce the use of toxic and hazardous materials, but estimating their sustainability is more important. Future applications of these NMs are based on improvements of the economic viability, biocompatibility, and interaction mechanisms of NPs. Future studies will be based on their applications in environmental, energy usage, and biomedical fields. In the future, we have to overcome some of the limitations of NMs.

There is a need to find innovative approaches to reducing the cost of NP production and also to develop novel methods to synthesize NMs from biowaste. Further studies should be carried out to find novel NMs that are non-toxic, abundant, and easy to functionalize.

NMs are becoming high-performance materials for the diagnosis of diseases with applications as antimicrobial agents. They are used in cancer diagnostics and therapy. However, compared to conventional molecular scale contrast, NPs are used more for *in vivo* detection. NMs are used in MRI and CT scanning. NPs are used in the detection and treatment of viral infections. Surface functionalization of NPs helps to enhance their properties, leading to the development of more novel nanovaccines, multifunctional NMs, and biomimetic agents.

Abbreviations

AlNP	Aluminium nanoparticle
AMP	Antimicrobial peptide
CNT	Carbon nanotube
CT	Computed tomography
CuNP	Copper nanoparticle
CVD	Chemical vapour decomposition
DNA	Deoxyribonucleic acid
AgNP	Silver nanoparticle
AuNP	Gold nanoparticle
GO	Graphene oxide
MRI	Magnetic resonance imaging
NM	Nanomaterial
NP	Nanoparticle
PEG	Poly(ethylene glycol)
PLA	Poly(lactic acid)
PS	Polystyrene
TiO_2	Titanium dioxide
TNT	Titania nanotube
ZnO	Zinc oxide

References

1. R. Sridharan, B. Monisha, P. Kumar and K. Gayatahri, *Chemosphere*, 2022, 133731.
2. W. Park, H. Shin, B. Choi, W. K. Rhim, K. Na and D. K. Han, *Prog. Mater. Sci.*, 2020, **114**, 100686.
3. M. M. Khin, A. S. Nair, V. J. Babu and R. Murugan, *Energy Environ. Sci.*, 2012, **5**(8), 8075.
4. S. H. Pham, Y. Choi and J. Choi, *Pharmaceutics*, 2020, **12**(7), 630.
5. X. Zheng, P. Zhang, Z. Fu, S. Meng, L. Dai and H. Yang, *RSC Adv.*, 2021, **11**(31), 19041.

6. J. Zhang, H. Chen, M. Zhao, G. Liu and J. Wu, *Nano Res.*, 2020, **13**, 2019.
7. K. Loh, D. Ho, G. N. C. Chiu, D. T. P. Leong, G. Pastorin and E. K.-H. Chow, *Adv. Mater.*, 2018, **30**(47), 1802368.
8. Y. He, C. Hu, Z. Li, C. Wu, Y. Zeng and C. Peng, *Mater. Today Bio*, 2022, 100231.
9. X. Y. Wong, A. Sena-Torralba, E. Alvarez-Diduk, K. Muthoosamy and A. Merkoçi, *ACS Nano*, 2020, **14**(3), 2585.
10. M. R. Khan, Z. A. Siddiqui and X. Fang, *Chemosphere*, 2022, 134114.
11. L. Chen, Z. Liu, Z. Guo and X. J. Huang, *J. Mater. Chem.*, 2020, **8**(34), 17326.
12. A. B. Sengul and E. Asmatulu, *Environ. Chem. Lett.*, 2020, **18**, 1659–1683.
13. M. Vallet-Regí, B. González and I. Izquierdo-Barba, *Int. J. Mol. Sci.*, 2019, **20**(15), 3806.
14. M. Nasrollahzadeh, M. Sajjadi, G. J. Soufi, S. Iravani and R. S. Varma, *Nanomaterials*, 2020, **10**(6), 1072.
15. L. C. W. Lin, S. Chattopadhyay, J. C. Lin and C. M. J. Hu, *Adv. Healthcare Mater.*, 2018, **7**(13), 1701395.
16. J. Yang, L. Yue, Z. Yang, Y. Miao, R. Ouyang and Y. Hu, *Nanomaterials*, 2021, **11**(8), 2129.
17. A. Waris, A. Ali, A. U. Khan, M. Asim, D. Zamel, K. Fatima and M. A. Bourehab, *Nanomaterials*, 2022, **12**(13), 2140.
18. M. T. Amin, A. A. Alazba and U. Manzoor, *Adv. Mater. Sci. Eng.*, 2014, 1–24.
19. K. K. Chenab, B. Sohrabi, A. Jafari and S. Ramakrishna, *Mater. Today Chem.*, 2020, **16**, 100262.
20. R. S. Devan, R. A. Patil, J. H. Lin and Y. R. Ma, *Adv. Mater. Sci. Eng.*, 2012, **22**(16), 3326.
21. M. Parashar, V. K. Shukla and R. Singh, *J. Mater. Sci.: Mater.*, 2020, **31**, 3729–3749.
22. W. Cai, J. Yu, C. Anand, A. Vinu and M. Jaroniec, *Chem. Mater.*, 2011, **23**(5), 1147.
23. J. G. Lu, P. Chang and Z. Fan, *Mater. Sci. Eng.*, 2006, **52**(1–3), 49.
24. E. Comini, C. Baratto, G. Faglia, M. Ferroni, A. Vomiero and G. Sberveglieri, *Prog. Mater. Sci.*, 2009, **54**(1), 1.
25. A. Mirzaei and G. Neri, *Sens. Actuators, B*, 2016, **237**, 749.
26. R. Krakowiak, J. Musial, P. Bakun, M. Spychała, B. Czarczynska-Goslinska, D. T. Mlynarczyk and T. Goslinski, *Appl. Sci.*, 2021, **11**(18), 8674.
27. A. Mirzaei, Z. Chen, F. Haghighat and L. Yerushalmi, *Sustainable Cities Soc.*, 2016, **27**, 407.
28. M. Umar and H. A. Aziz, *Risk Treat.*, 2013, **8**, 196.
29. M. M. El-Kady, I. Ansari, C. Arora, N. Rai, S. Soni, D. K. Verma, P. Singh and A. E. D. Mahmoud, *J. Mol. Liq.*, 2023, **370**, 121046.
30. S. Iijima, *Nature*, 1991, **354**, 56–58.
31. N. Abid, A. M. Khan, S. Shujait, K. Chaudhary, M. Ikram, M. Imran and M. Maqboo, *Adv. Colloid Interface Sci.*, 2021, 102597.
32. A. Kumar and K. M. Gangawane, *Powder Technol.*, 2022, **401**, 117298.
33. J. D. Mackenzie and E. P. Bescher, *Acc. Chem. Res.*, 2007, **40**(9), 810.

34. J. Khan, H. Ullah, M. Sajjad, A. Ali and K. H. Thebo, *Inorg. Chem. Commun.*, 2018, **98**, 132.
35. J. Bi, L. Wu, Z. Li, Z. Ding, X. Wang and K. Fu, *J. Alloys Compd.*, 2009, **480**(2), 684.
36. Y. M. Manawi, A. Samara, T. Al-Ansari and M. A. Atieh, *Materials*, 2018, **11**(5), 822.
37. N. Devi, S. Sahoo, R. Kumar and R. K. Singh, *Nanoscale*, 2021, **13**(27), 11679.
38. A. S. Mukasyan and P.Dinka, *Int. J. Self-Propag. High-Temp. Synth.*, 2007, **16**, 23.
39. S. Govindaraju, M. Ramasamy, R. Baskaran, S. J. Ahn and K. Yun, *Int. J. Nanomed.*, 2015, **10**(sup1), 67.
40. Q. W. Zhu and M. G. Ou, *Appl. Mech. Mater.*, 2014, **670**, 3.
41. S. J. Amina and B. Guo, *Int. J. Nanomed.*, 2020, 9823–9857.
42. B. Rivas-Santiago, C. J. Serrano and J. A. Enciso-Moreno, *Infect. Immun.*, 2009, **77**(11), 4690.
43. A. M. Manea, B. S. Vasile and A. Meghea, *C. R. Chim.*, 2014, **17**(4), 331.
44. M. Alavi, N. Karimi and M. Safaei, *Adv. Pharm. Bull.*, 2017, **7**(1), 3.
45. N. E. Mushi, T. Nishino, L. A. Berglund and Q. Zhou, *ACS Sustainable Chem. Eng.*, 2018, **7**(1), 1692.
46. W. S. Cho, R. Duffin, C. A. Poland, S. E. Howie, W. MacNee, M. Bradley and K. Donaldson, *Environ. Health Perspect.*, 2010, **118**(12), 1699.
47. A. Nazarizadeh, A. H. Staudacher, N. L. Wittwer, T. Turnbull, M. P. Brown and I. Kempson, *Int. J. Mol. Sci.*, 2022, **23**(9), 4707.
48. Z. Wu, Y. Wang, X. Liu, C. Lv, Y. Li, D. Wei and Z. Liu, *Adv. Mater.*, 2019, **31**(9), 1800716.
49. W. Liang, C. E. Bunker and Y. P. Sun, *ACS Omega*, 2020, **5**(2), 965.
50. H. W. Yoon, Y. H. Cho and H. B. Park, *Proc.: Math., Phys. Eng.*, 2016, **374**(2060), 20150024.
51. A. Al-Nahain, S. Y. Lee, I. In, K. D. Lee and S. Y. Park, *Int. J. Pharm.*, 2013, **450**(1–2), 208.
52. N. Song, X. Gao, Z. Ma, X. Wang, Y. Wei and C. Gao, *Desalination*, 2018, **437**, 59.
53. U. Shashikumar, B. C. Jha, S. Chawla, C. M. Hussain, G. Andaluri, Y. C. Lin and V. K. Ponnusamy, *Fuel*, 2023, **336**, 126829.
54. D. P. Stephen and S. B. Palanisamy, Advances in biopolymer composites and biomaterials for the removal of emerging contaminants, *Phys. Sci. Rev*, 2021.
55. Y. Bu, H. X. Xu, X. Li, W. J. Xu, Y. X. Yin, H. L. Dai, X. B. Wang, Z. J. Huang and P. H. Xu, *RSC Adv.*, 2018, **8**(20), 10806–10817.
56. T. H. Kim, H. L. Jiang and C. Chong-Su, *Prog. Polym. Sci.*, 2007, **2**(7), 726.
57. E. Oyervides-Muñoz, E. Pollet, G. Ulrich, G. de Jesús Sosa-Santillán and L. Avérous, *Carbohydr. Polym.*, 2017, **157**, 1922.
58. S. K. Sohaebuddin, P. T. Thevenot, D. Baker, J. W. Eaton and L. Tang, *Part. Fibre Toxicol.*, 2010, 7, 1.

59. T. R. Kyriakides, A. Raj, T. H. Seng, T. H. Xiao, H. Nguyen, F. Mohammed and W. C. Sheu, *Biomed. Mater.*, 2021, **16**(4), 042005.
60. N. A. Nikitin, M. V. Arkhipenko, O. V. Dement'eva, M. E. Kartseva and V. N. Morozov, *Part. Part. Syst. Charact.*, 2022, **39**(11), 2200074.
61. Y. L. Wu, N. Putcha, K. W. Ng, D. T. Leong, C. T. Lim, S. C. J. Loo and X. Chen, *Acc. Chem. Res.*, 2013, **46**(3), 782.
62. A. Naskar and K. S. Kim, *Microb. Pathog.*, 2019, **137**, 103800.

CHAPTER 9

Nanomaterial-based Narrowband UVB Light-emitting Devices for Phototherapy Treatment of Infectious Skin Diseases

AVULA BALAKRISHNA, AMANDA PERCY SEFAGE AND LEELAKRISHNA REDDY*

Department of Physics, University of Johannesburg, Johannesburg ZA-2006, South Africa
*Email: krishr@uj.ac.za

9.1 Introduction

Ultraviolet (UV) radiation is a fractional component of the electromagnetic spectrum with a specific wavelength range of 200–400 nm. This specific spectral wavelength range is further divided into three ranges, namely UVA, UVB, and UBC at 320–400, 280–320, and 200–280 nm, respectively. Within the broad UVB spectrum mentioned above, production of light in the narrowband (NB) spectrum of wavelength 310–313 nm is regarded as the most effective, safest, and least phototoxic in treating more than 40 skin diseases and disorders, such as psoriasis, vitiligo, and other skin diseases using phototherapy treatments.[1] The basis of phototherapy is the interaction of ~313 nm UVB radiation with cell tissues, which induces an immune

Nanoscience & Nanotechnology Series No. 62
Smart Nanomaterials for Infectious Diseases
Edited by Suvardhan Kanchi, Nidhi Sharotri, Rajasekhar Chokkareddy, Deepali Sharma and Falah H. Hussein

Published by the Royal Society of Chemistry, www.rsc.org

response.[2] Among the rare earth ions, the trivalent gadolinium (Gd^{3+}) ion with a half-filled ($4f^7$) electronic configuration has been consistently used as an activator in fluorescent lamps because of its narrow single emission line spectrum that originates due to a transition from an excited metastable ($6P_{7/2}$) state to its ground ($8S_{7/2}$) state, which falls within the action spectrum of phototherapy.[3] The emission line produced by Gd^{3+} ions that falls within the UVB spectral range is necessary to serve as a light source to treat skin diseases in phototherapy treatments. However, this UVB spectral emission produced by Gd^{3+} ions is too low owing to its parity 4f–4f forbidden transitions, and this affects its performance efficiency in fluorescent phototherapy lamps. To overcome this shortcoming, it is necessary for the Gd^{3+} ions to be sensitized through co-doping with sensitizers such as Pr^{3+}, Ce^{3+}, Bi^{3+}, and Pb^{2+} ions, whereby high UVB emissions are achieved through energy transfer mechanisms from the sensitizer ions to the Gd^{3+} ions. The UVB emission by Gd^{3+} ions is mainly influenced by the type of host material used. Among various inorganic host materials, wide-bandgap materials such as fluorides, phosphates, and borates have been considered for dopant embedding. This chapter is focused on highlighting the efficiencies of different sensitizers in pairing with Gd^{3+}-activated inorganic hosts in producing enhanced ~313 nm UVB emissions for resourceful phototherapeutic applications.

9.2 Reasons for Choosing Specific Phosphor Host Materials and Sensitizers for Gd^{3+}-activated NB-UVB Emissions

Many researchers have reported on Gd^{3+} emissions using a few specific sensitizers prepared with different host materials to achieve narrowband emission in the UVB region for the treatment of multiple types of skin diseases and disorders, which can be summarized as follows:

1. appropriate choice of inorganic phosphor host materials;
2. appropriate choice of gadolinium as an activator;
3. appropriate choice of sensitizer ions;
4. Gd^{3+} emissions through energy transfer mechanisms.

These topics are discussed in the following sections.

9.2.1 Appropriate Choice of Inorganic Phosphor Host Materials

9.2.1.1 Borates

Borate compounds are excellent inorganic phosphor host materials for phototherapy applications, owing to the nature of their planar sp^2 or tetrahedral sp^3 hybridization, which creates complex structures when borate atoms bond to either three or four oxygen atoms, respectively.[4] Further, inorganic borates can easily be synthesized at low temperatures and have a relatively large bandgap (3.0–3.5 eV), making them ideal for the embedding of multiple dopants.[5] In addition, borates also have additional properties such as diverse crystal structures, good thermal and chemical stability, supreme optical qualities, and a high optical damage threshold, and are transparent to a wide range of wavelengths,[5,6] which make them more appropriate compounds for phototherapeutic applications.

9.2.1.2 *Phosphates*

Phosphates are amazing inorganic host phosphor materials because of their excellent properties such as high chemical and thermal stability, high efficiency, wide bandgap, high refractive index, low sintering temperature, good charge stabilities, facile methods of preparation, and excellent optical properties, making these materials ideal hosts for phototherapeutic applications.[7,8] UVB emissions from Gd^{3+}-activated phosphate materials are found to be a useful model to fabricate fluorescent phototherapy lamps.

9.2.1.3 *Fluorides/Oxyfluorides*

Fluoride compounds are good phosphor host materials for phototherapy applications because of their low phonon energy (<350 cm^{-1}), wide bandgap (9–10 eV), wide transparent region, and ease of oxidization. However, fluoride materials have some drawbacks, namely, they are chemically unstable, have low chemical durability, and have low probabilities of material optimization using various synthesis techniques. On the other hand, oxides may provide the solution to these issues because of their good stability and relatively high phonon energy. A combination of the properties of both fluorides and oxides to form oxyfluorides is also a better option for an inorganic host phosphor material because of new advantages such as moderate phonon energy and better chemical stability.[9] Fluorides/oxyfluorides doped with rare earth metal ions have found attractive applications in the optics and phototherapy fields because of their high quantum efficiencies and favorable chemical and mechanical properties.[10]

9.2.2 Appropriate Choice of Gadolinium as an Activator

Activation of phosphor materials with the trivalent gadolinium ion, Gd^{3+}, is consistently used for the production of NB-UVB emissions that can be integrated into fluorescent phototherapy lamps for the treatment of various skin diseases and skin disorders.[11] The main interest in the use of Gd^{3+} ions is because of the existence of the energy transfer process, leading to a high luminescent efficiency of the ~313 nm UVB band that forms a vital part of phototherapy lamps.[12] The Gd^{3+} ion with a $4f^7$ electronic configuration has a stable ground state $^8S_{7/2}$ and high excited states such as ^6P_J, ^6I_J, ^6D_J, and ^6G_J

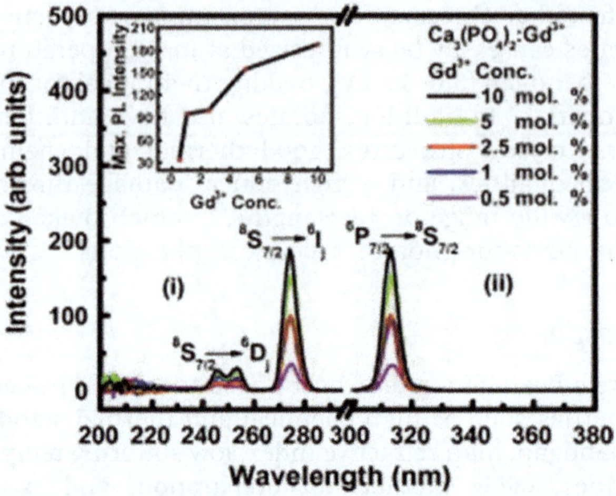

Figure 9.1 (i and ii): Full-length photoluminescence excitation (PLE) and photoluminescence (PL) spectra of $Ca_3(PO_4)_2:Gd^{3+}$ nanophosphors under excitation ($\lambda_{em} = 313$ nm) and emission ($\lambda_{ex} = 274$ nm) wavelengths, respectively. Reproduced from ref. 13 with permission from Elsevier, Copyright 2014.

(see the energy level diagram in Figure 9.3b). The energy difference between the first excited state and the ground state is 32 000 cm^{-1} in the UV region, which makes this ion applicable for phototherapy applications through the excitation of a wider UV band.[12]

Figure 9.1 shows the excitation and emission spectra of a typical Gd^{3+} activated $Ca_3(PO_4)_2$ host material with an excitation wavelength of 274 nm and monitored at an emission wavelength of 313 nm.[13] The excitation spectrum shows two peaks, one at 245 nm ($^8S_{7/2} \rightarrow {}^6D_J$) and the other at 274 nm ($^8S_{7/2} \rightarrow {}^6I_J$), with that at 274 nm being more prominent. The emission spectrum shows a prominent UVB peak at 313 nm ($^6P_{7/2} \rightarrow {}^8S_{7/2}$). The development of NB-UVB emission from Gd^{3+}-activated phosphor materials, with emission at around 313 nm, due to the $^6P_{7/2} \rightarrow {}^8S_{7/2}$ transition, falls within the action spectrum of phototherapy skin treatments.[3] Owing to the parity-forbidden nature of its 4f–4f transitions, its emission intensity is sharp but very low, hence sensitizers are required to enhance its emission intensity by absorbing sufficient energy from the excitation source through energy transfer from sensitizers to the activator ions. For this to happen, the emitting levels of sensitizers should be placed very close to the excited states of Gd^{3+} activator ions, commonly known as spectral overlap. According to the literature, possible sensitizers to couple with the Gd^{3+} activator ions are Ce^{3+}, Pb^{2+}, Bi^{3+}, and Pr^{3+} ions.

9.2.3 Appropriate Choice of Sensitizers

The addition of the sensitizer ion creates channels for the possible energy transfer mechanism between the sensitizer and activator ions: Pr^{3+}(sensitizer) → Gd^{3+}(activator). Upon UV excitation, Pr^{3+} with a $4f^2$ electronic configuration exhibits a cascade of photons ranging from the UV to the IR regions of the electromagnetic spectrum. The excitation spectrum of a typical $Ca_3(PO_4)_2:Pr^{3+}$ phosphor material shows a few narrow peaks between 400 and 500 nm, which are attributed to Laporte-forbidden f–f transitions as shown in Figure 9.2. The narrow band located within this region is due to transitions from the ground state ($^3H_4$) of the Pr^{3+} ion to various excited states such as $^3P_2$ and ^2P_J ($J = 0, 1$).[8,13] A broad excitation band located at around 300 nm could be due to the charge-transfer state or 4f to 4f–5d transition.[8,13] The emission observed at 603 nm is due to a red photon emission, arising from the radiative $^3P_0 \rightarrow {}^3H_6$ transition.

Addition of Pr^{3+} to the $Ca_3(PO_4)_2:Gd^{3+}$ phosphor material as a co-dopant leads to suppression of Gd^{3+} excitation bands at 276 nm.[8,13] Figure 9.3a shows the emission spectra of Gd^{3+} singly doped and Gd^{3+}–Pr^{3+} co-doped $Ca_3(PO_4)_2$ powder phosphors. A broad, intense emission located at around 227 nm is due to the 4f → 4f5d transition of the Pr^{3+} ion, while a minor peak at 274 nm is due to the $^8S_{7/2} \rightarrow {}^6I_J$ transition of the Gd^{3+} ion.[8,13] An intense peak at 313 nm (five times more intense than for the singly doped Gd^{3+} ion) is due to the $^8P_{7/2} \rightarrow {}^8S_{7/2}$ transition of the Gd^{3+} ion (see the inset in Figure 9.3a).[8,13] The energy transfer mechanism from the Pr^{3+} to the Gd^{3+} ion is shown in Figure 9.3b. The Pr^{3+} ion absorbs energy from the high-energy source and transfers it non-radiatively to the Gd^{3+} ion, thereby producing enhanced UVB emission at a wavelength of 313 nm. Energy is transferred from the Pr^{3+} ion to either the ^6D_J or ^6I_J state of Gd^{3+} and thereafter it populates the ^6P_J level before making a downward transition to the $^8S_{7/2}$ ground state. This implies that the Pr^{3+} ion is a good sensitizer for the Gd^{3+} ion in UVB emissions.

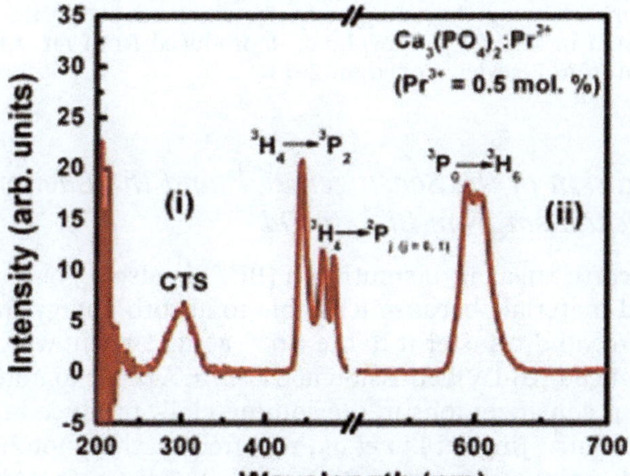

Figure 9.2 (i), (ii): Full-length PLE and PL spectra of $Ca_3(PO_4)_2:Pr^{3+}$ nanophosphors under excitation ($\lambda_{em} = 603$ nm) and emission ($\lambda_{exc} = 444$ nm) wavelengths, respectively. Reproduced from ref. 13 with permission from Elsevier, Copyright 2014.

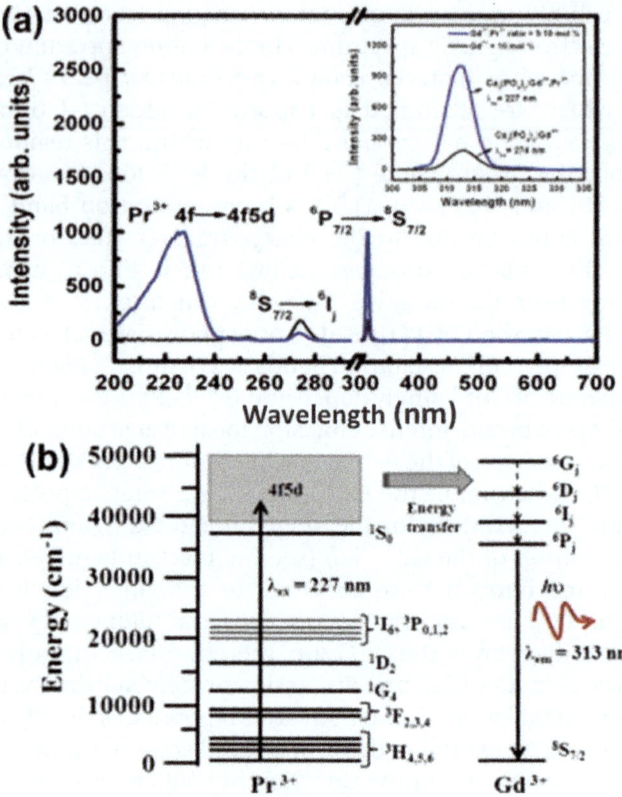

Figure 9.3 (a) Comparison of the PLE and PL spectra of Gd^{3+} singly doped and Gd^{3+}–Pr^{3+} co-doped $Ca_3(PO_4)_2$ phosphors. (b) Partial energy level diagrams showing the energy transfer mechanism for the $Pr^{3+} \rightarrow Gd^{3+}$ system in $Ca_3(PO_4)_2$ phosphors. Reproduced from ref. 13 with permission from Elsevier, Copyright 2014.

9.2.3.1 *Addition of the Sensitizer Bi^{3+} and the Energy Transfer Mechanism from Bi^{3+} to Gd^{3+}*

The non-rare earth trivalent bismuth ion (Bi^{3+}) is also a good sensitizer for Gd^{3+}-activated materials because it is able to absorb energy from the high excitation source and transfer it to the Gd^{3+} activator ion, which eventually results in enhanced NB-UVB emission at 311 nm. A point to note is that Bi^{3+} ions are used as sensitizer ions in the commercially produced LaB_3O_6:Gd,Bi phosphor material.[14] Jingwei Liu *et al.*, reported on the photoluminescence excitation and emission spectra of a typical $(Y,Gd)_3Ga_5O_{12}$ phosphor material using the parameters *i.e.*, λ_{exc} wavelength of 275 nm and λ_{emi} wavelength of 313 nm of Gd^{3+} ions.[15] The excitation spectrum of this phosphor material co-doped with Bi^{3+} ions reveals sharp and broad excitation peaks at 275 and 289 nm.[15] The peak at 275 nm is due to the $^8S_{7/2} \rightarrow {}^6I_J$ transition of

Gd^{3+} ions and that at 289 nm is due to the $^1S_0 \rightarrow {}^3P_1$ transition of Bi^{3+} ions. A single-line UVB emission at 313 nm is produced due to the non-radiative energy transfer from the excited $^3P_1$ levels of Bi^{3+} to ^6P_J of Gd^{3+} through the resonant energy mechanism. The energy emission at 313 nm originates from Gd^{3+} ions due to the $^6P_{7/2} \rightarrow {}^8S_{7/2}$ transition. This confirms that the Bi^{3+} ion is a good sensitizer in producing enhanced NB-UVB emissions in Gd^{3+}-activated compounds.

9.2.3.2 Addition of the Sensitizer Pb^{2+} and the Energy Transfer Mechanism from Pb^{2+} to Gd^{3+}

The Pb^{2+} ion has an [Xe]6s^2 electronic configuration, which, together with its non-shielded s and p orbitals, implies that it depends on the crystal field of the various host materials.[1] All of the properties of the Pb^{2+} ion (excitation, emission, Stokes shift, and quenching temperatures) will be affected by choosing different host matrices.[1] Co-doping of Gd^{3+}-activated phosphor materials with Pb^{2+} ions produces an excitation spectrum that has a peak maximum at 252 nm (due to Pb^{2+} ions) along with two more emission peaks, one a broad peak at 313 nm (corresponding to the $^6P_J \rightarrow {}^8S_{7/2}$ transition of Gd^{3+} ions) and the other a sharp peak at 330 nm (corresponding to the $^3P_1 \rightarrow {}^1S_0$ transition of Pb^{2+} ions).

Figure 9.4a and b show the photoluminescence emission and excitation spectra of $Sr_2Mg(BO_3)_2$:Pb^{2+},Gd^{3+} phosphors. In the spectra of a typical sample of $Sr_2Mg(BO_3)_2$:Pb^{2+},Gd^{3+}, if the concentration of Gd^{3+} is varied while Pb^{2+} is kept constant or *vice versa*, the peak emission pattern of Gd^{3+} and Pb^{2+} ions remains the same, except for a change in the Stokes shift values.[1] The fact that an NB-UVB emission at 313 nm occurs in both cases implies that energy has been absorbed by Pb^{2+} ions, and a portion of it was transferred to Gd^{3+} ions. This also confirms that Pb^{2+} is a reasonably good

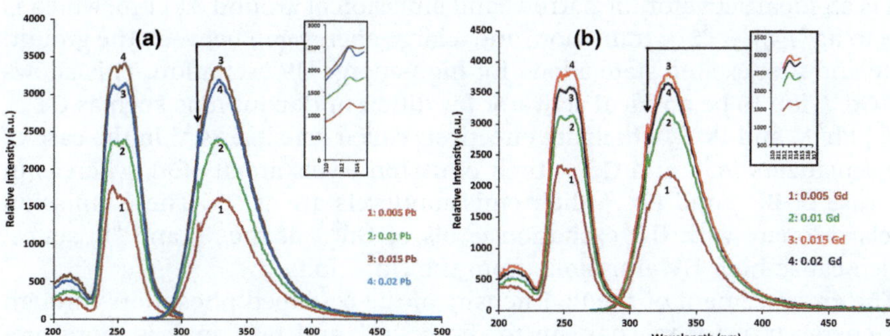

Figure 9.4 (a) PLE and PL spectra of $Sr_2Mg(BO_3)_2{:}_xPb^{2+}{,}_{0.015}Gd^{3+}$ ($x = 0.005$, 0.01, 0.015, 0.02) nanophosphors. (b) PLE and PL spectra of $Sr_2Mg(BO_3)_2{:}_{0.015}Pb^{2+}{,}_xGd^{3+}$ ($x = 0.005$, 0.01, 0.015, 0.02) nanophosphors. Reproduced from ref. 1 with permission from Elsevier, Copyright 2015.

sensitizer for Gd^{3+}-activated host materials. The energy transfer mechanism between Pb^{2+} and Gd^{3+} ions has scarcely been reported.

9.2.3.3 *Addition of the Sensitizer Ce^{3+} and the Energy Transfer Mechanism from Ce^{3+} to Gd^{3+} Ions*

When the Ce^{3+} ion with an electronic configuration of [Xe]4f^1 is excited, there exists only one electron in the 5d shell that interacts strongly with the crystal field of the host, while the 4f shell remains empty.[16] The 4f ground states of the Ce^{3+} ion are split into two by spin–orbit coupling, namely $^2F_{5/2}$ and $^2F_{7/2}$, while the 5d excited state is split into five configurations.[16] The energy difference between the ground state and the excited state is around 51 000 cm^{-1}, which implies that Ce^{3+} must be UV excited. The role of the Ce^{3+} ion is rather complicated in different host materials: sometimes it acts as a sensitizer and sometimes it acts as an activator in Gd^{3+}-activated phosphor host materials, and in many cases there is an energy transfer crossover between Ce^{3+} and Gd^{3+} ions.

Figure 9.5a and b display typical excitation and emission spectra of a $Ca_3(PO_4)_2$:1 mol% $Ce^{3+}/_xGd^{3+}$ (where $x = 0.5$, 1.0, 3.0, and 5.0 mol%) phosphors where mostly Ce^{3+} is found to sensitize Gd^{3+} ions, as a case scenario.[17] The excitation peaked at 271 nm and the emission band consists of two peaks, one broad peak at 326 nm and a narrow peak at 314 nm which is due to the $^6P_{7/2} \rightarrow {}^8S_{7/2}$ transition of the Gd^{3+} ion. The broad peak at 370 nm could be due to the 5d → 4f transition of the Ce^{3+} ion. This scenario is a case where Ce^{3+} plays a dual role, on the one hand being a sensitizer and on the other being an activator.

Despite the dual role played by Ce^{3+} ions, it is certain in one scenario that the enhanced emission at 314 nm is due to energy absorbed by Ce^{3+} and transferred to the Gd^{3+} ion, as can be seen from the energy mechanism diagram in Figure 9.6.[17,28] In summary, we can say that the trivalent Gd^{3+} ion is an ideal activator for narrowband emission at around 313 nm, which is due to its $^6P_{7/2} \rightarrow {}^8S_{7/2}$ transition, and a large energy gap between the ground state and first excited state allows for high-energy UV excitation. This allows the Gd^{3+} ion to be an ideal activator for different dopant ions such as Ce^{3+}, Bi^{3+}, Pb^{2+}, and Pr^{3+}, which are effectively called sensitizers.[15] In the case of the sensitizers Pr^{3+} and Ce^{3+}, their emission levels are at 4f5d, whereas in the case of Bi^{3+} and Pb^{2+}, their emission levels are at $^3P_1$. These emission levels resonate with the excitation levels of Gd^{3+} at the ^6I_J and ^6P_J states, which cause high UV emissions from the Gd^{3+} ions.[15]

The enhancement of the PL intensity of the co-doped phosphors through the energy transfer mechanism for both Pr^{3+} and Bi^{3+} ions is more pronounced than that for singly doped Gd^{3+} ions.[5] This proves that the Pr^{3+} ion is the best sensitizing ion of Gd^{3+} ions for producing NB-UVB emission at around 311 nm. Reports of enhanced NB-UVB emissions through the energy transfer mechanism using various sensitizers (Pb^{2+}, Bi^{3+}, Pr^{3+}, Ce^{3+}) with

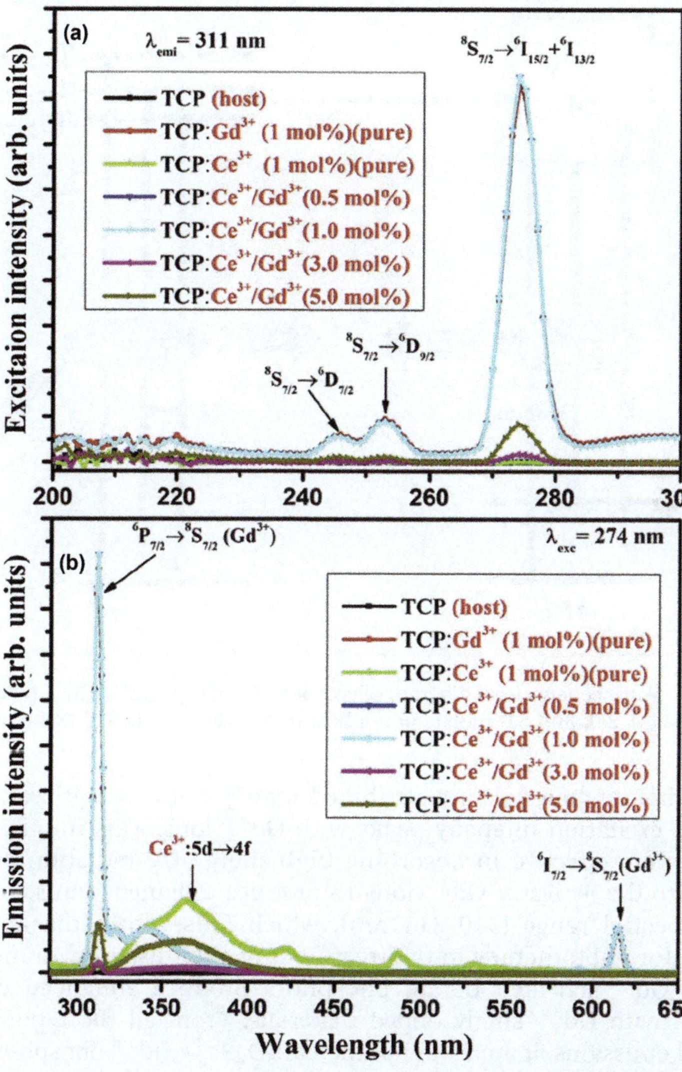

Figure 9.5 (a and b) PLE and PL emission spectra for undoped $Ca_3(PO_4)$ and co-doped $Ca_3(PO_4)_2$:Ce^{3+}/$_x Gd^{3+}$ (where $x = 0.5$, 1.0, 3.0, and 5.0 mol%) nanophosphors with $\lambda_{em} = 311$ nm and $\lambda_{exc} = 274$ nm, respectively. Reproduced from ref. 28 with permission from Elsevier, Copyright 2020.

Gd^{3+} as an activator ion between the various dopants in different host materials, such as borates, phosphates, and fluorides/oxyfluorides, are mainly considered in this section. Table 9.1 presents the results for Gd^{3+}-sensitized emissions. From the "Comments" in the table, Gd^{3+}-sensitized borate host materials appear to show enhanced NB-UVB emissions around the 313 nm wavelength range. The commonly used methods of synthesis are combustion and solid-state reaction methods. Enhanced excitation peaks of

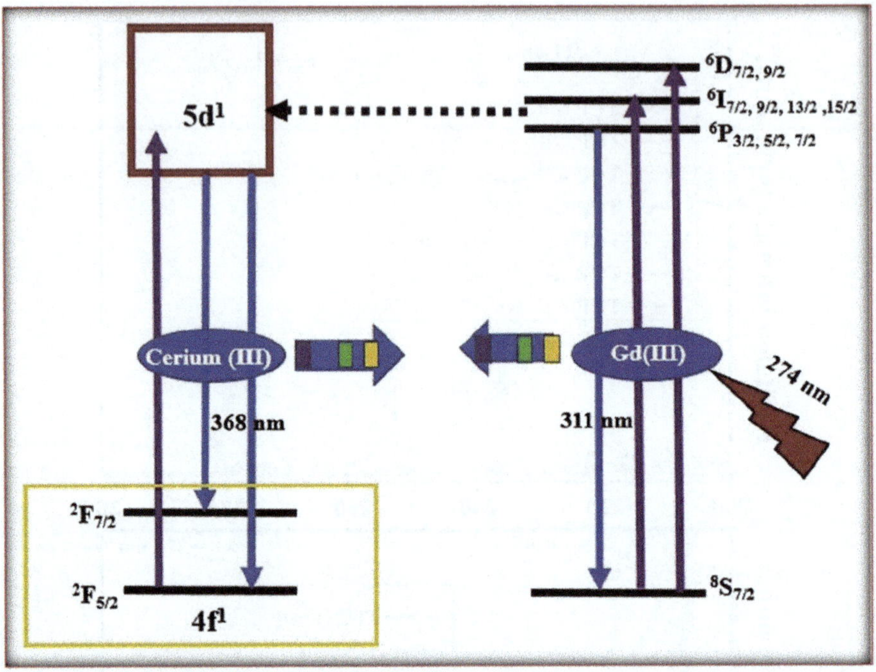

Figure 9.6 Partial energy level diagram of co-doped $Ca_3(PO_4)_2$:Ce^{3+}/$_x$$Gd^{3+}$ (where $x = 0.5$, 1.0, 3.0, and 5.0 mol%) nanophosphors with 274 nm UV excitation.

co-doped borate materials are attributed mainly to the sensitizers, followed by weaker excitation intensity peaks with Gd^{3+} ions. This implies that the sensitizers are effective in absorbing high-energy UV radiation and transferring it to the activator Gd^{3+} ions to produce enhanced emissions in the narrow spectral range (310–313 nm), which falls within the appropriate spectrum for phototherapy in the treatment of skin diseases. In most cases, co-doped Gd^{3+}-activated borate phosphors produce enhanced emissions compared with Gd^{3+} singly doped materials. From all the reported Gd^{3+}-sensitized emissions, it appears that the LaB_5O_9:Pr^{3+},Gd^{3+} phosphor material is close to the commercially produced LaB_3O_6:Bi^{3+},Gd^{3+} phosphor material used in the Philips TL01 phototherapy lamps. Table 9.2 presents the results for Gd^{3+}-activated emissions with phosphates as the host material. Phosphates are considered the next best host materials after borates because of their high chemical and thermal stability and their excellent charge stabilities. The complex structures that borate-based materials form with oxygen make better host materials than other more stable host compounds.

Phosphate host materials for NB-UVB emissions are not as useful as borate host materials. Combustion is the most commonly used method of synthesis of phosphor materials. Enhanced excitation of co-doped phosphate host materials is likewise dependent on the sensitizers, while minor excitation peaks due to Gd^{3+} ions are observed. The only exception is when

Table 9.1 Comparison of different Gd^{3+} singly doped and co-doped borate-based compounds for narrowband UVB emissions.

No.	Host sample	Synthesis technique	Excitation wavelength/nm	Emission wavelength/nm	Comments
1	$(LaY)BaB_9O_{16}$: Bi^{3+},Gd^{3+}[18]	Combustion	260 (Bi^{3+}/Hg)	312	The excitation spectrum consists of a broad spectrum at around 260 nm due to Bi^{3+} ions ($^1S_0 \rightarrow {}^3P_1$) and the emission spectrum consists of a narrow band at around 312 nm due to Gd^{3+} ions. Efficient energy transfer between Bi^{3+} and Gd^{3+} ions was responsible for the enhanced emission observed. The Stokes shift was calculated to be 6410 cm^{-1}[6]
2a	YBO_3:Pr^{3+},Gd^{3+}[4]	Combustion	260 (Bi^{3+})	312	When Bi^{3+} is co-doped with Gd^{3+} in YBO_3, the emission intensity is four times greater than that of singly doped YBO_3:Gd^{3+} phosphor materials, but when Pr^{3+} is co-doped with YBO_3:Gd^{3+} phosphor, the emission intensity is around 30 times greater than that of YBO_3:Gd^{3+} phosphors at 312 nm. This proves that Pr^{3+} is a much better sensitizer of Gd^{3+} than Bi^{3+} ions. The Stokes shift in the first case is 10 000 cm^{-1} and in the second case it is 7949 cm^{-1}[6]
2b	YBO_3:Bi^{3+},Gd^{3+}[4]	Combustion	250 (Bi^{3+})	312	
3	$Sr_2Mg(BO_3)_2$: Pr^{3+},Gd^{3+}[19]	Combustion	248 (Pr^{3+})	313	When $Sr_2Mg(BO_3)_2$ is doped with Pr^{3+} ions, it shows an excitation spectrum at around 250 nm, but when doped with Gd^{3+} alone, the emission spectrum occurs at around 248 and 276 nm. When Pr^{3+} is co-doped with $Sr_2Mg(BO_3)_2$:Gd^{3+} phosphor materials, the Pr^{3+} excitation band at 248 nm is enhanced, while the Gd^{3+} excitation is reduced. The emission intensity of Gd^{3+} at 313 nm is almost doubled compared with singly doped Gd^{3+} ions. This means that Pr^{3+} effectively absorbs the excitation energy and transfers it non-radiatively to Gd^{3+} ions. No trace of Pr^{3+} ions was observed in the emission spectra
4	$Ba_2B_2O_5$:Pr^{3+}, Gd^{3+}[20]	Combustion	248 (Pr^{3+})	313	When $Sr_2Mg(BO_3)_2$ is doped with the Pr^{3+} ion, it shows an excitation spectrum at around 250 nm, but when doped with Gd^{3+} alone, the excitation spectra occur at around 248 and 276 nm. When Pr^{3+} is co-doped with $Sr_2Mg(BO_3)_2$:Gd^{3+} phosphor materials, the Pr^{3+} excitation intensity is

Table 9.1 (*Continued*)

No.	Host sample	Synthesis technique	Excitation wavelength/nm	Emission wavelength/nm	Comments
					enhanced at 248 nm (more than double), while the Gd^{3+} excitation is reduced at 276 nm (almost half) with others suppressed. The emission intensity of Gd^{3+} at 313 nm is almost doubled compared with singly doped Gd^{3+} ions. This means that Pr^{3+} ions effectively absorb the excitation energy and transfer it non-radiatively to Gd^{3+} ions. No trace of Pr^{3+} ions was observed in the emission spectra
5	$Sr_2Mg(BO_3)_2$:Pb^{2+},Gd^{3+} [1]	Combustion	254 (Pb^{2+}/Hg)	313	When Pb^{2+} ions are co-doped with $Sr_2Mg(BO_3)_2$:Gd^{3+} phosphor materials, only absorption bands due to Pb^{2+} ions are observed, with no trace of Gd^{3+} ions. The emission spectra consist of a sharp peak at 313 nm and a broad peak of higher intensity at 330 nm due to Pb^{2+} ions. The excitation spectra were monitored using a wavelength of 330 nm. In this case, the Stokes shift was 7420 cm^{-1}. If the excitation was monitored at 313 nm, the Stokes shift was 9070 cm^{-1}. When the Gd^{3+} ion concentration was kept constant while the Pb^{2+} ion concentration was varied, and *vice versa*, the curve patterns were almost the same, except that the intensity was slightly higher for Gd^{3+} variations. Here, Pb^{2+} ions absorb some of the excitation energy and transfer it to Gd^{3+} ions, showing effective energy transfer between them, with enhanced emission at 313 nm
6	$LiSr_4(BO_3)_2$:Pr^{3+}, Gd^{3+} [21]	Solid-state reaction	245 (Pr^{3+})	313	When Pr^{3+} is co-doped with $LiSr_4(BO_3)_2$:Gd^{3+} phosphor materials, the excitation peak of Pr^{3+} is at 245 nm (almost double in size), while the Gd^{3+} excitation peak at 276 nm is reduced, but the emission intensity of the peak of Gd^{3+} is enhanced at 313 nm and almost double that when singly doped with Gd^{3+} ions. This confirms that energy is effectively transferred from Pr^{3+} ions to Gd^{3+} ions. The emission spectrum shows no trace of Pr^{3+} ions

7	$SrZr(BO_3)_2:Pr^{3+}$, Gd^{3+} [22]	Combustion	230 (Pr^{3+})	313	When Pr^{3+} is co-doped with $SrZr(BO_3)_2:Gd^{3+}$ phosphor materials, it shows an intense excitation spectrum at 230 nm due to Pr^{3+} ions and a minor excitation peak at 276 nm due to Gd^{3+} ions. The emission spectrum shows an enhanced peak at 313 nm due to Gd^{3+} ions, with no trace of Pr^{3+} ions. This spectrum is like the Philips commercial TL01 lamp. This confirms that Pr^{3+} is an effective sensitizer of the Gd^{3+}-activated borate host material
8	$BaB_8O_{13}:Pr^{3+}$, Gd^{3+} [23]	Solid-state reaction	220 (Pr^{3+})	313	When Pr^{3+} is co-doped in $BaB_8O_{13}:Gd^{3+}$ phosphor materials, it reveals an excitation spectrum with a broad, intense peak at 220 nm due to Pr^{3+} ions ($4f \rightarrow 4f5d$ transition) and a slightly less intense peak at 274 nm due to Gd^{3+} ions ($^8S_{7/2} \rightarrow {}^6I_J$ transition), monitored with an emission wavelength of 313 nm. The emission peak at 313 nm at an excitation wavelength of 220 nm has twice the intensity compared with an excitation wavelength of 274 nm. The enhanced emission intensity is due to spectral overlap or effective energy transfer from Pr^{3+} ions to Gd^{3+} ions
9a	$LaB_3O_6:Bi^{3+}$, Gd^{3+} [24]	Combustion	254 (Bi^{3+})	310	When Bi^{3+} ions are co-doped with $LaB_3O_6:Gd^{3+}$ phosphor materials, it shows a broad excitation band at 254 nm due to Bi^{3+} ions and an intense emission band at 310 nm due to Gd^{3+} ions. This confirms that Bi^{3+} is an effective sensitizer for Gd^{3+}-activated borate materials. When Pr^{3+} ions are co-doped with $YBO_3:Gd^{3+}$ and $LaBO_3:Gd^{3+}$ phosphor materials, it shows a broad excitation band at around 250 nm due to Pr^{3+} ions and an identical emission band of high intensity at 312 nm. When Pr^{3+} ions are co-doped with $LaB_5O_9:Gd^{3+}$ phosphor materials, it shows an intense excitation spectrum at 276 nm and an intense emission band at 311 nm due to Gd^{3+} ions. Comparison of the above co-doped phosphor materials shows that the best emission intensity is produced for the LaB_5O_9 phosphor material, followed by the LaB_3O_6 and $LaBO_3$ phosphor materials, in that order
9b	$YBO_3:Pr^{3+}$, Gd^{3+} [24]	Combustion	250 (Pr^{3+})	312	
9c	$LaB_5O_9:Pr^{3+}$, Gd^{3+} [24]	Combustion	276 (Gd^{3+})	311	

Table 9.2 Comparison of different Gd^{3+} singly doped and co-doped phosphate-based compounds for narrowband UVB emissions.

No.	Host sample	Synthesis technique	Excitation wavelength/nm	Emission wavelength/nm	Comments
1	$Ca_5(PO_4)_3OH$: Pr^{3+},Gd^{3+} [8,25]	Coprecipitation method	220 (Pr^{3+})	316	When Pr^{3+} is co-doped with $Ca_5(PO_4)_3OH:Gd^{3+}$, the excitation band shows an enhanced peak at 220 nm that is due to Pr^{3+}, while a minor peak at 275 nm is due to Gd^{3+} ions. The emission spectra, under an excitation wavelength of 220 nm, show enhanced emission at 316 nm, almost double the emission intensity for a Gd^{3+} singly doped phosphate material, but with traces of suppressed Pr^{3+} red emissions at 610 nm. This implies that Pr^{3+} is a good sensitizer in absorbing energy from the UV source and transferring it non-radiatively to Gd^{3+} ions
2	$Ca_3(PO_4)_2$:Pr^{3+}, Gd^{3+} [8,13]	Combustion	227 (Pr^{3+})	313	When Pr^{3+} is co-doped with $Ca_3(PO_4)_2:Gd^{3+}$ phosphor materials, the excitation spectrum shows an enhanced peak at 227 nm due to Pr^{3+} ions and a very low-intensity band at 276 nm due to Gd^{3+} ions. The emission band monitored at an excitation wavelength of 227 nm shows *ca.* fivefold enhanced emission compared with singly doped Gd^{3+} ions. This confirms that Pr^{3+} is a good sensitizer in absorbing excitation energy and transferring it to Gd^{3+} to produce enhanced UVB emission at 313 nm
3	$Ca_9Mg(PO_4)6F_2$: Pr^{3+},Gd^{3+}, Li^{+} [8,26]	Combustion	229 (Pr^{3+})	313	When Pr^{3+} is co-doped with $Ca_9Mg(PO_4)_6F_2:Gd^{3+}$ phosphor materials, an enhanced excitation band is observed at 229 nm due to Pr^{3+} ions and a low-intensity excitation band at 275 nm due to Gd^{3+} ions. The emission at 313 nm is attributed to the energy transfer between Pr^{3+} and Gd^{3+} ions and also to the charge compensator ion effect, which further enhances the emission intensity (three times larger than Gd^{3+} singly doped emission) at 313 nm, with no trace of Pr^{3+} ions

4	$LiSrPO_4:Pb^{2+}$, Gd^{3+} [27]	Solid-state reaction	247 (Pb^{2+})	312	When Pb^{2+} ions are co-doped with $LiSrPO_4:Gd^{3+}$ phosphor materials, a broad, intense excitation band is observed at 247 nm due to Pb^{2+} ions and a low-intensity excitation band at 274 nm due to Gd^{3+} ions. An interesting aspect of Pb^{2+} co-doping is that the UVB emission at 312 nm is 10 times greater than that with singly doped Gd^{3+} ions. This implies that Pb^{2+} is an excellent sensitizer to Gd^{3+}-activated phosphate materials in absorbing the high excitation energy and transferring it to Gd^{3+} ions, leaving no trace of Pb^{2+} ions in the emission spectra
5	$Ca_3(PO_4)_2$: Gd^{3+},Ce^{3+} [28]	Combustion	274 (Gd^{3+})	313	When Gd^{3+} ions are co-doped with $Ca_3(PO_4)_2:Ce^{3+}$ phosphor materials, an intense excitation band at 274 nm is observed, which is due to Gd^{3+} ions, when monitored using an emission wavelength of 313 nm. The emission spectrum of co-doped compounds shows an intense emission at 313 nm, which is due to Gd^{3+} ions, and a second, smaller emission at 365 nm, which is due to Ce^{3+} ions, and a third, smaller emission at 615 nm, which is due to Gd^{3+} ions. Interestingly, in this case, the high-intensity emission at 313 nm is due to the energy transfer from Gd^{3+} to Ce^{3+} ions. In this case, Gd^{3+} is a sensitizer to Ce^{3+}, yet it produces enhanced UVB emission at 313 nm. The same compound, when monitored with an excitation wavelength of 265 nm, shows a dominant blue-shifted Ce^{3+} emission at around 360 nm with no trace of Gd^{3+} ions (Gd^{3+} transfers all of its energy to Ce^{3+} ions)

Table 9.3 Comparison of different Gd^{3+} singly doped and co-doped fluoride-based compounds for narrowband UVB emissions.

No.	Host sample	Synthesis technique	Excitation wavelength/nm	Emission wavelength/nm	Comments
1	$LiSrAlF_6$:Ce^{3+}, Gd^{3+} [29]	Wet chemical method	274 (Gd^{3+})	312	For there to be efficient energy transfer from Ce^{3+} to Gd^{3+} ions, the absorption spectrum must lie anywhere in the region between 274 and 305 nm (due to absorption peaks of Gd^{3+} lying in this region). Therefore, efficient energy transfer is achieved when the absorption spectrum of Gd^{3+} ions overlaps with the emission spectrum of Ce^{3+} ions. In this scenario, Ce^{3+} is a sensitizer, and the emission produced is enhanced owing to the energy transfer from Ce^{3+} ions to Gd^{3+} ions. Interestingly, this emission at 312 nm is not far away from the commercial Philips TL01 phototherapy lamp and can be adjusted in some way to exceed that emission intensity
2	$Y_7O_6F_9$:Pr^{3+}, Gd^{3+} [9]	Solid state reaction	238 (Pr^{3+})	314	There are two ways of producing UVB Gd^{3+} emissions at 314 nm: one is by exciting the co-doped material using an emission wavelength of 314 nm and the other is by exciting Pr^{3+} ions using a wavelength of 450 nm (visible excitation), which can produce enhanced emission at 314 nm through the up-conversion process. In the first process, the excitation band produces a high-intensity excitation peak at 276 nm due to Gd^{3+} ions and a broad, low-intensity peak at 238 nm due to Pr^{3+} ions. In the second case, Pr^{3+} absorbs two blue energy photons and transfers them to Gd^{3+} *via* the up-conversion process to produce super-enhanced emission at 314 nm. The sample is an oxyfluoride compound

3a	LaF_3:Ce^{3+},Gd^{3+} [30]	Solid-state reaction	229 (Pr^{3+})	311	When LaF_3 is doped with Ce^{3+} ions, emission is produced at 307 nm, but when it is co-doped with Gd^{3+} ions, a broad excitation band is observed at 254 nm, and the emission at 307 nm is red shifted to 311 nm and enhanced due to the energy transfer from Ce^{3+} to Gd^{3+} ions. This emission at 311 nm is far greater than that at 307 nm. The Ce^{3+} ion acts as a sensitizer for Gd^{3+} co-doped fluoride host materials.
3b	$SrAlF_5$:Ce^{3+}, Gd^{3+} [30]	Solid-state reaction	254 (Ce^{3+})	311	When Ce^{3+} is co-doped with $SrAlF_5$:Gd^{3+} phosphor materials, it reveals a broad excitation band at 254 nm due to Ce^{3+} ions, and enhanced emission is produced at 311 nm, which is due to the energy transfer from Pr^{3+} to Gd^{3+} ions. The emission intensity is 10 times lower than for the above LaF_3:Ce^{3+},Gd^{3+} phosphor material. In both cases, Ce^{3+} ions act as a sensitizer to Gd^{3+}-activated fluoride materials
4	(Gd,Ce):SrF_2 [31]	Coprecipitation method	273 (Ce^{3+})	312	When SrF_2 is doped with Ce^{3+} ions, it produces emission at 330 nm due to the 5d → ^2F_J transition, when monitored at an excitation wavelength of 273 nm. On the other hand, when Gd^{3+} is co-doped with this fluoride material, Ce^{3+} ions, which absorb the majority of the excitation energy, transfer some of it to Gd^{3+} ions, leading to a narrow, low-intensity peak at 312 nm. This also confirms that energy has been transferred radiatively from Ce^{3+} to Gd^{3+} ion, proving that Ce^{3+} has effectively sensitized Gd^{3+}-activated fluoride materials

Table 9.3 *(Continued)*

No.	Host sample	Synthesis technique	Excitation wavelength/nm	Emission wavelength/nm	Comments
5	$Lu_6O_5F_8:Pr^{3+}$, Gd^{3+} [28,32]	Hot-injected protocol method	238 (Pr^{3+})	315	When Pr^{3+} is co-doped with oxyfluoride $Lu_6O_5F_8:Gd^{3+}$ phosphor materials, monitored at an emission wavelength of 315 nm, an enhanced excitation band is observed at 238 nm due Pr^{3+} ions and a low-intensity excitation band at 276 nm due to Gd^{3+} ions. When these co-doped materials are monitored using an excitation wavelength of 238 nm, enhanced emission is produced at 315 nm due to Gd^{3+} ions, with no trace of Pr^{3+} ions. However, when these co-doped materials are monitored using an emission wavelength of 315 nm, an enhanced excitation band is observed at 450 nm due to Pr^{3+} ions. Likewise, when these co-doped materials are monitored using an excitation wavelength of 450 nm, enhanced emission is produced at 315 nm due to Gd^{3+} ions that resulted from the up-conversion process whereby Pr^{3+} ions absorb blue photons and transfer them radiatively to Gd^{3+} ions

phosphate materials are doped with Ce^{3+} ions, where the role played by Gd^{3+} ions is reversed. The emission peaks for Gd^{3+}-activated phosphate materials lie in the wavelength range 312–316 nm. Mostly, Gd^{3+}-activated phosphate materials produce enhanced NB-UVB emissions through an energy transfer mechanism that is better compared with singly doped Gd^{3+} ions. Table 9.3 presents the results for Gd^{3+}-activated emissions with fluorides/oxyfluorides in the last category of phosphor host materials. Fluorides are considered the next best host materials after phosphates because their oxygen component generates moderate phonon interactions and due to their high chemical stability.

Gd^{3+}-activated fluoride (or oxyfluoride) phosphor materials have attracted renewed interest in phototherapy applications owing to their enhanced NB-UVB emissions in the 311–315 nm wavelength range. Researchers have employed a variety of synthesis methods for the preparation of these phosphor materials, such as wet chemical methods, solid-state reactions, a hot injected protocol, and coprecipitation methods, but not many reports of the application of the combustion synthesis method were found. The excitation peaks are mainly concerned with the selection of sensitizer ions, and this can impact the activator ion emission properties through energy transfer mechanisms. Of particular interest is the dual way of producing NB-UVB emissions from Gd^{3+} ions, either by UV absorption and energy transfer from different sensitizers to Gd^{3+} ions or through visible absorption of blue photons and transferring them through the up-conversion process. The activation of Gd^{3+} ions in different host materials such as borates, phosphates, and fluorides is currently showing remarkable properties for the creation of NB-UVB emissions in the 311–315 nm wavelength range, which can be integrated into fluorescent lamps for phototherapeutic treatment of various skin diseases and skin disorders. In particular, the interest in these well-pursued inorganic host materials is due to the possibility of well-established energy transfer mechanisms that yield amazingly high luminescent efficiencies that are leading the way for the next generation of fluorescent lamps for phototherapeutic treatments of dermatological maladies.

9.3 Conclusion

Gd^{3+}-activated host materials are mostly excitable in the UV region in producing NB-UVB emissions, which can be achieved through up-conversion processes from the visible (blue) region. These inorganic hosts are ideal for phototherapeutic applications because they are facile to synthesize and have unique structural features that allow for easy sensitization of Gd^{3+} ions to improve their limited parity forbidden transitions. Each of the sensitizers (Ce^{3+}, Pb^{2+}, Pr^{3+}, Bi^{3+}) has a unique role to play in different host materials in enhancing NB-UVB emissions. This chapter has provided insights into various Gd^{3+}-activated luminescent inorganic host materials that are potential candidates for the treatment of various skin diseases because they lie in the appropriate spectrum for phototherapy.

Acknowledgements

The South African National Nanoscience Postgraduate Teaching and Training Platform (NNPTTP) of the Department of Science and Technology supports this work financially.

References

1. S. P. Bhagat, A. B. Gawande and S. K. Omanwar, *Opt. Mater.*, 2015, **40**, 36–40.
2. V. V. Shinde, R. G. Kunghatkar and S. J. Dhoble, *Luminescence*, 2015, **30**, 1257–1262.
3. X. Wang, Y. Chen, P. A. Kner and Z. Pan, *Dalton Trans.*, 2021, **50**, 3499.
4. A. B. Gawande, R. P. Sonekar and S. K. Omanwar, *Combust. Sci. Technol.*, 2014, **186**, 785–791.
5. S. Ebraheem and A. EL-Saied, *Mater. Sci. Appl.*, 2013, **4**, 31475–31476.
6. M. M. Lanje, M. M. Yawalkar, J. S. Dahegaonkar and S. J. Dhoble, *J. Phys.: Conf. Ser.*, 2021, **1913**, 012031.
7. S. Tamboli, G. B. Nair, S. J. Dhoble and H. C. Swart, *Luminescent Materials in display and biomedical applications*, CRC Press, 1st edn, 2020.
8. M. Nimbalkar, M. Yawalkar, N. Mahajan and S. J. Dhoble, *Photodiagn. Photodyn. Ther.*, 2021, **33**, 102082.
9. Y. Sun, Y. Wang, C. Hu, X. Zhou, J. Hao, W. Li and H. Li, *Materials*, 2022, **15**, 7680.
10. S. P. Puppalwar, S. J. Dhoble, G. Singh and A. Kumar, *J. Mod. Phys.*, 2011, **2**, 1560–1566.
11. V. Singh, G. Sivaramaiah, N. Singh, M. S. Pathak, J. L. Rao, P. K. Singh and A. S. Nagpure, *Bull. Mater. Sci.*, 2019, **42**, 19.
12. V. V. Shinde, R. G. Kunghatkar and D. J. Dhoble, *Luminescence*, 2015, **30**, 1257–1262.
13. P. P. Mokoena, M. Gohain, V. Kumar, B. C. B. Bezuidenhoudt, H. C. Swart and O. M. Ntwaeaborwa, *J. Alloys Compd.*, 2014, **595**, 33–38.
14. J. A. Waje, P. N. Dandle and A. B. Gawande, *J. Biol. Chem. Chron.*, 2019, **5**, 88–90.
15. J. Liu, Y. Liang, S. Yan, D. Chen, S. Miao and W. Wang, *Dalton Trans.*, 2021, **50**, 15413–15421.
16. P. D. Belsare, C. P. Joshi, S. V. Moharil, V. K. Kondawar, P. L. Muthal and S. M. Dhopte, *Radiat. Eff. Defects Solids*, 2009, **164**, 197–203.
17. U. Manik, S. C. Gedam and S. J. Dhoble, *Luminescence*, 2016, **31**, 911–914.
18. A. B. Gawande, R. P. Sonekar and S. K. Omanwar, *Int. J. Curr. Res.*, 2018, **5**(5), 176–179.
19. A. B. Gawande, R. P. Sonekar and S. K. Omanwar, *Mater. Res. Bull.*, 2014, **60**, 285–291.
20. A. B. Gawande, R. P. Sonekar, S. D. Kapse and N. S. Mahaele, *Int. J. lumin. appl.*, 2016, **6**, 82–84.
21. A. O. Chauhan, N. S. Bajaj and S. K. Omanwar, *Bull. Mater. Sci.*, 2017, **40**, 1–6.

22. P. A. Nagpure and S. K. Omanwar, *Indian J. Pure Appl. Phys.*, 2015, **53**, 77–81.
23. S. Tamboli, G. B. Nair, S. J. Dhoble and D. K. Burghate, *Phys. B*, 2018, **535**, 232–236.
24. R. P. Sonekar, S. K. Omanwar, S. V. Moharil, S. M. Dhopte, P. L. Muthal and V. K. Kondawar, *Opt. Mater.*, 2007, **30**, 622–625.
25. P. P. Mokoena, I. M. Nagpure, V. Kumar, H. C. Swart and O. M. Ntwaeaborwa, *J. Phys. Chem. Solids*, 2014, **75**, 998–1003.
26. S. Tamboli and S. J. Dhoble, *Spectrochim. Acta, Part A*, 2017, **184**, 119–127.
27. A. O. Chauhan, A. B. Gawande and S. K. Omanwar, *J. Inorg. Organomet. Polym. Mater.*, 2016, **26**, 1023–1027.
28. L. Reddy, T. J. Nkosi, P. L. Masiteng, A. Balakrishna, H. C. Swart and O. M. Ntwaeaborwa, *Curr. Appl. Phys.*, 2020, **20**, 696–702.
29. P. D. Belsare, S. V. Moharil, C. P. Joshi and S. V. Dhoble, *AIP Conf. Proc.*, 2011, **1391**, 194.
30. P. S. Hemne and S. V. Dhoble, *Int. J. Lumin. Appl.*, 2015, **Vol I**(II), 31.
31. I. Milasavljevic, Y. Wu and E. Zych, *J. Lumin.*, 2020, **224**, 117243.
32. Y. Du, Y. Wang, Z. Deng, X. Chen, X. Yang, T. Sun, X. Zhang, G. Zhu, S. F. Yu and F. Wang, *Adv. Opt. Mater.*, 2020, **8**, 1900968.

CHAPTER 10

Bioengineered Nanomaterials: Recent Trends and Bottlenecks in the Management of Infectious Diseases

SAMEER NADAF,*[a] DURGACHARAN BHAGWAT[b] AND SHAILENDRA GURAV*[c]

[a] Bharati Vidyapeeth College of Pharmacy, Palus, Maharashtra 416310, India; [b] Bharati Vidyapeeth College of Pharmacy, Kolhapur, Maharashtra 416013, India; [c] Department of Pharmacognosy, Goa College of Pharmacy, Goa University, Panaji, Goa 403001, India
*Emails: sam.nadaf@rediffmail.com; shailendra.gurav@nic.in

10.1 Introduction

The majority of diseases that exist in the world are caused by infections. Lower respiratory tract infections, diarrhea, tuberculosis (TB), HIV infection, and malaria are all conditions with high death rates. In developing nations, where commodities such as vaccines and antibiotics may be scarce, death rates are higher.[1] The prevalence of respiratory disorders is rising, especially among children and older people with impaired immune systems, as a result of rising pollutant levels, rapid variations in lifestyle, and recurrent outbreaks of microbial infections.[2] By 2050, infection-related mortality is expected to be comparable to cancer mortality, according to a 2019 World Health Organization (WHO) report.[3] Infectious diseases that are arising and

Nanoscience & Nanotechnology Series No. 62
Smart Nanomaterials for Infectious Diseases
Edited by Suvardhan Kanchi, Nidhi Sharotri, Rajasekhar Chokkareddy, Deepali Sharma and Falah H. Hussein

Published by the Royal Society of Chemistry, www.rsc.org

resurging cause enormous costs in terms of both human life and economic output. The onset of diseases is influenced by a variety of causes, the majority of which are related to human activity. Anticipating the possibility of major outbreaks, the WHO has given several infectious illnesses top priority status for research and development.[4]

To have a fundamental grasp of infectious diseases, it may be said that they are disorders brought on by the existence of pathogenic agents, such as viruses, bacteria, and fungi. Dengue, malaria, and TB are examples of infectious diseases that are considered communicable diseases since they may be spread from one person to another; influenza (flu), in contrast, can be spread from one species to another.[5] Because of the accessibility and use of a large range of antibiotics and antimicrobial medicines, the consequences associated with infectious diseases have decreased considerably, especially in industrialized nations. Despite this, anti-infective therapy is still not regarded as being ideal.

Over the last few centuries, people have become more vulnerable to unexpected outbreaks and new infectious diseases. Overuse of antibiotic and antimicrobial agents has led to an increase in drug-resistant bacteria.[6] To date, each recognized category of antibiotics has indeed been proven to experience resistance. The scenario of antibiotic resistance has emerged over more than three decades as a challenge due to the absence of sufficient new types of antibiotics.[7] There is currently a pressing demand for new cutting-edge treatment approaches. Major barriers to effective therapy include poor patient adherence to medicines and the requirement for ongoing patient monitoring. Access to appropriate therapies is hampered by inefficient purchasing procedures, the incapacity to purchase medications, and the poor durability of prescription items at high temperatures and humidity.[1] These shortcomings have prompted scientists to investigate advances in the creation of new vaccinations and antibiotics, with a focus on nanomedicine- and nanotechnology-based treatments at the moment.[5]

Both the diagnosis and treatment of numerous diseases could be improved upon with the use of nanotechnology. The development, characterization, and use of therapeutic systems on a nano- or micron-sized scale are all aspects of nanomedicine.[8] Nanoparticles (NPs) have great potential for the rapid and accurate detection of bacteria, viruses, and fungi owing to their distinctive optical, magnetic, and electrical properties. By generating reactive oxygen species (ROS) and metal ions, inducing photothermal effects, and rupturing cell membranes, NPs demonstrate impressive antibacterial activity. Drug permeation can be further enhanced by nano-based delivery methods, which can also lessen adverse effects, increase the systemic circulation time, and extend the half-life of medications.[3]

Modern methods of antimicrobial treatment will tackle the numerous drawbacks and shortcomings of conventional antibiotics by leveraging nanoscale carriers. In this regard, a wide range of nanomaterials have been investigated to increase therapeutic effectiveness and minimize adverse effects during the treatment of infectious disorders.[9] There are several potential benefits to including anti-infective medications in NPs. It is possible to target specific organs, tissues, cells, or bacteria with manufactured particles. Targeting offers the advantage of concentrating elevated amounts of antimicrobials at the location where bacterial death is required. This should result in a decrease in the overall effective antibiotic dose and, consequently, a reduction in adverse effects. Furthermore, the NP formulation has the power to alter how medications that fight bacteria become soluble. Additionally, there is a chance of changing the pharmacokinetics, which would provide desired pharmacological effects. Owing to the advantages of nanocarrier formulations, numerous antibiotics may be used in clinical settings and existing resistance issues may be resolved.[7]

As a result, several NPs based on bacteria, viruses, and mammalian cells are first introduced in this chapter. The most recent advances in curative strategies that employ bioengineered and biomimetic nanomaterials to treat infectious illnesses are then reviewed. The development of bioengineered nanomaterials at the present time for the cure of infectious diseases is discussed. Additionally, the ability of "nanovaccines" to treat or prevent infectious diseases is also emphasized. Finally, significant difficulties relating to the therapeutic uses and translation of these bioengineered nanotherapies are explored.

10.2 Bioengineered Nanoparticles

To create materials with specified properties, engineered NPs are created in laboratories or industries. Bioinspired nanomaterials are artificial NPs whose characteristics resemble those of biological or natural nanomaterials.[10] In the recent past, diagnostic or therapeutic substances for the treatment of infectious diseases have been delivered using NPs made from bacteria, viruses, and mammalian cells. Numerous biomimetic nanomaterials with particular functions can be created by altering their architectures using cutting-edge nanofabrication technology.

10.2.1 Virus-based Nanoparticles

The production of virus-based nanoparticles (VNPs) uses viruses that infect bacteria, animals, or plants. Viruses are abundant in the environment. Since viruses have evolved naturally to transmit nucleic acids, they are a perfect entity to start with since they may be manipulated to transport other compounds, such as medications and imaging agents.[11] The exteriors of virus particles can be changed to showcase large and tiny components in clearly defined arrays, while the interiors can enclose and shield sensitive chemicals. Owing to these characteristics of viruses and their inherent biocompatibility, they have been developed as actively targeted delivery systems for drugs that supplement and enhance the present pharmaceutical alternatives.[12] Virion proteins interact specifically with lipids, membrane proteins, or carbohydrate moieties on the cell surface to help viruses bind to host cells effectively. In most cases, this attachment is followed by the virus entering the host cell through endocytosis, pinocytosis, fusion, or penetration.[13] Genetic

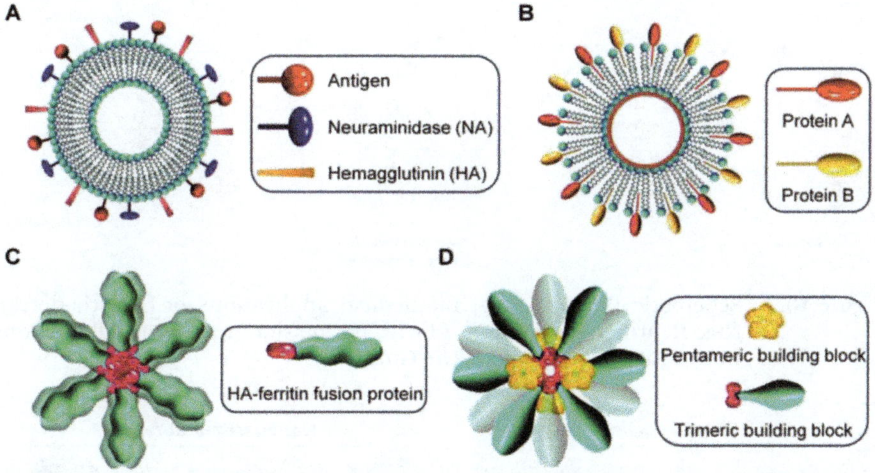

Figure 10.1 Schematic illustration of virus-mimetic NPs. (A) Virosomes; (B) virus-like particles (VLPs); (C) self-assembling NPs displaying antigens on their surface; (D) fully synthetic NPs mimicking viruses. Reproduced from ref. 13, https://doi.org/10.3389/fphar.2019.00751, under the terms of the CC BY 4.0 license https://creativecommons.org/licenses/by/4.0/.

engineering, encapsulation, biomineralization, infusion, and bioconjugation are a few of the methods that can be utilized to tailor and change virus-based products.[12] A schematic illustration is shown in Figure 10.1.

10.2.2 Bacterial or Fungal Pathogen-derived Nanoparticles

The health and wellbeing of their mammalian hosts are maintained by microbes, which are a crucial component of natural ecosystems. Studies on materials made from bacteria have evolved in the past few decades from proofs of concept to ways to enhance conventional biomaterials. It is possible to create biomimetic nanoplatforms using both Gram-negative and Gram-positive bacteria, with functionality mostly drawn from the characteristics of the parent organism.[14] Gram-negative and Gram-positive bacteria, comparable to mammalian cells, release nanoscale membrane vesicles into the extracellular medium either constitutively or in a controlled way. Spherical bilayered proteolipids comprising bioactive proteins, lipids, nucleic acids, and virulence factors make up these bacterial extracellular vesicles.[15] Applications of bacteria-derived NPs are shown in Figure 10.2.

10.2.3 Nanoparticles Mimicking Mammalian Cells

Natural cell membranes are applied to synthetic nanostructures to give them specific cell surface antigens and functionalities.[16] Because of their characteristics of biocompatibility, biodegradability, and long circulation half-life, erythrocytes [red blood cells (RBCs)], which are the majority of

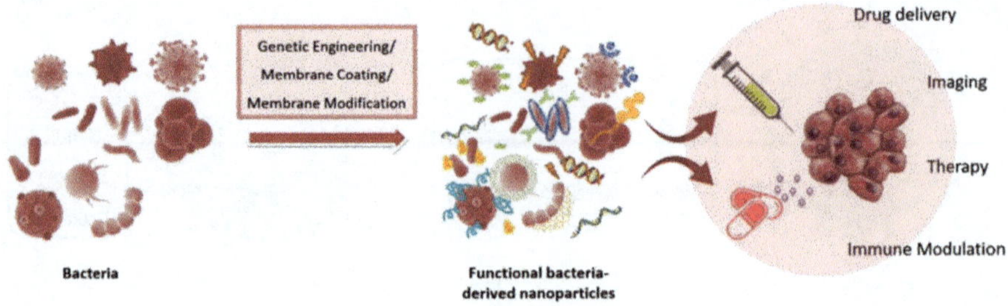

Figure 10.2 Schematic illustration of biomedical applications of bacteria-derived NPs. Reproduced from ref. 14 with permission from John Wiley & Sons, Copyright © 2020 Wiley-VCH GmbH.

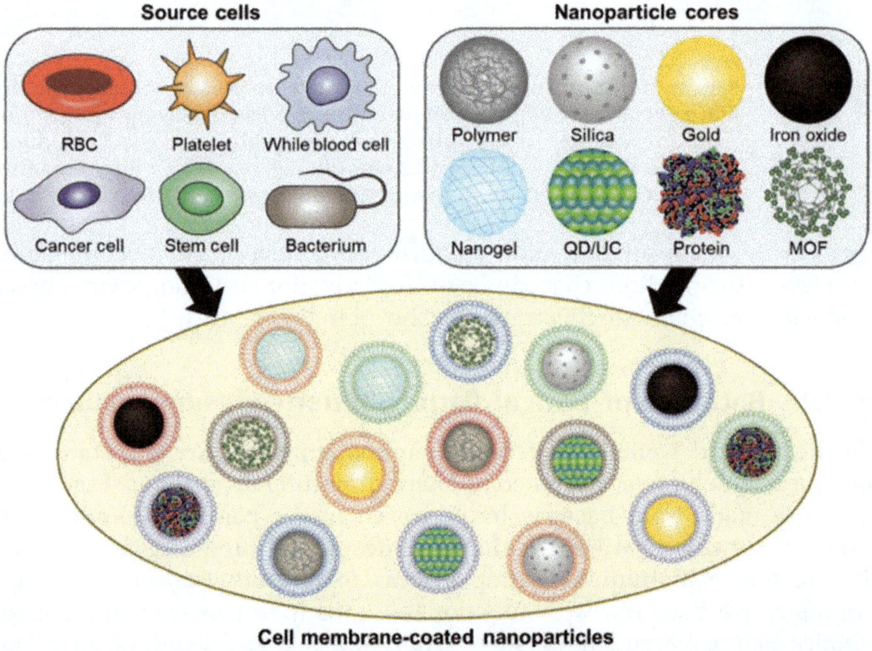

Figure 10.3 Cell membrane-coated NPs. Reproduced from ref. 18 with permission from John Wiley & Sons, Copyright © 2018 Wiley-VCH Verlag GmbH & Co. KGaA, Weinheim.

circulating blood cells, have been extensively applied in drug delivery systems.[17] Cell membrane-coated NPs can be fabricated using numerous methods.[18] For the delivery of a wide range of medications, blood cell-based delivery systems have been developed. They can increase targeted release, decrease the carrier's immunogenicity, lengthen blood circulation duration, and improve the cytocompatibility of the developed drug delivery system.[19] The coating of the NPs can be carried out with a variety of cells, including RBCs, platelets, blood cells, and so on (see Figure 10.3).

10.3 Application of Bioengineered Nanoparticles in Treating Infections

10.3.1 Bacterial Infections

Human health is seriously endangered by bacterial infections, which are made worse by rising antibiotic resistance. It is crucial to detect and eradicate pathogenic microorganisms swiftly and efficiently since they can cause severe morbidity and mortality.[20] Toxins produced by bacteria can be lethal if not treated correctly. Targeting toxins provides a reasonable method of treating ailments on a causal level owing to their harmful nature and the substantial role they can play in many pathogeneses.

10.3.1.1 Toxin Neutralization

Diverse organisms use toxins as a long-range attack strategy to help them survive. The more lethal toxins frequently result in mortality and can have a substantial and immediate influence on human health. Toxins must first get past cell membranes to exert their pathogenic effect on living things. Owing to their molecular similarities, these NPs can be used to create broadly applied antitoxin compositions (see Figure 10.4).[21] Endotoxins and exotoxins are two types of microbial toxins that bacteria commonly create.

10.3.1.1.1 Neutralization of Exotoxins. Gram-positive bacteria are predominantly responsible for secreting bacterial exotoxins. Some exotoxins exhibit superantigen-like behavior, which prompts T lymphocytes to release cytokines. Pre-exposure to endotoxins typically leads to tolerance, but pre-exposure to exotoxins characteristically increases the production of cytokines generated by lipopolysaccharides and associated lethality.[22] Traditional detoxification methods depend on antigen–antibody-specific binding, with an antibody focusing on a single antigen; these methods

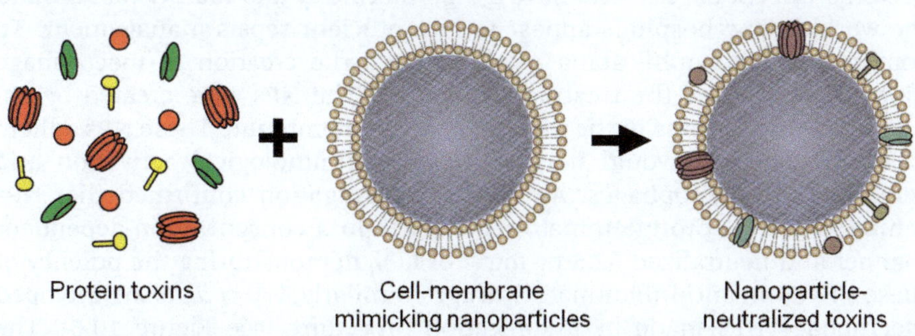

Figure 10.4 Engineered NPs for detoxification aim to mimic natural cellular membrane structures. Reproduced from ref. 21 with permission from Elsevier, Copyright 2015.

necessitate knowing the structure of the toxin and the kind of pathogenic bacteria, which presents several difficulties for clinical interventions.[23]

10.3.1.1.2 Neutralization of Endotoxins. Lipopolysaccharides (LPS), or endotoxins, are elements of the cell walls of Gram-negative bacteria that are liberated from the inner membrane at division, upon bacterial death, or after antibiotic therapy.[24] The systemic inflammatory response syndrome, which includes widespread intravascular coagulation, sepsis, shock, and multiorgan failure, can be brought on by the upregulation of the mammalian immune system caused by LPS.[23] Therefore, effective LPS removal is a key to treating these diseases.

The use of cell membrane-coated NPs as an antivirulence intervention against bacterial infections has shown potential. Since *Escherichia coli* produce numerous bacterial toxins, particularly endotoxins, and exotoxins that are difficult to eradicate concurrently, ways to treat *E. coli* infections remain a challenge. To address these problems, Jiang *et al.* developed a polymyxin B (PMB)-modified, RBC-mimetic hybrid liposome (P-RL). *In vitro* and *in vivo* outcomes established that P-RL had a substantial anchoring effect on *E. coli*. Furthermore, compared with the existing RBC vesicles or PMB-modified liposomes, P-RL showed a higher therapeutic effect against RBC hemolysis, macrophage activation, and a mixed-toxin infection in mice.[25]

According to reports, photothermal therapy is a successful antibacterial strategy. Although certain antibacterial substances are harmful, photothermal reagents cannot remove bacterial toxins. In light of this, Chen *et al.* developed RBC membrane-coated Fe_3O_4 (RBC@Fe_3O_4) NPs as effective antibacterial agents. The toxin of methicillin-resistant *Staphylococcus aureus* (MRSA) was absorbed by the NPs, according to an *in vitro* investigation. *In vivo* research further confirmed this fact. Additionally, it was discovered that NPs with laser irradiation had a superior wound-healing effect.[26] An overall representation is shown in Figure 10.5.

Sepsis, which is brought on by unmanaged inflammatory reactions to bacterial infections, tends to have a high morbidity and fatality rate around the world. Many hospitals appear to lack efficient sepsis management. To combat this, Thamphiwatana *et al.* reported the creation of macrophage biomimetic NPs for the treatment of sepsis. The NPs were created by encasing polymeric cores in the macrophage cell membrane. These NPs adhere to endotoxins that would normally cause immunological activation and serve as fake macrophages. An *in vitro* investigation confirmed that NPs inhibited several proinflammatory cytokines in a concentration-dependent manner and neutralized 62.5 $ng\,mg^{-1}$ of LPS, demonstrating the potency of these NPs as an anti-inflammatory drug.[27] Similarly, Wu *et al.* also developed macrophage NPs made by a sonication procedure (see Figure 10.6). The findings showed that in BALB/c mice, the fabricated NPs significantly reduced the production of nitric oxide and LPS promoted the cytokines TNF-α and IL-6. Additionally, it decreased the expression level of CD40.[28]

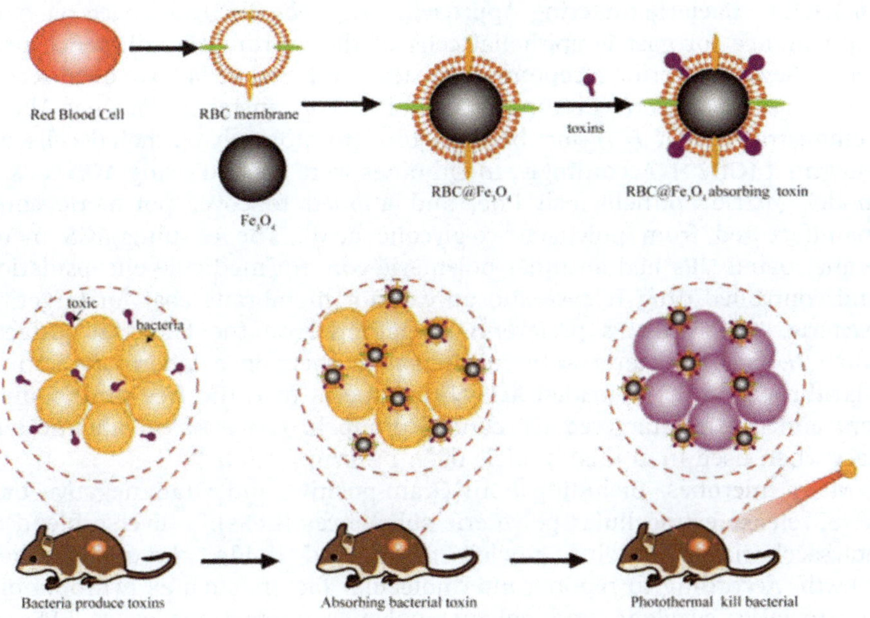

Figure 10.5 Schematic diagram of photothermal therapy for bacterial infections with RBC@Fe_3O_4 NPs. Reproduced from ref. 26 with permission from the Royal Society of Chemistry.

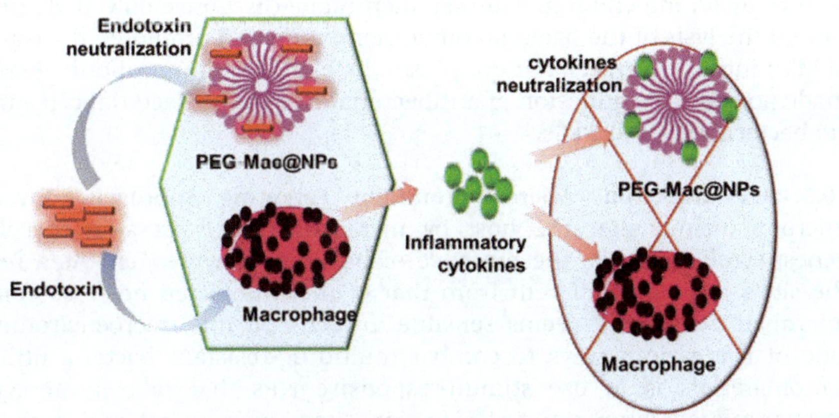

Figure 10.6 Formulation of macrophage membrane-coated poly(ethylene glycol) NPs (PEG-Mac@NPs). Reproduced from ref. 28 with permission from Springer Nature, Copyright 2020.

In situ gelation using a membrane vesicle template was used by Zhang *et al.* to generate RBC membrane-coated nanogels for the treatment of MRSA infection. In contrast to free antibiotics and non-responsive nanogel analogs, RBC-nanogels greatly reduced bacterial growth when given to macrophages treated with intracellular MRSA.[29]

10.3.1.1.3 Bacteria-targeting Approach. *Helicobacter pylori* bacteria have a preference for gastric epithelial cells of the antrum *via* adherence pedestals. Several exterior receptors on gastric epithelial cells have been recognized as *H. pylori* binding receptors. For instance, it has been demonstrated that *H. pylori* bacteria bind to stomach epithelial cells *via* integrin 1 (CD29). Accordingly, membranes were created using AGS cells, a model gastric epithelia cell line, and applied to cover polymeric cores manufactured from poly(lactic-*co*-glycolic acid). The ensuing AGS membrane-coated NPs had an inner polymeric core for medicine encapsulation and controlled drug release and an exterior membrane coat for targeting bacteria. The AGS-NPs preferentially gathered on the bacterial surfaces when *H. pylori* bacteria were cultured with them in a laboratory setting. Clarithromycin (CLR)-loaded AGS-NPs outperformed the free drug equivalent and the non-targeted NP control group in terms of therapeutic efficacy when used in animal models of *H. pylori* infection.[30]

Many microbes, including both Gram-positive and Gram-negative bacteria, release extracellular polymeric substances (EPSs), a diverse blend of polysaccharides, proteins, nucleic acids, and lipids, throughout their growth. According to reports, intermolecular factors such as hydrophobic, electrostatic, covalent, and polymer–polymer interactions cause EPSs to adsorb on minerals. Based on this knowledge, Ping *et al.*, developed a bioinspired design technique to create membrane-disrupting and bacteria-targeting NPs for the efficient administration of metronidazole (MTZ) to treat *H. pylori* infection. In contrast, membrane-disruptive polyethylenimine caused the lysis of the bacterial outer membrane and promoted the transfer of MTZ into the intracellular cytoplasm. Exfoliated montmorillonite (eMMT) made possible the adhesion of antibacterial NPs to the bacterial cell surface *via* bacteria-secreted EPSs.[31]

10.3.1.1.4 Infectious Microenvironment Targeting Approach. Invading microorganisms alter the host by increasing the expression of certain genes, cytokine levels, the presence of bacterial enzymes, *etc.* As a result, the site's milieu is different from that of the uninfected host. To combat microbial diseases, it seems sensible to focus on this microenvironment. One of the various ways to combat multidrug-resistant bacteria utilizing nanomaterials is to use stimuli-responsive NPs that take advantage of certain microenvironments at infection sites, such as pH and pathogen-derived metabolites.[32] Biofilm microenvironment stimuli-responsive nanoagents not only discharge bactericidal composition efficiently upon specific stimuli, but also alter the physicochemical characteristics of the nanoagents upon stimulation, increasing the accumulation of antibacterial agents in biofilm-infected sites, enhancing the permeation of biofilms, and killing bacteria.[33] The stimulation of nanoagents includes some endogenous triggering factors (see Figure 10.7).

In view of this, Li *et al.* developed RBC-coated supramolecular gelatin nanoparticles (GNPs) to deliver vancomycin in the presence of gelatinase at

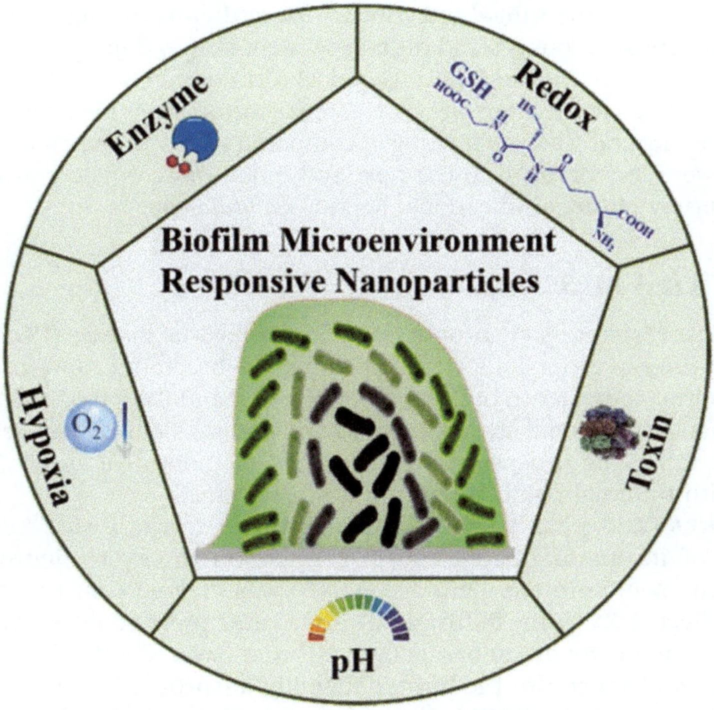

Figure 10.7 Biofilm microenvironment-responsive NPs for biofilm eradication. Reproduced from ref. 33 with permission from Elsevier, Copyright 2022.

the bacterial infection site. This delivery strategy has an extremely lethal effect on Gram-positive and gelatinase positive pathogenic bacteria with minimal systemic toxicity owing to the site-specific breakdown of the NPs.[34] In another study, Lin *et al.* prepared RBC-based GNPs to detect and inhibit MRSA. Due to the RBC membrane, the immune system clearance was reduced and exotoxins were neutralized efficiently. The developed GNPs were found to be degraded by gelatinase in pathogen-infected areas *in situ* and facilitated the wound-healing process in MRSA-infected mouse models.[35]

10.3.2 Fungal Infections

Currently, the foremost antifungal therapeutics comprise polyenes, azoles, and echinocandins. However, fatality rates continue to be very high even when antifungals are administered properly. Nanotechnology is one promising path that is being actively studied to create effective vaccinations and new antifungal treatments because it allows focused medication and antigen transfer, which lowers toxicity and therapy expenses.[36] Since its discovery in 1955, the antibiotic amphotericin B (AmB) has been favored for the treatment of systemic fungal infections.[37] Nevertheless, owing to infusion-associated febrile responses and dose-dependent kidney damage in the host, the clinical relevance of AmB and its typical formulation is constrained. Zia *et al.* used *Aloe vera* leaf extract to create biomimetically designed NPs of AmB to get around this. Interestingly, compared with the drug's free form, the NPs were less harmful to the host and much more efficient at eliminating a variety of fungal infections, notably *Candida* spp.[38]

10.3.3 Viral Infections

Middle East respiratory syndrome (MERS), coronavirus disease (COVID-19), and other highly contagious viral infections are becoming more prevalent and constitute a danger to both human wellbeing and the world economy.[39] In contrast to traditional chemical-based antiviral medicines, the use of NPs offers an intriguing prospect for the creation of innovative antiviral treatments with minimal risk of acquiring drug resistance.

Zika virus (ZIKV) has become a serious threat to worldwide health as a result of its unanticipated causal connection to severe neurological conditions. A biomimetic nanodecoy (ND) was created that can capture ZIKV, deflect ZIKV from its original target, and prevent ZIKV infection. The ND, which is made up of a gelatin NP core covered with the host cell membranes of mosquito medium, efficiently adsorbs ZIKV and prevents ZIKV reproduction in ZIKV-vulnerable cells. The chance of survival of mice infected with ZIKV was increased by NDs because they considerably reduced the inflammatory reactions and degenerative alterations that ZIKV causes.[40]

SARS-CoV-2 specifically attacks angiotensin-converting enzyme 2 (ACE2) in the lungs and goblet secretory cells in the nasal mucosa by the interaction of its spike protein with ACE2. Li *et al.* developed ACE2-NDs derived from human lung spheroid cells (LSCs) that can bind and neutralize SARS-CoV-2 and protect the host lung cells from infection. Inhalation treatment was used to administer these LSC-nanodecoys to mice, where they remained in the lungs for more than 72 h after administration. In addition, SARS-CoV-2 mimics were cleared from the lungs faster after being inhaled by LSC-NDs, and no toxicity was noted.[41] Similar to this, Wang *et al.* produced NPs using the membranes of human embryonic kidney-239T cells that express high levels of ACE2. Spike protein S1 was entrapped by developed NPs in a dose-dependent manner. Further research showed that ACE2-NPs effectively prevented viral infection *in vitro* and *in vivo* and suppressed the entry of SARS-CoV-2 S pseudovirions into host cells.[42] Figure 10.8 displays a detailed mechanism.

Rao *et al.* created NDs by combining ACE2 receptor-expressing genetically modified cells with human monocyte-derived cellular membrane nanovesicles. These NDs efficiently shielded host cells from SARS-CoV-2 infection by fighting with host cells for virus binding. Additionally, the NDs effectively attached and neutralized inflammatory cytokines such as interleukin 6 (IL-6)

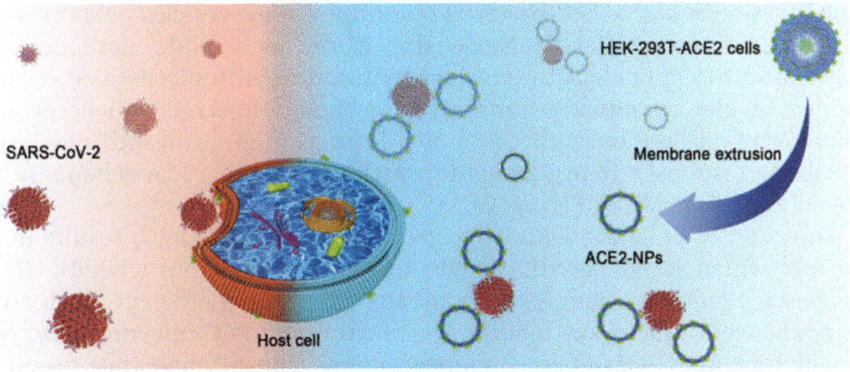

Figure 10.8 Diagram depicting the preparation and function of ACE2-NPs. Reproduced from ref. 42 with permission from American Chemical Society, Copyright 2021.

and granulocyte–macrophage colony-stimulating factor (GM-CSF), and they greatly decreased immune disorder and lung injury in an animal model of acute pneumonia by reliance on the plentiful cytokine receptors on their surface.[43]

SARS-CoV-2 infectivity is dependent on its ability to bind to either known or unknown protein receptors on target cells. Zhang *et al.* were inspired by this fact and developed cellular nanosponges made of plasma membranes derived from human lung epithelial type II cells or human macrophages. The nanosponges exhibited the same receptors that viruses use for cellular entry. As a result of incubation of the nanosponges, SARS-CoV-2 was neutralized and did not infect cells.[44]

In particular, HIV entry/fusion is started by an encounter between the viral envelope glycoprotein (gp120) and CD4+ on T helper cells, resulting in a reduction in T helper cells. In light of this, Wei *et al.* created T-cell membrane-coated nanoparticles (TNPs). TNPs acted as decoys for the viral attack, diverting viruses aside from their intentional host targets and neutralizing HIV. TNPs specifically bind to gp120 and inhibit gp120-induced CD4+ T-cell killing.[45] In related research, TNPs were created that can allegedly neutralize a wide range of HIV-1 strains.[46]

10.3.4 Vaccine Development

10.3.4.1 Antibacterial Vaccines

The best method for reducing morbidity and mortality from infectious diseases is vaccination. Antibacterial vaccinations have significantly decreased worldwide morbidity and death in contemporary practice. One cutting-edge approach to NP vaccination platforms is to encapsulate synthetic NPs with natural cell membranes.[47] Using cell membrane fragments to decorate a nanoparticulate adjuvant as the antigen-presenting source is a promising tactic.[48] Gao *et al.* created bacterial outer-membrane vesicles (OMVs) coated with gold nanoparticles (AuNPs) using *E. coli* as a model pathogen. The generated NPs quickly activated and matured dendritic cells in the lymph nodes of the immunized animals after being injected subcutaneously. Moreover, vaccination with these NPs formed more avid and long-lasting antibody responses than vaccination with OMVs alone.[49] A schematic representation is given in Figure 10.9.

Infection of chickens with avian pathogenic *E. coli* (APEC) results in significant financial losses and danger to the general public's health. In the large-scale chicken business of today, the adoption of efficient vaccinations is acknowledged as a key method for controlling APEC infections. Hu *et al.* created an APEC–OMV-based nanovaccine as a result, and they tested the vaccine's efficacy and safety in broiler chicks. The findings demonstrated that the average diameter of the membrane-enclosed structures in the APEC–OMVs was 89 nm. A 50 g dose of APEC–OMV vaccine effectively protected chicks against homologous infection with minimal adverse effects.[50] In another study, self-assembled glycyrrhizic acid nanoparticles (GANs) were coated with OMVs to create a stable OMV vaccine. The subcutaneous GAN–OMV vaccine greatly increased the generation of antibodies against *Bordetella bronchiseptica* (Bb) and the proliferation of lymphocytes. The ratio of CD4+/CD8+ T cells to CD19+ B cells was greater in the splenic lymphocytes of mice inoculated with GAN–OMVs, and these cells also produced considerably more Th1/Th2/Th17 cytokines. Additionally, GAN–OMVs successfully stopped Bb reinfection.[51] These results imply that coating synthetic nanomaterials with bacterial OMVs may provide useful information for the development of an extremely potent and secure antibacterial vaccine.

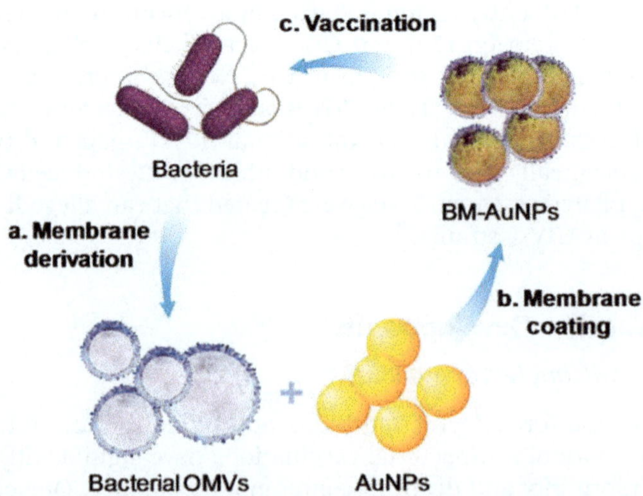

Figure 10.9 Schematic illustration of modulating antibacterial immunity *via* bacterial membrane-coated NPs. Reproduced from ref. 49 with permission from American Chemical Society, Copyright 2015.

10.3.4.2 Antiviral Vaccines

Nanomedicine was advocated by researchers as a new option in the development of a new generation of vaccines because of the low immunogenicity of viral antigens. A lipidic nanomaterial with potential bioinspiration and biomimetic effectiveness against viral infections, virosomes are a new class of US Food and Drug Administration (FDA)-approved nanocarriers. Production of vaccines is based on the crucial viral fusion proteins that modify the virosome surface.[52] The reconstituted virosome, as shown in Figure 10.10, is mostly comparable to the enveloped virus structure that elicited strong humoral and cellular immune responses. In a murine model system,

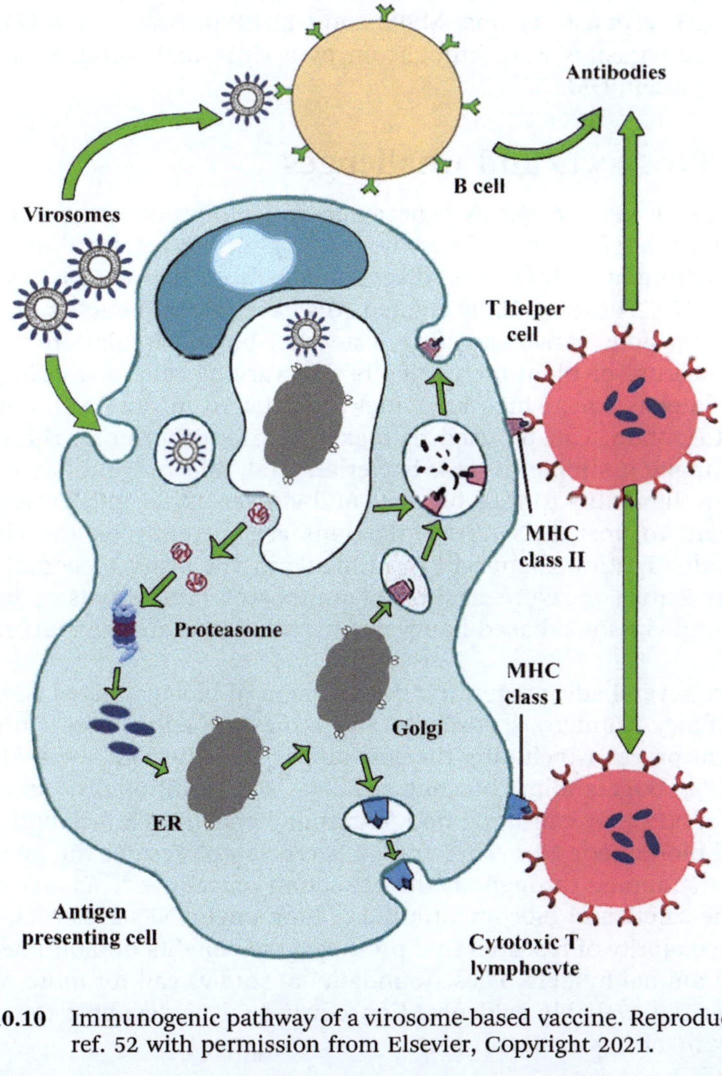

Figure 10.10 Immunogenic pathway of a virosome-based vaccine. Reproduced from ref. 52 with permission from Elsevier, Copyright 2021.

virosomes induce a more balanced T helper 1 *versus* T helper 2 response than a conventional subunit vaccine, with virosomes specificaly evoking a greater T helper 1 response than a subunit vaccine. Additionally, a portion of the virosomal antigen gains entry to the major histocompatibility class I presentation pathway as a consequence of the fusion of the virosomes with the endosomal membrane, priming cytotoxic T lymphocyte activity. Conclusively, virosomes provide an excellent platform for the addition of lipophilic adjuvants to further stimulate vaccine immunogenicity.[53]

Another intriguing report described the immunogenicity of a recombinant virosomal fusion protein made *via* the fusion of the MSP3 and GLURP domains from two malaria vaccine candidate antigens. The highly pure recombinant protein GMZ2 was linked to phosphatidylethanolamine and the conjugates integrated into the membrane of influenza virosomes. Cross-reactive IgG responses against MSP3 and GLURP in *Plasmodium falciparum* blood-stage parasites were brought on by a virosomal formulation of the chimeric protein GMZ2.[54]

10.4 Prospects and Challenges

Conventional NPs frequently experience rapid immune system clearance and have unsatisfactory medication-targeting properties. Until now, drugs for the treatment of infectious diseases have been delivered using bioengineered NPs. These specially created NPs have several benefits, including immune evasion, active targeting, extended blood circulation time, *etc.* The key feature of biomimetic NPs is that various cells, including RBCs, WBCs, platelets, *etc.*, which are widely distributed in the body and hence simple to obtain, can be used to make them. Furthermore, these nanosystems make it simple to treat bacterial, viral, and fungal diseases since they have the ability to lead bacteria and viruses astray and prevent their attachment to host cells. Viral infections are currently on the rise and, as a result, fighting them will be difficult in the years to come. As the main preventive measure against a number of fatal diseases, bacteria-derived and virosome-based nanovaccines will dominate the market in the future.

Despite several advantages, the development of bioengineered NPs is still in its infancy. Numerous obstacles and issues are still present in the development process, including the cell supply, the laborious method of creating NPs, targetability, biocompatibility, *etc.* Additionally, an uneven coating of NPs may cause the body's immune system to attack and destroy supplied biomimetic NPs. Therefore, it is crucial to preserve the integrity of the cell membrane throughout the extraction procedure.[55] As stated previously, the safety and biocompatibility of biomimetic NPs are crucial. However, the majority of research that promotes the benefits of biomimetic NPs relies on animal models. These foundational studies call for more research to bolster the available evidence. Consequently, it is desirable to continue the work involving clinical trials.

10.5 Conclusion

In a nutshell, bioengineered nanoparticulate systems have contributed significantly to the treatment of infectious disorders. Owing to the outstanding results that have been obtained and the inherent benefits that these systems possess, biomimetic NPs have attracted considerable attention in the past 10 years. Owing to their innate capacity to strengthen human immunity, bacteria-derived and virosomal nanovaccines have taken center stage. However, in order for the area to advance further, some issues, including laborious fabrication procedures, scaling-up difficulties, low yields, *etc.*, must be resolved. In conclusion, biomimetic nanosystems will soon be a potent weapon against a wide range of infectious diseases.

References

1. A. R. Kirtane, M. Verma, P. Karandikar, J. Furin, R. Langer and G. Traverso, *Nat. Nanotechnol.*, 2021, **16**, 369.
2. M. X. Luo, S. Hua and Q. Y. Shang, *Mol. Med. Rep.*, 2021, **23**(5), 325.
3. H. Liu, W. Zhong, X. Zhang, D. Lin and J. Wu, *J. Mater. Chem. B*, 2021, **9**(38), 7878.
4. K. Watkins, *Curr. Emerg. Hosp. Med. Rep.*, 2018, **6**(3), 86.
5. A. Vashist, A. Kaushik, A. Vashist, R. D. Jayant, A. Tomitaka, S. Ahmad, Y. K. Gupta and M. Nair, *Biomater. Sci.*, 2016, **4**(11), 1535.
6. P. Basnet and N. Škalko-Basnet, *Curr. Pharm. Des.*, 2013, **19**(41), 7237.
7. K. M. Rubey and J. S. Brenner, *Adv. Drug Delivery Rev.*, 2021, **179**, 113996.
8. W. Yang, J. I. Peters and R. O. Williams, *Int. J. Pharm.*, 2008, **356**(1–2), 239.
9. M. H. Xiong, Y. Bao, X. Z. Yang, Y. H. Zhu and J. Wang, *Adv. Drug Delivery Rev.*, 2014, **78**, 63.
10. A. Barhoum, M. L. García-Betancourt, J. Jeevanandam, E. A. Hussien, S.A. Mekkawy, M. Mostafa, M. M. Omran, M. S. Abdalla and M. Bechelany, *Nanomaterials*, 2022, **12**(2), 177.
11. K. J. Koudelka, A. S. Pitek, M. Manchester and N. F. Steinmetz, *Annu. Rev. Virol.*, 2015, **2**(1), 379.
12. L. Chen, W. Hong, W. Ren, T. Xu, Z. Qian and Z. He, *Signal Transduction Targeted Ther.*, 2021, **6**, 225.
13. G. Yang, S. Chen and J. Zhang, *Front. Pharmacol.*, 2019, **10**, 751.
14. L. Liu, J. Wu, J. Gao and X. Lu, *Adv. Healthcare Mater.*, 2020, **9**(22), e2000893.
15. J. H. Kim, J. Lee, J. Park and Y. S. Gho, *Semin. Cell Dev. Biol.*, 2015, **40**, 97.
16. W. Chen, Q. Zhang, B. T. Luk, R. H. Fang, Y. Liu, W. Gao and L. Zhang, *Nanoscale*, 2016, **8**(19), 10364.
17. Q. Xia, Y. Zhang, Z. Li, X. Hou and N. Feng, *Acta Pharm. Sin. B*, 2019, **9**(4), 675.
18. R. H. Fang, A. V. Kroll, W. Gao and L. Zhang, *Adv. Mater.*, 2018, **30**(23), e1706759.

19. A. Jha, A. N. Nikam, S. Kulkarni, S. P. Mutalik, A. Pandey, M. Hegde, B. S. S. Rao and S. Mutalik, *J. Controlled Release*, 2021, **329**, 413.
20. C. Deusenbery, Y. Wang and A. Shukla, *ACS Infect. Dis.*, 2021, **7**(4), 695.
21. R. H. Fang, B. T. Luk, C. M. Hu and L. Zhang, *Adv. Drug Delivery Rev.*, 2015, **90**, 69.
22. J. M. Cavaillon, *Toxicon*, 2018, **149**, 45.
23. J. Ma, L. Jiang and G. Liu, *Wiley Interdiscip. Rev.: Nanomed. Nanobiotechnol.*, 2022, **14**(5), e1825.
24. S. Shen, F. Han, A. Yuan, L. Wu, J. Cao, J. Qian, X. Qi, Y. Yan and Y. Ge, *Biomaterials*, 2019, **189**, 60.
25. L. Jiang, Y. Zhu, P. Luan, J. Xu, G. Ru, J. G. Fu, N. Sang, Y. Xiong, Y. He, G. Q. Lin, J. Wang, J. Zhang and R. Li, *ACS Nano*, 2021, **15**(3), 4173.
26. B. Chen, F. Li, X. K. Zhu, W. Xie, X. Hu, M. H. Zan, X. Li, Q. Y. Li, S. S. Guo, X. Z. Zhao, Y. A. Jiang, Z. Cao and W. Liu, *Biomater.Sci.*, 2021, **9**(3), 826.
27. S. Thamphiwatana, P. Angsantikul, T. Escajadillo, Q. Zhang, J. Olson, B. T. Luk, S. Zhang, R. H. Fang, W. Gao, V. Nizet and L. Zhang, *Proc. Natl. Acad. Sci. U. S. A.*, 2017, **114**(43), 11488.
28. B. Wu, L. Lin, F. Zhou and X. Wang, *Bioprocess Biosyst. Eng.*, 2020, **43**(11), 2065.
29. Y. Zhang, J. Zhang, W. Chen, P. Angsantikul, K. A. Spiekermann, R. H. Fang, W. Gao and L. Zhang, *J. Controlled Release*, 2017, **263**, 185.
30. P. Angsantikul, S. Thamphiwatana, Q. Zhang, K. Spiekermann, J. Zhuang, R. H. Fang, W. Gao, M. Obonyo and L. Zhang, *Adv.Ther.*, 2018, **1**(2), 1800016.
31. Y. Ping, X. Hu, Q. Yao, Q. Hu, S. Amini, A. Miserez and G. Tang, *Nanoscale*, 2016, **8**(36), 16486.
32. J. M. V. Makabenta, A. Nabawy, C. H. Li, S. Schmidt-Malan, R. Patel and V. M. Rotello, *Nat. Rev. Microbiol.*, 2021, **19**, 23.
33. Y. Hu, X. Ruan, X. Lv, Y. Xu, W. Wang, Y. Cai, M. Ding, H. Dong, J. Shao, D. Yang and X. Dong, *Nano Today*, 2022, **46**, 101602.
34. L. L. Li, J. H. Xu, G. B. Qi, X. Zhao, F. Yu and H. Wang, *ACS Nano*, 2014, **8**(5), 4975.
35. A. Lin, Y. Liu, X. Zhu, X. Chen, J. Liu, Y. Zhou, X. Qin and J. Liu, *ACS Nano*, 2019, **13**(12), 13965.
36. B. Kischkel, S. A. Rossi, S. R. Santos, J. D. Nosanchuk, L. R. Travassos and C. P. Taborda, *Front. Cell. Infect. Microbiol.*, 2020, **10**, 463.
37. P. Balabathula, S. G. Whaley, D. R. Janagam, N. K. Mittal, B. Mandal, L. A. Thoma, P. D. Rogers and G. C. Wood, *Pharmaceutics*, 2020, **12**(3), 247.
38. Q. Zia, O. Mohammad, M. A. Rauf, W. Khan and S. Zubair, *Sci. Rep.*, 2017, **7**(1), 11873.
39. S. Gurunathan, M. Qasim, Y. Choi, J. T. Do, C. Park, K. Hong, J. H. Kim and H. Song, *Nanomaterials*, 2020, **10**(9), 1645.
40. L. Rao, W. Wang, Q. F. Meng, M. Tian, B. Cai, Y. Wang, A. Li, M. Zan, F. Xiao, L. L. Bu, G. Li, A. Li, Y. Liu, S. S. Guo, X. Z. Zhao, T. H. Wang, W. Liu and J. Wu, *Nano Lett.*, 2019, **19**(4), 2215.

41. Z. Li, Z. Wang, P. C. Dinh, D. Zhu, K. D. Popowski, H. Lutz, S. Hu, M. G. Lewis, A. Cook, H. Andersen, J. Greenhouse, L. Pessaint, L. J. Lobo and K. Cheng, *Nat. Nanotechnol.*, 2021, **16**, 942.
42. C. Wang, S. Wang, Y. Chen, J. Zhao, S. Han, G. Zhao, J. Kang, Y. Liu, L. Wang, X. Wang, Y. Xu, S. Wang, Y. Huang, J. Wang and J. Zhao, *ACS Nano*, 2021, **15**(4), 6340.
43. L. Rao, S. Xia, W. Xu, R. Tian, G. Yu, C. Gu, P. Pan, Q. F. Meng, X. Cai, D. Qu, L. Lu, Y. Xie, S. Jiang and X. Chen, *Proc. Natl. Acad. Sci. U. S. A.*, 2020, **117**(44), 27141.
44. Q. Zhang, A. Honko, J. Zhou, H. Gong, S. Downs, J. Vasquez, R. Fang, W. Gao, A. Griffiths and L. Zhang, *Nano Lett.*, 2020, **20**(7), 5570.
45. X. Wei, G. Zhang, D. Ran, N. Krishnan, R. H. Fang, W. Gao, S. A. Spector and L. Zhang, *Adv. Mater.*, 2018, **30**(45), e1802233.
46. G. Zhang, G. R. Campbell, Q. Zhang, E. Maule, J. Hanna, W. Gao, L. Zhang and S. A. Spector, *mBio*, 2020, **11**(5), e00903-20.
47. P. Angsantikul, S. Thamphiwatana, W. Gao and L. Zhang, *Vaccines*, 2015, **3**(4), 814.
48. H. Hu, C. Yang, F. Zhang, M. Li, Z. Tu, L. Mu, J. Dawulieti, Y. H. Lao, Z. Xiao, H. Yan, W. Sun, D. Shao and K. W. Leong, *Adv. Sci.*, 2021, **8**(15), 2002020.
49. W. Gao, R. H. Fang, S. Thamphiwatana, B. T. Luk, J. Li, P. Angsantikul, Q. Zhang, C. M. Hu and L. Zhang, *Nano Lett.*, 2015, **15**(2), 1403.
50. R. Hu, H. Liu, M. Wang, J. Li, H. Lin, M. Liang, Y. Gao and M. Yang, *Nanomaterials*, 2020, **10**(11), 2293.
51. Y. Huang, L. Nan, C. Xiao, J. Dong, K. Li, J. Cheng, Q. Ji, Q. Wei, G. Bao and Y. Liu, *Int. J. Nanomed.*, 2022, **17**, 647.
52. K. Asadi and A. Gholami, *Int. J. Biol.Macromol.*, 2021, **182**, 648.
53. A. Huckriede, L. Bungener, T. Stegmann, T. Daemen, J. Medema, A. M. Palache and J. Wilschut, *Vaccine*, 2005, **23**, S26.
54. M. Tamborrini, S. A. Stoffel, N. Westerfeld, M. Amacker, M. Theisen, R. Zurbriggen and G. Pluschke, *Malar. J.*, 2011, **10**, 359.
55. X. Zhao and C. Yan, *Nanoscale Res. Lett.*, 2022, **17**, 36.

CHAPTER 11

Ex Vivo *Immunization: A Strategy for Immunization Against SARS-CoV-2*

HARPREET SINGH,[a] POOJITH NUTHALAPATI,[b] PREETHI YENDAPALLI[b] AND DEBASIS SAHU*[c,d]

[a] School of Biosciences and Bioengineering, Lovely Professional University, Phagwara, Punjab 144411, India; [b] Sri Ramachandra Institute of Higher Education and Research, Chennai, Tamil Nadu 600116, India; [c] Amity Institute of Biotechnology, Amity University, Noida, Uttar Pradesh 201303, India; [d] Ubioquitos Inc., Markham, Canada
*Email: debasissahu.phd@gmail.com

11.1 Introduction

Given the nature of highly contagious novel coronavirus 2 (nCoV-2) worldwide, a collaborative effort among all researchers is vital to prevent the further spread of the virus and to develop every possible route for the treatment of coronavirus COVID-19. It caused an outbreak of severe acute respiratory syndrome (SARS) in Wuhan, China,[1] which left its neighboring countries and trading partners vulnerable to COVID-19. The virus is believed to have originated from bats when the whole genome sequence similarity of 96% was confirmed, and the sequence was also found to be 79.6% identical with SARS-CoV-2.[2]

The virulence of this virus is mediated *via* the receptor binding domain (RBD) of spike protein (S protein), having a core receptor binding motif (RBM) at the S1 region of S protein which complements the angiotensin-

Nanoscience & Nanotechnology Series No. 62
Smart Nanomaterials for Infectious Diseases
Edited by Suvardhan Kanchi, Nidhi Sharotri, Rajasekhar Chokkareddy, Deepali Sharma and Falah H. Hussein

Published by the Royal Society of Chemistry, www.rsc.org

converting enzyme 2 (ACE2) and facilitates the endocytosis of the nCoV-2 into the host cell.[3–5] The S2 region of the S protein of nCoV-2 then undergoes acidic cleavage by some proteases such as cathepsin, transmembrane protease serine-2 (TMPRRS2), or any other protease,[6] which facilitates the separation of the RBD domain and fusion domain and exposes the fusion peptide in the S2 region to embed into the cell membrane and promote the fusion of the virion and lysosomal membrane, thereby releasing the viral genome into the cytosol of the cell followed by replication and release of new virion particles.[7]

Certain drugs, such as hydroxychloroquine, were the first choice to treat COVID-19, working on the principle of increasing the pH of the intracellular vacuoles, *i.e.* lysosome, thus preventing the hydrolytic cleavage of regions of S protein from acid hydrolases and thereby preventing the release of the viral genome into the cytosol of the cell. However, this would also mean that the antigen-presenting cells (APCs) would be compromised in providing immunity against the virus and presenting antigens.[8] Other drugs under consideration are some nucleoside and nucleotide analogs that are supposed to target the viral-specific RNA-dependent RNA polymerase to prevent the replication of the viral genome and suppress its pathogenicity.[9]

The last outbreak of SARS-CoV in 2003 was reported to have originated in Foshan City, Guangdong Province, China.[10] The virus was traced back to a live animal market where *Paguma larvata* (masked palm civet) and *Nyctereutes procyonoides* (raccoon dog) harbored the same virus with 99.8% sequence similarity to palm civet.[11] SARS-CoV was noted to cause lymphocytopenia, and in a study of 38 patients, 84% suffered from lymphocytopenia, 100% of whom were deficient in CD4+ T cells, 87% were deficient in CD8+ T cells, 76% were deficient in B cells, and 55% were deficient in natural killer (NK) cells.[12]

An alternative to these drugs and other types of vaccination is the *ex vivo* immunization for patients infected with nCoV-2 and naïve individuals. It promises a fast and effective means of treatment and prophylactic measures for the time being.

11.2 *Ex Vivo* Immunization: A Hypothetical Attempt to Treat COVID-19

Ex vivo antigen-pulsed APCs and other leukocytes may provide a safe and efficient immunization method for a wide variety of populations with the least adverse effects. A few reports suggest that *ex vivo* immunization has been successful against certain bacterial and viral diseases in the past, and it has also been used to alleviate cancer.[13] In the chlamydial genital tract infection caused by *Chlamydia trachomatis*, dendritic cells (DCs) isolated from the patient were pulsed with the killed bacterial cells, which resulted in their phagocytosis by the DCs with subsequent secretion of IL-12 p40 and antigen presentation to naïve CD4+ T cells. Such pulsed DCs were able to elicit a similar immune response in the host as in naturally infected individuals but without any associated pathogenesis.[14]

As far as the viral infection is concerned, SARS-CoV spike-specific S1203 and S978 epitopes were able to induce interferon gamma (IFN-γ) secreting T-cell response in an HLA-A2$^+$ SARS-CoV-infected host.[15] DCs coated with spike peptide of SARS-CoV induced activation of IFN-γ secreting CD8+ T lymphocytes in draining lung lymph nodes and protected mice from SARS and increased the survival rate of infected mice from 20% to >90%, and a combination of the peptide of all structural proteins from SARS-CoV was also reported to induce cell-mediated immunity (CMI).[16] Epstein–Barr virus (EBV), or the kissing virus, is known to undergo contact spreading and causes mononucleosis. In such cases, cytotoxic T lymphocytes (CTLs) of the infected individual show high reactivity to the EBV surface antigens gp85 and gp350, resulting in the recuperation of patients. It was observed that such individuals show the retention of memory CTLs in their blood over a long period. This supports the use of CTLs for *ex vivo* immunization. It was reported that when individuals were injected with the gp85- and gp350-specific CTL epitopes, they developed an effective immune response, thus providing protection against the vaccinia virus expressing the respective antigens.[17] Similarly, DCs infected with PR8 and X31 influenza virus, when injected into mice, induced a Th1-like immune response, but without releasing any progeny viral particles. These mice showed an early recovery from viral infection in comparison with non-vaccinated mice.[18]

Kumar *et al.* demonstrated that the Prevnar 13 vaccine, which contains the antigens Ft from *Francisella tularensis* and Sp from *Streptococcus pneumoniae*, elicited a lower immune response to the infection compared with individuals who were administered peripheral blood mononuclear cells (PBMCs) pulsed with Prevnar 13 *ex vivo*. Intranasal administration was found to be highly effective.[19]

In addition, *ex vivo* immunization has also been reported to be safe, primarily because of the absence of other biological interferences in the host body. This could be applied for nCoV-2 immunization mainly in a susceptible population. Such a strategy can be of great importance as a prophylactic and also in a therapeutic approach against COVID-19, thereby curbing its further spread.

11.3 Effective Immunogens of SARS-CoV-2

As the exposed surface antigens of the coronavirus are easily accessible to immunoglobulins, these are the most favorable candidates for vaccine development. One such component of nCoV-2 is the spike (S) protein. Thereby the main components of nCoV-2 are its structural components which facilitate attachment and entry into the host cell.[7] Envelope (E) protein is less abundant with ion channel activity which facilitates the assembly and release of viruses from the cell.[20] The membrane (M) protein gives curvature to the virion and binds to the nucleocapsid for proper packaging,[21] and the nucleocapsid (N) protein, in turn, is capable of binding and packaging RNA to give a beads-on-a-string conformation for complete packaging.[22,23]

11.3.1 Importance of S Protein in the Immunization Approach

It was found that the adenoviral (Ad5)-associated vector carrying the SARS-CoV spike S1 fragment, membrane, and nucleocapsid coding genes, when injected intramuscularly, was capable of inducing antibody generation against the S1 fragment and T-cell activity against nucleocapsid protein in rhesus macaques.[24]

In another study, *Cercopithecus aethiops* was immunized intranasally with a recombinant attenuated parainfluenza virus expressing the complete SARS-CoV S protein (3768 bp of cDNA) and was found to develop antiserum against the S protein when challenged with SARS-CoV and prevented viral shedding compared with the control.[25]

The recombinant fusion protein of poly-γ-glutamic acid synthetase A from *Bacillus subtilis* and segments of spike protein SA (residues 2–114) and SB (residues 264–596) from SARS-CoV were expressed in *Lactobacillus casei* and administered to mice *via* the nasal and oral routes. High levels of serum IgG and mucosal IgA antibodies were induced, thereby neutralizing the viral particles.[26]

11.3.1.1 M Protein

M1–31 and M132–161 residues from N- and C-termini, respectively, of SARS-CoV M protein were found to elicit high reactivity against the convalescent phase sera of 40 SARS-positive patients with a negative response against sera of control healthy patients. It was found to secrete high titers of antibodies against these regions of M protein and thereby was termed an immunodominant epitope.[27]

11.3.1.2 N Protein

The pCI-N plasmid carrying the full-length SARS-CoV N protein coding gene was transfected intramuscularly into BALB/c mice. It was observed to induce IgG-type antibodies along with high levels of IFN-γ and IL-2 from induced splenocytes against N protein, respectively. Activation of N protein-specific CD8+ T cells was also observed, thereby stimulating both humoral immunity and CMI.[28] CMI was also observed in a lymphocyte proliferation test against the infectious bronchitis virus (IBV) N protein immunogen when BALB/c mice were injected subcutaneously with pEX-N vector carrying the gene for N protein mixed with dimethyldioctadecylammonium bromide.[29]

N protein is believed to actively induce both arms of immunity. A further result was obtained when an adenoviral (Ad5-N-V) vector carrying the SARS-CoV-specific N protein gene expressed and induced both humoral immunity and CMI in C57BL/6 mice.[30]

11.3.1.3 E Protein and Others

Although insufficient data are available to support the use of E protein as an immunogen, the deletion of E protein in SARS-CoV renders the virus attenuated. It inhibited its growth in hACE2 Tg mice and protected them from severe respiratory disease.[31] A lack of E protein prevents the assembly of virion particles.[32] Since E protein also acts as a cation-selective ion channel in a planar lipid bilayer, the drug hexamethylene amiloride was found to block the ion channel conductance and prevent the replication of human coronavirus 229E (hCoV-229E) and murine hepatitis virus E (MHV-E) in cultured cells.[33] This channel could be targeted to design antivirals and vaccines. It was also reported that the non-structural protein 1 (nsp1) of SARS-CoV is capable of interfering with type 1 interferon and that upon deletion of the nsp1 gene the virus was rendered inactive *in vivo*, and low doses of nsp1 mutant were capable of eliciting a high cytotoxic T lymphocyte response and protected mice against infection from homologous and heterologous viral invasion. Hence this candidate could also be targeted to treat COVID-19.[34]

11.4 Feasibility of *Ex Vivo* Immunization Against nCoV-2

The standard protocol for the isolation of blood components is straightforward. The common density gradient centrifugation of the blood layer on Histopaque (Sigma-Aldrich Cat. No. 10771) or Percoll (Sigma-Aldrich Cat. No. GE17-0891-02) to separate the blood components can be utilized.[35] These PBMCs can be standardized for pulsing with heat-killed or inactivated nCoV-2 under the required proportions and conditions. Complete attenuation or killing is necessary to avoid retrogressing of the virus for safety, and it can be verified by viability assay on cell cultures such as Vero cells.[36–39]

These stimulated PBMCs could be screened *in vitro via* enzyme-linked immunosorbent spot (ELISpot),[40] dot enzyme-linked immunosorbent assay (dot ELISA), lymphocyte proliferation assay,[41] ELISA,[42] *etc.* The data might vary among the nCoV-2 positive and negative hosts. The pulsed PBMCs can be readministered into the host *via* different routes after the conditioning of PBMCs for human standards. The nCoV-2 immune host can be screened for increased antisera titer and chemokines. A simple flow chart is presented in Figure 11.1 to explain the workflow of *ex vivo* immunization, and Figure 11.2 provides an overview of the mechanism of the immune response.

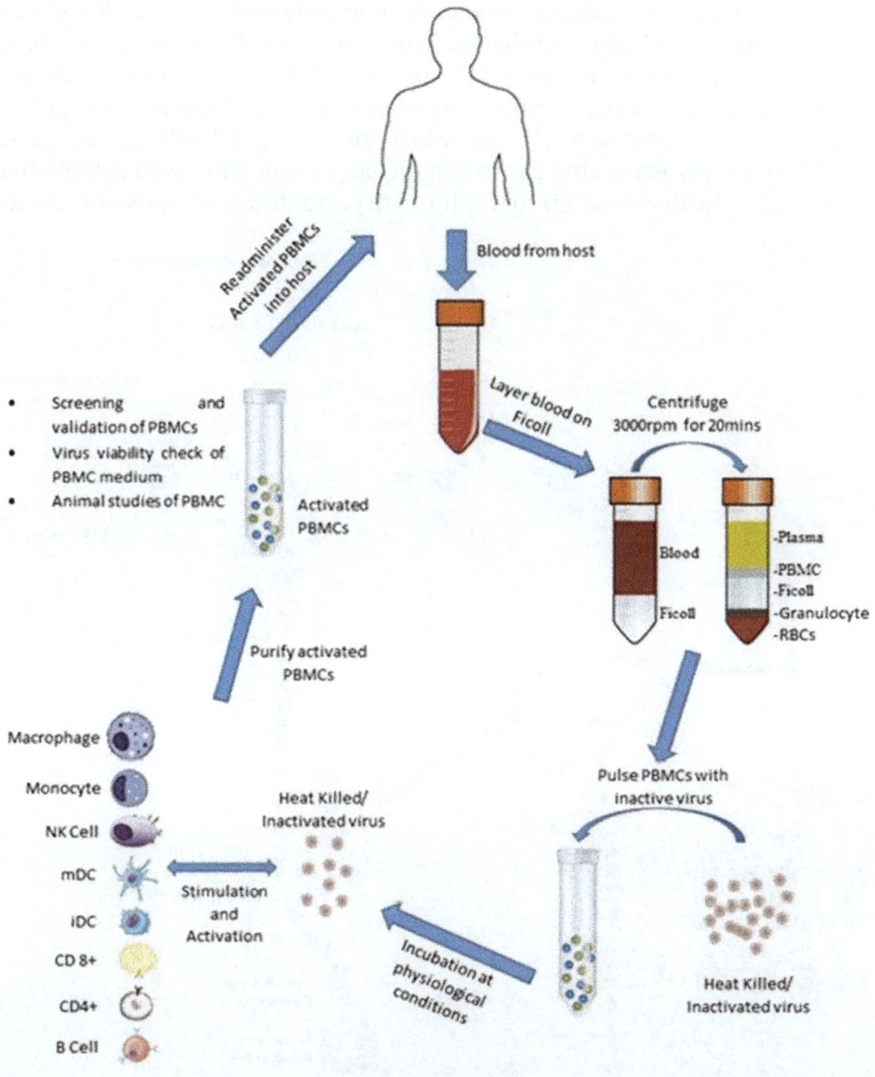

Figure 11.1 Diagrammatic representation of PBMC activation against heat-killed or inactivated nCoV-2. iDC, immature dendritic cell; mDC, mature dendritic cell; NK cell, natural killer cell.

11.5 Future Prospects

The urgency of the situation regarding COVID-19 and other future possible cases of lethal outbreaks demands the rapid development of safe and effective drugs and vaccines. Research organizations worldwide are putting great efforts into the development of vaccines and repurposing old drugs to analyze their potency, and this could be a time-consuming and lengthy process in case of emergencies. However, *ex vivo* immunization can be carried out with any new causative agents effectively to deliver quick and safe treatment to individuals worldwide. One can identify many pre-existing antivirals from closely related viral families or protein families that are conserved throughout the evolutionary changes *via* molecular docking,[43,44] surface plasmon resonance,[45] homology modeling of related antigens, *etc.*,[46–48] or simply extracting the antigenic sequences from well-established databases for the development of high-quality and different types of vaccines

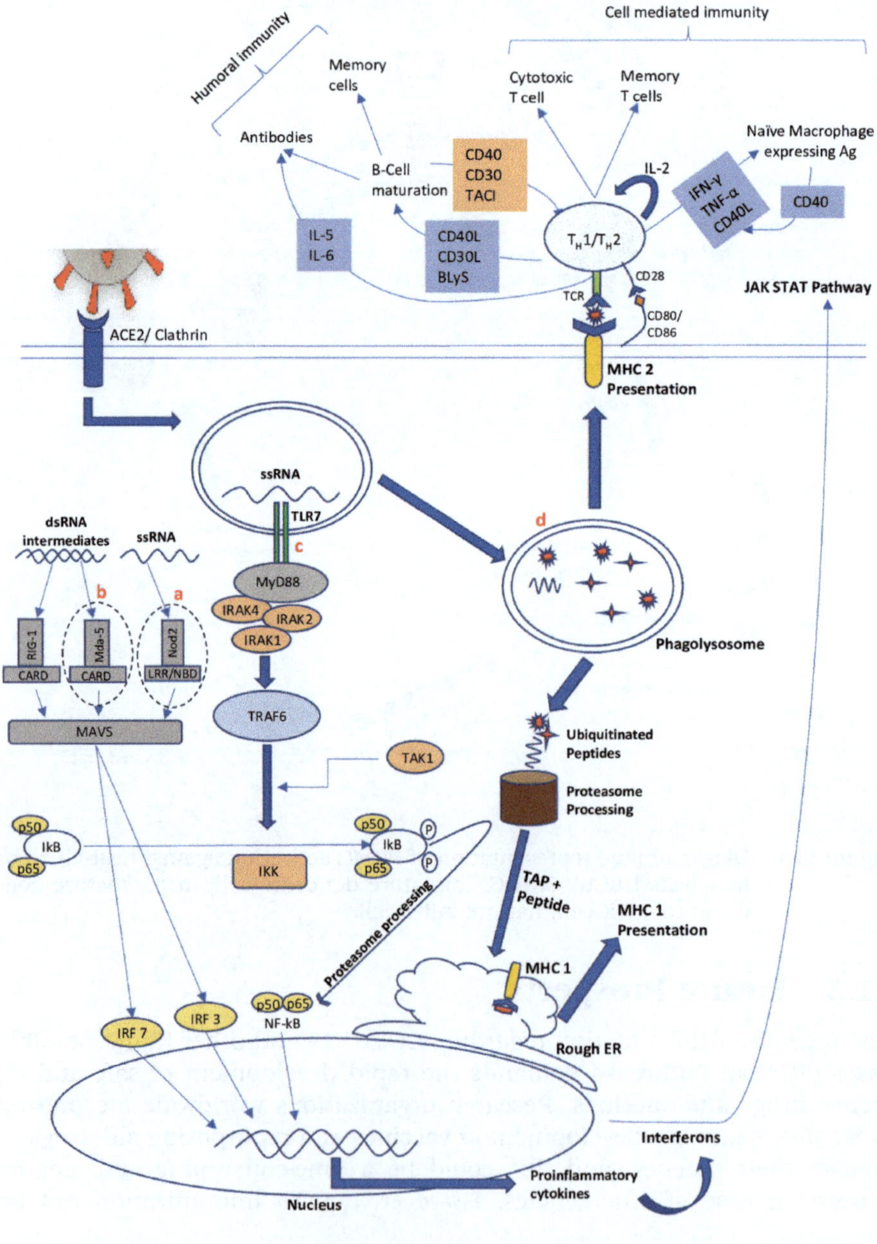

such as subunit,[49,50] DNA,[51] killed or attenuated,[52,53] recombinant,[54,55] live,[56] *etc.*, which will prove to be effective for *ex vivo* treatments. Given the advances in state-of-the-art technologies, vaccination strategies, and drug-targeting approaches, there is still scope for new ideas and techniques for rapid and effective responses to such emergencies.

11.6 Conclusion

Ex vivo immunization has earlier been carried out with many immunogenic compounds under conditions where the causative agent has proved to be difficult to handle or administer to the host. Similarly, in epidemics, when safety and treatment are of the utmost priority, and the need for treatment is urgent worldwide, all the available ideas and strategies should be assessed for the betterment of society. Certain proofs have been provided to support the feasibility of *ex vivo* immunization for nCoV-2. With the interaction of the immune system (PBMCs) and immunogens outside the host, the combinatorial advantage results in the development of immunity against nCoV-2 without any adverse effect on the host. The near-natural environment for PBMCs and immunogens allows natural interaction and signaling, thereby

---

Figure 11.2 The intracellular pattern recognition receptor (PRR) such as NOD-like receptor (NLR), RIG-like receptor (RLR), and Toll-like receptor (TLR) recognizes the pathogen-associated molecular patterns (PAMP) to induce the production of proinflammatory cytokines to control and restrict infection. (a) NOD2 has two N-terminal caspase recruitment domains (CARDs), one central nucleotide binding domain (NBD), and a C-terminal leucine-rich region (LRR) where NBD interacts with ssRNA of the virus and induces an immune cascade by interacting its CARD/LRR/NBD domain with MAVS (mitochondrial antiviral signaling), thereby activating TBK1–IRF3-induced expression of IFN1.[57] (b) Mda5 has two N-terminal CARD domains and one central helicase and C-terminal domain (CTD); the last two bind to dsRNA, dsDNA, and ssRNA (+ve/−ve) depending on the length and secondary structure of the nucleotide stretch whereas the CARD domain interacts with MAVS and triggers the activation of TBK1–IRF3-mediated induction of IFN1 response.[58] (c) TLR7 dimerizes upon ssRNA stimulation in the endosome and interacts with myeloid differentiation primary response gene 88 (Myd88) *via* the Toll/IL-1 receptor (TIR) domain to recruit IL-1 receptor-associated kinase-4/2/1 (IRAK-4/2/1) to activate TNF receptor-associated factor 6 (TRAF6) by its K63 polyubiquitination to activate the transforming growth factor β-activated kinase-1 (TAK1), which in turn activates the inhibitor of nuclear factor κB kinase dB (IKK) to phosphorylate the inhibitor of nuclear factor κB kinase (IκB) which degrades *via* the ubiquitination pathway to release the free NFκB. (d) MHC 1/2 plays a crucial role in antigen presentation *via* a cytosolic or endogenous pathway, respectively, to initiate the activation of T effector cells (T_H1/T_H2) for further recruitment and activation of B cells, macrophages, dendritic cells, neutrophils, and cytotoxic T cells. T_H1 and T_H2 effector cells provide cell-mediated and humoral immunity against intracellular and extracellular pathogens, respectively.

boosting the resistivity of immunity against closely related viruses by cross-reactivity. S, M, and N proteins are the major components in the *ex vivo* immunization process. However, proper pretreatment or inactivation of the immunogen is required for the major immunogenic sites to be exposed for fast and strong antigen–antibody interactions, which leads to improved and heightened immune responses (both humoral and cell mediated).

Abbreviations

ACE2	Angiotensin-converting enzyme 2
APCs	Antigen-presenting cells
BALB	Bagg albino
CMI	Cell-mediated immunity
COVID-19	Coronavirus disease-19
CTLs	Cytotoxic T lymphocytes
DCs	Dendritic cells
EBV	Epstein–Barr virus
ELISA	Enzyme-linked immunosorbent assay
ELISpot	Enzyme-linked immunosorbent spot
hACE2	Human angiotensin-converting enzyme 2
hCoV-229E	Human coronavirus-229E
HLA	Human leukocyte antigen
IBV	Infectious bronchitis virus
IFN	Interferon
Ig	Immunoglobulin
IL	Interleukin
MHV-E	Murine hepatitis virus E
NK	Natural killer
Nsp1	Non-structural protein 1
PBMCs	Peripheral blood mononuclear cells
RBD	Receptor binding domain
SARS-nCoV-2	Severe acute respiratory syndrome-novel coronavirus-2
Tg	Transgenic
TMPRRS2	Transmembrane protease serine-2

Acknowledgements

H. S. acknowledges the contributions of P. N. and P. Y. and the corresponding author D. S. for their diligence and efforts in compiling this chapter.

References

1. F. Wu, S. Zhao, B. Yu, Y. M. Chen, W. Wang and Z. G. Song, *et al.*, A new coronavirus associated with human respiratory disease in China, *Nature*, 2020, **579**(7798), 265–269.

2. P. Zhou, X. L. Yang, X. G. Wang, B. Hu, L. Zhang and W. Zhang, *et al.*, A pneumonia outbreak associated with a new coronavirus of probable bat origin, *Nature*, 2020, **579**(7798), 270–273.
3. H. Kubo, Y. K. Yamada and F. Taguchi, Localization of neutralizing epitopes and the receptor-binding site within the amino-terminal 330 amino acids of the murine coronavirus spike protein, *J. Virol.*, 1994, **68**(9), 5403–5410.
4. K. Wu, G. Peng, M. Wilken, R. J. Geraghty and F. Li, Mechanisms of host receptor adaptation by severe acute respiratory syndrome coronavirus, *J. Biol. Chem.*, 2012, **287**(12), 8904–8911.
5. Y. Wan, J. Shang, R. Graham, R. S. Baric and F. Li, Receptor Recognition by the Novel Coronavirus from Wuhan: an Analysis Based on Decade-Long Structural Studies of SARS Coronavirus, *J. Virol.*, 2020, **94**(7), e00127-20.
6. A. R. Fehr and S. Perlman, Coronaviruses: An overview of their replication and pathogenesis, in *Coronaviruses: Methods and Protocols*, Springer New York, 2015, pp. 1–23.
7. B. J. Bosch, R. van der Zee, C. A. M. de Haan and P. J. M. Rottier, The Coronavirus Spike Protein Is a Class I Virus Fusion Protein: Structural and Functional Characterization of the Fusion Core Complex, *J. Virol.*, 2003, **77**(16), 8801–8811.
8. R. I. Fox, Mechanism of action of hydroxychloroquine as an anti-rheumatic drug, *Semin. Arthritis Rheum.*, 1993, **23**(2), 82–91.
9. H. Lu, Drug treatment options for the 2019-new coronavirus (2019-nCoV), *BioSci. Trends*, 2020, **14**(1), 69–71.
10. J. D. Cherry and P. Krogstad, SARS: The first pandemic of the 21st century, *Pediatr. Res.*, 2004, **56**, 1–5.
11. Y. Guan, Isolation and Characterization of Viruses Related to the SARS Coronavirus from Animals in Southern China, *Science*, 2003, **302**(5643), 276–278.
12. W. Cui, Y. Fan, W. Wu, F. Zhang, J. Wang and A. Ni, Expression of Lymphocytes and Lymphocyte Subsets in Patients with Severe Acute Respiratory Syndrome, *Clin. Infect. Dis.*, 2003, **37**(6), 857–859.
13. M. Yazdani, M. R. Jaafari, J. Verdi, B. Alani, M. Noureddini and A. Badiee, Ex vivo- generated dendritic cell-based vaccines in melanoma: the role of nanoparticulate delivery systems, *Immunotherapy*, 2020, **12**(5), 333–349.
14. H. Su, R. Messer, W. Whitmire, E. Fischer, J. C. Portis and H. D. Caldwell, Vaccination against chlamydial genital tract infection after immunization with dendritic cells pulsed *ex vivo* with nonviable Chlamydiae, *J. Exp. Med.*, 1998, **188**(5), 809–818.
15. Y.-D. Wang, W.-Y. F. Sin, G.-B. Xu, H.-H. Yang, T. Wong and X.-W. Pang, *et al.*, T-Cell Epitopes in Severe Acute Respiratory Syndrome (SARS) Coronavirus Spike Protein Elicit a Specific T-Cell Immune Response in Patients Who Recover from SARS, *J. Virol.*, 2004, **78**(11), 5612–5618.

16. J. Zhao, J. Zhao and S. Perlman, T Cell Responses Are Required for Protection from Clinical Disease and for Virus Clearance in Severe Acute Respiratory Syndrome Coronavirus-Infected Mice, *J. Virol.*, 2010, **84**(18), 9318–9325.
17. R. Khanna, M. Sherritt and S. R. Burrows, EBV structural antigens, gp350 and gp85, as targets for *ex vivo* virus-specific CTL during acute infectious mononucleosis: potential use of gp350/gp85 CTL epitopes for vaccine design, *J. Immunol.*, 1999, **162**(5), 3063–3069.
18. C. B. López, A. Fernandez-Sesma, S. M. Czelusniak, J. L. Schulman and T. M. Moran, A mouse model for immunization with *ex vivo* virus-infected dendritic cells, *Cell. Immunol.*, 2000, **206**(2), 107–115.
19. S. Kumar, R. Sunagar, G. Pham, E. J. Gosselin and D. Nalin, *Ex vivo* antigen-pulsed PBMCs generate potent and long lasting immunity to infection when administered as a vaccine, *Vaccine*, 2017, **35**(7), 1080–1086.
20. J. L. Nieto-Torres, M. L. DeDiego, C. Verdiá-Báguena, J. M. Jimenez-Guardeño, J. A. Regla-Nava and R. Fernandez-Delgado, *et al.*, Severe Acute Respiratory Syndrome Coronavirus Envelope Protein Ion Channel Activity Promotes Virus Fitness and Pathogenesis, *PLoS Pathog.*, 2014, **10**(5), e1004077.
21. B. W. Neuman, G. Kiss, A. H. Kunding, D. Bhella, M. F. Baksh and S. Connelly, *et al.*, A structural analysis of M protein in coronavirus assembly and morphology, *J. Struct. Biol.*, 2011, **174**(1), 11–22.
22. L. S. Sturman, K. V. Holmes and J. Behnke, Isolation of coronavirus envelope glycoproteins and interaction with the viral nucleocapsid, *J. Virol.*, 1980, **33**(1), 449–462.
23. C. K. Chang, S. C. Sue, T. H. Yu, C. M. Hsieh, C. K. Tsai and Y. C. Chiang, *et al.*, Modular organization of SARS coronavirus nucleocapsid protein, *J. Biomed. Sci.*, 2006, **13**(1), 59–72.
24. W. Gao, A. Tamin, A. Soloff, L. D'Aiuto, E. Nwanegbo and P. D. Robbins, *et al.*, Effects of a SARS-associated coronavirus vaccine in monkeys, *Lancet*, 2003, **362**(9399), 1895–1896.
25. A. Bukreyev, E. W. Lamirande, U. J. Buchholz, L. N. Vogel, W. R. Elkins and M. St Claire, *et al.*, Mucosal immunisation of African green monkeys (Cercopithecus aethiops) with an attenuated parainfluenza virus expressing the SARS coronavirus spike protein for the prevention of SARS, *Lancet*, 2004, **363**(9427), 2122–2127.
26. J.-S. Lee, H. Poo, D. P. Han, S.-P. Hong, K. Kim and M. W. Cho, *et al.*, Mucosal Immunization with Surface-Displayed Severe Acute Respiratory Syndrome Coronavirus Spike Protein on Lactobacillus casei Induces Neutralizing Antibodies in Mice, *J. Virol.*, 2006, **80**(8), 4079–4087.
27. Y. He, Y. Zhou, P. Siddiqui, J. Niu and S. Jiang, Identification of immunodominant epitopes on the membrane protein of the severe acute respiratory syndrome-associated coronavirus, *J. Clin. Microbiol.*, 2005, **43**(8), 3718–3726.

28. P. Zhao, J. Cao, L. J. Zhao, Z. L. Qin, J. S. Ke and W. Pan, *et al.*, Immune responses against SARS-coronavirus nucleocapsid protein induced by DNA vaccine, *Virology*, 2005, **331**(1), 128–135.
29. A. M. H. Boots, J. G. Kusters, B. A. M. van der Zeijst and E. J. Hensen, The nucleocapsid protein of IBV comprises immunodominant determinants recognized by T-cells, in *Advances in Experimental Medicine and Biology*, Springer, Boston, MA, 1990, pp. 189–200.
30. A. N. Zakhartchouk, S. Viswanathan, J. B. Mahony, J. Glaudei and L. A. Babiuk, Severe acute respiratory syndrome coronavirus nucleocapsid protein expressed by an adenovirus vector is phosphorylated and immunogenic in mice, *J. Gen. Virol.*, 2005, **86**(1), 211–215.
31. J. Netland, M. L. DeDiego, J. Zhao, C. Fett, E. Álvarez and J. L. Nieto-Torres, *et al.*, Immunization with an attenuated severe acute respiratory syndrome coronavirus deleted in E protein protects against lethal respiratory disease, *Virology*, 2010, **399**(1), 120–128.
32. F. Fischer, C. F. Stegen, P. S. Masters and W. A. Samsonoff, Analysis of Constructed E Gene Mutants of Mouse Hepatitis Virus Confirms a Pivotal Role for E Protein in Coronavirus Assembly, *J. Virol.*, 1998, **72**(10), 7885–7894.
33. L. Wilson, P. Gage and G. Ewart, Hexamethylene amiloride blocks E protein ion channels and inhibits coronavirus replication, *Virology*, 2006, **353**(2), 294–306.
34. R. Züst, L. Cervantes-Barragán, T. Kuri, G. Blakqori, F. Weber and B. Ludewig, *et al.*, Coronavirus non-structural protein 1 is a major pathogenicity factor: Implications for the rational design of coronavirus vaccines, *PLoS Pathog.*, 2007, **3**(8), 1062–1072.
35. A. J. Ulmer, W. Scholz, M. Ernst, E. Brandt and H. D. Flad, Isolation and Subfractionation of Human Peripheral Blood Mononuclear Cells (PBMC) by Density Gradient Centrifugation on Percoll, *Immunobiology*, 1984, **166**(3), 238–250.
36. N. C. Ammerman, M. Beier-Sexton and A. F. Azad, Growth and maintenance of vero cell lines, *Current Protocols in Microbiology*, NIH Public Access, 2008, Vol. APPENDIX, p. Appendix.
37. M. Kaye, J. Druce, T. Tran, R. Kostecki, D. Chibo and J. Morris, *et al.*, SARS-associated Coronavirus Replication in Cell Lines, *Emerging Infect. Dis.*, 2006, **12**(1), 128–133.
38. W. B. Park, N. J. Kwon, S. J. Choi, C. K. Kang, P. G. Choe and J. Y. Kim, *et al.*, Virus isolation from the first patient with SARS-CoV-2 in Korea, *J. Korean Med. Sci.*, 2020, **35**(7), e84.
39. M. L. Ng, S. H. Tan, E. E. See, E. E. Ooi and A. E. Ling, Proliferative growth of SARS coronavirus in Vero E6 cells, *J. Gen. Virol.*, 2003, **84**(12), 3291–3303.
40. C. C. Czerkinsky, L. Å. Nilsson, H. Nygren, Ö. Ouchterlony and A. Tarkowski, A solid-phase enzyme-linked immunospot (ELISPOT) assay for enumeration of specific antibody-secreting cells, *J. Immunol. Methods*, 1983, **65**(1–2), 109–121.

41. C. Lopez and R. J. O'Reilly, Cell-mediated immune responses in recurrent herpesvirus infections: I. Lymphocyte proliferation assay, *J. Immunol.*, 1977, **118**(3), 895–902.
42. E. Engvall and P. Perlmann, Enzyme-linked immunosorbent assay (ELISA) quantitative assay of immunoglobulin G, *Immunochemistry*, 1971, **8**(9), 871–874.
43. B. K. Shoichet, I. D. Kuntz and D. L. Bodian, Molecular docking using shape descriptors, *J. Comput. Chem.*, 1992, **13**(3), 380–397.
44. E. Jenwitheesuk and R. Samudrala, Identifying inhibitors of the SARS coronavirus proteinase, *Bioorg. Med. Chem. Lett.*, 2003, **13**(22), 3989–3992.
45. B. Liedberg, C. Nylander and I. Lunström, Surface plasmon resonance for gas detection and biosensing, *Sens. Actuators*, 1983, **4**(C), 299–304.
46. O. Spiga, A. Bernini, A. Ciutti, S. Chiellini, N. Menciassi and F. Finetti, *et al.*, Molecular modelling of S1 and S2 subunits of SARS coronavirus spike glycoprotein, *Biochem. Biophys. Res. Commun.*, 2003, **310**(1), 78–83.
47. Z. Xu and D. Shi, *et al.*, *Signal Transduction Targeted Ther.*, 2020, **8**, 169.
48. S. Liu, G. Xiao, Y. Chen, Y. He, J. Niu and C. R. Escalante, *et al.*, Interaction between heptad repeat 1 and 2 regions in spike protein of SARS-associated coronavirus: Implications for virus fusogenic mechanism and identification of fusion inhibitors, *Lancet*, 2004, **363**(9413), 938–947.
49. Y. He, Y. Zhou, S. Liu, Z. Kou, W. Li and M. Farzan, *et al.*, Receptor-binding domain of SARS-CoV spike protein induces highly potent neutralizing antibodies: Implication for developing subunit vaccine, *Biochem. Biophys. Res. Commun.*, 2004, **324**(2), 773–781.
50. W. H. Chen, P. J. Hotez and M. E. Bottazzi, Potential for developing a SARS-CoV receptor-binding domain (RBD) recombinant protein as a heterologous human vaccine against coronavirus infectious disease (COVID)-19, *Hum. Vaccines Immunother.*, 2020, **16**(6), 1239–1242.
51. A. N. Zakhartchouk, Q. Liu, M. Petric and L. A. Babiuk, Augmentation of immune responses to SARS coronavirus by a combination of DNA and whole killed virus vaccines, *Vaccine*, 2005, **23**(35), 4385–4391.
52. S. U. Kapadia, J. K. Rose, E. Lamirande, L. Vogel, K. Subbarao and A. Roberts, Long-term protection from SARS coronavirus infection conferred by a single immunization with an attenuated VSV-based vaccine, *Virology*, 2005, **340**(2), 174–182.
53. M. Spruth, O. Kistner, H. Savidis-Dacho, E. Hitter, B. Crowe and M. Gerencer, *et al.*, A double-inactivated whole virus candidate SARS coronavirus vaccine stimulates neutralising and protective antibody responses, *Vaccine*, 2006, **24**(5), 652–661.
54. P. C. Y. Woo, S. K. P. Lau, H. W. Tsoi, Z. W. Chen, B. H. L. Wong and L. Zhang, *et al.*, SARS coronavirus spike polypeptide DNA vaccine priming with recombinant spike polypeptide from *Escherichia coli* as booster induces high titer of neutralizing antibody against SARS coronavirus, *Vaccine*, 2005, **23**(42), 4959–4968.

55. Y. Chen, C. Qin, Q. Wei, R. Li, H. Gao, H. Zhu, *et al.*, Protection of Rhesus Macaque from SARS-Coronavirus challenge by recombinant adenovirus vaccine, *bioRxiv*, 2020, DOI: 10.1101/2020.02.17.951939.
56. E. W. Lamirande, M. L. DeDiego, A. Roberts, J. P. Jackson, E. Alvarez and T. Sheahan, *et al.*, A Live Attenuated Severe Acute Respiratory Syndrome Coronavirus Is Immunogenic and Efficacious in Golden Syrian Hamsters, *J. Virol.*, 2008, **82**(15), 7721–7724.
57. A. Sabbah, T. H. Chang, R. Harnack, V. Frohlich, K. Tominaga and P. H. Dube, *et al.*, Activation of innate immune antiviral responses by Nod2, *Nat. Immunol.*, 2009, **10**(10), 1073–1080.
58. B. Wu, A. Peisley, C. Richards, H. Yao, X. Zeng and C. Lin, *et al.*, Structural basis for dsRNA recognition, filament formation, and antiviral signal activation by MDA5, *Cell*, 2013, **152**(1–2), 276–289.

CHAPTER 12

Neuroinvasion by SARS-CoV-2 Influences the Prognosis of Neurological Disorders

POOJITH NUTHALAPATI,[a] MOHAN KRISHNA GHANTA,[b] BATOUL FARRAN,[c] SANTENNA CHENCHULA,[d] VENKATA AKSHEENA VARAHI VEDAM,[e] VIGNESHWARAN PUKALENDIRAN,[f] KRISHNA CHAITANYA AMERNENI[g] AND L. V. K. S. BHASKAR*[h]

[a] Sri Ramachandra Institute of Higher Education and Research, Chennai, Tamil Nadu 600116, India; [b] Department of Pharmacology, MVJ Medical College and Research Hospital, Hoskote, Karnataka 562114, India; [c] Department of Hematology and Medical Oncology, Winship Cancer Institute, Emory University, Atlanta, GA 30322, USA; [d] Department of Pharmacology, All India Institute of Medical Sciences, Bhopal, Madhya Pradesh 462020, India; [e] Department of Pharmacology, ESIC Medical College and PGIMSR, K. K Nagar, Chennai, Tamil Nadu 600078, India; [f] Department of General Surgery, Mahatma Gandhi Medical College and Research Institute, Pillaiyarkuppam, Pondicherry 607402, India; [g] Department of Internal Medicine, Western Michigan University, Kalamazoo, MI 49008, USA; [h] Department of Zoology, Guru Ghasidas Vishwavidyalaya, Bilaspur, Chhattisgarh 495009, India
*Email: lvksbhaskar@gmail.com

Nanoscience & Nanotechnology Series No. 62
Smart Nanomaterials for Infectious Diseases
Edited by Suvardhan Kanchi, Nidhi Sharotri, Rajasekhar Chokkareddy, Deepali Sharma and Falah H. Hussein

Published by the Royal Society of Chemistry, www.rsc.org

12.1 Introduction

On March 11, 2020, the World Health Organization (WHO) declared the emergence and spread of the novel coronavirus (COVID-19) outbreak to be a global pandemic.[1] A large number of patients suffering from pneumonia following travel to seafood markets were reported in China in December 2019. Subsequently, rapid transmission to other parts of the world was observed.[2] The current pandemic is characterized by high morbidity and mortality rates, highlighting the urgent need for accurate diagnostic, therapeutic, and prevention strategies to curb the number of infections worldwide.[3,4] Coronaviruses are RNA viruses that cause respiratory illnesses, including mild symptoms such as the common cold or severe complications such as pneumonia.[2] In addition to respiratory symptoms, emerging cases of various neurological manifestations caused by SARS-CoV-2 show that symptoms can range from headaches to seizures and strokes.[5] In total, seven coronavirus types have been reported to infect vital organs. SARS-CoV-2 is one of these coronaviruses and can affect the central nervous system (CNS), leading to several neurological disorders, further aggravating its health burden. However, the exact mechanism by which SARS-CoV-2 infects the CNS remains unclear. This chapter explores the most likely mechanisms of neurological injury and examines the effects of COVID-19 and other coronaviruses on the nervous system, describing its neurological manifestations and relevance to other neurological disorders.

12.2 Methods

We searched the literature using the MEDLINE and EMBASE databases, and specifically screened studies that were published between December 2019 and November 2020. The following search terms were used: COVID-19, SARS-CoV-2 relevance with neurological injury, neurological manifestations, stroke, epilepsy, Guillain–Barré syndrome, and Parkinson's disease. We imposed restrictions to exclude studies without detailed methodological reporting. Abstracts and titles were screened for the relevance of COVID-19 to neurology and its manifestations; the research papers considered in this study were case reports, systemic reviews, and meta-analyses or studies relevant to neurology and coronavirus. We restricted the literature review to articles available in English only.

12.3 Neuroinvasive Potential of SARS-CoV-2

In March 2020, the Beijing Ditan Hospital reported the first case of COVID-19 (Coronavirus Disease 2019), diagnosed with encephalitis and severe acute respiratory syndrome-coronavirus 2 (SARS-CoV-2) in the cerebrospinal fluid (CSF) using genome sequencing. Since then, neurological symptoms of patients with COVID-19 have been closely monitored. The virus displays neuroinvasive tendencies and can infect the CNS through the transneuronal and hematogenous routes, resulting in various neurological disorders. A study showed that 78 out of 214 COVID-19 patients experienced neurological symptoms of the CNS (headache, dizziness), peripheral nervous system (PNS) (taste and smell impairment), and skeletal muscular injury,[6] indicating the importance of investigating potential long-term neurological effects of the infection.

12.3.1 Viral Infection in the Nervous System

Research indicates that viruses invade the CNS when host defense mechanisms fail to limit the infection at the peripheral site. Some viruses can directly enter the CNS by bypassing peripheral barriers. At peripheral sites, nerve endings are in direct contact with the tissues, which allows the virus to access the PNS through infected circulating cells that carry the infection from the blood–brain barrier (BBB) into the CNS.[7]

12.3.2 Invasion of Sensory Nerve Endings and Neuromuscular Junctions (NMJs)

Viral entry into the PNS occurs mainly through receptors in axon termini, particularly sensory neurons and autonomic neurons, such as HSV-1, HSV-2, and varicella zoster viruses (VZV).[7] NMJs can be an entry route for many viruses invading the CNS [*e.g.* poliovirus, rabies virus (RABV) infection].

12.3.3 Invasion of Olfactory Neurons

The olfactory epithelium of the nasopharyngeal cavity (roof) is innervated by olfactory nerves. It functions in odor perception and remains the major entry point for viruses to the olfactory system in the CNS. The olfactory nerve (PNS) piercing the olfactory epithelium terminates in the olfactory bulb (CNS), which facilitates infection of the CNS *via* this olfactory route.[8] For instance, zoonotic infections can traverse the olfactory pathway to reach the CNS. Other examples include the Hendra virus, Nipah virus, Chikungunya virus, La Crosse virus, *etc.*,[9] highlighting their key role in CNS infections.

Owing to the similar features elicited by SARS-CoV and SARS-CoV-2, it has been postulated that COVID-19 could also result in neurological symptoms. In fact, SARS-CoV can infect brain cells owing to its neuroinvasive and neurotropic properties, which further aggravate its effects. A recent study demonstrated the importance of analyzing neurological symptoms in COVID-19 patients, such as viral piercing of the olfactory bulb, brain entry, thalamus and brainstem entry by *trans*-synaptic migration, and disruption of the pre-Bötzinger complex, leading to collapse in the respiratory centers, respiratory failure, and subsequent death. The effects of this infection can be observed within a week of infection in the CNS, resulting in major disruption of the respiratory centers. The study also showed that MERS-CoV infects only the brain and lungs, with increased chances of mortality on testing with low doses in mice,[10] which highlights the importance of elucidating the mechanisms of infections to develop successful prevention and therapeutic regimens.

12.3.4 Neuroinvasive Potential of SARS-CoV-2 to Cause Respiratory Failure

ACE2 receptors are commonly detected in vascular endothelia, lung parenchyma, kidney cells, airway epithelial cells, and the small intestine.[11] However, some cells, such as the hepatocytes, which do not express ACE2, can also be infected by SARS-CoV-2. Likewise, infection can also occur in the CNS although it displays lower levels of ACE2. Viral antigens have been detected in the brainstems of the solitary tract and nucleus ambiguus. The nucleus of the solitary tract receives sensory information from the lungs *via* mechanoreceptors and chemoreceptors. Furthermore, the nucleus of the solitary tract and the efferent fibers of the nucleus ambiguus innervate the airway smooth muscle, blood vessels, and glands. These various anatomical features favor viral infection, which disrupts the cardiorespiratory center in the brain.[12] These neurological complications further aggravate the health burden posed by these viruses, including COVID-19.

12.3.5 Invasion of Microvascular Endothelium in the Brain

Some viruses, such as the West Nile virus (WNV), can infect the brain microvascular endothelial cells (BMVECs) and disrupt the BBB, leading to loss of tight junction proteins such as the zonula occludens, claudins, and occludin, which maintain the BBB integrity. They also increase the production of matrix metalloproteinases (MMPs), which degrade the basement membrane and disrupt the BBB, inducing the influx of uncontrolled immune cells in the brain parenchyma and inflammation, resulting in neurological disease.[13]

12.3.6 Hematogenous Route

Research has elucidated some of the mechanisms of viral entry into the CNS, including the hematogenous route. For instance, respiratory viruses can enter the CNS through this hematogenous route, either by infecting the endothelial cells of the BBB in the ventricles or through leucocytes that act as a vector for infiltrating the CNS. The human coronavirus (HCoV) can enter the CNS through this hematogenous route,[14] leading to potential severe neurological complications as found in various studies that will be considered in later sections of this chapter.

12.3.7 Lymphatic Route

The lymphatic pathway in the brain has olfactory and cervical lymphatic vessels through which viruses can penetrate the CNS. Recent studies have reported paranasal sinusitis caused by obstruction of paranasal lymph vessels following viral infection. Importantly, SARS-CoV-2 can infect lymph endothelial cells by

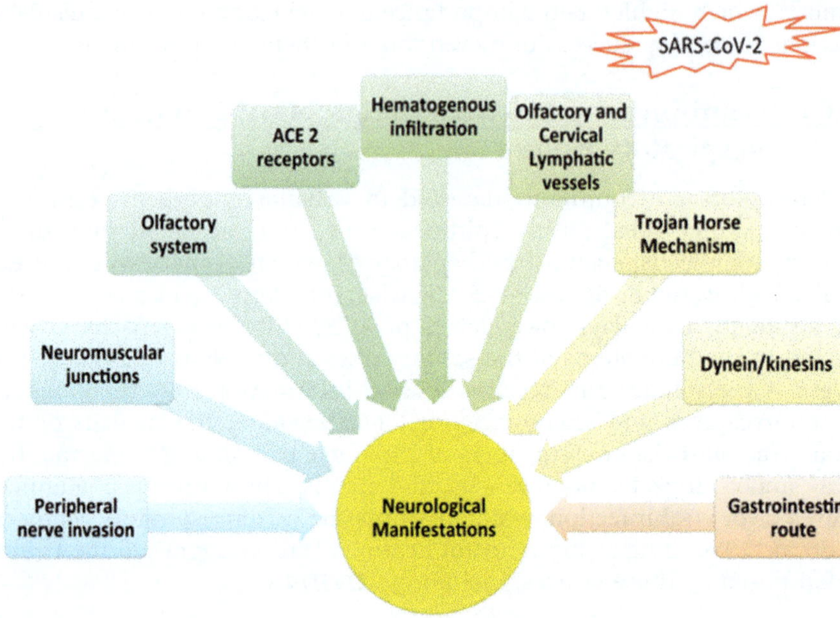

Figure 12.1 Mechanisms of neuroinvasion by SARS-CoV-2.

crossing the nasal cavity through the cervical lymph nodes to reach the brain.[15] The mechanisms of infection that account for the neuroinvasive and neurotropic effects of coronaviruses in humans are depicted in Figure 12.1. Several resulting neurological disorders include encephalitis, acute flaccid paralysis, Guillain–Barré syndrome (GBS) and development or aggravation of other chronic diseases. For example, the respiratory syncytial virus (RSV) can affect sensory neurons locally and spread to the CNS from the lungs, causing long-term behavioral and cognitive impairments. The entry of the virus to the brain *via* the cribriform plate adjacent to the olfactory bulb provides an additional route of entry to reach the brain, further aggravating its symptoms.[16]

12.4 Illuminating Viral Infections in the Nervous System

Viruses can indirectly affect the CNS by infecting immune cells, circulating in the blood and infiltrating the brain through the "Trojan Horse" mechanism.[17] The following sections examine these various routes of infection and their clinical effects.

12.4.1 Neuronal Pathway

Coronaviruses can invade the CNS *via* infected proteins (dynein, kinesins) of the sensory and motor nerve endings through retrograde/anterograde transmission.[18]

12.4.2 Neurological Alterations Due to Respiratory Virus Infection

Typically, CNS infections are due to neurotropic viruses such as maedi-visna (MV), HIV, and Herpes virus. Respiratory viruses such as the human RSV, influenza virus, coronavirus, and human metapneumovirus are also known to produce neuropathological features.[19]

12.4.3 Nervous System Association with SARS-CoV-2 and Other Coronaviruses

SARS-CoV elicits common symptoms of fever, chills, dry cough, and difficulty in breathing. SARS-CoV can also induce neurological diseases such as polyneuropathy, encephalitis, and aortic ischemic strokes. In most SARS-infected cases, autopsy findings have revealed signs of cerebral edema, meningeal vasodilation, infiltration of monocytes and lymphocytes, ischemic changes, demyelination of nerve fibers, *etc.* Some SARS-infected individuals have presented with symptoms of red degeneration of neurons in selective areas, indicating neuronal hypoxia or ischemic changes.[20]

12.4.3.1 MERS-CoV

Middle East respiratory syndrome coronavirus (MERS-CoV) is a type of coronavirus causing respiratory disease. It was first identified in Saudi Arabia in 2021 and has now spread to other countries, including other Middle East countries, Africa, Europe, Asia, and the USA.[21] MERS-CoV infections are commonly associated with fever, cough, and shortness of breath; pneumonia is one of the less prevalent symptoms. MERS-CoV can also display neuroinvasive behavior. For instance, a study conducted in Saudi Arabia showed that 25.7% of infected individuals displayed confusion and 8.6% developed seizures. Furthermore, up to one-fifth of MERS-CoV patients can develop neurological symptoms.[22] A study in transgenic mice also showed inflammatory changes in the lungs and brain, revealing the involvement of the thalamus and brainstem in the neurological effects induced by MERS-CoV.[23]

12.4.3.2 SARS-CoV-2

SARS-CoV and SARS-CoV-2 have a genetic similarity of about 79.5%. SARS CoV-2 is the seventh known coronavirus found to infect humans. Other viruses include HCoV-229E, HCoV-OC43, HCoV-NL63, HCoV-HKU1, SARS-CoV, and MERS-CoV. SARS-CoV-2 can result in fever and cough in mild cases and pneumonia and other organ dysfunctions in severe cases. A few studies have reported olfactory and taste disorders following SARS-CoV-2 infection, in addition to neurological symptoms.[24] Data from EEG and neuroimaging procedures have additionally indicated occurrences of hemorrhagic necrotizing encephalopathy, epileptogenicity, and encephalomalacia.[25] As these various studies show, one of the aggravating factors linked to SARS-CoV-2 is the extensive range of symptoms that it can elicit in different patients, including neurological lesions, which complicates diagnostic and therapeutic regimens.

12.4.3.3 HCoV-OC43

Encephalitis is a neurological disorder characterized by fatal brain inflammation. The virus is widely known to produce respiratory tract infections, but it can also spread to other organs such as the CNS, causing viral encephalitis.[26]

12.4.3.4 HCoV-229E

Post-mortem findings from brain samples of multiple sclerosis patient have indicated the presence of HCoV-229E particles in these patients. HCoV-229E is neuroinvasive and can increase the production of proinflammatory cytokines, leading to multiple sclerosis, a neurological disease characterized by demyelination and infiltration of inflammatory cells. Hence HCoV-OC43 and HCoV-229E can cause multiple sclerosis.[27] HEV-67N CoV is the first known coronavirus to infect the brain. It has been reported that HEV-67N CoV-infected mice show neurological symptoms,[28] highlighting its neuroinvasive characteristics.

12.4.4 Neuropathology

12.4.4.1 Neurological Alteration Due to Respiratory Virus Infection

A recent study showed that infectious respiratory viruses can be detected in the CNS after 7 days of infection, indicating their rapid spread to the CNS, leading to encephalitis.[19]

12.4.4.2 Hypoxia and Nervous System Damage

SARS-CoV can enter the lower respiratory tract, causing severe lung injury, impaired gas exchange, and hypoxia in the CNS. In response to hypoxic states, several changes occur, including anaerobic metabolism and acid accumulation, causing increased vasodilation and edema, which produces ischemic changes and intracranial hypertension. Patients displaying these symptoms have a higher risk of developing acute cerebrovascular diseases.[18]

12.4.4.3 Angiotensin-converting Enzyme (ACE)

ACE is a cardio-cerebrovascular protection protein distributed evenly in the nervous system. It regulates blood pressure and various other functions. One of the key mechanisms of action of COVID-19 is the binding of SARS-CoV-2 to ACE2 receptors. Furthermore, SARS-CoV-2 can attack ACE2 receptors in the brain, as observed in glial cells and neurons. After the binding of SARS-CoV-2 to ACE2, increased blood pressure can aggravate the risk of cerebral hemorrhage and BBB disruption by damaging the vascular system.

Autopsy findings from COVID 19 patients revealed SARS-CoV-2 infection in the endothelium of the cerebral microcirculation and in cerebrospinal fluids, glial cells, and neuronal tissues, indicating the wide neurological effects of COVID-19.[16]

12.4.4.4 Gastrointestinal Tract

SARS-CoV-2 binds to ACE2 receptors in the gastrointestinal tract (GIT), destroying the epithelial cells and producing increased amounts of proinflammatory cytokines, also known as the "cytokine storm", which disrupts gastric absorption and increases mucosal secretions. The virus further invades GIT mucosa and infects host cells, producing more virions. The spike protein in the virus is activated by furin in human tissues, mainly in the intestine and lungs. Owing to increased inflammatory responses, which damage the intestinal barrier, the virus can enter the bloodstream, or lymphatic pathway, and invade other organs. The disrupted GIT system favors the entry of the virus into the CNS. Following infection, the vomiting reflex is activated by signals from the sensory neurons of the digestive tract and afferent fibers of the vagus nerve. The mechanisms underlying these effects remain unknown, indicating that further studies examining the connection between the nervous system and the GIT in the context of SARS-CoV infection are required to explain the exact pathophysiological process.[29]

12.4.4.5 Systemic Inflammatory Response

Severe viral infection with increased systemic inflammatory response allows cytokines to disrupt the BBB, which aggravates neuroinflammation, leading to encephalitis.[30] A recent study demonstrated that lymphocytes can access the brain through the nose by disrupting the BBB, activating microglial cells with neurophysiological defects in odor processing *via* Th17 lymphocytes along with entry of autoantibodies in postinfectious autoimmune encephalitis.[31]

12.4.5 Neurological Manifestations

Recent studies indicated that the CNS effects of SARS-CoV include headache, dizziness, acute cerebrovascular disease, ataxia, impaired consciousness, and seizures. PNS manifestations include smell and taste disorders and vision impairments. A study in Wuhan, China, reported that 78 out of 214 patients had neurological symptoms, as depicted in Figure 12.2.[5]

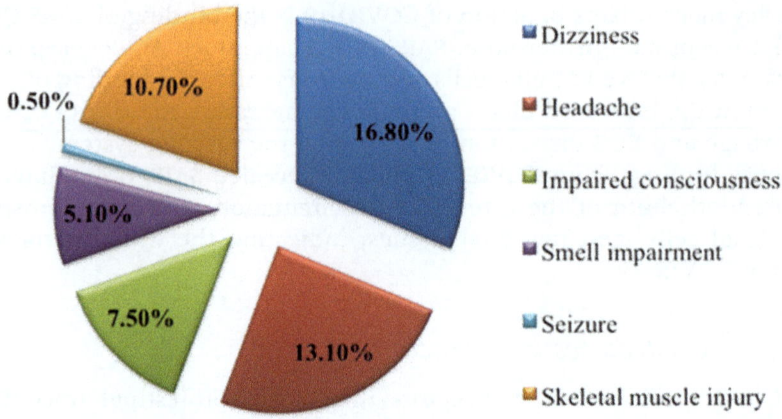

Figure 12.2 Incidence of neurological manifestations in COVID-19 patients.

12.4.5.1 Headache

Headache is one of the most common symptoms of COVID patients along with fever, cough, and dyspnea. Headaches are a symptom of brain infections such as meningitis, encephalitis, intracranial hypertension, *etc.* Headache was observed to be one of the co-manifestations of other CNS-associated symptoms in COVID-19 patients. It was characterized as tension-type headache, migraine without aura, and migraine-like headache types. Certain neurological inflammatory processes can cause headache syndromes *via* cytokines and chemokines. The release of proinflammatory mediators in COVID-19 infection may also cause pain.[32] A meta-analysis study by Islam *et al.* identified a pooled prevalence of headache in 10% and 8.3% of Chinese and American adult COVID-19 patients, respectively. The study also concluded that the risk of headache was similar in severe and non-severe and in intensive care unit (ICU) and non-ICU patients.[33] In a study by Caronna *et al.* conducted at a hospital in Barcelona, Spain, 130 adult patients were studied for their symptoms and 97 of the patients had headache as one of the COVID-19 symptoms. The headache reported was moderate in the majority of the patients and was migraine type in mostly female patients, who consisted of one-quarter of the total patients. It was concluded that headache can be a good early-stage COVID-19 prognosis symptom enabling patients to seek medical attention. It was also reported that in a significant number of people headache persisted even after 6 weeks of recovery from COVID-19.[34]

12.4.5.2 Olfactory and Gustatory Dysfunction

Olfaction and multisensory integration give vital information about the environment. Odorants enter the nasal cavity through the superior aspect, which is lined by the olfactory epithelium (OE).[35] Olfactory sensory neurons (OSNs) perceive these odorants through cilia on odorant receptors.

A conductive loss can be caused by sinonasal abnormalities that influence airflow and make it difficult for odorants to reach the intact OE. Nasal congestion or edema of the nasal respiratory epithelium caused by a variety of factors can produce transient anosmia, leading to olfactory dysfunction.[36] It was found that 85.6% and 88.0% of SARS-CoV-2 patients experience olfactory and gustatory dysfunction, respectively.[37] Previous studies identified the presence of SARS-CoV-2 in the nasal discharge of patients with olfactory dysfunction.[38] However, the pathophysiological mechanisms of COVID infection leading to olfactory and gustatory dysfunction remain unclear. A study in Italy *via* telephone interviews reported that 130 (64.4%) patients had altered smell and taste among 202 subjects.[39] Further research is needed to elucidate the mechanisms leading to olfactory dysfunction following SARS-CoV-2 infection.

12.4.5.3 *Impaired Consciousness*

Several factors may affect the level of consciousness of COVID patients through direct injury affecting the brain parenchyma, toxic-metabolic encephalopathy, encephalitis, or seizures.[40] The most frequent symptom among COVID 19 patients is "isolated sudden onset anosmia".[41]

12.4.5.4 *Encephalitis*

A study in Japan reported a case of meningitis/encephalitis in a SARS-CoV-2 patient with a negative nasopharyngeal sample, diagnosed with headache, fever, convulsions, *etc.*[42] A study in Iran reported acute encephalitis in COVID-19 patients with fever, cough, tonic–clonic seizures, and impaired consciousness.[43]

A study of COVID-19 patients at a hospital in France highlighted the presence of 67% of diffuse symptoms of corticospinal tract exhibiting altered tendon reflexes, bilateral extensor plantar reflexes, and ankle clonus; 33% of the patients had dysexecutive syndrome (alterations in cognitive and behavioral patterns, disorientation, poor response of organized movements to commands).[44]

12.4.5.5 *Miller Fisher Syndrome and Polyneuritis*

The Miller Fisher syndrome (MFS) is a variant of GBS and is an acquired nerve disease. Its symptoms show lack of tendon reflexes, poor muscular coordination, and paralysis of the ocular muscles.[45] Polyneuritis, as the name suggests, is an abnormality in many nerves simultaneously; there is collective inflammation in many cranial nerves.[46] Both MFS and polyneuritis are rare neurological manifestations and are preceded by an infection similar to GBS *para*- or postviral manifestations. A recent study reported a case of polyneuritis with COVID-19. The pathophysiology of Miller Fisher syndrome and polyneuritis could be due to over-response of proinflammatory cytokines and the presence of GD1b–IgG antibodies for MFS.[47]

MFS is mostly linked to problems with the third, fourth, and sixth cranial nerves. In MFS, the activation of the adaptive immune system is considered to cause molecular mimicry between peripheral neurons and viral antigens. Studies have suggested antibody IgG anti-GQ1b against gangliosides to be the main characteristic of MFS. Anti-GQ1b antibodies are hypothesized to operate directly on the neuromuscular junction between the cranial nerves and the ocular muscles, causing ophthalmoparesis in MFS patients.[48] An enzyme-linked immunosorbent assay (ELISA) will yield positive results in around 70–90% of individuals.[49]

Polyneuritis is characterized by several cranial nerve palsies leaving the spinal cord unaffected. This illness is a combination of GBS and MFS. Demyelination of lower cranial nerves is a hallmark of polyneuritis etiology having a pathogenic mechanism similar to MFS.[46] Here, the olfactory neurons are assumed to be the major site of direct viral inoculation.[50] To infect the neurons, the SARS-CoV-2 viral spike protein interacts with sialic acids connected to the patient's cell surface gangliosides. The interplay between host-cell proteases and viral S protein spikes explains the neurotropic invasion of the SARS-CoV-2 virus.[18] The existence of neurological symptoms in individuals with severe COVID-19 illness and a link between IL-6 and disease severity suggests that neurological damage is caused by the immune system. The SARS-CoV-2 virus, which is neurotropic, has been shown *in vitro* to induce a proinflammatory state in glial cells, resulting in an increase in inflammatory factors such as interleukins.[19]

Abnormally high blood pressure increases the risk of intracranial hemorrhage in COVID-19 patients. Patients with existing neurological conditions such as amyotrophic lateral sclerosis (ALS), Parkinson's disease, Alzheimer's disease, myasthenia gravis, and multiple sclerosis should be advised to follow specific recommendations during this pandemic as COVID-19 will aggravate their existing conditions.[51]

Further, some patients can develop neurological syndromes such as GBS and acute inflammatory conditions such as myelitis, necrotizing encephalopathy, disseminated encephalomyelitis, and ischemic stroke. Neurological effects of SARS-CoV-2 can cause long-term or permanent neurological disabilities,[52] indicating the importance of conducting detailed research to understand the mechanisms underlying these symptoms.

12.4.5.6 Cerebrovascular Manifestations

Acute stroke is an emergency condition caused by SARS-CoV-2 infection. A retrospective study presented five case reports of acute ischemic stroke in COVID-19 patients requiring immediate hospitalization.[53]

A study in Singapore reported five cases of large artery cerebral infarction among 206 severe acute respiratory syndrome (SARS) patients. Computed tomography (CT) scans revealed infarctions involving large cerebral arteries, especially the middle cerebral artery. Patients exhibited symptoms of cardiac dysfunction, disseminated intravascular coagulation (DIC), and hypotension prior to the development of the stroke. Moreover, some patients developed a venous thromboembolism even after receiving low molecular weight heparin (LMWH).[54]

A study in New York City reported five cases of large vessel stroke in COVID-19 patients below 50 years of age. All of the patients showed new-onset symptoms of stroke with COVID-19, except one patient who had a previous history of stroke and others with few risk factors. The average NIHSS (National Institutes of Health stroke scale) for all patients was 17.[55] These findings highlight the importance of understanding the mechanisms of infection with SARS-CoV-2 in order to devise preventive measures and suitable treatments to avoid fatal strokes in infected patients.

12.4.5.6.1 COVID-19-related Stroke. The main pathological action of COVID-19 is the binding of SARS-CoV-2 to ACE2 receptors, as noted above. In normal conditions, ACE2 acts as a cardioprotective and neuroprotective agent by regulating the harmful effects of ACE1, angiotensin-II, and the AT1 axis. Disruption of ACE2 by SARS-CoV-2 results in its downregulation and shedding, leading to dysfunctions in the renin–angiotensin mechanism, and increased release of proinflammatory mediators that can facilitate thrombotic events leading to stroke.[56] Further research is required to elucidate the signaling pathways involved in this viral action in order to develop potential personalized molecular-based treatments to prevent and/or alleviate these symptoms.

12.4.5.6.2 Ischemic Stroke Associated with COVID-19. A recent study exploring six case reports found that all six SARS-CoV-2 patients studied displayed large-vessel occlusion with increased D-dimer levels. Furthermore, three patients had multi-territory infarcts, two had concurrent venous thrombosis, and two had ischemic stroke despite anticoagulation therapy,[57] indicating the large repertoire of symptoms that the virus can elicit and the need for further understanding of its actions and effects to restrain the pandemic.

12.4.5.6.3 Formation of Thrombus and Thrombotic Events. Inflammation mechanisms trigger coagulation activity (and *vice versa*). Coagulation is a well-known defense mechanism of innate immunity aimed at confining infection in thrombi and limiting the proliferation and dissemination of pathogens.[58] The overactivation of proinflammatory mediators in response to SARS-CoV-2 binding to ACE2 favors the coagulation process. Proinflammatory mediators influence platelet activation, thrombin generation, and fibrin formation by inhibiting anticoagulation systems such as the antithrombin system, the activated protein C system, and tissue factor pathway inhibitors leading to the production of microvascular clots in vital organs,[59] further aggravating infection symptoms.

12.4.5.6.4 Hypoxia-induced Thrombogenesis. COVID-19 patients tend to develop hypoxia due to lung injury and myocardial infarction. Hypoxia in

COVID-19 patients causes vascular occlusion, which increases blood viscosity and stimulates thrombogenesis through hypoxia-inducible factor (HIF) agents or directly *via* proinflammatory cytokines.[60,61]

12.4.5.6.5 D-dimer Levels and CVA. A recent study reported that nearly 5.7% of severe COVID-19 patients had cerebrovascular disease with elevated D-dimer and C-reactive protein (CRP) levels, and an average onset duration of 12 days.[53] The D-dimer level indicates fibrin deposition and thrombus formation, and its value depends on various conditions.[62] For instance, an elevated D-dimer level denotes hypercoagulation in COVID-19 patients. Data from the pandemic epicenter in Wuhan, China, highlighted the presence of acute cerebrovascular disease in 88 severe COVID-19 patients.[5]

12.4.5.7 Epilepsy

SARS-CoV-2 spreads from the respiratory tract to the CNS, which could lead to symptoms of seizures.[63] In fact, primary manifestations with status epilepticus were observed in some COVID-19 patients.[64,65] A case report revealed respiratory failure with episodes of epilepsy in a COVID-19 patient with no history of epilepsy.[66] A 20-years-old man who was brought to hospital due to seizures had CSF with lymphocytosis and tested positive for COVID-19,[67] indicating that SARS-CoV-2 can lead to epilepsy in patients with no underlying history of the condition.

In this pandemic age, the Italian Society of Clinical Neurophysiology (SINC), Italian League Against Epilepsy (LICE), and Italian Association of Neurophysiology Technologists (AITN) were formed with the goal of providing clinical and technical recommendations for electroencephalography (EEG) indications and recording standards. According to these guidelines, during the various stages of the COVID-19 pandemic, EEG should be used only on people who would benefit from it in terms of diagnosis, treatment, prognosis, and avoiding emergency department visits.[68] A case report addressed the importance of EEG in identifying encephalopathies in COVID-19 patients.[69] Another study discussed epileptiform discharges in the EEG of a severe COVID-19 patient with encephalopathy or suspected clinical seizures. The most common findings from EEG studies are that SARS-CoV-2 infection can elicit the onset of encephalopathy and seizure-like symptoms in patients with no previous medical history. It can also cause frontal fluctuations similar to epileptiform discharges, as observed in many COVID-19 patients. Additionally, fewer than 50% of COVID-19 patients showed sporadic epileptiform discharges,[70] indicating the importance of monitoring the progress of the infection to prevent such aggravating symptoms.

A case report of two patients with coronavirus infection in 2019 found that their EEG traces displayed unique patterns.[71] These unusual patterns should be verified in a larger sample of COVID-19 patients.[72] The International League Against Epilepsy (ILAE) has provided valuable resources to tackle some of the important issues related to this topic.[72]

Further, an important review offers recommendations on how to provide the best possible healthcare for people with epilepsy during the coronavirus pandemic.[73] The main challenge is to handle the interactions between drugs used for COVID-19 treatment and antiepileptic drugs (AEDs).[73–75] In the context of the challenges faced by healthcare workers during the COVID-19 pandemic, telemedicine may aid in the reduction of disease transmission.[76] Preferably, an efficient telemedicine service provided to epilepsy patients should include all possible technologies for seizure recordings, daily reminders, and audio/visual conversation facilities to interact with physicians.[77]

12.4.5.8 Guillain–Barré Syndrome (GBS)

GBS is an autoimmune disorder characterized by an acute reflex paralytic state. It is an after-disease syndrome in which the immune system developed antibodies against pathogens start to cross-react with gangliosides at nerve membranes of the host itself. The nature of the preceding infection and the specificity of such antibodies influence the subtype and severity of the illness.[78] The general manifestations include progressive limb weakness, often with sensory and cranial deficits. These symptoms develop between 3 and 6 weeks following a viral infection, with respiratory or diarrheal-type illness, and can result in significant morbidity and mortality.

A study revealed the association of GBS with COVID-19 infection. It also indicated that GBS coupled with SARS-CoV-2 infection may present as a parainfectious pattern as opposed to a postinfectious pattern as for the Zika virus.[79]

Another study reported a case in which a COVID-19 patient with no previous medical history developed neurological symptoms such as paresthesia and diminished tendon reflexes in all limbs. Later electrodiagnostic reports showed a demyelinating pattern with GBS.[80]

A further study also reported a number of SARS-CoV-2 cases with a classical postinfectious picture of GBS after recovery from SARS-CoV-2, similarly to the Zika virus. The symptoms identified in this study included albuminocytological dissociation in the CSF, demyelinating neuropathy from electrodiagnostic testing, and subacute peripheral neurological disorders. The mechanism responsible for the development of polyradiculoneuritis is considered to be autoimmune cross-reactivity.[81]

Another study performed in northern Italy examined five patients presenting with GBS in association with SARS-CoV-2, with a 5–10-days interval between the onset of the infection and the emergence of neurological symptoms.[82] The infection-associated etiology of GBS includes the Epstein–Barr virus, HIV virus, Zika virus, and cytomegalovirus.[83,84] As research has shown, the proteins related to autoimmune reactions can also be detected in the PNS, which supports neurological involvement.[84]

These various findings indicate that clinicians should pay attention to the neurological presentations of GBS, which are likely associated with SARS-CoV-2 infection. While bacterial and viral infections are known to trigger

GBS, emerging evidence points to a link between COVID-19 and GBS. Hence clinicians should consider and incorporate this link in their diagnostic/therapeutic regimen to avoid delays in diagnosis and treatment. Currently, the management of COVID-19-induced GBS is the same as the standard recognized guidelines for GBS.[85]

12.4.5.9 Critical Illness Neuropathy in COVID-19 Infection

SARS-CoV-2 infection causing COVID-19 and leading to serious illness in a few cases requires intensive care admittance, which often causes ICUAW (intensive care unit-acquired weakness) that manifests as symmetric and generalized weakness in patients. In one study, 12 SARS-CoV-2-positive patients were referred with the suspicion of critical illness myopathy (CIM) or polyneuropathy (CIP) between March and May 2020. During their stay in the hospital, all patients underwent nerve conduction and concentric needle electromyography. Three patients had muscle biopsies taken. It was found that four of the patients had sensory-motor axonal polyneuropathy, whereas the other seven had myopathy. One muscle sample revealed a mixture of necrotic and regenerating fibers with no evidence of inflammation. There is a lack of understanding of any specific features in COVID-19-related CIM/CIP.[86,87]

12.4.5.10 Encephalopathy

Encephalopathy is a major life-threatening complication of the nervous system, Hypoxic and metabolic shocks cause diffuse brain damage.[88] Toxic metabolic encephalopathy (TME) is due to endocrine or metabolic disturbances, exogenous toxins, renal dysfunction, liver dysfunction, and infections. TME results in altered consciousness, delirium, cognitive impairment, coma, *etc.* Altered consciousness, which is one of the major manifestations of encephalopathy, was reported to occur mostly in older patients with severe disease symptoms. COVID-19 infection triggers increased production of proinflammatory cytokines, chemokines, tumor necrosis factor, and interleukins, which may lead to a cytokine storm, sepsis, and renal dysfunction. This facilitates the development of encephalopathy in critically ill patients. Sepsis, inflammatory processes, and cytokine storms contribute to encephalopathy along with other risk factors and metabolic derangements.[89]

12.4.5.11 Parkinson's Disease

A study examined SARS-CoV-2 infection as a potential etiological factor in the development of Parkinson's disease (PD) and other neurodegenerative diseases, highlighting its clinical relevance for long-term complications.[90] The causative physiology of PD is the death of the dopamine-generating cells in the substantia nigra pars compacta (SNpc) in the midbrain region, although the exact etiology of the disease is still unknown.[91] Aggregates of the proteon α-synuclein, called Lewy bodies, accumulate in the SNpc and also necrotic regions of the brain.[92] There is a continuous progression in the severity of the symptoms in the disease, four primary symptoms being tremor, muscle rigidity, postural instability and bradykinesia.[93] A few PD symptoms, such as hyposmia, overlapped with COVID-19 symptoms, showing its link with SARS-CoV-2.[94] This effect was also substantiated by another study that showed vast expression of angiotensin-converting enzyme 2 (ACE2) receptors in dopaminergic neurons.[95] A third study also revealed that ACE2 and TMPRSS2 protease are highly expressed in the cortex and substantia nigra of the CNS. Hence these studies could explain the possible presence of coronaviruses in the CNS of patients with neurodegenerative diseases, as supported by a number of studies.[96]

Patient with existing PD can be severely affected by COVID-19 infection owing to the rigidity of their respiratory muscles, diminished cough reflex, and pre-existing dyspnea. Moreover, indirect effects such as stress and anxiety caused by lockdown could worsen these symptoms. Additional complications such as fatigue, orthostatic hypotension, cognitive impairment, and psychosis could be protein aggravated by SARS-CoV-2 infection in PD patients.[97]

Furthermore, COVID-19 may worsen PD either through pharmacodynamic changes such as reciprocal interactions between dopaminergic and renin–angiotensin systems in the substantia nigra and striatum, or through the systemic inflammatory response. A case–control study in Italy found that several motor and non-motor symptoms such as urinary incontinence and fatigue are the most prominent in these patients and are aggravated by COVID-19.[98]

PD diminishes motor and respiratory activities, which could lead to poor outcomes following COVID-19 infection.[99–101] As physical activity has a protective role in PD patients, reduced physical activity is a major concern in the PD population during the COVID-19 pandemic and will negatively affect it.[102,103] Self-management strategies and supplementations may aid in overcoming these complications and are therefore of the utmost importance. Moreover, a study reported the neuroprotective potential role of vitamin D_3 uptake in the elderly to promote resistance to COVID and slow progression of PD.[104] These various factors should be taken into consideration by healthcare professionals and can guide personalized management techniques aimed at helping and protecting vulnerable populations during the COVID-19 pandemic.

12.5 Post-COVID-19 Effects on Incidence and Prognosis of Neurological Disease

The above mechanism elucidated the involvement of COVID-19 infections causing neurological disease or worsening the prognosis of existing disease. The literature reporting the effects of COVID-19 on the incidence and prognosis of neurological diseases is documented in Table 12.1.

Table 12.1 Effects of COVID-19 on the incidence and prognosis of various neurological diseases.

No.	No. of cases/ patients	Location	Patient characteristics	Symptoms	Treatment and response	Ref.
Parkinson's disease						
1	3	Mangalore, India	Case 1: 72 years old with no past comorbidities	Presented with symptoms of fever, chills, cough, and dyspnea for 4 days. Developed acute kidney injury and was referred to a higher center. Superficial thrombophlebitis of the basilic vein. Three episodes of fall, no mobility	Levodopa 110 mg, half tablet four times per day. Symptoms improved, no falls, rigidity subsided, and gait speed increased. After 4 months: resolution of symptoms and daily activities performed independently	126
			Case 2: 66 years old with history of (h/o) diabetes mellitus (DM), hypertension, and epilepsy	Presented with cough, hoarseness of voice for 2 weeks, and one episode of a generalized tonic–clonic seizure. Immobile and mute. Rigidity in right upper and lower limbs with severe bradykinesia. Gliosis in bilateral temporal lobes, periventricular punctate white matter, ischemic changes in bilateral frontal and parietal lobes, and age-related cerebral and cerebellar atrophy	Intravenous dexamethasone and anticoagulant. Levodopa–carbidopa. Improvement in mobility and speech. After 1 month: optimization of levodopa, complete mobility, and independence in daily life	
			Case 3: 74 years old with h/o COVID-19	Presented with decreased mobility developed during the previous hospitalization 8 weeks ago, diagnosed with COVID-19. Ischemic changes in periventricular white matter	Levodopa–carbidopa and physiotherapy. Decreasing rigidity, improved mobility, speech, and swallowing. At 6 months: able to walk independently with a TUG (timed up and go) score of 18 s	

2	27	UK, Italy, Romania, and Mexico	H/o Parkinson's disease	Breathlessness, cough, chest tightness, chest pain, palpitations, fever pain, neurological symptoms. "Long COVID". Motor worsening, cognitive disturbances including "brain fog", loss of concentration, and memory deficits, and sleep disturbances	Increased levodopa equivalent daily dose requirements within the long-COVID spectrum	118
3	20	Literature review (global)		Mild parkinsonian symptoms in some patients. Symmetrical bradykinesia of the right limbs. Alterations in consciousness, ocular abnormalities, myoclonus, and epileptic phenomena, hyposmia or anosmia. Encephalopathy, occasionally including alterations in consciousness, ocular abnormalities, myoclonus, and epileptic phenomena	1. Immunosuppressive treatment in six cases: no effect in two 2. Levodopa or dopamine agonists given to 12 patients: no effect in three 3. Convalescent plasma therapy administration: in one patient, full recovery after 24 h 4. Levodopa–carbidopa and modafinil for 1 month: in one patient, full resolution of symptoms after 1 month follow-up 5. Two cycles of intravenous immunoglobulin (IVIg) and high doses of methylprednisolone: two patients showed full recovery 6. One death 30 days post-discharge	127
4	1	Jerusalem, Israel	45-years-old man with h/o hypertension and asthma	Dry cough and muscle pain. A few days before admission, he had also noticed a loss of smell, shortness of breath, and chest pain without fever. Difficulties	Salbutamol inhalations as needed for mild asthma symptoms. 0.375 mg extended release pramipexole, once daily: quick improvement in all clinical signs.	128

Table 12.1 (*Continued*)

No.	No. of cases/ patients	Location	Patient characteristics	Symptoms	Treatment and response	Ref.
				speaking and writing text messages. Episodes of tremor in right hand. Moderate cogwheel rigidity in the neck and in the right arm, mild cogwheel rigidity in the left arm, moderate bradykinesia in the right extremities, mild bradykinesia in the left extremities, and no tremor. Gait was slightly slow	On discharge, still had unreadable handwriting, hypomimia, bradykinesia, and cogwheel rigidity, mostly on the right side. Started on a course of 5 days intravenous high-dose methylprednisolone. Due to worsening of tremor in his right extremities at follow-up visit, biperiden added (2 mg daily, increased to 4 mg daily after 1 week): improvement of tremor	
Epilepsy						
1	1	West Bengal, India	27-years-old female with premorbid h/o generalized anxiety disorder	Presented first with an episode of generalized tonic–clonic seizure, about 1 week prior to ictus diagnosed with SARS-CoV-2. Post-ictal confusion and lethargy that persisted for about 1 h. Focal cerebritis	Continuation on selective serotonin reuptake inhibitors (SSRIs): escitalopram 10 mg, levetiracetam 500 mg twice daily, midazolam nasal spray on a seen on symptoms basis	129
2	860 934	UK	Two cohorts: COVID+ and influenza+	COVID and influenza positive, absence of pre-existing diagnosis of epilepsy or recurrent seizures. Patients with anxiety and mood disorders. Lower respiratory tract diseases		130

3	212 678	Korea	3919 (1.8%) with h/o epilepsy. Other comorbidities: hypertension, diabetes, stroke, heart failure, coronary artery disease, atrial fibrillation, asthma, chronic kidney disease, malignancy	Patients with a diagnostic code for epilepsy or seizure: epilepsy [ICD-10 code G40], status epilepticus [ICD-10 code G41], Landau–Kleffner syndrome [ICD-10 code F803], and convulsions	Patients on AEDs	107
4	17 278 392	Different countries ILAE database				109
Alzheimer's disease (AD)						
1	1	Spain	67-years-old female with h/o type 2 DM, rheumatic fever in childhood, benign uterine tumor removed. No relevant past medical or psychiatric history	Fever, cough, breathing difficulties, and myalgias. Bilateral pneumonia. Verbal episodic memory deficit. Hypometabolism in regions linked to AD. Decreased Aβ_{1-42} and elevated tau and phosphorylated tau. AD at a prodromal stage	Oxygen therapy with a nasal cannula, heparin, and hydroxychloroquine. Metformine for DBS and zolpidem on demand for insomnia	131
2	6 245 282	USA	Older adults (age ≥65 years), 410 748 in the COVID-19 cohort and 5 834 534 in the non-COVID-19 cohort. No prior diagnosis of AD	No prior diagnosis of AD	No treatment mentioned	132

Table 12.1 (*Continued*)

No.	No. of cases/ patients	Location	Patient characteristics	Symptoms	Treatment and response	Ref.
3	10 de-identified human heart, lung, and brain tissues			1. Acute hypoxic–ischemic injury in the hippocampus, pons, and cerebellum 2. Hypoxic ischemic encephalopathy, severe, global 3. Hypoxic/ischemic injury, global, widespread astrogliosis/microgliosis 4. Dementia. β-Amyloid plaques noted in cortex and cerebellum 5. Severe hypoxic–ischemic encephalopathy, severe global astrogliosis and microgliosis. Mild Alzheimer-type pathology 6. Acute hypoxic–ischemic encephalopathy, global, moderate to severe. Arteriolosclerosis, mild. Metabolic gliosis, moderate 7. Left frontal subacute hemorrhagic infarct. Multifocal subacute infarcts in pons and left cerebral peduncle. Global astrogliosis and microgliosis (see microscopic description). Alzheimer's pathology	No treatment mentioned	133

					8. Hypoxic–ischemic encephalopathy, moderate. Alzheimer's pathology. Atherosclerosis, moderate. Arteriolosclerosis, moderate 9. Hypoxic–ischemic injury, acute to subacute, involving hippocampus, medulla, and cerebellum. Mild atherosclerosis. Mild arteriolosclerosis 10. Hypoxic–ischemic encephalopathy, acute, global, mild to moderate. Diffuse Lewy body disease, neocortical type, consistent with Parkinson's disease dementia. Atherosclerosis, severe. Arteriolosclerosis, mild		
Guillain–Barré syndrome (GBS)							
1	5		Lucknow, India	Case 1: 27-years-old male	Complaints of fever, sore throat, and myalgia for 5 days. Weakness in upper limbs, unable to walk, required support for routine chores	Initiated on IVIg at a dose of 0.4 $g\,kg^{-1}$ per day for 5 days; ivermectin and remdesivir for COVID-19. Responded to treatment (walking with minimal support), discharged on 14th day	134
				Case 2: 35-years-old female with no comorbidities	Presented with complaints of fever and anosmia for 4 days, developed tingling in both lower limbs followed by mild weakness	Discharged on fifth day with advice to report urgently in the event of any deterioration. Improved completely 6 weeks post-discharge	

Table 12.1 (*Continued*)

No.	No. of cases/ patients	Location	Patient characteristics	Symptoms	Treatment and response	Ref.
			Case 3: 40-years-old female with areflexic quadriparesis, bifacial and bulbar involvement	Areflexic quadriparesis with suggestion of bifacial and bulbar involvement. After having developed fever and breathlessness, experienced paresthesias in her lower limbs associated with weakness rapidly progressing from lower to upper limbs, followed by weakness of the respiratory muscles	Put on ventilatory support, IVIg, antibiotics, methylprednisolone. Expired on eighth day	
			Case 4: 48-years-old female with h/o constrictive pericarditis	Presented with complaints of both lower limb paresthesias and weakness for 3 days, preceded by a history of fever and sore throat for 1 day	IVIg at 0.4 $g\,kg^{-1}$ bodyweight per day for 5 days: good improvement. 10th day: buckling of knees improved but foot dorsiflexors were still weak	
			Case 5: 50-years-old male	Presented with complaints of both lower limb paresthesias and weakness for 3 days, preceded by a history of fever and sore throat for 1 day	Flavipiravir, hydroxychloroquine, IVIg, antibiotics, steroids, oxygen support. Day 36: discharged	
2	1	Iran	65-years-old male with type 2 DM diagnosed with COVID-19 2 weeks prior	Two weeks prior to hospitalization, the patient suffered from cough, fever, and sometimes dyspnea. Symptoms of acute progressive symmetric ascending quadriparesis.	Hydroxychloroquine, lopinavir/ritonavir (LPV/RTV), and azithromycin; 0.40 $g\,kg^{-1}$ per day IVIg for 5 days	135

				Neurological manifestations began with acute progressive weakness of distal lower extremities. Symptoms progressed from distal limbs to proximal limbs and he had been quadriplegic 1 day before admission. Facial paresis bilaterally		
3	1	Manchester, UK	9-years-old boy with h/o low vitamin D levels	Unbalanced gait, back pain, and lower limb weakness. Presented due to worsening back pain and unsteadiness over the last 3 weeks. Decreased appetite, some mild weight loss. Absent lower limb reflexes and CSF showed albuminocytological dissociation. Brain and spine: abnormal enhancement	Analgesics along with an increasing dose of gabapentin, 10 $mg\,kg^{-1}$ once per day on day 1, 10 $mg\,kg^{-1}$ twice per day on day 2, 10 $mg\,kg^{-1}$ three times per day on day 3, and then stopped. Regular physiotherapy. Able to walk independently with good reflexes	136
4	34	Italy	34 patients	GBS and COVID symptoms	Intravenous immunoglobulin or plasma exchange treatment in both COVID-19-positive and COVID-19-negative patients	116
5	5	Northern Italy	Case 1: 77-years-old female	Neurological symptoms: 7 days after fever, cough, and ageusia, flaccid areflexic tetraplegia evolving to facial weakness, upper-limb paresthesia (36 h), and respiratory failure (day 6)	Received two cycles of IVIg: poor outcome with severe upper-limb weakness, dysphagia, and lower-limb paraplegia	82

Table 12.1 (*Continued*)

No.	No. of cases/ patients	Location	Patient characteristics	Symptoms	Treatment and response	Ref.
			Case 2: 23-years-old male	Neurological symptoms: 10 days after fever and pharyngitis, facial diplegia and generalized areflexia evolving to lower-limb paresthesia with ataxia (day 2)	Received IVIg: improvements, with decrease in ataxia and a mild decrease in facial weakness	
			Case 3: 55-years-old male	Neurological symptoms: 10 days after fever and cough, flaccid tetraparesis and facial weakness evolving to areflexia (day 2) and respiratory failure (day 5)	Two cycles of IVIg: poor outcomes, ICU admission due to flaccid tetraplegia and neuromuscular respiratory failure	
			Case 4: 76-years-old male	Neurological symptoms: 5 days after cough and hyposmia, flaccid areflexic tetraparesis and ataxia (day 4)	IVIg: slight improvement but not able to stand 1 month post-onset	
			Case 5: 61-years-old male	Neurological symptoms: 7 days after cough, ageusia, and anosmia, facial weakness, flaccid areflexic paraplegia (days 2–3), and respiratory failure (day 4)	IVIg and plasma exchange: contracted bacterial pneumonia on IVIg treatment, hence delayed plasma exchange	
6	1	Italy	57-years-old male	Dysgeusia, cough, and fever lasting for 5 days. At 12 days after the resolution of the symptoms, numbness and tingling in the feet and, a few days later, also in the hands. Over 10 days, the patient developed distal limb weakness and severe gait impairment; referred to the emergency department	An IVIg cycle at 0.4 $g\,kg^{-1}$ per day over 5 days: significant improvement of the weakness in the upper limbs and the left foot but a poor benefit on the right foot and gait ataxic. Improved through physiotherapy: after 1 month, able to walk without support	137

12.5.1 Epilepsy

Many epilepsy patients struggle with one or more additional health issues and the conditions may worsen when these patients contract COVID-19.[105] In addition, some undiagnosed comorbid conditions may also arise, leading to further exacerbation of symptoms.[106] In COVID-19 patients, their epilepsy history may impact clinical outcomes.[107] Although a precise mechanism has not been identified, increased systemic inflammation brought on by COVID-19 infection may contribute to an increase in the incidence of status epilepticus in patients.[108] One of the key conclusions drawn from a PWE (persons with epilepsy) and caregiver questionnaire by the ILAE and a Telemedicine Task Forces survey showed an increase in the frequency of seizures in PWE after COVID-19 infection.[109] The vulnerability of epilepsy patients to COVID-19 infection is poorly understood, however, with conflicting results.[73] In Iranian healthcare facilities, it was found that PWE were not more susceptible to COVID-19 infection compared with those without epilepsy.[110] In contrast, hospitalized people with active epilepsy in Spain showed an increased incidence of COVID-19.[106] However, Yoo *et al.* showed that in contrast to the COVID-19 susceptibility, serious complications were more common in PWE infected with COVID-19.[107] Additionally, the prognosis for epilepsy patients may be worsened by COVID-19 treatment. Administration of antiepileptic medications may also interact with COVID-19 treatment, decreasing the effectiveness.[111,112] In PWE and COVID-19 patients, worsening or breakthrough seizures may be a cause of multiple factors. Owing to the neurotropism exhibited by SARS-CoV-2, and as shown in numerous earlier studies, COVID-19 has been linked to seizures even in individuals without a history of epilepsy.[113] The accessibility and quality of PWE care has been severely impacted by COVID-19, and the consequent rise in rates of stress and social disconnection has paved the way to increases in seizures along with other factors such as drug–drug interactions and organ failures.[114]

12.5.2 Guillain–Barré Syndrome (GBS)

The incidence of GBS ranges from 1.1 to 1.8 per 100 000 people annually on average, but it rises to 3.3 per 100 000 people annually in subjects over 50 years of age.[115] Many studies have shown a pathogenic connection between GBS and COVID-19 infection. A study in two regions of Italy showed that in contrast to COVID-19-negative cases, COVID-19-positive patients with GBS had worse total MRC muscle strength scores, increased ICU admissions, and recurrent arterial hypotension, thus giving rise to clinically more severe cases.[116]

12.5.3 Parkinson's Disease

The impact that COVID-19 may have on patients with PD is immense. Especially older people with advanced PD may be more at risk because COVID-19 can become more severe owing to cough reflex dysfunction, respiratory muscle rigidity, and underlying dyspnea.[97,117] Dopamine neurons have high levels of ACE2 receptors; however, their levels are decreased in PD owing to the degenerative changes. Hence the brain penetration associated with SARS-CoV-2 could aggravate symptoms and cause more damage.[95] In a recent study, it was found that COVID-19 had a 40% mortality rate and was fatal to PD patients who were older, *i.e.* with a mean age of more than 78 years, and had had the disease for more than 12 years.[97] In another study, which enrolled 27 PD patients from different health centers, 23 of these showed post-COVID-19 symptoms, of whom 51.9% showed decreasing motor functions, 48.2% had increased levodopa dosage requirements, 22.2% experienced brain fog, and 22.2% also reported loss of focus and had sleep disorders.[118] Numerous studies have suggested that PD patients who have COVID-19 with progressive disease, frailty, and comorbid conditions may have worsening of their motor and non-motor symptoms and also a possibility of an increase in mortality rate.[119,120] A further study concluded that patients with PD may not always be at an increased risk for contracting COVID-19, but they may not have a good disease outcome, especially if they have been living with PD for a long time.[121]

12.5.4 Alzheimer's Disease

The possible neurological damage caused by SARS-CoV-2 that may result in the emergence of neurodegenerative illnesses such as Alzheimer's disease (AD) has been explained by several different mechanisms. On the one hand patients with COVID-19 appear to be more likely to develop AD, and on the other AD individuals may also be more vulnerable to severe COVID-19 illness, thus leading to a mutualistic relationship between the two.[122] *Post-mortem* investigations revealed that ACE2 expression is elevated in the brains of AD patients compared with controls.[123] Furthermore, ACE2 gene expression is increased in the brain tissue of patients with more severe forms of AD, as shown by genome-wide association studies (GWAS).[124] As a result, this increased ACE2 expression in AD patients may serve as a potential risk for COVID-19 virus transmission. Patients with COVID-19 have been shown to have elevated levels of CSF biomarkers of neurodegeneration, which are in turn are linked to both neurological symptoms and the severity of the disease.[125] COVID-19 and AD have also been linked by the presence of several inflammatory biomarkers, such as IL-1, IL-6, and Gal-3.

12.6 Conclusion

SARS-CoV-2 viruses exhibit a wide range of neuroinvasive mechanisms compared with other viruses. The disease pathology highlights nervous system involvement and complications leading to various neurological disorders. This chapter may be helpful in understanding the various neurological mechanisms of SARS-CoV-2 infection to prevent and control neurological disease severity and mortality. Future studies are required to elucidate the mechanisms of neurological manifestations and management, as the impact of neurological involvement may precipitate long-term complications and disabilities. This chapter provides an essential framework.

References

1. D. Cucinotta and M. Vanelli, WHO Declares COVID-19 a Pandemic, *Acta Bio Med. Atenei Parmensis*, 2020, **91**, 157–160.
2. Y. Jin, H. Yang, W. Ji, W. Wu, S. Chen, W. Zhang and G. Duan, Virology, Epidemiology, Pathogenesis, and Control of COVID-19, *Viruses*, 2020, **12**, 372.
3. H. K. Verma, N. Merchant, M. K. Verma, C. İ. Kuru, A. N. Singh, F. Ulucan, P. Verma, A. Bhattacharya and L. V. K. S. Bhaskar, Current updates on the European and WHO registered clinical trials of coronavirus disease 2019 (COVID-19), *Biomed. J.*, 2020, **43**, 424–433.
4. H. K. Verma, B. Farran and L. V. K. S. Bhaskar, Convalescent plasma transfusion a promising therapy for coronavirus diseases 2019 (COVID-19): current updates, *Antibody Ther.*, 2020, **3**, 115–125.
5. L. Mao, H. Jin, M. Wang, Y. Hu, S. Chen, Q. He, J. Chang, C. Hong, Y. Zhou and D. Wang, *et al.*, Neurologic Manifestations of Hospitalized Patients With Coronavirus Disease 2019 in Wuhan, China, *JAMA Neurol.*, 2020, **77**, 683–690.
6. T. Sun, J. Guan and C. You, The neuroinvasive potential of severe acute respiratory syndrome coronavirus 2, *Brain, Behav., Immun.*, 2020, **88**, 59.
7. O. O. Koyuncu, I. B. Hogue and L. W. Enquist, Virus infections in the nervous system, *Cell Host Microbe*, 2013, **13**, 379–393.
8. D. van Riel, R. Verdijk and T. Kuiken, The olfactory nerve: a shortcut for influenza and other viral diseases into the central nervous system, *J. Pathol.*, 2015, **235**, 277–287.
9. I. Mori, Y. Nishiyama, T. Yokochi and Y. Kimura, Olfactory transmission of neurotropic viruses, *J. Neurovirol.*, 2005, **11**, 129–137.
10. S. Gandhi, A. K. Srivastava, U. Ray and P. P. Tripathi, Is the Collapse of the Respiratory Center in the Brain Responsible for Respiratory Breakdown in COVID-19 Patients?, *ACS Chem. Neurosci.*, 2020, **11**, 1379–1381.
11. R. Valizadeh, A. Baradaran, A. Mirzazadeh and L. V. Bhaskar, Coronavirus-nephropathy; renal involvement in COVID-19, *J. Renal Inj. Prev.*, 2020, **9**, e18.
12. Y.-C. Li, W.-Z. Bai and T. Hashikawa, The neuroinvasive potential of SARS-CoV-2 may play a role in the respiratory failure of COVID-19 patients, *J. Med. Virol.*, 2020, **92**, 552–555.
13. S. Verma, M. Kumar, U. Gurjav, S. Lum and V. R. Nerurkar, Reversal of West Nile virus-induced blood-brain barrier disruption and tight junction proteins degradation by matrix metalloproteinases inhibitor, *Virology*, 2010, **397**, 130–138.

14. M. Desforges, A. Le Coupanec, P. Dubeau, A. Bourgouin, L. Lajoie, M. Dubé and P. J. Talbot, Human Coronaviruses and Other Respiratory Viruses: Underestimated Opportunistic Pathogens of the Central Nervous System?, *Viruses*, 2019, **12**, 14.
15. M. Bostancõklõoğlu, SARS-CoV-2 entry and spread in the lymphatic drainage system of the brain, *Brain, Behav., Immun.*, 2020, **87**, 122–123.
16. A. M. Baig, A. Khaleeq, U. Ali and H. Syeda, Evidence of the COVID-19 Virus Targeting the CNS: Tissue Distribution, Host-Virus Interaction, and Proposed Neurotropic Mechanisms, *ACS Chem. Neurosci.*, 2020, **11**, 995–998.
17. D. B. McGavern and S. S. Kang, Illuminating viral infections in the nervous system, *Nat. Rev. Immunol.*, 2011, **11**, 318–329.
18. Y. Wu, X. Xu, Z. Chen, J. Duan, K. Hashimoto, L. Yang, C. Liu and C. Yang, Nervous system involvement after infection with COVID-19 and other coronaviruses, *Brain, Behav., Immun.*, 2020, **87**, 18–22.
19. K. Bohmwald, N. M. S. Gálvez, M. Ríos and A. M. Kalergis, Neurologic Alterations Due to Respiratory Virus Infections, *Front. Cell. Neurosci.*, 2018, **12**, 386.
20. J. Gu, E. Gong, B. Zhang, J. Zheng, Z. Gao, Y. Zhong, W. Zou, J. Zhan, S. Wang and Z. Xie, *et al.*, Multiple organ infection and the pathogenesis of SARS, *J. Exp. Med.*, 2005, **202**, 415–424.
21. A. Mostafa, A. Kandeil, M. Shehata, R. El Shesheny, A. M. Samy, G. Kayali and M. A. Ali, Middle East Respiratory Syndrome Coronavirus (MERS-CoV): State of the Science, *Microorganisms*, 2020, **8**, 991.
22. M. Saad, A. S. Omrani, K. Baig, A. Bahloul, F. Elzein, M. A. Matin, M. A. Selim, M. Al Mutairi, D. Al Nakhli and A. Y. Al Aidaroos, *et al.*, Clinical aspects and outcomes of 70 patients with Middle East respiratory syndrome coronavirus infection: a single-center experience in Saudi Arabia, *Int. J. Infect. Dis.*, 2014, **29**, 301–306.
23. K. Li, C. Wohlford-Lenane, S. Perlman, J. Zhao, A. K. Jewell, L. R. Reznikov, K. N. Gibson-Corley, D. K. Meyerholz and P. B. McCray Jr., Middle East Respiratory Syndrome Coronavirus Causes Multiple Organ Damage and Lethal Disease in Mice Transgenic for Human Dipeptidyl Peptidase 4, *J. Infect. Dis.*, 2016, **213**, 712–722.
24. A. Giacomelli, L. Pezzati, F. Conti, D. Bernacchia, M. Siano, L. Oreni, S. Rusconi, C. Gervasoni, A. L. Ridolfo and G. Rizzardini, *et al.*, Self-reported Olfactory and Taste Disorders in Patients With Severe Acute Respiratory Coronavirus 2 Infection: A Cross-sectional Study, *Clin. Infect. Dis.*, 2020, **71**, 889–890.
25. A. Sepehrinezhad, A. Shahbazi and S. S. Negah, COVID-19 virus may have neuroinvasive potential and cause neurological complications: a perspective review, *J. Neurovirol.*, 2020, **26**, 324–329.
26. S. Morfopoulou, J. R. Brown, E. G. Davies, G. Anderson, A. Virasami, W. Qasim, W. K. Chong, M. Hubank, V. Plagnol and M. Desforges,

et al., Human Coronavirus OC43 Associated with Fatal Encephalitis, *N. Engl. J. Med.*, 2016, **375**, 497–498.
27. N. Arbour, R. Day, J. Newcombe and P. J. Talbot, Neuroinvasion by human respiratory coronaviruses, *J. Virol.*, 2000, **74**, 8913–8921.
28. H. C. Yashavantha Rao and C. Jayabaskaran, The emergence of a novel coronavirus (SARS-CoV-2) disease and their neuroinvasive propensity may affect in COVID-19 patients, *J. Med. Virol.*, 2020, **92**, 786–790.
29. Z. Li, T. Liu, N. Yang, D. Han, X. Mi, Y. Li, K. Liu, A. Vuylsteke, H. Xiang and X. Guo, Neurological manifestations of patients with COVID-19: potential routes of SARS-CoV-2 neuroinvasion from the periphery to the brain, *Front. Med.*, 2020, **14**, 533–541.
30. F. G. De Felice, F. Tovar-Moll, J. Moll, D. P. Munoz and S. T. Ferreira, Severe Acute Respiratory Syndrome Coronavirus 2 (SARS-CoV-2) and the Central Nervous System, *Trends Neurosci.*, 2020, **43**, 355–357.
31. M. P. Platt, K. A. Bolding, C. R. Wayne, S. Chaudhry, T. Cutforth, K. M. Franks and D. Agalliu, Th17 lymphocytes drive vascular and neuronal deficits in a mouse model of postinfectious autoimmune encephalitis, *Proc. Natl. Acad. Sci. U. S. A.*, 2020, **117**, 6708–6716.
32. P. Conti, C. D'Ovidio, C. Conti, C. E. Gallenga, D. Lauritano, A. Caraffa, S. K. Kritas and G. Ronconi, Progression in migraine: Role of mast cells and pro-inflammatory and anti-inflammatory cytokines, *Eur. J. Pharmacol.*, 2019, **844**, 87–94.
33. M. A. Islam, S. S. Alam, S. Kundu, T. Hossan, M. A. Kamal and C. Cavestro, Prevalence of Headache in Patients With Coronavirus Disease 2019 (COVID-19): A Systematic Review and Meta-Analysis of 14,275 Patients, *Front. Neurol.*, 2020, **11**, 562634.
34. E. Caronna, A. Ballvé, A. Llauradó, V. J. Gallardo, D. M. Ariton, S. Lallana, S. López Maza, M. Olivé Gadea, L. Quibus and J. L. Restrepo, *et al.*, Headache: A striking prodromal and persistent symptom, predictive of COVID-19 clinical evolution, *Cephalalgia*, 2020, **40**, 1410–1421.
35. M. C. Whitman and C. A. Greer, Adult neurogenesis and the olfactory system, *Prog. Neurobiol.*, 2009, **89**, 162–175.
36. A. Y. Han, L. Mukdad, J. L. Long and I. A. Lopez, Anosmia in COVID-19: Mechanisms and Significance, *Chem. Senses*, 2020, bjaa040.
37. J. R. Lechien, C. M. Chiesa-Estomba, D. R. De Siati, M. Horoi, S. D. Le Bon, A. Rodriguez, D. Dequanter, S. Blecic, F. El Afia and L. Distinguin, *et al.*, Olfactory and gustatory dysfunctions as a clinical presentation of mild-to-moderate forms of the coronavirus disease (COVID-19): a multicenter European study, *Eur. Arch. Oto-Rhino-Laryngol.*, 2020, **277**, 2251–2261.
38. M. Suzuki, K. Saito, W.-P. Min, C. Vladau, K. Toida, H. Itoh and S. Murakami, Identification of Viruses in Patients With Postviral Olfactory Dysfunction, *Laryngoscope*, 2007, **117**, 272–277.

39. G. Spinato, C. Fabbris, J. Polesel, D. Cazzador, D. Borsetto, C. Hopkins and P. Boscolo-Rizzo, Alterations in Smell or Taste in Mildly Symptomatic Outpatients With SARS-CoV-2 Infection, *JAMA*, 2020, **323**, 2089–2090.
40. A. S. Zubair, L. S. McAlpine, T. Gardin, S. Farhadian, D. E. Kuruvilla and S. Spudich, Neuropathogenesis and Neurologic Manifestations of the Coronaviruses in the Age of Coronavirus Disease 2019: A Review, *JAMA Neurol.*, 2020, **77**, 1018–1027.
41. S. B. Gane, C. Kelly and C. Hopkins, Isolated sudden onset anosmia in COVID-19 infection. A novel syndrome?, *Rhinology*, 2020, **58**, 299–301.
42. T. Moriguchi, N. Harii, J. Goto, D. Harada, H. Sugawara, J. Takamino, M. Ueno, H. Sakata, K. Kondo and N. Myose, *et al.*, A first case of meningitis/encephalitis associated with SARS-Coronavirus-2, *Int. J. Infect. Dis.*, 2020, **94**, 55–58.
43. N. Karimi, A. Sharifi Razavi and N. Rouhani, Frequent Convulsive Seizures in an Adult Patient with COVID-19: A Case Report, *Iran. Red Crescent Med. J.*, 2020, **22**, e102828.
44. J. Helms, S. Kremer, H. Merdji, R. Clere-Jehl, M. Schenck, C. Kummerlen, O. Collange, C. Boulay, S. Fafi-Kremer and M. Ohana, *et al.*, Neurologic Features in Severe SARS-CoV-2 Infection, *N. Engl. J. Med.*, 2020, **382**, 2268–2270.
45. Z. Li, X. Li, J. Shen, M. T. V. Chan and W. K. K. Wu, Miller Fisher syndrome associated with COVID-19: an up-to-date systematic review, *Environ. Sci. Pollut. Res. Int.*, 2021, **28**, 20939–20944.
46. A. Polo, P. Manganotti, G. Zanette and D. De Grandis, Polyneuritis cranialis: clinical and electrophysiological findings, *J. Neurol., Neurosurg. Psychiatry*, 1992, **55**, 398–400.
47. C. Gutiérrez-Ortiz, A. Méndez-Guerrero, S. Rodrigo-Rey, E. San Pedro-Murillo, L. Bermejo-Guerrero, R. Gordo-Mañas, F. de Aragón-Gómez and J. Benito-León, Miller Fisher syndrome and polyneuritis cranialis in COVID-19, *Neurology*, 2020, **95**, e601–e605.
48. S. Kusunoki, Diagnosis, pathogenesis and treatment of Miller Fisher syndrome and related disorders: clinical significance of antiGQ1b IgG antibody, *Expert Rev. Neurother.*, 2003, **3**, 133–140.
49. K. Kaida, M. Kanzaki, D. Morita, K. Kamakura, K. Motoyoshi, M. Hirakawa and S. Kusunoki, Anti-ganglioside complex antibodies in Miller Fisher syndrome, *J. Neurol., Neurosurg. Psychiatry*, 2006, **77**, 1043–1046.
50. G. D. Vavougios, Potentially irreversible olfactory and gustatory impairments in COVID-19: Indolent *vs.* fulminant SARS-CoV-2 neuroinfection, *Brain, Behav., Immun.*, 2020, **87**, 107–108.
51. N. Calcagno, E. Colombo, A. Maranzano, J. Pasquini, I. J. Keller Sarmiento, F. Trogu and V. Silani, Rising evidence for neurological involvement in COVID-19 pandemic, *Neurol. Sci.*, 2020, **41**, 1339–1341.
52. A. O. Payus, C. Liew Sat Lin, M. Mohd Noh, M. S. Jeffree and R. A. Ali, SARS-CoV-2 infection of the nervous system: A review of the literature on neurological involvement in novel coronavirus disease-(COVID-19), *Bosnian J. Basic Med. Sci.*, 2020, **20**, 283–292.

53. A. Avula, K. Nalleballe, N. Narula, S. Sapozhnikov, V. Dandu, S. Toom, A. Glaser and D. Elsayegh, COVID-19 presenting as stroke, *Brain, Behav., Immun.*, 2020, **87**, 115–119.
54. T. Umapathi, A. C. Kor, N. Venketasubramanian, C. C. Lim, B. C. Pang, T. T. Yeo, C. C. Lee, P. L. Lim, K. Ponnudurai and K. L. Chuah, *et al.*, Large artery ischaemic stroke in severe acute respiratory syndrome (SARS), *J. Neurol.*, 2004, **251**, 1227–1231.
55. T. J. Oxley, J. Mocco, S. Majidi, C. P. Kellner, H. Shoirah, I. P. Singh, R. A. De Leacy, T. Shigematsu, T. R. Ladner and K. A. Yaeger, *et al.*, Large-Vessel Stroke as a Presenting Feature of Covid-19 in the Young, *N. Engl. J. Med.*, 2020, **382**, e60.
56. D. C. Hess, W. Eldahshan and E. Rutkowski, COVID-19-Related Stroke, *Transl. Stroke Res.*, 2020, **11**, 322–325.
57. R. Beyrouti, M. E. Adams, L. Benjamin, H. Cohen, S. F. Farmer, Y. Y. Goh, F. Humphries, H. R. Jäger, N. A. Losseff and R. J. Perry, *et al.*, Characteristics of ischaemic stroke associated with COVID-19, *J. Neurol., Neurosurg. Psychiatry*, 2020, **91**, 889–891.
58. H. Minasyan and F. Flachsbart, Blood coagulation: a powerful bactericidal mechanism of human innate immunity, *Int. Rev. Immunol.*, 2019, **38**, 3–17.
59. M. Levi and T. van der Poll, Coagulation and sepsis, *Thromb. Res.*, 2017, **149**, 38–44.
60. N. Gupta, Y. Y. Zhao and C. E. Evans, The stimulation of thrombosis by hypoxia, *Thromb. Res.*, 2019, **181**, 77–83.
61. P. Nuthalapati, M. K. Ghanta, N. S. Natesh and L.V.K.S. Bhaskar, Association of hypercoagulation with severe acute respiratory syndrome coronavirus 2 infection, *Blood Res.*, 2021, **56**, 61–64.
62. A. Tripodi, D-dimer testing in laboratory practice, *Clin. Chem.*, 2011, **57**, 1256–1262.
63. A. A. Asadi-Pooya and L. Simani, Central nervous system manifestations of COVID-19: A systematic review, *J. Neurol. Sci.*, 2020, **413**, 116832.
64. C. Vollono, E. Rollo, M. Romozzi, G. Frisullo, S. Servidei, A. Borghetti and P. Calabresi, Focal status epilepticus as unique clinical feature of COVID-19: A case report, *Seizure*, 2020, **78**, 109–112.
65. G. Balloy, L. Leclair-Visonneau, Y. Péréon, A. Magot, A. Peyre, P.-J. Mahé and P. Derkinderen, Non-lesional status epilepticus in a patient with coronavirus disease 2019, *Clin. Neurophysiol.*, 2020, **131**, 2059–2061.
66. S. Sohal and M. Mansur, COVID-19 Presenting with Seizures, *IDCases*, 2020, **20**, e00782.
67. S. Lyons, B. O'Kelly, S. Woods, C. Rowan, D. Brady, G. Sheehan and S. Smyth, Seizure with CSF lymphocytosis as a presenting feature of COVID-19 in an otherwise healthy young man, *Seizure*, 2020, **80**, 113–114.
68. A. Grippo, G. Assenza, M. Scarpino, L. Broglia, R. Cilea, C. A. Galimberti, G. Lanzo, R. Michelucci, L. Tassi and M. Vergari, *et al.*, Electroencephalography during SARS-CoV-2 outbreak: practical

recommendations from the task force of the Italian Society of Neurophysiology (SINC), the Italian League Against Epilepsy (LICE), and the Italian Association of Neurophysiology Technologists (AITN), *Neurol. Sci.*, 2020, **41**, 2345–2351.

69. M. Flamand, A. Perron, Y. Buron and W. Szurhaj, Pay more attention to EEG in COVID-19 pandemic, *Clin. Neurophysiol.*, 2020, **131**, 2062–2064.
70. A. S. Galanopoulou, V. Ferastraoaru, D. J. Correa, K. Cherian, S. Duberstein, J. Gursky, R. Hanumanthu, C. Hung, I. Molinero and O. Khodakivska, *et al.*, EEG findings in acutely ill patients investigated for SARS-CoV-2/COVID-19: A small case series preliminary report, *Epilepsia Open*, 2020, **5**, 314–324.
71. G. Vellieux, A. Rouvel-Tallec, P. Jaquet, A. Grinea, R. Sonneville and M.P. d'Ortho, COVID-19 associated encephalopathy: Is there a specific EEG pattern?, *Clin. Neurophysiol.*, 2020, **131**, 1928–1930.
72. N. Kuroda, Epilepsy and COVID-19: Associations and important considerations, *Epilepsy Behav.*, 2020, **108**, 107122.
73. J. A. French, M. J. Brodie, R. Caraballo, O. Devinsky, D. Ding, L. Jehi, N. Jette, A. Kanner, A. C. Modi and C. R. Newton, *et al.*, Keeping people with epilepsy safe during the COVID-19 pandemic, *Neurology*, 2020, **94**, 1032–1037.
74. C. Santenna, K. Vidyasagar, K. C. Amarneni, S. N. Ghanta, B. Sadasivam, S. Pathan and R. Padmavathi, The safety, tolerability and mortality reduction efficacy of remdesivir; based on randomized clinical trials, observational and case studies reported safety outcomes: an updated systematic review and meta-analysis, *Ther. Adv. Drug Saf.*, 2021, **12**, 20420986211042517.
75. S. Chenchula, A. Ray and B. Sadasivam, Famotidine Repurposing for Novel Corona Virus Disease of 2019: A Systematic Review, *Drug Res.*, 2021, **71**, 295–301.
76. R. Ohannessian, T. A. Duong and A. Odone, Global Telemedicine Implementation and Integration Within Health Systems to Fight the COVID-19 Pandemic: A Call to Action, *JMIR Public Health Surveill.*, 2020, **6**, e18810.
77. C. Escoffery, R. McGee, J. Bidwell, C. Sims, E. K. Thropp, C. Frazier and E. D. Mynatt, A review of mobile apps for epilepsy self-management, *Epilepsy Behav.*, 2018, **81**, 62–69.
78. B. van den Berg, C. Walgaard, J. Drenthen, C. Fokke, B. C. Jacobs and P. A. van Doorn, Guillain–Barré syndrome: pathogenesis, diagnosis, treatment and prognosis, *Nat. Rev. Neurol.*, 2014, **10**, 469–482.
79. H. Zhao, D. Shen, H. Zhou, J. Liu and S. Chen, Guillain–Barré syndrome associated with SARS-CoV-2 infection: causality or coincidence?, *Lancet Neurol.*, 2020, **19**, 383–384.
80. J. P. Camdessanche, J. Morel, B. Pozzetto, S. Paul, Y. Tholance and E. Botelho-Nevers, COVID-19 may induce Guillain–Barré syndrome, *Rev. Neurol.*, 2020, **176**, 516–518.

81. S. Arnaud, C. Budowski, S. Ng Wing Tin and B. Degos, Post SARS-CoV-2 Guillain–Barré syndrome, *Clin. Neurophysiol.*, 2020, **131**, 1652–1654.
82. G. Toscano, F. Palmerini, S. Ravaglia, L. Ruiz, P. Invernizzi, M. G. Cuzzoni, D. Franciotta, F. Baldanti, R. Daturi and P. Postorino, *et al.*, Guillain–Barré Syndrome Associated with SARS-CoV-2, *N. Engl. J. Med.*, 2020, **382**, 2574–2576.
83. L. Barbi, A. V. C. Coelho, L. C. A. Alencar and S. Crovella, Prevalence of Guillain–Barré syndrome among Zika virus infected cases: a systematic review and meta-analysis, *Braz. J. Infect. Dis.*, 2018, **22**, 137–141.
84. N. Yuki and H. P. Hartung, Guillain–Barré syndrome, *N. Engl. J. Med.*, 2012, **366**, 2294–2304.
85. S. Webb, V. C. J. Wallace, D. Martin-Lopez and M. Yogarajah, Guillain–Barré syndrome following COVID-19: a newly emerging post-infectious complication, *BMJ Case Rep.*, 2020, **13**, e236182.
86. L. Cabañes-Martínez, M. Villadóniga, L. González-Rodríguez, L. Araque, A. Díaz-Cid, I. Ruz-Caracuel, H. Pian, S. Sánchez-Alonso, S. Fanjul and M. Del Álamo, *et al.*, Neuromuscular involvement in COVID-19 critically ill patients, *Clin. Neurophysiol.*, 2020, **131**, 2809–2816.
87. H. Tankisi, Critical illness myopathy and polyneuropathy in Covid-19: Is it a distinct entity?, *Clin. Neurophysiol.*, 2021, **132**, 1716–1717.
88. R. K. Garg, V. K. Paliwal and A. Gupta, Encephalopathy in patients with COVID-19: A review, *J. Med. Virol.*, 2021, **93**, 206–222.
89. V. Krishnan, L. Y. Leung and L. R. Caplan, A neurologist's approach to delirium: diagnosis and management of toxic metabolic encephalopathies, *Eur. J. Intern. Med.*, 2014, **25**, 112–116.
90. E. Fazzini, J. Fleming and S. Fahn, Cerebrospinal fluid antibodies to coronavirus in patients with Parkinson's disease, *Mov. Disord.*, 1992, **7**, 153–158.
91. L. V. Kalia and A. E. Lang, Parkinson's disease, *Lancet*, 2015, **386**, 896–912.
92. C. Klein and A. Westenberger, Genetics of Parkinson's disease, *Cold Spring Harbor Perspect. Med.*, 2012, **2**, a008888.
93. A. J. Lees, J. Hardy and T. Revesz, Parkinson's disease, *Lancet*, 2009, **373**, 2055–2066.
94. T. Goerttler, E. H. Kwon, M. Fleischer, M. Stettner, L. Tönges and S. Klebe, SARS-CoV-2, COVID-19 and Parkinson's Disease-Many Issues Need to Be Clarified-A Critical Review, *Brain Sci.*, 2022, **12**, 456.
95. A. I. Rodriguez-Perez, P. Garrido-Gil, M. A. Pedrosa, M. Garcia-Garrote, R. Valenzuela, G. Navarro, R. Franco and J. L. Labandeira-Garcia, Angiotensin type 2 receptors: Role in aging and neuroinflammation in the substantia nigra, *Brain, Behav., Immun.*, 2020, **87**, 256–271.
96. J. Matías-Guiu, U. Gomez-Pinedo, P. Montero-Escribano, P. Gomez-Iglesias, J. Porta-Etessam and J. A. Matias-Guiu, Should we expect neurological symptoms in the SARS-CoV-2 epidemic?, *Neurologia*, 2020, **35**, 170–175.

97. A. Antonini, V. Leta, J. Teo and K. R. Chaudhuri, Outcome of Parkinson's Disease Patients Affected by COVID-19, *Mov. Disord*, 2020, **35**, 905–908.
98. R. Cilia, S. Bonvegna, G. Straccia, N. G. Andreasi, A. E. Elia, L. M. Romito, G. Devigili, E. Cereda and R. Eleopra, Effects of COVID-19 on Parkinson's Disease Clinical Features: A Community-Based Case-Control Study, *Mov. Disord.*, 2020, **35**, 1287–1292.
99. R. C. Helmich, E. Aarts, F. P. de Lange, B. R. Bloem and I. Toni, Increased dependence of action selection on recent motor history in Parkinson's disease, *J. Neurosci.*, 2009, **29**, 6105–6113.
100. T. W. Robbins and R. Cools, Cognitive deficits in Parkinson's disease: a cognitive neuroscience perspective, *Mov. Disord.*, 2014, **29**, 597–607.
101. E. H. Douma and E. R. de Kloet, Stress-induced plasticity and functioning of ventral tegmental dopamine neurons, *Neurosci. Biobehav. Rev.*, 2020, **108**, 48–77.
102. M. Schenkman, C. G. Moore, W. M. Kohrt, D. A. Hall, A. Delitto, C. L. Comella, D. A. Josbeno, C. L. Christiansen, B. D. Berman and B. M. Kluger, *et al.*, Effect of High-Intensity Treadmill Exercise on Motor Symptoms in Patients With De Novo Parkinson Disease: A Phase 2 Randomized Clinical Trial, *JAMA Neurol.*, 2018, **75**, 219–226.
103. N. M. van der Kolk, N. M. de Vries, R. P. C. Kessels, H. Joosten, A. H. Zwinderman, B. Post and B. R. Bloem, Effectiveness of home-based and remotely supervised aerobic exercise in Parkinson's disease: a double-blind, randomised controlled trial, *Lancet Neurol.*, 2019, **18**, 998–1008.
104. C. A. Hribar, P. H. Cobbold and F. C. Church, Potential Role of Vitamin D in the Elderly to Resist COVID-19 and to Slow Progression of Parkinson's Disease, *Brain Sci.*, 2020, **10**, 284.
105. R. D. Thijs, R. Surges, T. J. O'Brien and J.W. Sander, Epilepsy in adults, *Lancet*, 2019, **393**, 689–701.
106. P. Cabezudo-García, N. L. Ciano-Petersen, N. Mena-Vázquez, G. Pons-Pons, M. V. Castro-Sánchez and P. J. Serrano-Castro, Incidence and case fatality rate of COVID-19 in patients with active epilepsy, *Neurology*, 2020, **95**, e1417–e1425.
107. J. Yoo, J. H. Kim, J. Jeon, J. Kim and T. J. Song, Risk of COVID-19 Infection and of Severe Complications Among People With Epilepsy: A Nationwide Cohort Study, *Neurology*, 2022, **98**, e1886–e1892.
108. F. Dono, B. Nucera, J. Lanzone, G. Evangelista, F. Rinaldi, R. Speranza, S. Troisi, L. Tinti, M. Russo and M. Di Pietro, *et al.*, Status epilepticus and COVID-19: A systematic review, *Epilepsy Behav.*, 2021, **118**, 107887.
109. J. H. Cross, C. S. Kwon, A. A. Asadi-Pooya, G. Balagura, P. Gómez-Iglesias, A. Guekht, J. Hall, A. Ikeda, N. A. Kishk and P. Murphy, *et al.*, Epilepsy care during the COVID-19 pandemic, *Epilepsia*, 2021, **62**, 2322–2332.
110. A. A. Asadi-Pooya, A. Emami, A. Akbari and F. Javanmardi, COVID-19 presentations and outcome in patients with epilepsy, *Acta Neurol. Scand.*, 2021, **143**, 624–628.
111. S. Jain, H. Potschka, P. P. Chandra, M. Tripathi and D. Vohora, Management of COVID-19 in patients with seizures: Mechanisms of action of

potential COVID-19 drug treatments and consideration for potential drug-drug interactions with anti-seizure medications, *Epilepsy Res.*, 2021, **174**, 106675.

112. A. A. Asadi-Pooya, A. Attar, M. Moghadami and I. Karimzadeh, Management of COVID-19 in people with epilepsy: drug considerations, *Neurol. Sci.*, 2020, **41**, 2005–2011.
113. A. A. Asadi-Pooya, L. Simani, M. Shahisavandi and Z. Barzegar, COVID-19, *de novo* seizures, and epilepsy: a systematic review, *Neurol. Sci.*, 2021, **42**, 415–431.
114. A. A. Asadi-Pooya, Seizures associated with coronavirus infections, *Seizure*, 2020, **79**, 49–52.
115. A. McGrogan, G. C. Madle, H. E. Seaman and C. S. de Vries, The epidemiology of Guillain-Barré syndrome worldwide. A systematic literature review, *Neuroepidemiology*, 2009, **32**, 150–163.
116. M. Filosto, S. Cotti Piccinelli, S. Gazzina, C. Foresti, B. Frigeni, M. C. Servalli, M. Sessa, G. Cosentino, E. Marchioni and S. Ravaglia, *et al.*, Guillain-Barré syndrome and COVID-19: an observational multicentre study from two Italian hotspot regions, *J. Neurol., Neurosurg. Psychiatry*, 2021, **92**, 751–756.
117. D. J. van Wamelen, V. Leta, J. Johnson, C. L. Ocampo, A. M. Podlewska, K. Rukavina, A. Rizos, P. Martinez-Martin and K. R. Chaudhuri, Drooling in Parkinson's Disease: Prevalence and Progression from the Non-motor International Longitudinal Study, *Dysphagia*, 2020, **35**, 955–961.
118. V. Leta, M. Rodríguez-Violante, A. Abundes, K. Rukavina, J. T. Teo, C. Falup-Pecurariu, L. Irincu, S. Rota, R. Bhidayasiri and A. Storch, *et al.*, Parkinson's Disease and Post-COVID-19 Syndrome: The Parkinson's Long-COVID Spectrum, *Mov. Disord*, 2021, **36**, 1287–1289.
119. Q. Zhang, J. L. Schultz, G. M. Aldridge, J. E. Simmering and N. S. Narayanan, Coronavirus Disease 2019 Case Fatality and Parkinson's Disease, *Mov. Disord*, 2020, **35**, 1914–1915.
120. D. Sulzer, A. Antonini, V. Leta, A. Nordvig, R. J. Smeyne, J. E. Goldman, O. Al-Dalahmah, L. Zecca, A. Sette and L. Bubacco, *et al.*, COVID-19 and possible links with Parkinson's disease and parkinsonism: from bench to bedside, *NPJ Parkinsons Dis.*, 2020, **6**, 18.
121. C. A. Artusi, A. Romagnolo, G. Imbalzano, A. Marchet, M. Zibetti, M. G. Rizzone and L. Lopiano, COVID-19 in Parkinson's disease: Report on prevalence and outcome, *Parkinsonism Relat. Disord.*, 2020, **80**, 7–9.
122. M. Ciaccio, B. Lo Sasso, C. Scazzone, C. M. Gambino, A. M. Ciaccio, G. Bivona, T. Piccoli, R. V. Giglio and L. Agnello, COVID-19 and Alzheimer's Disease, *Brain Sci.*, 2021, **11**, 305.
123. Q. Ding, N. V. Shults, B. T. Harris and Y. J. Suzuki, Angiotensin-converting enzyme 2 (ACE2) is upregulated in Alzheimer's disease brain, *bioRxiv*, 2020, 2020.10.08.331157.
124. K. H. Lim, S. Yang, S. H. Kim and J. Y. Joo, Elevation of ACE2 as a SARS-CoV-2 entry receptor gene expression in Alzheimer's disease, *J. Infect.*, 2020, **81**, e33–e34.

125. J. Virhammar, A. Nääs, D. Fällmar, J. L. Cunningham, A. Klang, N. J. Ashton, S. Jackmann, G. Westman, R. Frithiof and K. Blennow, *et al.*, Biomarkers for central nervous system injury in cerebrospinal fluid are elevated in COVID-19 and associated with neurological symptoms and disease severity, *Eur. J. Neurol.*, 2021, **28**, 3324–3331.
126. A. R. Rao, S. M. Hidayathullah, K. Hegde and P. Adhikari, Parkinsonism: An emerging post COVID sequelae, *IDCases*, 2022, **27**, e01388.
127. I. Boura and K. R. Chaudhuri, Coronavirus Disease 2019 and Related Parkinsonism: The Clinical Evidence Thus Far, *Mov. Disord. Clin. Pract.*, 2022, **9**, 584–593.
128. M. E. Cohen, R. Eichel, B. Steiner-Birmanns, A. Janah, M. Ioshpa, R. Bar-Shalom, J. J. Paul, H. Gaber, V. Skrahina and N. M. Bornstein, *et al.*, A case of probable Parkinson's disease after SARS-CoV-2 infection, *Lancet Neurol.*, 2020, **19**, 804–805.
129. S. Chattopadhyay, S. Basu and K. Majumdar, Seizure after recovery from Covid-19, *J. Fam. Med. Primary Care*, 2021, **10**, 2713–2715.
130. M. Taquet, O. Devinsky, J. H. Cross, P. J. Harrison and A. Sen, Incidence of Epilepsy and Seizures Over the First 6 Months After a COVID-19 Diagnosis: A Retrospective Cohort Study, *Neurology*, 2023, **100**, e790–e799.
131. J. A. Matias-Guiu, C. Delgado-Alonso, M. Yus, C. Polidura, N. Gómez-Ruiz, M. Valles-Salgado, I. Ortega-Madueño, M. N. Cabrera-Martín and J. Matias-Guiu, "Brain Fog" by COVID-19 or Alzheimer's Disease? A Case Report, *Front Psychol.*, 2021, **12**, 724022.
132. L. Wang, P. B. Davis, N. D. Volkow, N. A. Berger, D. C. Kaelber and R. Xu, Association of COVID-19 with New-Onset Alzheimer's Disease, *J. Alzheimer's Dis.*, 2022, **89**, 411–414.
133. S. Reiken, L. Sittenfeld, H. Dridi, Y. Liu, X. Liu and A. R. Marks, Alzheimer's-like signaling in brains of COVID-19 patients, *Alzheimer's Dementia*, 2022, **18**, 955–965.
134. F. Khan, P. Sharma, S. Pandey, D. Sharma, V. Vijayavarman, N. Kumar, S. Shukla, H. Dandu, A. Jain and R. K. Garg, *et al.*, COVID-19-associated Guillain-Barre syndrome: Postinfectious alone or neuroinvasive too?, *J. Med. Virol.*, 2021, **93**, 6045–6049.
135. Z. Sedaghat and N. Karimi, Guillain Barre syndrome associated with COVID-19 infection: A case report, *J. Clin. Neurosci.*, 2020, **76**, 233–235.
136. S. Kanou, L. Wardeh, S. Govindarajan and K. Macnay, Guillain-Barre syndrome (GBS) associated with COVID-19 infection that resolved without treatment in a child, *BMJ Case Rep.*, 2022, **15**, e245455.
137. A. Zito, E. Alfonsi, D. Franciotta, M. Todisco, M. Gastaldi, M. Cotta Ramusino, M. Ceroni and A. Costa, COVID-19 and Guillain-Barré Syndrome: A Case Report and Review of Literature, *Front. Neurol.*, 2020, **11**, 909.

CHAPTER 13

Socioeconomics: Involving Communities in the Fight Against Infectious Illnesses for Stronger Health Systems

RAUNAK KUMAR TAMRAKAR,*[a] ABHISHEK CHAKRABORTY,[b] SUSHMA SINGH[b] AND SUNIL KUMAR[b]

[a] Department of Applied Physics, Bhilai Institute of Technology, Near Bhilai Power House, Durg, Chhattisgarh 49100, India; [b] Department of Management, Bhilai Institute of Technology, Near Bhilai Power House, Durg, Chhattisgarh 49100, India
*Emails: rounak.tamrakar@bitdurg.ac.in; raunak.ruby@gmail.com

13.1 Overview of Socioeconomic Factors

How well and how long we live can be greatly influenced by social and economic factors, including money, education, work, neighbourhood safety, and social supports. These elements have an impact on our capacity to make healthy decisions, pay for housing and healthcare, control stress, and other things. The non-medical elements that affect health outcomes are referred to as social determinants of health (SDH). In addition to the larger group of factors and systems that influence the conditions of daily life, these are the circumstances in which people are born, develop, work, live, and age. These factors and systems include political systems, societal norms, social policies, economic policies and systems, and development objectives. Long-term

Nanoscience & Nanotechnology Series No. 62
Smart Nanomaterials for Infectious Diseases
Edited by Suvardhan Kanchi, Nidhi Sharotri, Rajasekhar Chokkareddy, Deepali Sharma and Falah H. Hussein

Published by the Royal Society of Chemistry, www.rsc.org

research has been carried out on the link between socioeconomic status (SES) and health. Academics largely held the belief in the 1960s that advances in medical technology and economic growth would diminish health inequality, at least in industrialized nations.[1] Black discovered that health disparities in Britain not only did not decrease but actually significantly worsened during the 1980s.[2] The health of the group with higher SES is unquestionably better than that of the group with lower SES, according to studies conducted in the USA and Europe.[3,4] These investigations demonstrated that SES has a significant negative impact on health; however, the exact mechanism underlying this phenomenon is still open to debate. Social causality theory and health selection theory are two ideas put forth by academics.[5] According to the former, SES disparities are what ultimately drive health inequities.[6] On the other hand, the latter suggests that persons with good health have a higher tendency to advance in society and have a higher SES.[7] Despite these disagreements, there seems to be a growing consensus that a person's lifestyle has a significant impact on their SES.[8] In a more practical setting, people's efforts and decisions to live a healthy lifestyle help to maintain and improve their health.[9] A healthy lifestyle is a set of behavioural patterns that people use to preserve and promote good health based on their motives, social norms, skills, and understanding of what habits are healthful, stress relieving, or enjoyable.[10] Lifestyle includes both unhealthy activities such as smoking, drinking, and being sedentary[11,12] and healthy behaviour such as physical activity, social contact, stress reduction, and spiritual development.[13] Health and lifestyle are closely associated in a variety of social circumstances, according to research findings. For instance, a study carried out in the USA revealed that behavioural risk factors including smoking, a poor diet, and inactivity are the real primary causes of mortality. According to the study's findings, over 40% of deaths are caused by behavioural factors connected to lifestyle.[14] Lifestyles could also be passed down the generations. According to a study of mother–child couples, if a mother of a child between the ages of 0 and 3 years leads a healthy lifestyle, the youngster is 27% more likely to follow suit.[15] Lifestyle has also been linked to psychological health in addition to physical health. People who have unhealthy lifestyles typically experience worse mental health than those who lead healthy lifestyles. For instance, risky habits, such as smoking, have been linked to poor mental health,[16] whereas in contrast, health-promoting behaviour, such as exercising, have been suggested as a successful treatment for depression.[17] A comprehensive social study of individuals in New Zealand found that adopting a healthy lifestyle increases the likelihood of experiencing optimal wellbeing.[18] Notably, as shown in Figure 13.1, lifestyle is not solely a matter of personal preference, although being intimately tied to health. In actuality, a multitude of social factors, particularly a person's SES, have an impact on lifestyle. SES and way of life are closely intertwined. One can select a specific lifestyle from the options available, as Max Weber remarked, but the range of options is primarily influenced by one's SES and other social variables.[19] Similarly to this,

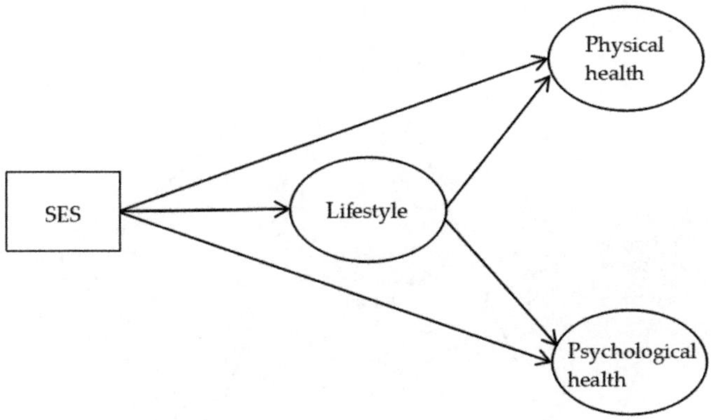

Figure 13.1 Conceptual framework.

Cockerham asserted that social rank and a person's living circumstances are important constraints on lifestyle.[20] In addition, a study found that the total impact of socioeconomic position on health would be much less if lifestyle factors were taken out.[21] As a result, one of the intermediary mechanisms connecting SES and health could be lifestyle. Although the literature indicates a close association between SES, lifestyle, and health, it is noteworthy that the specifics of this relationship are not well defined. On the one hand, prior research frequently concentrated on a single habit or small group of lifestyle activities (particularly risk behaviours, such as drinking and smoking), although doing so has limited implications.[22] However, a large proportion of the research on SES and lifestyle to date has focused on people's physical health,[16] but the definition of health used today encompasses both mental and physical wellbeing.

13.2 Social Determinants of Health

As shown in Figure 13.2, there are five main social and economic factors that make up the social determinants of health (SDOH): health and medical treatment, neighbourhood and constructed environment, social and community context, education, and financial security.[23] Access to primary care, health insurance, and health literacy are all components of health and healthcare.[24] Patients who lack health literacy may find it challenging to navigate the complex healthcare system and understand medical advice or medications. Without health insurance, people are less likely to use or even have access to primary care, which makes it challenging to diagnose and treat chronic illnesses such as cancer, diabetes, asthma, and cardiovascular disease (CVD).

The environments in which a person lives, learns, and works are called the social and communal context. Participation in the community and prejudice are part of this SDOH domain. With social and communal support and coherence come lower mortality rates. Housing, community, transportation,

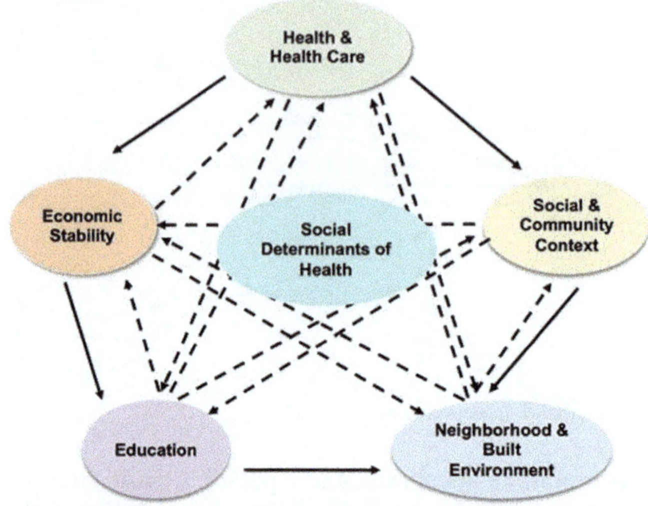

Figure 13.2 The five domains of social determinants of health (SDOH).

access to wholesome foods, air quality, water quality, and availability of green space are all parts of the neighbourhood and built environment.[24] It has been demonstrated that asthma attacks are correlated with air pollution. According to the US Centers for Disease Control and Prevention (CDC), those who have asthma are more likely to contract a serious COVID-19 condition.[25] Health plays a large part in safety. When a person feels safe in their community, they are more inclined to go for a walk or a run outside. People can maintain good mental health in safe surroundings without having to worry about crime and risk. Stress on the mind has an impact on immune function. According to Algren *et al.*, people in deprived neighbourhoods were seen to experience higher levels of stress than people in non-poor communities. "Overcrowding, high crime rates, perceived danger, bad transportation, poor housing, disrepair, limited amenities, poor infrastructure, and a lack of social support" are some of the stressors that affect those who live in underprivileged areas.[26] Education comprises completing high school, enrolling in post-secondary programmes, and mastering a language and literacy. One's life expectancy increases with increased levels of education.[24] It is crucial to provide health information in a patient-specific way that takes into consideration their level of education. Employment, poverty, food security, and housing stability are all aspects of economic stability. According to the American Medical Association (AMA), the proportion of persons aged 25 years and older with an activity-limiting chronic condition rises as poverty levels rise.[24] Since unemployment is linked to physical sickness, domestic abuse, substance abuse, and depression, it has a variety of negative effects on a person's health.

Income, education, job, and social support are some examples of SDOH.[27] They are the circumstances that one is born into, grows up in, lives in, works in, and ages in.[28] They consider the individual as a whole. These issues collectively have an effect on people's and communities' health. Inequalities in any of these factors are translated into a socioeconomic status indicator, which represents the social hierarchy (SES). People will experience worse health results the lower they are on the SES range. Life expectancy decreases for persons at the bottom end of the range as a result of poor results.[27] Socioeconomic disparity adds health issues to the financial difficulties that already beset underprivileged groups of the community.

The five SDOH are interconnected and had a significant impact on the COVID-19 epidemic. In the recent COVID-19 pandemic, for instance, those factors played a significant role. For instance, an individual's education level can affect their occupation, which determines economic stability and income level, which can impact the type of healthcare that the individual is eligible for and what neighbourhood they live in. As a result, it is clear that socioeconomic factors significantly affect infection and mortality rates. Examples include the fact that particular counties in New York, including the Bronx, Brooklyn, and Queens, had higher death rates than other counties, indicating that a sizable population of people of poor socioeconomic level resided there. Another illustration to take into account is from the viewpoint of a child growing up in a household with little financial security. The child's parents are compelled to live in low-income neighbourhoods with perhaps subpar school systems because of their low-paying occupations. This child will not receive an education of the same calibre as a child who lives in a wealthy area with a wealthier school system, since local, state, and federal governments in the USA all contribute to funding public schools.[29]

Property and income taxes provide funding. Richer districts and areas have greater revenues because they collect more taxes. Low-income areas receive less financing, have inadequate school infrastructure, and hire the least qualified instructors.[29] As a result, if a child decides to pursue a college degree, low-quality schooling will not result in high test scores for college admission, keeping them out of the best schools. The youngster may not have many possibilities for high-paying jobs even with a low-tier college diploma. The child will then be in the same situation as their parents, with a low-paying job and housing in a depressed area. According to van Ham *et al.*,[30] children who grow up in impoverished communities with their parents have a higher likelihood of experiencing poverty as adults. Even in the COVID-19 epidemic, the five determinants can be viewed as a cycle of interconnected occurrences rather than as distinct entities.

13.3 Assessments of the Economic Impact of Disease

Socioeconomic factors, such as poverty, influence infection risk and disease outcomes. People who are poor frequently experience living in cramped accommodation, having limited access to high-quality healthcare, being forced to work while ill, eating less nutritively, sleeping less, experiencing more stress, and being more likely than others to abuse drugs and alcohol.

All of these things lower immunity and make people more vulnerable to illness and death.[31]

The majority of studies in developed nations such as in Europe concentrate on the socioeconomic factors that influence chronic diseases such as diabetes, cancer, and cardiovascular disease because infectious diseases account for only 7% of the global burden of disease.[32] Conversely, communicable diseases in Africa cause 63% of all fatalities.

The majority of infectious diseases in South Asia are directly related to poverty and its effects on poor nutrition, crowded living conditions, access to healthcare, poor hygiene, and poor sanitation. The World Health Organization (WHO) estimates that diseases brought on by poverty account for 45% of all illnesses in the world's poorest nations, and that tuberculosis, malaria, and HIV/AIDS combined account for nearly 18% of all diseases.[33] Poverty, contaminated air and water, and other socioeconomic and environmental factors are recognized as major risk factors for the spread of tuberculosis (TB) infection.[34]

The basic framework of the cyclical relationship between social determinants and infectious diseases is shown in Figure 13.3.[35] According to studies,[36,37]

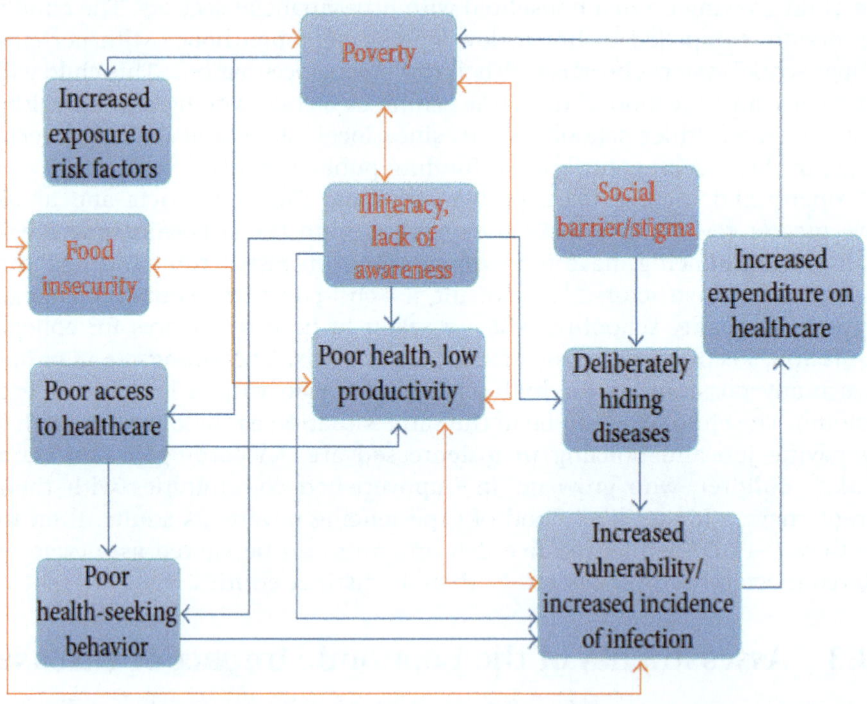

Figure 13.3 The connection between social determinants and an infectious disease. Four main social determinants of the infectious diseases are shown in red.[35] Reproduced from ref. 35, https://doi.org/10.1155/2014/135243, under the terms of the CC BY 3.0 license https://creativecommons.org/licenses/by/3.0/.

a lack of access to food in the home makes people more likely to engage in behaviours that increase their risk of contracting HIV. Another crucial factor is health literacy, as those who are poor and illiterate do not always respond to health programmes as well as those who are literate.[38]

Economic growth is linked to improved longevity and health. However, this might not be true in the short term, as economic booms have been shown to increase mortality rates and economic busts have been shown to decrease them. There are numerous hypotheses as to the cause, including some data implicating the healthcare system. Better nutrition, improvements in public health infrastructure (such as sanitation and cleanliness of water supplies), and more effective medical technologies have all been linked to economic growth in the long term (such as antibiotics and vaccines). The classic wealth–health relationship is evident in the findings of many studies, both between and within countries. People in countries where the median income is higher have a lower rate of disability and a longer average life expectancy.[39]

Fewer cases of anaemia are seen among females in countries with higher per capita incomes. In those countries, babies typically weigh more at birth. Within nations, those who have higher wages also enjoy a longer life expectancy. Children from affluent backgrounds tend to be physically and mentally stronger. Rates of infant and child mortality, and also childhood disease, have decreased. The economic booms of the 1960s and 1970s led to an increase in life expectancy in Japan. The Great Depression led to a decrease in life expectancy. Gordon and Sommers discovered a similar relationship between economic downturns and increased mortality in their recent analysis of the USA from 1993 to 2012.[40] Thus, it can be concluded that there is a correlation between a larger chance of death and unemployment, rising poverty, lower income, and employment status.

13.3.1 Consequences of Economic Growth

Nonetheless, a sizable body of literature suggests that, in the short term, growth in the economy can have negative effects on people's health. In addition, certain people may benefit from the adverse effects of a recession on the economy. Unemployment skyrocketed worldwide during the recession of 2007–2012, although to varying extents depending on country. One study in Europe found that for every 1% increase in the unemployment rate, the death rate decreased by 0.5%. There has been a correlation between the two in the past, both in Europe and the USA. In the USA, this equates to around 13 000 fewer fatalities for every percentage point increase in the unemployment rate.[39]

How can a sluggish economy help people live longer while also improving their health? The link between the economy and people's wellbeing can be different in different settings, shift over time, and have distinct effects in the short and long term. Owing to the dynamic nature of the economy, this could shift (such as agrarian *versus* industrial *versus* service). There is also a correlation with recession severity and boom intensity. Lastly, the demographic make-up matters since different people are hit harder or easier by economic shifts.

Air pollution is thought to worsen the effects of several diseases and increase mortality rates during economic expansions in industrial economies. According to research, the rise in the unemployment rate during the recession referred to above was linked to lower rates of death from air pollution-sensitive diseases such as asthma, heart disease, and chronic obstructive pulmonary disease (COPD). One study suggested that air pollution is responsible for two-thirds of the positive effects of a robust economy on mortality over the long term. Short-term economic difficulties can influence another major cause of death, namely driving. Fewer persons are killed in car accidents as a result of a slowing economy since fewer people are driving (both to and from work and transporting fewer things). Those who do drive during recessions are more cautious, according to research by Maheshri and Winston.[41] They found that during economic downturns, higher-risk drivers (those who had been in an accident in the past, were over the age of 60 years, or lived in households with no more than one other person) cut back on their mileage more than other drivers.

13.3.2 Who Will be Impacted?

A dwindling or expanding economy can have varied health repercussions on different people, and it is vital to keep that in mind. An intriguing study by Stevens *et al.*[42] proved this to be the case. It was discovered that the elderly, and especially elderly women residing in nursing homes, are disproportionately affected by the rise in mortality that occurs during prosperous economic times. One possible explanation for this occurrence is that as the unemployment rate decreases, skilled nursing institutions lay off workers, especially nurses. This may occur if nurses who would normally work in those facilities find better employment elsewhere, or if other members of their households find work. Also, it is possible that the quality of the nursing personnel that hospitals are able to keep increases during economic expansions. This link between employment rates and nursing home mortality seems reasonable given that quality is favourably connected with staffing numbers in nursing homes. These results agree with those of other studies. One study, for instance, indicated that cardiovascular, pulmonary, and degenerative brain disorders, in addition to infections, are the primary causes of increased mortality among the elderly during prosperous times.[43] The results are supported by those of another recent study that looked at the effects of a publicly subsidized parental leave programme on nurses' employment and patient outcomes in Denmark.[44] As a result of the scheme, the number of nurses employed by nursing homes decreased by 12%, and the death rate for people aged 85 years and older in those facilities increased by 13%.

These studies aid in making sense of the paradox that rising earnings benefit health but flourishing economies can lead to an increase in deaths.

If those who feel the impacts of a strong economy on their mortality tend to be different from those who benefit from it, then both statements can be true. Consider that while you and your family may benefit in the long term from a healthy economy, your grandparents may feel the effects in the immediate term.[39]

13.4 Structural Determinants of Health

Health is a state of physical, mental, and social wellbeing and not merely the absence of disease or infirmity, *i.e.* the ability to lead a socially and economically productive life (Table 13.1). The longitudinal curriculum thread on structural determinants of health focuses on the circumstances of our birth, development, employment, and existence, including the social, political, economic, and environmental forces and systems that influence and inform the health of individuals, groups, and populations.[45] Determinants are factors that can influence a person's health. While the focus of health interventions has typically been on who people are and what they do, the conditions in which they are born, grow, live, work, and age are critically important in determining the health of individuals and communities. Health is influenced by many factors, which may generally be organized into five broad categories known as determinants of health: genetics, behaviour, environmental and physical influences, medical care, and social factors. These five categories are interconnected. A novel framework for comprehending and addressing the disparities that cause illness in people and populations is called structural competency. In order to address how these factors contribute to illness and disease, this paradigm examines institutional, structural, and racial hierarchies and discrimination.[46]

This section uncovers the determinants of health and when they are equities. The underlying health inequities that exist in the USA have become more obvious as a result of the COVID-19 outbreak. The conventional biomedical approach attributes individual decisions as the cause of these discrepancies. The model makes the assumption that one's eating, exercising, and smoking behaviours are based on personal preferences. Therefore, a person's poor decisions are to blame for the prevalence of morbidities, including high blood pressure, obesity, and diabetes. This section examines how people's health outcomes are influenced by the environments in which they live, work, play, and pray. The structural determinants of health (SDH) are the umbrella name for these factors, and they frequently

Table 13.1 Structural determinants of health.

Structural determinants of health
Socioeconomic status
Access to healthcare
Environment and safety
Culture and language

highlight the structural injustices that disproportionately afflict underprivileged populations.[46]

13.4.1 Socioeconomics and Status

The SDH are the conditions in which people are born, grow, live, and age, and they have a large influence on our health. They also determine health inequities, which are the unfair and avoidable health differences between different groups of people within a country or between different countries. In this section, we take a look at the social determinants of health, which include who they are, *i.e.* age, gender, and genetic factors, what they are, *i.e.* smoking, physical activities, alcohol consumption, and diet pattern, and are also largely influenced by the conditions in which people are born, grow, live, and work, how it impacts health, *i.e.* their social and community networks, socioeconomic culture, and environmental conditions that people live in and the available health system, and a useful framework to understand it. These collectively form the determinants of health. "How can we change these determinants?" is a good question to begin with. But next, we need to ask which determinants would have a stronger impact on the overall health of society. This second question can help policy makers prioritize what needs to be done, because usually it is not feasible to focus on all determinants adequately. The SDH are ultimately shaped by the distribution of money, power, and resources at international, national, and local levels, and have marked influences on health inequities, as defined above.[39]

There are many different SDH at many levels and with complex interactions between them. Therefore, to understand and explain how these determinants influence and interact with each other to affect wellbeing, several models have been proposed. There are many different frameworks developed by the WHO that influence health and can lead to health inequities. The framework highlights two broad determinants, the first being structural determinants that are in socioeconomic and political contexts, *i.e.* governance, policies, value system, and socioeconomic position describing a person's place in society, this might alter their susceptibility to situations that affect their health state, which is affected by a number of attributes such as education, occupation, gender, income level, race or ethnicity, and social class. The second determinant is intermediary and includes material circumstances such as the quality of housing, the financial means to buy healthy food and clothing, or requirements for healthy living and also work environments and psychological circumstances such as stressful living, relationships, and social support. A system dynamics approach to understanding the problem is necessary. System dynamics simulations take into account different cause–effect paths, both positive and negative, in addition to feedback loops that can help expedite or negate changes in outcomes.[36]

13.4.2 Access to Healthcare

Access to healthcare, which is one of the determinants of health access, refers to a person's ability to obtain the appropriate healthcare services when they are needed. There are five dimensions to access to healthcare: availability, accessibility, accommodation, affordability, and acceptability. Availability of healthcare services refers to the match between the services' capacity and an individual's requirements such as sufficient healthcare specialists, social services, and even primary care that are available to patients when they are needed. Accessibility means the fit between the location of service providers and patients, *i.e.* convenience, design, and payment options. Improving access to healthcare is too great a problem to solve alone, as there is a good chance that the medical supplies and medicines that one needs will not be available and low-cost treatments do not exist. This is the challenge of improving access to healthcare in the developing world. As recently as 2008 diarrhoea was one of the leading causes of child deaths worldwide. The most advanced therapy available for tuberculosis was developed in the 1960s, but still as of today tuberculosis kills nearly 2 million people every year. The infrastructure, the availability of doctors, and the supply of medicines and vaccines are all in the process of improvement yet advances in these areas are only part of the solution. There is huge scope for research and development to find better treatments and infrastructure for all the players in global healthcare, and there is a need to encourage cooperation, innovation, and investment in disease prevention. Great things can happen when we focus on all relevant stakeholders. For example, the WHO 20 years ago focused on eradicating leprosy as a priority, working with global companies, national health systems, and local workers. This coordinated effort expanded access to health education, diagnosis, and free medicines and the result was that 14 million people were cured of leprosy. A similar comprehensive approach also helped to fight malaria. Countries that have increased the supply and use of mosquito bed nets while increasing the availability of newer and more effective medicines have seen malaria cases and deaths fall by 50%, and this is just the beginning of what we can achieve. No single organization can solve all the access to healthcare problems, they are too large to solve alone, but with effective coordination, continued innovation and research, and shared commitment, we can make a big difference together.[38]

13.4.3 Environment and Safety

Environmentally constituted health risks are disproportionately experienced by the world's poorest communities, which frequently bear the brunt of the health risks of climate change. Despite this, these communities are predominately marginalized in policy discourse and are frequently forced to foot the bill for environmental harms.[47] Experts positioned inside global power structures suggest solutions to environmental challenges that are frequently divorced from the lived experiences of communities. For instance, despite the fact that the poor in the global south bear a disproportionate amount of the costs associated with human-induced climate change, these groups are frequently left out of discussion and policy-making forums. Similarly, the poor are frequently removed from discursive venues where policies are discussed and decisions are made, despite the fact that they bear a disproportionate share of the dangers associated with pollution, chemical exposure, and poisoning.[48]

To "guarantee healthy lifestyles and promote wellbeing for everybody at all ages," a healthy environment is essential.[5] Addressing physical, chemical, and biological factors that are external to an individual as well as all related behaviours, environmental public health is defined as the environmental intersection and health of the public and includes local, national, regional, and global environmental factors that have an impact on human health.

13.4.3.1 Key Facts

- Around 83 million people still lack access to proper sanitation facilities, including 15.6 million who continue to use open defecation and 28 million without improved supplies of safe drinking water, which leads to over 30 000 avoidable deaths annually.
- Children and expectant mothers are disproportionately affected by dangers from hazardous chemicals, such as toxic pesticides, lead, and mercury exposure.
- Frequent and chronic irreversible concerns about health, for instance issues with neurological development, birth malformations, and diseases linked to endocrine disruption, can be brought on by exposure to toxic chemicals.
- Owing to the global disruption of physical, biological, and ecological systems, environmental phenomena such as climate change are having a growing negative influence on the health and wellbeing of people in poorer regions.
- The appearance of new environmental dangers, such as water scarcity, microplastics, nanoparticles, toxins that affect the endocrine system, and electronic trash, is a concern.
- Management of complex difficulties brought on by pollution in poorer regions (*e.g.* air pollution over country borders and polluted basins shared between countries) is a problem.

Environmental public health programmes must assess potential health issues linked to environmental causes, create comprehensive and reasonable policies for the public so as to safeguard everyone from risks from the environment, and guarantee that they are followed in order to achieve these goals. Interprogramme, intersectoral, multisectoral, subnational, national, and supranational methodologies are used to accomplish this. Environmental

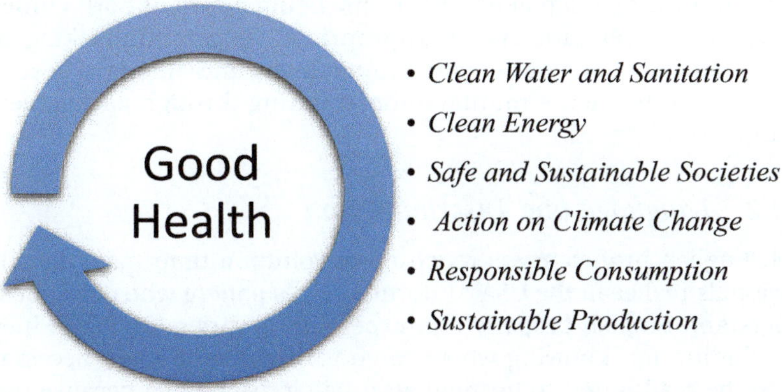

Figure 13.4 Determinants for good health.

public health initiatives are crucial for promoting resilient communities and a health sector that is environmentally conscious and resilient (Figure 13.4).[49]

13.4.4 Culture and Language

Often containing health-promoting activities held within communal norms and beliefs, culture is a dynamic network of shared morals, practices, definitions, and symbolic manifestations.[50] For example, cultural traditions such as dances and songs enhance the happiness and health of the neighbourhood. There are games and physical activities that are culturally specific in communities all over the world that promote health and wellbeing. Unhealthy western lifestyle promotions and the advertising and public relations industries' global hegemony frequently directly endanger human health while also endangering the great range of health procedures practiced around the world. The theory of the free market is used to reduce regulation of these industries and to let fast food and sugary drinks into nation states.[51]

Similar to this, the militarization and privatization of health treatments are caused by the cultural portrayal as being health threatening and backward in the global discursive constructs of infectious disease.[52]

13.4.4.1 Cultural Awareness, Humility, and Sensitivity

Healthcare professionals and educators began collectively to acknowledge in the 1990s that patients come from a variety of cultural backgrounds that may affect how they perceive health and sickness, interact with institutions and providers of healthcare, and respond to treatment recommendations.[53]

Instead of "competence," which incorrectly implies that culture is a talent that can be learned, there are various ways to realize that culture matters in some therapeutic encounters, including cross-cultural caring, cultural humility, and respect.[54] These strategies include the clinician being modest about acknowledging the limitations of their knowledge of a patient's situation, refraining from generalizations, being aware of both clinicians' and patients' biases, and, when appropriate, respectfully asking open-ended questions about patients' circumstances and values. These strategies are meant to ensure mutual understanding through patient-centred communication.[55]

13.4.4.2 Language and Discrimination

Limited English proficiency is much more common than many healthcare professionals realize in the USA. Unfortunately, a patient who does not speak or understand English properly may experience a variety of health inequalities, including not knowing where to go for treatment when necessary or having to be readmitted to hospital after being discharged because they do not understand the aftercare arrangements. Different groups of individuals may interpret discrimination differently. For instance, some patients might not have access to adequate healthcare because of overt bias against them based on their colour, culture, or sexual orientation. In a similar vein, some patients come to distrust doctors as a result of how their race has been perceived in the past. The patient may have a smaller chance of achieving health equity in any setting where discrimination is present or may be present, whether as a result of actual or potential prejudice.[56]

13.5 Intermediary Determinants of Health

The intermediary determinants of healthcare are the material, social, psychological, and environmental parameters that are required to support and improve the wellbeing of an individual, which include essential basic needs such as food, shelter, and clothing, *i.e.* an adequate standard of living and physical, mental, spiritual, and moral development in a healthy body with a healthy mind are real healthcare determinants (Table 13.2).[57]

13.5.1 Psychological Factors

The factors that affect our health have no limit; they start with everything from diet to genetics to different lifestyle factors. One of the most important factors that have an impact on health is primarily the brain, with psychological factors that could impact our digestive system because everything related to our health is controlled by the brain. There is a

Table 13.2 Intermediary determinants of health.

Intermediary determinants of health
Psychological factors
Biological factors
Health behaviour
Others such as awareness and education

complex interconnection between the body and mind for communication through nerves in the nervous system and neurochemical signalling, and that pathway can become dysregulated or dysfunctional so, in essence, it stops working in the way that it is supposed to and we need to help in order to address and improve the functioning of the brain–body connection. One of the factors that have adverse effects on the body is psychological, and its response may stem from environmental or psychological factors, so anyone is susceptible to the effects of these factors on their health. Maybe we all have felt that our stomach "ties up in knots" in an emotional situation or felt "stomach churning" when under stress or in fear, and this shows the connection between psychological factors and our overall wellbeing. Hence in many cases medical intervention may not be sufficient and psychological treatment is needed along with medicines. Therefore, it is necessary to carry out a psychological evaluation of an individual whether they are under medication or not, and only then can actual wellbeing be defined.[58]

Mental wellbeing is a major challenge for communities and keeping it fine-tuned requires huge efforts, but unfortunately mental health is degrading day by day. A serious problem is how to cope with a stressful lifestyle, which affects individual abilities and the ability of the community to grow. The problem with mental health is that it is different for each individual and its intensity is also different. Hence fighting this type of illness is crucial at present.

Our mental health is undermined by different challenges: long working hours, changing structure of society, greed, Internet and mobile phone addiction, high ambition, unrealistic deadlines, and misleading media. Unfavourable socioeconomic and geopolitical circumstances such as poverty, violence, inequality, and deprivation are increasing mental disorder-like conditions. The problem is worldwide and responding to it is becoming very important in the present scenario.

The problem of mental health is a community problem at different scales and involving different age groups. One problem is that if it affects those in their developing years, especially adolescents, it will be a big threat to any growing economy. The use of the Internet (Internet addiction) is a major issue worldwide and a large contributor to mental health disorders. To reduce psychological disorders, it very important to identify structural and social determinants of mental health in different communities.

Building capacity for improving mental health equity and initiating strategies that are actionable and realistic are the needs of the day. Some suggestions for mental wellbeing for individual and communities include the following:

- *Connect with others:* mental wellbeing is directly related to relationships, so one should try to develop good relationships with others and have an open mindset, thinking that the universe banks on diversity.

- *Physically activity:* regular routines of exercise, walking, and fitness activities reduce stress and help to develop confidence, which is important for mental wellbeing.
- *Updating with new things and skills:* this directly engages the mind in creative activity and thus boosts us psychologically, which is very important in a fast-moving lifestyle.
- *Focus on the present moment:* this helps us to understand ourselves and keeps us focused on our present work, which helps us to achieve our goals and creates a sense of happiness, leading to mental wellbeing.
- *Tech-free time:* giving ourselves technology-free time is very important, as we are very much heading towards a virtual world, and this is a common cause of stress, *e.g.* with messages received on social media that are not needed at all and have no direct influence on our life but may be a cause of stress.

13.5.2 Biological Factors

A healthy body is a combination of behaviour, feelings, thought processes, and motivation that define an individual. Heredity and genetic factors have a remarkable influence on the physical and mental health of an individual. The genetic endowment sets the boundaries of the abilities and potential of the individual. The genetic heritage is less modifiable than the other determinants. Some metabolic diseases, chromosomal abnormalities, and intellectual disability have genetic origins. Most mental disorders are found to have responsible biological factors, hence researchers have concentrated on biological causal factors, and most of these diseases or dysfunctions are caused by some kind of problems in the central nervous system, the automatic nervous system, or the endocrine system.

Heart-related diseases, diabetes, and cancers are very prevalent in the community nowadays. Poor living conditions, food habits, and working conditions, obesity, and lack of awareness about health are leading societies towards health problems. A major issue is the increase in diabetes among individuals of younger age, which is a great threat to social immunity as diabetes causes many other problems, including lower cognitive levels. Hormonal and metabolic responses to a new lifestyle also cause many types of problems in the community and strategies to improve these conditions are required globally or else they may lead to chronic diseases among the community. Studies are under way to identify the impact of a changing environment on our genes. The COVID-19 pandemic had a great impact on our overall health and many new problems arose after the treatment of the population who contracted the disease.

Genetic disorders will be the next major challenge in the near future. These problems may be due either to certain biological factors or to some psychological factors. Health workers, psychologists, and medical experts have the responsibility to provide genetic counselling to people who are risk of having impaired children.[59]

13.5.3 Health Behaviour

Health is the mirror of a person's lifestyle. Health behaviour is an individual-level effort to prevent disease, prolong life, and promote the dimensions of health. Theories of health behaviour, aimed at both the individual level and the societal level, are useful for providing an understanding of health education. Behaviour includes lifestyle, which is a person's way of living, *i.e.* habits, attitude, social values, eating patterns, rest, sleep, and exercise, and understanding and making adjustments under different circumstances are learnt through culture, family, friends, and peer groups. Major health issues include heart disease, cancer, drug addiction, smoking, and obesity. Health has behavioural and psychological dimensions, and we constantly acquire new knowledge about these aspects, giving us more and different ways of looking at health. Older people have to face greater challenges related to health issues along with the impact of environmental factors, hence the choices we make influence our own health and that of those around us. Therefore, it is necessary to understand the health interventions that each and every one needs to have, policies are required to be framed, and there is a need to create awareness and practices to be followed. Hence there is a need to interconnect health behaviour with health promotion, which is required to be connected with health education, and health education is passed through health screening, which is interconnected with disease prevention and is closed through the health pyramid.[60]

13.5.4 Others Such as Awareness and Education

An important question that has been raised is whether years of education can affect how long and how well one will live. It has been reported that Americans with fewer years of education have poorer health and shorter lives, and that is still true today. In fact, since the 1990s, life expectancy has decreased for people without a high school education. Education is important not only for higher paying jobs and economic productivity, but also for saving lives and saving money. Awareness and education have very strong relationships with wellbeing and a healthy body and mind. Mental health and behavioural problems such as depression, anxiety, drug abuse, *etc.*, are reported to be primary drivers of disability worldwide. Nearly half of those diagnosed with depression are also diagnosed with an anxiety disorder. Unlike the common misconception that stroke occurs only in the elderly, the truth is that stroke can affect anyone, including teenagers, at any time. What can we do about this? Sometimes just admitting that a situation is stressful and being prepared to deal with it can reduce anxiety. The following advice has been put forward. Get treatment from a healthcare professional or therapist. Live longer and prevent chronic diseases, have better cardiorespiratory and muscular fitness, maintain a healthier brain function, increase bone density, achieve better quality of sleep, and experience a better overall quality of life. Have healthy relationships, make good choices, maintain physical health and wellbeing, handle the natural ups and downs of life, and discover and grow towards your potential.[61]

13.6 Lessons Learned and Global Health Response

The world continues to grow smaller, with international travel, a global economy, and a changing environment contributing to the emergence of new diseases and the spread of existing ones. Understanding these connections and how they impact the furthest reaches of the globe is becoming an essential skill in international development, humanitarian assistance, business and commerce, and at all levels of healthcare. The Ebola outbreak offers a window through which global health can be examined in a variety of disciplines. A Toronto medical centre and a Dallas hospital are far removed from the outbreaks of SARS in China and Ebola in Guinea, yet those diseases arrived and created crises within those facilities and beyond. Research has shown that health security contributes to civil order, economic growth, and stable governments. It has also been demonstrated that every nation bears the economic and human burden of disease, illness, and injury. When COVID-19 hit, the world was not ready. This pandemic has revealed gaps in how we think about, implement, and measure epidemic preparedness. This has reinforced the need to work together on the reality with which we are all connected.

To address the inequalities in access to health-related rights, the goals of next-generation health research are to challenge societal health inequities and encourage health advocacy. A close examination of the assumed assumptions and values in these mainstream health communication frameworks/structures can help to facilitate theorising about health communication that addresses the disparities in health outcomes that we see around the world and examines claims about health and wellbeing embedded within local cultural contexts, lived experiences of communities, and community desires for health. Important aspects are lessons from past outbreaks, successful preparedness, and the need to invest in and prioritise funding to fight the next pandemic.[62]

13.7 Conclusion and Future Prospects

Pandemics mostly raise social issues, rather than medical issues. The poor and those who reside in overcrowded areas with poor maintenance and sanitation have borne the brunt of the COVID-19 pandemic. More assistance, such as financial support from government or social sectors, must be made available to low-income families. In the midst of a pandemic, this is of paramount importance because this demographic is particularly susceptible to infection. Being part of the vulnerable population at risk during outbreaks is mostly attributable to socioeconomic status and racial/ethnic group membership. Poor nutrition and eating habits have been linked to having a lower income.

Individual and structural discrimination puts people from minority groups such as Latinx and Black Americans at a disadvantage, making them more susceptible to worse health outcomes than their Caucasian counterparts. Because of the COVID-19 pandemic, it is clear that the SDOH have been neglected. According to a scholarly article, Richard Clarke Cabot in the USA was the first physician in the country to take into account patients' socioeconomic status, family dynamics, and mental health when providing medical care.[63] Low socioeconomic position was associated with increased mortality, as he had seen in his medical practice. Poor living conditions, unequal distribution of resources, and poor public health have all been demonstrated to contribute to the development of infectious diseases, according to research conducted over time. Disparities in morbidity and mortality may be exacerbated by health and healthcare inequalities. Existing studies of influenza pandemics, according to Quinn *et al.*, have not accounted for health inequalities or tried to investigate differences in socioeconomic determinants and their effect on health during times of health emergency.[64] Hence, in a medical emergency, it is crucial to act quickly and efficiently. To accomplish this goal, it is essential to learn about potential health and healthcare aspects before a disease epidemic starts. Those who are economically disadvantaged can benefit from increased access to medical and non-medical services if they have a better understanding of the issues that affect health and healthcare, such as SDOH. It is also crucial to educate the public and raise awareness of the seriousness of the illness. Individuals should be motivated to limit their exposure to others in order to halt the spread of infectious diseases through raising awareness of the disadvantaged community, which is more vulnerable than the typical individual, and the speed with which these diseases can spread. In order to combat an outbreak effectively, it is crucial to consider all of the variables that contribute to a country's prosperity. Effective responses to epidemics such as COVID-19 can include providing timely and relevant healthcare, social services, and educational opportunities for the affected populations.

Reductions in global illness may be attainable with the incorporation of SDOH into existing initiatives to end health and healthcare inequities. Healthcare providers, public health experts, anthropologists, sociologists, researchers, governments, the National Institutes of Health (NIH), the Centers for Disease Control and Prevention (CDC), the World Health Organization (WHO), and others can form an interdisciplinary team to understand better the causes of health disparities in vulnerable populations. In order to be better prepared for future pandemics, it is essential to evaluate the successes and failures of the responses that were implemented this time around. Each individual should have the same access to healthcare regardless of their location, race, or socioeconomic status. Disparities in social and economic outcomes can be mitigated by the use of SDOH in disease prevention and a more interdisciplinary approach to patient care.

Abbreviations

AMA	American medical association
CDC	Centers for disease control and prevention
COPD	Chronic obstructive pulmonary disease
CVD	Cardiovascular disease
IDH	Intermediary determinants of health
NIH	National institutes of health
SDH	Structural determinants of health
SDOH	Social determinants of health
SES	Socioeconomic status
TB	Tuberculosis
WHO	World health organization

Acknowledgements

Some say that research and writing are a lone profession, but this is not true. An amazing group of people have come together to make this chapter possible. We feel delight and pleasure in expressing our gratitude to every person who enabled this accomplishment with inspired dialogues. We especially express our earnest gratitude to Dr Sanjay Guha, Professor, Department of Management Studies, Bhilai Institute of Technology, for his guidance and help in initiating our research work. The beneficial viewpoints, insightfulness, and valuable discussions at various stages during this journey of research are gratefully acknowledged. We are greatly obliged to Dr Arun Arora, Principal, Bhilai Institute of Technology, for the research and administrative acumen that enabled this chapter to be completed. The awareness and vision of many people have provided lifetime benefits to our research through conferences and seminars that we attended during this endeavour.

References

1. K. Lutfey and J. Freese, Toward some fundamentals of fundamental causality: Socioeconomic status and health in the routine clinic visit for diabetes, *Am. J. Sociol.*, 2005, **110**, 1326–1372.
2. G. D. Smith, M. Bartley and D. Blane, The Black report on socioeconomic inequalities in health 10 years on, *Br. Med. J.*, 1990, **301**, 373–377.
3. S. Harper and J. Lynch, Trends in socioeconomic inequalities in adult health behaviors among US States, 1990–2004, *Public Health Rep.*, 2007, **122**, 177–189.
4. J. P. Mackenbach, S. Irina, A. J. R. Roskam, M. M. Schaap, M. Gwenn, L. Mall and A. E. Kunst, Socioeconomic inequalities in health in 22 European countries, *N. Engl. J. Med.*, 2008, **358**, 2468–2481.
5. E. Jon Ivar and K. Steinar, Social causation, health-selective mobility, and the reproduction of socioeconomic health inequalities over time: Panel study of adult men, *Soc. Sci. Med.*, 2003, **57**, 1475–1489.

6. E. Dahl, Social mobility and health: Cause or effect?, *Br. Med. J.*, 1996, **313**, 435–436.
7. P. West, Rethinking the health selection explanation for health inequalities, *Soc. Sci. Med.*, 1991, **32**, 373–384.
8. D. Simandan, Rethinking the health consequences of social class and social mobility, *Soc. Sci. Med.*, 2018, **200**, 258–261.
9. G. E. Fraser and D. J. Shavlik, Ten years of life: Is it a matter of choice?, *Arch. Intern. Med.*, 2001, **161**, 1645–1652.
10. W. C. Cockerham, Health lifestyle theory and the convergence of agency and structure, *J. Health Soc. Behav.*, 2005, **46**, 51–67.
11. N. P. Pronk, L. H. Anderson, A. L. Crain, B. C. Martinson, P. J. O'Connor, N. E. Sherwood and R. R. Whitebird, Meeting recommendations for multiple healthy lifestyle factors: Prevalence, clustering, and predictors among adolescent, adult, and senior health plan members, *Am. J. Prev. Med.*, 2004, **27**, 25–33.
12. E. Morawa and Y. Erim, Health-related lifestyle behavior and religiosity among first-generation immigrants of polish origin in Germany, *Int. J. Environ. Res. Public Health*, 2018, **15**, 2545.
13. S. N. Walker, K. R. Sechrist and N. J. Pender, The health-promoting lifestyle profile: Development and psychometric characteristics, *Nurs. Res.*, 1987, **36**, 76–81.
14. A. H. Mokdad, J. S. Marks, D. F. Stroup and J. L. Gerberding, Actual causes of death in the United States, 2000, *JAMA*, 2004, **291**, 1238–1245.
15. G. Ponthiere, Mortality, family and lifestyles, *J. Fam. Econ. Issue*, 2011, **32**, 175–190.
16. J. E. Rohrer, J. R. Pierce Jr and C. Blackburn, Lifestyle and mental health, *Prev. Med.*, 2005, **40**, 438–443.
17. D. A. Lawlor and S. W. Hopker, The effectiveness of exercise as an intervention in the management of depression: Systematic review and meta-regression analysis of randomised controlled trials, *Br. Med. J.*, 2001, **322**, 763–767.
18. K. B. Prendergast, G. M. Schofield and L. M. Mackay, Associations between lifestyle behaviours and optimal wellbeing in a diverse sample of New Zealand adults, *BMC Public Health*, 2015, **16**, 62.
19. W. C. Cockerham, T. Abel and G. Lüschen, Max Weber, formal rationality, and health lifestyles, *Sociol. Q.*, 1993, **34**, 413–428.
20. W. C. Cockerham, Health lifestyles: Bringing structure back, in *The New Blackwell Companion to Medical Sociology*, ed. W. C. Cockerham, Blackwell, Oxford, UK, 2010, pp. 159–183, ISBN 9781405188685.
21. P. Contoyannis and A. M. Jones, Socio-economic status, health and lifestyle, *J. Health Econ.*, 2004, **23**, 965–995.
22. K. L. Frohlich, E. Corin and L. Potvin, A theoretical proposal for the relationship between context and disease, *Sociol. Health Illness*, 2001, **23**, 776–797.
23. USDHHS, Social Determinants of Health, Washington, DC, 2020.

24. AMA, Social Determinants of Health, Health Systems Science Learning Series, Chicago, IL, 2020.
25. CDC, Coronavirus Disease 2019 (COVID-19): People Who Are at Higher Risk for Severe Illness, Atlanta, GA, 2020.
26. M. H. Algren, O. Ekholm, L. Nielsen, A. K. Ersboll, C. K. Bak and P. T. Andersen, Associations between perceived stress, socioeconomic status, and health-risk behaviour in deprived neighbourhoods in Denmark: a cross-sectional study, *BMC Public Health*, 2018, **18**, 250.
27. E. C. Lowcock, L. C. Rosella, J. Foisy, A. McGeer and N. Crowcroft, The social determinants of health and pandemic H1N1 2009 influenza severity, *Am. J. Public Health*, 2012, **102**, e51-8.
28. A. R. Chapman, The social determinants of health, health equity, and human rights, *Health Hum. Rights*, 2010, **12**, 17–30.
29. K. Watson, Why schools in rich areas get more funding than poor areas, Global Citizen, 2016. Available online at: https://www.globalcitizen.org/en/content/cost-of-education-in-us/.
30. M. van Ham, L. Hedman, D. Manley, R. Coulter and J. Osth, Intergenerational transmission of neighbourhood poverty: an analysis of neighbourhood histories of individuals, *Trans. Inst. Br. Geogr.*, 2014, **39**, 402–417.
31. Connecting the dots between social determinants and infectious diseases By Bara Vaida | September 27, 2018, Association of Health Care Journalists Center for Excellence in Health Care Journalism.
32. J. C. Semenza, Strategies to intervene on social determinants of infectious diseases, *Eurosurveillance*, 2010, **15**(27), 32–39.
33. G. T. Keusch, O. Fontaine, A. Bhargava *et al.*, Diarrheal diseases, in *Disease Control Priorities in Developing Countries*, ed. D. T. Jamison, J. G. Breman, A. R. Measham, *et al.*, World Bank, Washington, DC, USA, 2nd edn, 2006, ch. 19.
34. J. R. Hargreaves, D. Boccia, C. A. Evans, M. Adato, M. Petticrew and J. D. H. Porter, The social determinants of tuberculosis: from evidence to action, *Am. J. Public Health*, 2011, **101**(4), 654–662.
35. G. Bishwajit, S. Ide and S. Ghosh, Social Determinants of Infectious Diseases in South Asia, *Int. Scholarly Res. Not.*, 2014, 135243.
36. N. Rollins, Food insecurity—a risk factor for HIV infection, *PLoS Med.*, 2007, **4**(10), e301.
37. R. Frega, F. Duffy, R. Rawat and N. Grede, Food insecurity in the context of HIV/AIDS: a framework for a new era of programming, *Food Nutr. Bull. Suppl.*, 2010, **31**(4), S292–S312.
38. R. Beaglehole, International trends in coronary heart disease, mortality, morbidity, and risk factors, *Epidemiol. Rev.*, 1990, **12**, 1–15.
39. A. B. Frakt, *JAMA, J. Am. Med. Assoc.*, 2018, **319**(12), 1187–1188.
40. S. H. Gordon and B. D. Sommers, Recessions, poverty, and mortality in the United States: 1993–2012, *Am. J. Health Econ.*, 2016, **2**(4), 489–510.
41. V. Maheshri and C. Winston, Did the Great Recession keep bad drivers off the road?, *J. Risk. Uncertainty*, 2016, **52**, 255–280.

42. D. L. Miller, M. Page, A. H. Stevens and M. Filipski, Why Are Recessions Good for your Health? Understanding Pro-cyclical Mortality, 2009.
43. S. Sharma, J. Gomes, S. Sarkar, A. Chakraborty and S. Kumar, Analyzing the effective healthcare management model for a COVID-19 pandemic: A suggestive study, *Int. J. Health Sci.*, 2022, **6**(S3), 6245–6266.
44. T. Wiß and B. Greve, A comparison of the interplay of public and occupational work–family policies in Austria, Denmark, Italy and the United Kingdom, *J. Comp. Policy Anal. Res. Pract.*, 2020, **22**(5), 440–457.
45. Structural Competency Working Group, 2016, https://samuelmerritt.libguides.com/structuraldeterminants#s-lg-box-19216823.
46. M. Momplaisir, (n.d.), Structural Determinants of Health. https://scienceandsociety.columbia.edu/courses/bc3202-structural-determinants-health-Momplaisir.
47. A. J. McMichael, J. W. Powles, C. D. Butler and R. Uauy, Food, livestock production, energy, climate change, and health, *Lancet*, 2007, **370**, 1253–1263.
48. R. J. Brulle and D. N. Pellow, Environmental justice: human health and environmental inequalities, *Annu. Rev. Public Health*, 2006, **27**, 103–124.
49. Sustainable Development Goal 3: E. h.-b. (n.d.). The 2030 Agenda for Sustainable Development. Article. (n.d.). https://www.paho.org/en/topics/environmental-determinants-health.
50. B. J. Good, *Medicine, Rationality and Experience: an Anthropological Perspective*, Cambridge University Press, New York, NY, 1993.
51. C. O. Airhihenbuwa, *Health and Culture: Beyond the Western Paradigm*, SAGE, New York, 1995.
52. M. J. Dutta, *Communicating Social Change: A Culture-Centered Approach*, Polity, London, 2008.
53. A. D. Napier, C. Ancarno, B. Butler, J. Calabrese, A. Chater and H. Chatterjee, Culture and health, *Lancet*, 2014, **384**, 1607–1639.
54. M. J. Dutta, *Neoliberal Health Organizing: Communication, Meaning, and Politics*, New York, NY, Left Coast Press, 2015.
55. V. Navarro, *Neoliberalism, Globalization, and Inequalities: Consequences for Health and Quality of Life*, Amityville, NY, Baywood Pub, 2007.
56. S. Sastry and M. J. Dutta, Public health, global surveillance, and the "emerging disease" worldview: a postcolonial appraisal of PEPFAR, *Health Commun.*, 2012, 519–532.
57. S. Saha and M. C. Beach, Cooper LA. Patient centeredness, cultural competence and healthcare quality, *J. Natl. Med. Assoc.*, 2008, **2008**, 1275–1285.
58. M. Tervalon and J. Murray-Garcia, Cultural humility *versus* cultural competence: a critical distinction in defining physician training outcomes in multicultural education, *J. Health Care Poor Underserved*, 1998, **1998**, 117–125.
59. J. E. Carrillo, A. R. Green and J. R. Betancourt, Cross-cultural primary care: a patient-based approach, *Ann. Intern. Med.*, 1999, **1999**, 829–834.

60. Health N. I., (2017) Cultural respect, Bethesda (MD): NIH, 2017. Available at: https://www.nih.gov/institutes-nih/nih-office-director/office-communications-public-liaison/clear-communication/cultural-respect. Retrieved September 12, 2017.
61. G. S. Das, Narrative humility, *Lancet*, 2008, **2008**, 980–981.
62. Posted by GLOBO, Language & Social Determinants of Health: Impact on Health Equity, 2022, https://www.helloglobo.com/blog/language-social-determinants-of-health-impact-on-health-equity.
63. S. Singu, A. Acharya, K. Challagundla and S. N. Byrareddy, Impact of social determinants of health on the emerging COVID-19 pandemic in the United States, *Front. Public Health*, 2020, **8**, 406.
64. S. Kumar and S. C. Quinn, Existing health inequalities in India: informing preparedness planning for an influenza pandemic, *Health Policy Planning*, 2012, **27**(6), 516–526.

CHAPTER 14

Sustainable Nanomaterials for Mosquito-based Infectious Disease Control

GANESH GOLLAVELLI,*[a,b] GANGARAJU GEDDA[c] AND YONG-CHIEN LING[d]

[a] Department of Chemistry, Aurora Deemed to be University, Hyderabad, 500098, India; [b] Jawaharlal Nehru Technological University Kakinada, Kakinada, Andhra Pradesh 533003, India; [c] Department of Chemistry, Presidency University, Bangalore, Karnataka 560064, India; [d] Department of Chemistry, National Tsing Hua University, Hsinchu 30013, Taiwan
*Email: ganesh@aurora.edu.in

14.1 Introduction

Dengue, malaria, and cholera are infectious diseases spread by mosquitoes that pose a serious risk of illness to billions of people worldwide.[1–6] Since there is no specific treatment for some of these diseases, eliminating their related vectors is the only strategy to reduce the risk of infection. For the control of these diseases, nanotechnology-based therapeutic agents with nanoformulations are currently gaining in popularity.[7–11] Potential uses for nanoparticles (NPs) produced by an environmentally benign method from plant extracts have been suggested as an mosquitocides.[12–19]

Nanomosquitocides have a great advantage over other mosquito-inactivation strategies such as mosquito repellents, synthetic microbial and plant-borne mosquitocides, biological control agents, sterile insect technique (SIT), "boosted SIT", symbiont-based methods, and transgenic

Nanoscience & Nanotechnology Series No. 62
Smart Nanomaterials for Infectious Diseases
Edited by Suvardhan Kanchi, Nidhi Sharotri, Rajasekhar Chokkareddy, Deepali Sharma and Falah H. Hussein

Published by the Royal Society of Chemistry, www.rsc.org

mosquitoes.[20–27] Nanomedicine technology is an emerging approach that offers many solutions to the problems that exist in biology and medicine owing to the size, shape, high surface area, tuneable functionality, and biocompatibility or toxicity against pathogens and insects of nanomaterials (NMs).[28–38] Effectiveness of NMs against mosquitoes could be achieved by functionalizing them with selective surface functional groups.[39–42] Mostly the surface functional groups originating from plant extracts have shown remarkable mosquitocidal properties.[7,16,43–47] Plant-derived NMs have a great advantage over those obtained using other methods of fabrication owing to their ease of preparation, such as low temperature and pressure, single step, green attributes and availability at low cost.[48–51] An increasing number of plant-derived compounds have been reported for efficient and low concentration-dependent mosquitocides (Table 14.1). The majority of the studies were focused on mosquito larvicidal, pupicidal, and adulticidal toxicity, and extremely low LC_{50} values were calculated. Most of the mosquitocides are effective at 1–30 $mg\,L^{-1}$ owing to their small size and spherical shape. It has been assumed that the biotoxicity of the reported NMs against mosquitoes is due to exoskeleton penetration and the internalized NMs can bind to sulfur in proteins and DNA phosphates, finally leading to organelle damage and cellular death.[52] In order to elucidate the parasite action of *Plasmodium* parasites in humans as a malaria infection agent, it is very important to understand its life cycle[53] (Figure 14.1), and thereby we can effectively design the mosquitocides rationally.

In 2015, the Nobel Prize in Physiology or Medicine was awarded to Youyou Tu for the discovery of artemisinin, which was the first vaccine against *Plasmodium falciparum* malaria [*i.e.* RTS,S/AS01 (RTS,S)].[54,55] However, considerable work on malaria prevention and control is necessary as some parasites such as *Plasmodium vivax* are resistant to artemisin.[56] Hence there is a need to develop NM-based alternative mosquitocides *via* green synthetic methods. This chapter describes the preparation and characterization of Ag, Au, ZnO, TiO_2 and other NMs using various parts of plant extracts and their mechanism of killing mosquitoes.

14.2 Silver and Gold Nanoparticles as Nanomosquitocides

Silver, especially in nanoform, is a well-known microbial agent owing to its strong binding to proteins and genetic materials of cells.[57–63] Many workers have reported plant-derived AgNMs (Figure 14.2) for mosquito inactivation.[64] The killing action of AgNPs against microbes is depicted in Figure 14.3.

Govindarajan *et al.* fabricated simple polydisperse and stable AgNPs from leaf extracts of *Ichnocarpus frutescens* (Apocyanaceae).[65] The prepared AgNPs were tested against *Anopheles subpictus*, a larva of the malaria vector, *Aedes albopictus*, the dengue vector, and *Culex tritaeniorhynchus*, the Japanese

Table 14.1 Green nanomosquitocides prepared from various bioresources.

Bioresource used	Nanomaterial fabricated	Size of the nanomaterial (nm)	Effective against	LC_{50}	Ref.
Leaf extract of *Ichnocarpus frutescens* (Apocyanaceae)	AgNPs	1–5	*Anopheles subpictus* *Anopheles albopictus* *Culex tritaeniorhynchus*	14.22 μg mL^{-1} 15.84 μg mL^{-1} 17.26 μg mL^{-1}	65
Extract of *Hugonia mystax*	AgNPs	10–75	*Anopheles stephensi* *Aedes aegypti* *Culex quinquefasciatus*	14.45 μg mL^{-1} 15.86 μg mL^{-1} 17.46 μg mL^{-1}	66
Leaf extract *of Trichodesma indicum*	AgNPs	20–50	*Mythimna separata*	500 ppm (500 mg L^{-1})	67
Leaf extract of *Leucas aspera* and *Hyptis suaveolens*	Aspara-AgNPs Suaveolens-AgNPs	7–22 and 5–25	*Aedes aegypti* *Anopheles stephensi* *Culex quinquefasciatus*	4.02, 4.69, 5.06 mg L^{-1} and 4.63, 4.04, 3.52 mg L^{-1}	68
Dried root extract of *Salvia miltiorrhiza* and light	AgNMs/AgNPs	White light 18.5, green light 28.02, red light 50.22, blue light 16.26, and sunlight 10.12	*Aedes aegypti* vector	100 mg mL^{-1}	69

Table 14.1 (*Continued*)

Bioresource used	Nanomaterial fabricated	Size of the nanomaterial (nm)	Effective against	LC_{50}	Ref.
Leaf extract of *Nilgirianthus ciliatus*	AgNPs	100	*Aedes aegypti* larvae	0.49 mg L^{-1} (I instar) 1.25 mg L^{-1} (II instar) 1.67 mg L^{-1} (III instar) 4.95 mg L^{-1} (IV instar)	70
Essential oil of *Curcuma zedoaria* (ZEO)	ZEO-AgNPs	—	*Culex quinquefasciatus*	Susceptible: 0.57 ppm Resistant strain: 0.64 ppm	71
Aqueous stem extract of *Achyranthes aspera* (AASE)	Silver nanocomposites (AgNCs)	1–30	*Aedes aegypti*	1 μg mL^{-1} (3 mM AASE–$AgNO_3$ at 48 h)	72
Citrate-capped	AuNPs	15–30	*Blattella germanica*	65.58 mg L^{-1}	73
Citrate-capped	AuNPs	15	*Drosophila melanogaster*	1.9, 3.8, 19, 38, 190, and 380 pmol L^{-1}	74
Leaf powder aqueous extract of *Scadoxus multiflorus*	ZnONPs	2–31	*Aedes aegypti*	34.04 ppm	76
Murraya koenigii berries	*Mk*-ZnONPs	10–15	*Culex quinquefasciatus*	2.1 μg mL^{-1}	77
Zingiber offcinale rhizome extract	ZnONPs	~103 nm	*Spodoptera litura* and *Macrosiphum euphorbiae*	—	78
Thiamethoxam-impregnated castor leaves	ZnONPs	~34	4th instar larvae of *Spodoptera litura*	—	79

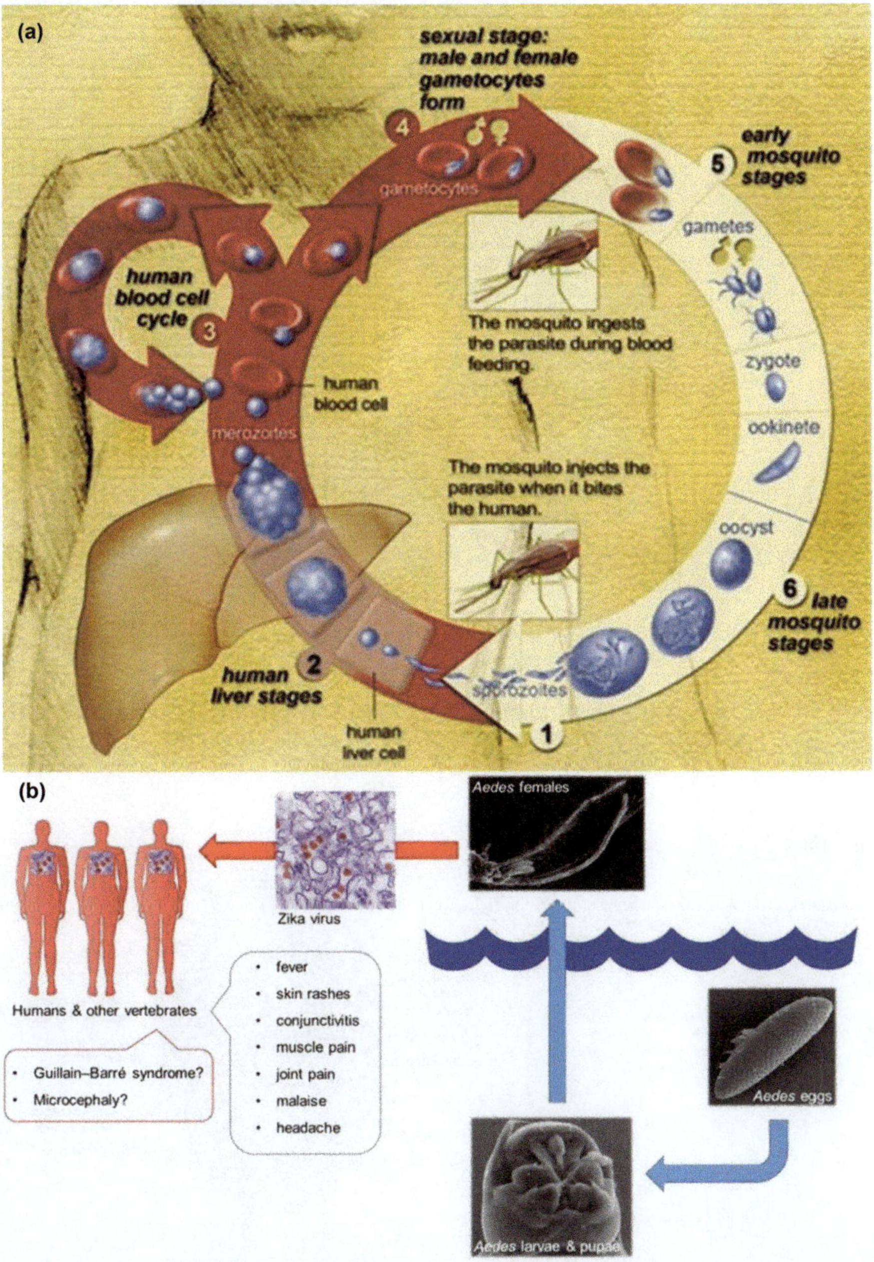

Figure 14.1 (a) *Plasmodium* parasites: life cycle and parasite development inside humans as host, more precisely inside the liver hepatocytes and red blood cells. (b) Zika virus infection in the human body through *Aedes* mosquito bites. Reproduced from ref. 53 with permission from Elsevier, Copyright 2017.

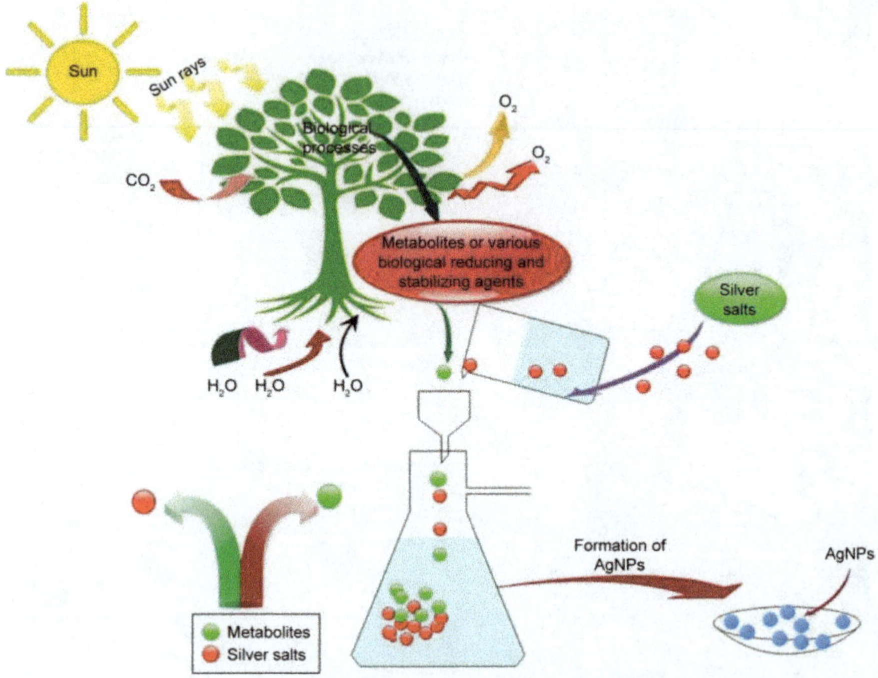

Figure 14.2 Green synthesis of plant-derived AgNPs. Reproduced from ref. 64 with permission from Dove Medical Press Ltd, Copyright 2018.

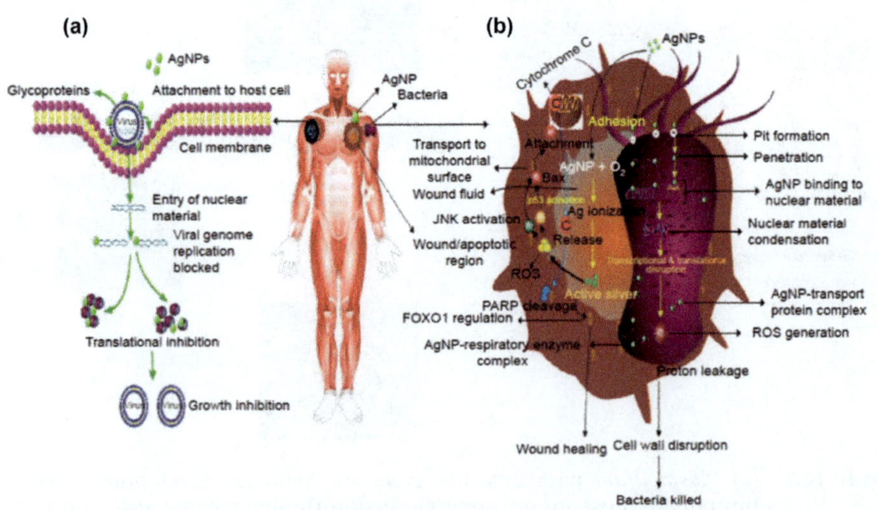

Figure 14.3 AgNPs in a biological system and action of killing mechanism. (a) Antiviral mechanism of AgNPs. (b) Antibacterial and wound-healing mechanism of AgNPs. Reproduced from ref. 64 with permission from Dove Medical Press Ltd, Copyright 2018.

encephalitis vector. The AgNPs showed higher toxicity than aqueous leaf extracts towards these three species and the respective LC_{50} values were 14.22, 15.84, and 17.26 $\mu g\,mL^{-1}$. The results prove that AgNPs are safe to non-target mosquito predators with LC_{50} values of 636.61–2098.61 $\mu g\,mL^{-1}$. This study is a good demonstration of the preparation of ultra-small AgNPs with a size of 1–5 nm using *I. frutescens* as a potential mosquitocide. The same group also suggested an alternative green fabrication of AgNPs from an extract of *Hugonia mystax*.[66] The mosquitocide activity of the AgNPs was tested against *Anopheles stephensi*, *Aedes aegypti*, and *Culex quinquefasciatus* larvae and the LC_{50} values were reported to be 14.45, 15.86, and 17.46 $\mu g\,mL^{-1}$, respectively, and they were non-toxic to other aquatic biocontrol organisms with LC_{50} values of 2567.15–829.63 $\mu g\,mL^{-1}$. The reported AgNPs had comparable toxicities to those reported previously, even though the size was larger (10–75 nm). This may be due to the changes in the plant extract and mosquito vectors.

Buhroo *et al.* explored the photosynthesis of AgNPs from *Trichodesma indicum* aqueous leaf extract at room temperature.[67] Its larvicidal capability towards *Mythimna separata* was tested and the reported LC_{50} value was 500 ppm (500 $mg\,L^{-1}$). The reported toxicity was lower than that in previous reports as there was a change in the different leaf extracts and larvae along with their size (20–50 nm). Elumalai *et al.* synthesized AgNPs using *Leucas aspera* and *Hyptis suaveolens* leaf extracts with sizes of 7–22 and 5–25 nm, respectively.[68] A larvicidal assay revealed that the AgNPs had 100% killing efficacy at 10 $mg\,mL^{-1}$. The LC_{50} values of the AgNPs prepared from *L. aspera* and *H. suaveolens* against three larvae (*A. aegypti*, *A. stephensi*, and *C. quinquefasciatus*) were 4.02, 4.69, 5.06 $mg\,L^{-1}$ and 4.63, 4.04, 3.52 $mg\,L^{-1}$, respectively. Even though the authors succeeded in preparing relatively smaller sized AgNPs than previously, the biocidal activity was not satisfactory.

Lee *et al.* carried out an interesting study on the fabrication of AgNPs from the dried roots of *Salvia miltiorrhiza* in the presence of sunlight and various LED lights[69] (Figure 14.4). The various preparation conditions, such as pH, light, extraction, and metal ion concentration, were optimized. High-resolution transmission electron microscopy (HR-TEM) revealed that the AgNMs were spherical, triangular, and oval (Figure 14.5). The average sizes of the AgNMs varied with the light used: white light (18.5 nm), green light (28.02 nm), red light (50.22 nm), blue light (16.26 nm), and sunlight (10.12 nm). The AgNMs prepared with exposure to sunlight had high toxicity towards the *A. aegypti* vector at 48 h of exposure with an LC_{50} of ~100 $mg\,mL^{-1}$. This study simply demonstrated another approach for the preparation of AgNM-based nanocides.

Meenambigai *et al.* prepared AgNPs using *Nilgirianthus ciliatus* leaf extract, which acted as a reducing and capping agent for AgNP formation.[70] The designed AgNPs were used to test the larvicidal activity against colonies of F3 mosquito larvae in the laboratory. The LC_{50} values of AgNPs against immature *A. aegypti* at different developmental stages were 0.49 (I instar), 1.25 (II instar), 1.67 (III instar), and 4.95 $mg\,L^{-1}$ (IV instar). In addition, the study also demonstrated inactivation against *Staphylococcus aureus*.

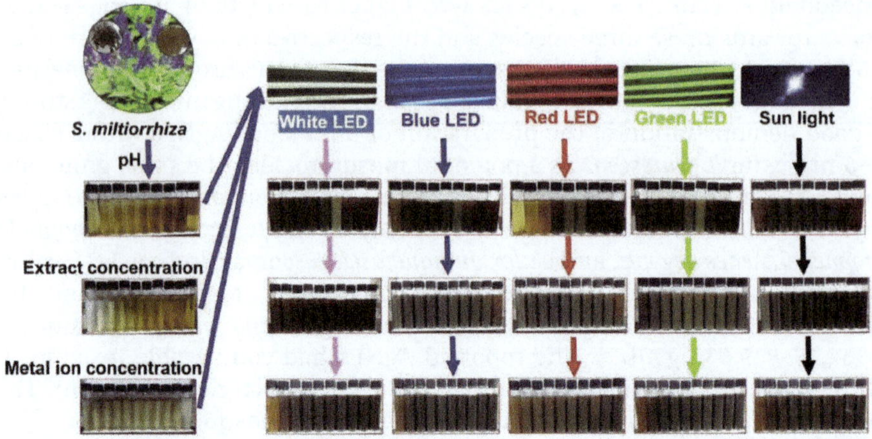

Figure 14.4 Fabrication of *Salvia miltiorrhiza*-synthesized AgNPs under sunlight and LED light. Effects of pH, reducing extract, and Ag ion concentration. Reproduced from ref. 69 with permission from Elsevier, Copyright 2018.

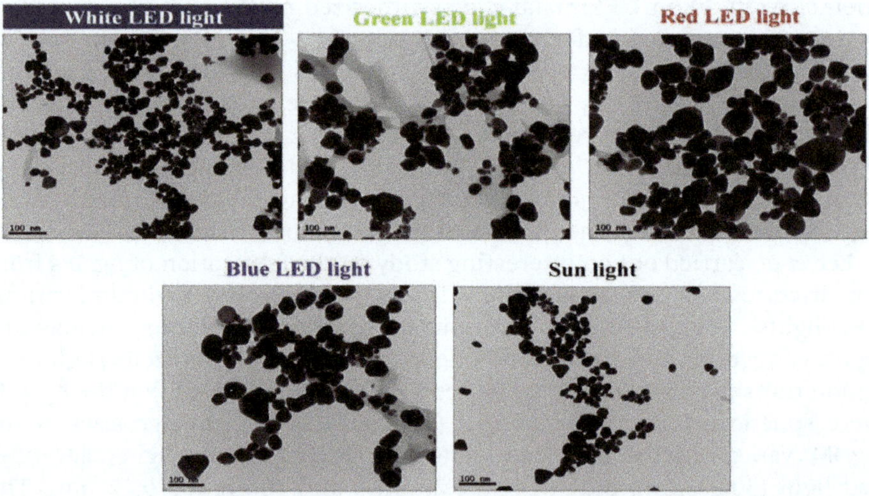

Figure 14.5 HR-TEM images of AgNPs prepared from aqueous suspensions of *Salvia miltiorrhiza* in the presence of sunlight and LED light. Reproduced from ref. 69 with permission from Elsevier, Copyright 2018.

Sutthanont *et al.* fabricated AgNPs based on *Curcuma zedoaria* essential oil (ZEO) to explore their larvicidal activity against both insecticide-susceptible and – resistant strains of *C. quinquefasciatus*.[71] The ZEO-AgNPs showed much better toxicity than ZEO biocidal oil alone towards both strains of *C. quinquefasciatus*. The LC_{50} values of ZEO-AgNPs were 0.57 and 0.64 ppm for susceptible and resistant strains, respectively, and caused maximum larval deaths within 24 h.

Sharma *et al.* reported silver nanocomposites (AgNCs) using an *Achyranthes aspera* (AASE) aqueous stem extract for mosquito control and they were safe towards non-target organisms with an average size of 1–30 nm.[72] The mosquitocidal activity of AgNCs against early fourth instars of *A. aegypti* revealed a notable toxicity that increased with $AgNO_3$ concentration. Time-dependent bioassays at 24, 48, and 72 h for different concentrations of $AgNO_3$ showed LC_{50} values of 26.693, 1.113, and 0.610 $\mu g\,mL^{-1}$, 9.119, 0.420, and 0.407 $\mu g\,mL^{-1}$, and 4.283, 0.3, and 0.248 $\mu g\,mL^{-1}$ for 3, 4 and 5 mM AASE–$AgNO_3$, respectively. These AgNCs showed better mosquitocidal activity than in previous literature reports. Moreover, this method of fabrication was claimed to be simple, economical, and green for mosquito control.

In addition to AgNPs and AgNCs, gold NMs (AuNPs, 15–30 nm) have also been demonstrated to be nanolarvicides against *Blattella germanica* at 65.58 $mg\,L^{-1}$.[73] Citrate-capped AuNPs with a size of 15 nm were also effective against *Drosophila melanogaster* at concentrations of 1.9, 3.8, 19, 38, 190, and 380 $pmol\,L^{-1}$.[74] The many potential applications of AuNPs were reviewed and illustrated (Figure 14.6) by Benelli.[75]

Figure 14.6 Biological applications of AuNPs towards various microbes and insects. The red dots symbolize toxicity to parasite and insect vectors. The development of novel detection tools is indicated by blue dots against parasites and a green dot against adjuvants for the induction of transmission-blocking immunity in malaria vaccine research. Reproduced from ref. 75 with permission from Elsevier, Copyright 2018.

14.3 Metal Oxide (ZnO and TiO_2) Nanomaterials

Metal oxide NMs are known to exert potential biocidal activity due to their small size, and by generating reactive oxygen species (ROS) that harm the cellular system (Figure 14.7). Al-Dhabi *et al.* produced 2–31 nm sized ZnONPs from a leaf powder aqueous extract of *Scadoxus multiflorus.*[76] The ZnONPs showed toxicity against the larvae and eggs of *A. aegypti*. The LC_{50} value obtained was 34.04 ppm. A higher mortality of about 96.4% at 120 ppm was observed for ovicidal activity with an LC_{50} of 32.73 ppm. In addition, the ZnONPs had antifungal activity towards *Aspergillus niger* and *Aspergillus flavus.*

Yazhiniprabha *et al.* used *Murraya koenigii* berries to prepare zinc oxide nanoparticles (*Mk*-ZnONPs) by a simple coprecipitation method.[77] The size of the *Mk*-ZnONPs determined by TEM was 10–15 nm and they were utilized for bacteriostatic and larvicidal inactivation. The *Mk*-ZnONPs showed potential toxicity against Gram-negative and Gram-positive bacteria and cell toxicity at 50 $\mu g\,mL^{-1}$. They had *in vitro* toxicity towards *Artemia nauplii* with an LC_{50} value of 78.73 $\mu g\,mL^{-1}$ and very effective larvicidal activity towards *C. quinquefasciatus* (Figure 14.8) with an LC_{50} value of 2.1 $\mu g\,mL^{-1}$, which is

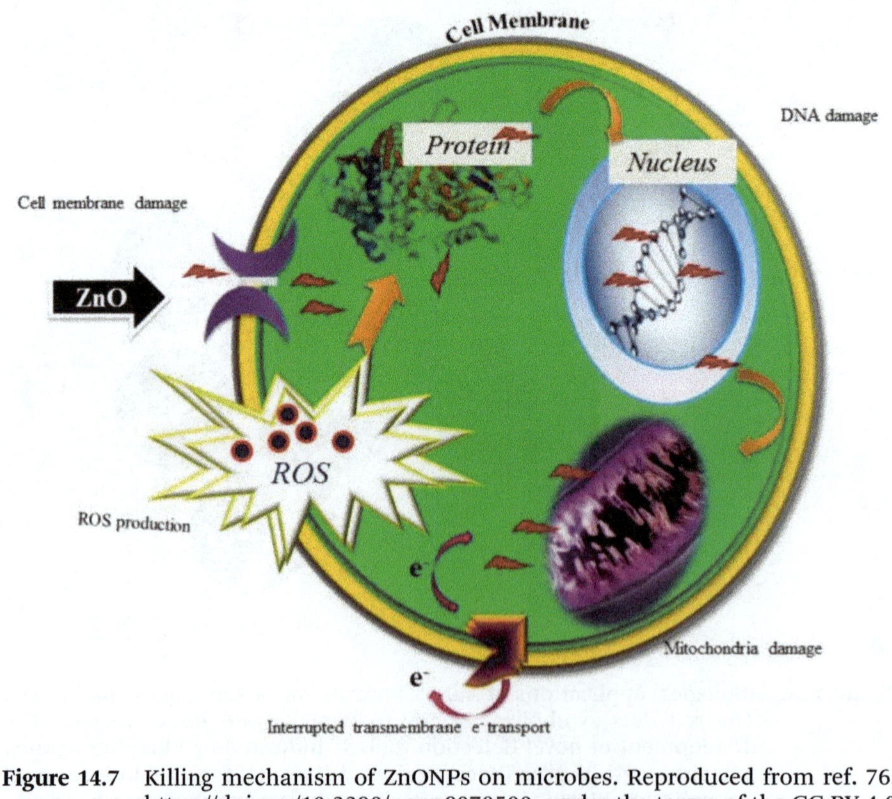

Figure 14.7 Killing mechanism of ZnONPs on microbes. Reproduced from ref. 76, https://doi.org/10.3390/nano8070500, under the terms of the CC BY 4.0 license https://creativecommons.org/licenses/by/4.0/.

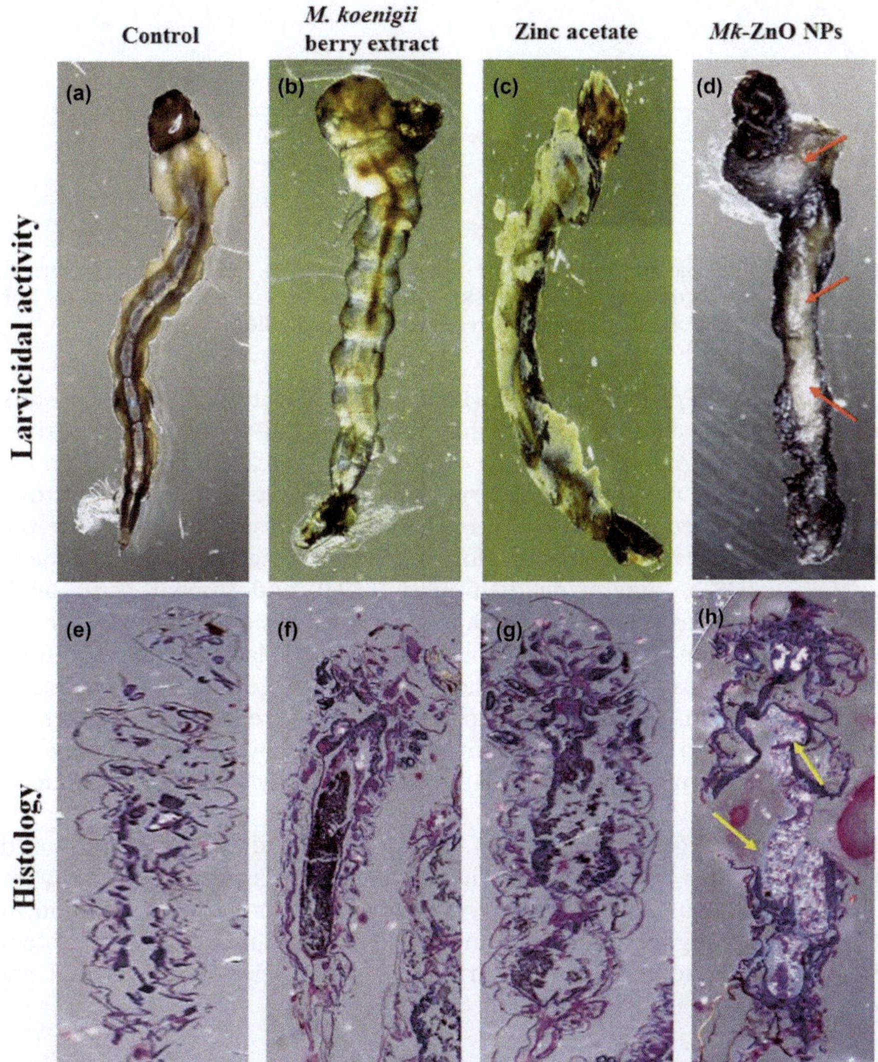

Figure 14.8 Stereomicroscopic images of *Culex quinquefasciatus* larvae treated with *Murraya koenigii* berry extract, zinc acetate, and *Mk*-ZnONPs. Images (a–d) depict morphological changes in larvae and the deposition of *Mk*-ZnONPs is indicated by red arrows in (d). Histology images of larvae are depicted in images (e–h) and the internalized *Mk*-ZnONPs are indicated by yellow arrows in (h). Reproduced from ref. 77 with permission from Elsevier, Copyright 2019.

very low compared with AgNPs. Overall, this study suggests that *Mk*-ZnONPs are very economical and highly potent nanomosquitocides.

Recently, Jameel *et al.* proposed that the synergetic effects of a thiamethoxam and ZnONP composite led to greater pesticide efficacy on fourth

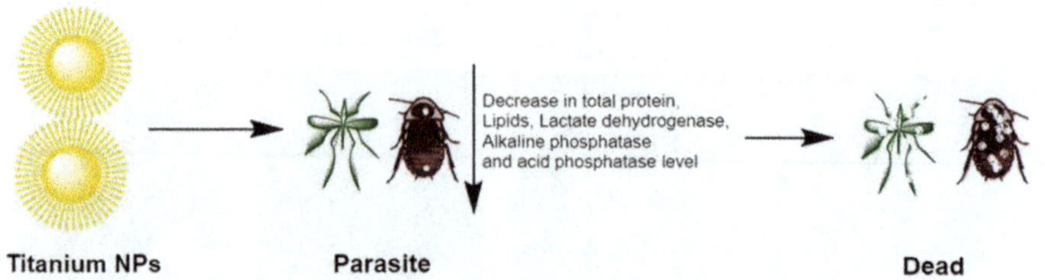

Figure 14.9 Biogenic TiO_2NPs and their action against larvae. Reproduced from ref. 80, https://doi.org/10.1080/17518253.2018.1538430, under the terms of the CC BY 4.0 license https://creativecommons.org/licenses/by/4.0/.

instar larvae of *Spodoptera litura* (Lepidoptera: Noctuidae).[79] The results from the ZnONP composite with thiamethoxam (10–90 $mg\,L^{-1}$) and thiamethoxam-impregnated castor leaves showed increased larval mortality (27% increase in mortality), malformations in pupae and adults, overdue emergence, and reduced fecundity and fertility. Bioassays performed on the treated larvae revealed a concentration-dependent variation of glutathione-*S*-transferase (GST), superoxide dismutase (SOD), and thiobarbituric acid-reactive substances (TBARS). The change in the physiological properties of the cellular system caused irrevocable damage to the larvae cells. Therefore, the ZnONP composite with thiamethoxam can be suggested as an alternative pest control agent to conventional products. Compared with the reports of ZnONPs discussed above, the *Mk*-ZnONPs proposed by Yazhiniprabha *et al.* have more economical fabrication and greater toxicity with an LC_{50} value of 2.1 $\mu g\,mL^{-1}$.[77]

Apart from the green synthesis of ZnONPs as biocides, many plant-based TiO_2NPs have also been reported. TiO_2NPs also possess a similar activity for killing microbes and parasites by generating ROS as they are well-known photocatalysts (Figure 14.9). The biogenic TiO_2NPs can damage the protein system, lipids, and cell wall and arrest other biological activities that are important for parasite survival.[80]

14.4 Conclusions and Future Perspectives

Plant products can be used to produce NMs in an economical and green fashion, and these NMs show much lower toxicity towards humans and non-targets. The plant products are well known to have medicinal value. As a result, the plant product-based synthesis of NMs has the great advantage that the NMs would be carriers in the targeted system. Plant products are useful for reduction and capping of NMs. Each plant material from leaves, roots, and stems contains unique bioactivity as repellents, ovicides, larvicides, and adulticides. As a result, the efficiency of the different plant product-based NMs differs from species to species. It has been observed that AgNPs show better toxicity than AuNPs and ZnONPs have better efficiency than the others as killing efficiency was observed at very low concentrations.

It is suggested that Ag/ZnONPs can be prepared with multiple functional groups on top and as a result a higher efficiency of killing mosquito larvae can be achieved.

References

1. N. M. Ganta, G. Gedda, B. Rathnakar, M. Satyanarayana, B. Yamajala, M. J. Ahsan, S. S. Jadav and T. Balaraju, *Eur. J. Med. Chem.*, 2019, **164**, 576–601.
2. H. F. Wertheim, P. Horby and J. P. Woodall, *Atlas of human infectious diseases*, John Wiley & Sons, 2012.
3. R. E. Baker, A. S. Mahmud, I. F. Miller, M. Rajeev, F. Rasambainarivo, B. L. Rice, S. Takahashi, A. J. Tatem, C. E. Wagner and L.-F. Wang, *Nat. Rev. Microbiol.*, 2022, **20**, 193–205.
4. S. Kumar, P. Bhushan and S. Bhattacharya, *Point-of-care diagnostics—new progresses and perspectives*, IAPC open book and monograph platform (OBP), Zagreb, Croatia, 2016, vol. 11, DOI: 10.5599/obp.11.2.
5. H. Han, B. Wu, Y. Song and R. Huang, *Dis. Surveill.*, 2019, **34**, 381–383.
6. H. Han, B. Wu, Y. Song and R. Huang, *Dis. Surveill.*, 2019, **34**, 287–289.
7. Y. K. Mohanta, D. Nayak, A. K. Mishra, I. Chakrabartty, M. K. Ray, T. K. Mohanta, K. Tayung, R. Rajaganesh, M. Vasanthakumaran and S. Muthupandian, *Int. J. Mol. Sci.*, 2022, **23**, 10626.
8. T. S. Hauck, S. Giri, Y. Gao and W. C. Chan, *Adv. Drug Delivery Rev.*, 2010, **62**, 438–448.
9. C. Sharma, M. Verma, S. Randhawa and A. Acharya, in *Handbook of Microbial Nanotechnology*, Elsevier, 2022, pp. 45–77.
10. P. Tallury, A. Malhotra, L. M. Byrne and S. Santra, *Adv. Drug Delivery Rev.*, 2010, **62**, 424–437.
11. M. Rai and K. Kon, *Nanotechnology in diagnosis, treatment and prophylaxis of infectious diseases*, Academic Press, 2015.
12. M. Govindarajan and G. Benelli, *J. Cluster Sci.*, 2017, **28**, 15–36.
13. C. Kulkarni, H. Mohanty, A. Bhagit, P. Rathod and R. P. Yadav, *Proc. Indian Natl. Sci. Acad.*, 2022, **88**, 576–591.
14. A. I. Hasaballah, H. A. El-Naggar, S. Abdelbary, M. A. Bashar and T. A. Selimd, *Bionanoscience*, 2022, 1–16.
15. A. Fouda, W. A. Al-Otaibi, T. Saber, S. M. AlMotwaa, K. S. Alshallash, M. Elhady, N. F. Badr and M. A. Abdel-Rahman, *J. Funct. Biomater.*, 2022, **13**, 157.
16. M. Suresh, S. Balu, S. C. Jose and J. Jeevanandam, in *Emerging Nanomaterials for Advanced Technologies*, Springer, 2022, pp. 369–396.
17. S. E.-D. Hassan, A. Fouda, E. Saied, M. M. Farag, A. M. Eid, M. G. Barghoth, M. A. Awad, M. F. Hamza and M. F. Awad, *J. Fungi*, 2021, 7, 372.
18. A. E. D. Mahmoud, *Handbook of nanomaterials and nanocomposites for energy and environmental applications*, 2020, pp. 1–21.
19. S. Iravani, *Green Chem.*, 2011, **13**, 2638–2650.

20. M. Suresh, J. Jeevanandam, Y. S. Chan, M. K. Danquah and J. Kalaiarasi, *Bionanoscience*, 2020, **10**, 292–310.
21. L. O. Zamora, A. E. M. de Faria Mota, J. C. T. Carvalho and C. P. Fernandes, in *Applications of nanobiotechnology for neglected tropical diseases*, Elsevier, 2021, pp. 493–507.
22. F. R. Formiga and P. Severino, *Applications of nanobiotechnology for neglected tropical diseases*, Academic Press, 2021.
23. G. Gollavelli and Y.-C. Ling, *Biomaterials*, 2012, **33**, 2532–2545.
24. G. Gollavelli and Y.-C. Ling, *Biomaterials*, 2014, **35**, 4499–4507.
25. G. Benelli, *Enzyme Microb. Technol.*, 2016, **95**, 58–68.
26. A. M. Partila, *Nanotechnology for Agriculture: Advances for Sustainable Agriculture*, 2019, pp. 19–36.
27. G. Benelli, F. Maggi, R. Pavela, K. Murugan, M. Govindarajan, B. Vaseeharan, R. Petrelli, L. Cappellacci, S. Kumar and A. Hofer, *Environ. Sci. Pollut. Res.*, 2018, **25**, 10184–10206.
28. G. Gedda and H.-F. Wu, *Sens. Actuators, B*, 2019, **288**, 667–677.
29. S. Dayalan, G. Gedda, R. N. Li, M. Zulfajri and G. G. Huang, *J. Chin. Chem. Soc.*, 2022, **69**, 2049–2060.
30. M. L. Bhaisare, M. S. Khan, S. Pandey, G. Gedda and H.-F. Wu, *RSC Adv.*, 2017, **7**, 23607–23614.
31. G. Gedda, G.-R. Chen, Y.-Y. Yao, W. M. Girma, J.-D. Li, C.-L. Yen, Y.-C. Ling and J.-Y. Chang, *New J. Chem.*, 2017, **41**, 14161–14170.
32. M. S. Khan, A. Talib, S. Pandey, M. L. Bhaisare, G. Gedda and H.-F. Wu, *Colloids Surf., B*, 2017, **159**, 564–570.
33. G. Gedda, S. Pandey, M. S. Khan, A. Talib and H.-F. Wu, *RSC Adv.*, 2016, **6**, 13145–13151.
34. M. S. Khan, G. Gedda, J. Gopal and H.-F. Wu, *Anal. Methods*, 2014, **6**, 5304–5313.
35. G. Gangaraju, G. Ganesh, R. Uma Devi and K. Balakrishna, in *Phytonanotechnology*, Springer, 2022, pp. 181–197.
36. J. Gopal, H.-F. Wu and G. Gangaraju, *J. Mater. Chem.*, 2011, **21**, 13445–13451.
37. G. Gollavelli, A. V. Ghule and Y.-C. Ling, *Molecules*, 2022, **27**, 5588.
38. A. Talib, M. S. Khan, G. Gedda and H.-F. Wu, *J. Mol. Liq.*, 2016, **220**, 463–467.
39. A. Sooresh, H. Kwon, R. Taylor, P. Pietrantonio, M. Pine and C. M. Sayes, *ACS Appl. Mater. Interfaces*, 2011, **3**, 3779–3787.
40. A. V. Samrot, K. S. Bhavya, J. L. A. Angalene, S. Roshini, R. Preethi, S. Steffi, P. Raji and S. S. Kumar, *Int. J. Biol. Macromol.*, 2020, **153**, 1024–1034.
41. N. Sultana, P. K. Raul, D. Goswami, D. Das, S. Islam, V. Tyagi, B. Das, H. K. Gogoi, P. Chattopadhyay and P. S. Raju, *RSC Adv.*, 2020, **10**, 9356–9368.
42. Y. Zhu, P. Prommana, N. S. Hosmane, P. Coghi, C. Uthaipibull and Y. Zhang, *ACS Omega*, 2022, **7**, 5864–5869.
43. G. Benelli, *Parasitol. Res.*, 2016, **115**, 23–34.

44. H. Mehlhorn, *Nanoparticles in the fight against parasites*, Springer, 2016.
45. V. Cittrarasu, D. Kaliannan, K. Dharman, V. Maluventhen, M. Easwaran, W. C. Liu, B. Balasubramanian and M. Arumugam, *Sci. Rep.*, 2021, **11**, 1032.
46. P. Amuthavalli, J.-S. Hwang, H.-U. Dahms, L. Wang, J. Anitha, M. Vasanthakumaran, A. D. Gandhi, K. Murugan, J. Subramaniam and M. Paulpandi, *Sci. Rep.*, 2021, **11**, 1–13.
47. V. Amutha, P. Deepak, C. Kamaraj, G. Balasubramani, D. Aiswarya, D. Arul, P. Santhanam, A. Ballamurugan and P. Perumal, *J. Cluster Sci.*, 2019, **30**, 797–812.
48. S. Kumar, V. Lather and D. Pandita, *Nanomedicine*, 2015, **10**, 2451–2471.
49. R. Trivedi, T. K. Upadhyay, M. H. Mujahid, F. Khan, P. Pandey, A. B. Sharangi, K. Muzammil, N. Nasir, A. Hassan and N. M. Alabdallah, *Processes*, 2022, **10**, 338.
50. A. C. M. Hacke, D. Lima and S. Kuss, *J. Electroanal. Chem.*, 2022, 116786.
51. M. M. Saleh, L. M. Alnaddaf, A. K. Almuhammady, K. F. Salem, M. T. Alloosh and J. M. Al-Khayri, *Nanobiotechnology: Mitigation of Abiotic Stress in Plants*, 2021, pp. 201–238.
52. K. Shahzad and F. Manzoor, *Drug Chem. Toxicol.*, 2021, **44**, 1–11.
53. G. Benelli, A. Caselli and A. Canale, *J. King Saud Univ., Sci.*, 2017, **29**, 424–435.
54. N. J. White, *N. Engl. J. Med.*, 2015, **373**, 1670–1671.
55. G. Benelli and H. Mehlhorn, *Parasitol. Res.*, 2016, **115**, 1747–1754.
56. W. H. Organization, Status report on artemisinin and ACT resistance, September 2015, World Health Organization, 2015.
57. V. Pareek, R. Gupta and J. Panwar, *Mater. Sci. Eng., C*, 2018, **90**, 739–749.
58. R. K. Das, V. L. Pachapur, L. Lonappan, M. Naghdi, R. Pulicharla, S. Maiti, M. Cledon, L. M. A. Dalila, S. J. Sarma and S. K. Brar, *Nanotechnol. Environ. Eng.*, 2017, **2**, 1–21.
59. L. Rizzello and P. P. Pompa, *Chem. Soc. Rev.*, 2014, **43**, 1501–1518.
60. Q. Li, S. Mahendra, D. Y. Lyon, L. Brunet, M. V. Liga, D. Li and P. J. Alvarez, *Water Res.*, 2008, **42**, 4591–4602.
61. A. Tauseef, F. Hisam, T. Hussain, A. Caruso, K. Hussain, A. Châtel and B. Chénais, *J. Cluster Sci.*, 2022, 1–26.
62. R. A. Omar, S. Afreen, N. Talreja, D. Chauhan, M. Ashfaq and W. Srituravanich, *Microbial nanobionics: Volume 1, State-of-the-Art*, 2019, pp. 141–158.
63. A. K. Suresh, D. A. Pelletier and M. J. Doktycz, *Nanoscale*, 2013, **5**, 463–474.
64. S. U. Khan, T. A. Saleh, A. Wahab, M. H. U. Khan, D. Khan, W. U. Khan, A. Rahim, S. Kamal, F. U. Khan and S. Fahad, *Int. J. Nanomed.*, 2018, **13**, 733.
65. M. Govindarajan, S. Hoti and G. Benelli, *Enzyme Microb. Technol.*, 2016, **95**, 155–163.
66. M. Govindarajan, S. Kadaikunnan, N. S. Alharbi and G. Benelli, *Artif. Cells, Nanomed., Biotechnol.*, 2017, **45**, 1317–1325.

67. A. A. Buhroo, G. Nisa, S. Asrafuzzaman, R. Prasad, R. Rasheed and A. Bhattacharyya, *J. Plant Prot. Res.*, 2017, **57**, 194–200.
68. D. Elumalai, M. Hemavathi, C. V. Deepaa and P. K. Kaleena, *Parasite Epidemiol. Control*, 2017, **2**, 15–26.
69. J.-H. Lee, P. Velmurugan, J.-H. Park, K. Murugan, N. Lovanh, Y.-J. Park, B.-T. Oh, P. Venkatachalam and G. Benelli, *Physiol. Mol. Plant Pathol.*, 2018, **101**, 178–186.
70. K. Meenambigai, R. Kokila and A. N. Kumar, *J. Entomol. Acarol. Res.*, 2018, **50**, 7450.
71. N. Sutthanont, S. Attrapadung and S. Nuchprayoon, *Insects*, 2019, **10**, 27.
72. A. Sharma, S. Kumar and P. Tripathi, *Saudi J. Biol. Sci.*, 2019, **26**, 698–708.
73. T. Small, M. A. Ochoa-Zapater, G. Gallello, A. Ribera, F. M. Romero, A. Torreblanca and M. D. Garcerá, *Sci. Total Environ.*, 2016, **565**, 882–888.
74. P. P. Pompa, G. Vecchio, A. Galeone, V. Brunetti, S. Sabella, G. Maiorano, A. Falqui, G. Bertoni and R. Cingolani, *Nano Res.*, 2011, **4**, 405–413.
75. G. Benelli, *Acta Trop.*, 2018, **178**, 73–80.
76. N. A. Al-Dhabi and M. Valan Arasu, *Nanomaterials*, 2018, **8**, 500.
77. M. Yazhiniprabha, B. Vaseeharan, A. Sonawane and A. Behera, *J. Photochem. Photobiol., B*, 2019, **192**, 158–169.
78. P. Thakur, S. Thakur, P. Kumari, M. Shandilya, S. Sharma, P. Poczai, A. A. Alarfaj and R. Sayyed, *Appl. Nanosci.*, 2022, **12**, 3835–3850.
79. M. Jameel, M. Shoeb, M. T. Khan, R. Ullah, M. Mobin, M. K. Farooqi and S. M. Adnan, *ACS Omega*, 2020, **5**, 1607–1615.
80. M. Nadeem, D. Tungmunnithum, C. Hano, B. H. Abbasi, S. S. Hashmi, W. Ahmad and A. Zahir, *Green Chem. Lett. Rev.*, 2018, **11**, 492–502.

CHAPTER 15

Sustainable Nanomaterials and Diagnosis for Arboviral Infections

VIGNESH MARIAPPAN,[a] RAJESH SRINIVASAN,[a] MANOHARAN PAUL SURENDRA[b] AND AGIESHKUMAR BALAKRISHNA PILLAI*[a]

[a] Mahatma Gandhi Medical Advanced Research Institute (MGMARI), Sri Balaji Vidyapeeth (Deemed to be University), Puducherry 607403, India; [b] Indira Gandhi Institute of Dental Science (IGIDS), Sri Balaji Vidyapeeth (Deemed to be University), Puducherry 607403, India

*Emails: agiesh.b@gmail.com; agieshkumarb@mgmari.sbvu.ac.in

15.1 Introduction

Viruses that are transmitted by the bite of an arthropod are classified as "arboviruses." To date, many arboviruses have been discovered, and about 500 arbovirus species are now officially classified into six families. Most arboviruses belong to Togaviridae, Flaviviridae, and Bunyaviridae; a few are listed in Sedoreoviridae (formerly Reoviridae), Orthomyxoviridae, and Rhabdoviridae.[1] They are known to infect humans and are considered the fastest-spreading pathogens across the globe (Figure 15.1). More than 17% of infectious diseases are caused by arthropod-borne infections, which are a result of re-emerging human pathogens and infect millions of individuals worldwide annually. An overview of common arbovirus infections in different geographical distributions, vectors that transmit the virus, and diagnostic methods is given in

Nanoscience & Nanotechnology Series No. 62
Smart Nanomaterials for Infectious Diseases
Edited by Suvardhan Kanchi, Nidhi Sharotri, Rajasekhar Chokkareddy, Deepali Sharma and Falah H. Hussein

Published by the Royal Society of Chemistry, www.rsc.org

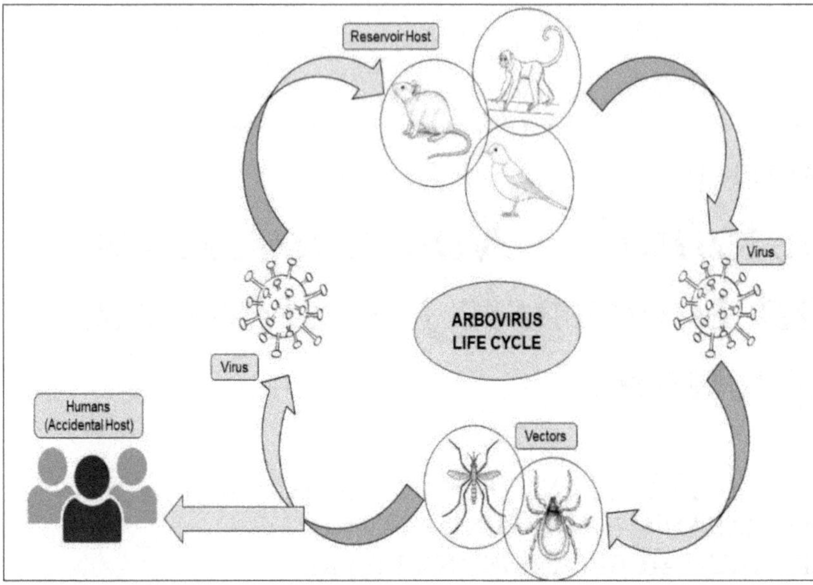

Figure 15.1 Life cycle of arboviruses.

Table 15.1 for a better understanding. Dengue, the most widespread arboviral infection, causes almost 90 million infections and 40 000 fatalities each year. The recent outbreak of Zika virus (ZIKV) has put the world's health infrastructure to the test in terms of diagnosing ZIKV infections, distinguishing them from infections caused by closely related arboviruses, and mitigating the disastrous effects of prenatal infection.[2] Further, other arbovirus infections, such as Rift Valley fever, Mayaro fever, West Nile fever, chikungunya fever, and tick-borne encephalitis, have drawn the attention of scientists as potential dangers to public health. Consequently, studies in terms of bio-surveillance, pandemic preparedness, virus transmission dynamics, epidemiology and ecology, molecular pathogenesis, diagnosis, and treatment of emerging and re-emerging viruses are currently being undertaken for the effective management of arboviral diseases.

Since there are no effective vaccines or specific antivirals for treating arbovirus infections, continuous monitoring and treatment of the infected patients are the only ways to alleviate disease symptoms and manage the disease.[3] In the above context, this chapter reviews (i) potential biomarkers for understanding the outcome of arboviral infection and (ii) current trends in developing nanosensor-based diagnostics and therapeutics for arboviral diseases.

15.2 Biomarkers for the Prediction of Arboviral Disease Severity

The virus enters the human body *via* the epidermis and is internalized by endocytes. The first line of defence is produced at the site of entry, which

Table 15.1 Overview of arboviral diseases.[a]

Virus	Geographical distribution	Vector	Clinical symptoms and signs	Traditional and current diagnostic methods
Togoviridae				
Genus: Alphavirus				
Barmah Forest virus (BFV)	Australia	Mosquitoes	Joint pain, joint swelling, skin rashes	IgM and IgG by HI, EIA, or FIA, PCR
Chikungunya virus (CHIKV)	Tanzania	Mosquitoes	Arthralgia, lymphopenia, hypocalcaemia, photophobia, lumbar back pain, chills, weakness, nausea, vomiting	PCR, IgM by EIA, HI, or N test
Western equine encephalitis virus (WEEV)	North America, Saskatchewan, Argentina, Brazil, Mexico, Guyana	Mosquitoes	Mild fever, malaise, headache, photophobia, nausea, vomiting, sore throat	PCR, HI, or N test, IgM by EIA
Eastern equine encephalitis virus (EEEV)	Eastern USA, Mexico, Panama, Brazil, Argentina, Guyana, Dominica, Jamaica, Venezuela	Mosquitoes	Encephalitis, headache, meningism, reduction of conscious state	PCR, HI, or N test, IgM by EIA
Venezuelan equine encephalitis virus (VEEV)	Venezuela, Trinidad, Colombia, Brazil, Panama, Mexico, southern USA	Mosquitoes	Fever, rigours, headache, myalgia, vomiting, conjunctivitis	RT-PCR, HI, or N test, IgM by EIA
Mayaro virus (MAYV)	Trinidad, Central America, northern South America, Amazon basin	Mosquitoes	The clinical condition closely resembles CHIKV	RT-PCR, IgM ELISA, PRNT
Rose River virus (RRV)	Australia, Fiji, American Samoa, Cook Islands, New Caledonia, New Guinea, Soloman Islands, The Moluccas, Vietnam	Mosquitoes	Joint pain, myalgia, fever, headache, photophobia, lymphadenopathy, sore throat	IgM and IgG by HI, EIA, or FIA, PCR
Sindbis virus (SINV)	Egypt, sub-Saharan Africa, Europe, Middle East, India, Asia, Philippines, Australia, New Zealand, Sweden, Finland, adjacent parts of Russia, South Africa	Mosquitoes, birds	Joint pain, rash, malaise, fatigue	PCR, IgM by EIA, FIA, HI

Table 15.1 (*Continued*)

Virus	Geographical distribution	Vector	Clinical symptoms and signs	Traditional and current diagnostic methods
Flaviviridae				
Genus: Flavivirus				
Dengue virus (DENV)	America, Africa, Middle East, Asia, Pacific Islands	Mosquitoes	Nausea, vomiting, rash, aches and pains	NAAT, IgM, and IgG by ELISA, PCR
Japanese encephalitis virus (JEV)	Japan, Siberia, North Korea, South Korea, Sri Lanka, India, Pakistan, Nepal, Indonesia, Papua New Guinea	Mosquitoes	Gastrointestinal disturbances, fever, headache, altered mental state, convulsions	RT-PCR, IgM by EIA, IFA, HI
Kyasanur Forest disease virus (KFDV)	South-western India, Saudi Arabia (Alkhurma virus subtype of KFDV)	Ticks	Fever, headache, myalgia, cough, vomiting, diarrhoea, dehydration, hypotension, bradycardia	IgM by EIA
Kunjin virus (KUNV)	Australia	Mosquitoes	Headache, myalgia, fatigue, maculopapular rash	IgM by ELISA
Louping Ill virus (LIV)	UK, Spain, Greece, Turkey	Ticks	Fever, headache, myalgia, nausea	RT-PCR
Murray Valley encephalitis virus (MVEV)	Australia, New Guinea	Mosquitoes	Fever, headache, arthralgia	RT-PCR, HI tests, EIA or IFA
OMSK haemorrhagic fever virus (OHFV)	Siberia	Ticks	The clinical condition closely resembles TFDV	CF, HI, N test
St Louis encephalitis virus (SLEV)	America, Trinidad, Brazil, Argentina	Mosquitoes	Fever, severe headache, neck stiffness, photophobia	IgM by EIA
Tick-borne encephalitis virus (TBEV)	Siberia, Europe	Ticks	Fever, headache, myalgia	RT-PCR, IgM
West Nile virus (WNV)	Africa, Middle East, southern Europe, Russia, southern India, parts of south-east Asia, North America, South America	Mosquitoes	Fever, headache, myalgia, backache, anorexia	PCR, IgM
Yellow fever virus (YFV)	Africa, South America	Mosquitoes	Fever, severe headache, nausea, vomiting, abdominal pain, distressing pain in the back, loins, limbs	RT-PCR, IgM by HI, EIA, or IFA, IgG by N titres

Zika virus (ZIKV)	East, central, and west Africa	Mosquitoes	Fever, headache, rash	RT-qPCR, IgG and IgM by ELISA
Bunyaviridae				
Genus: Bunyavirus				
Oropouche virus (OROV)	Amazon, Peru	Mosquitoes, *Culicoides* biting midges	Fever, chills, headache, myalgia, arthralgia, photophobia	Serology test
Rift Valley fever virus (RVFV)	Kenya, South Africa, Zimbabwe, Sudan, Egypt, Uganda, Tanzania, Zambia	Mosquitoes	Fever, headache, joint and muscle pains, conjunctivitis, photophobia	PCR, IgM by EIA or HI test
Sandfly fever (Naples virus, SFNV; Sicilian virus, SFSV)	Mediterranean, Middle East, Malta, Aegean Islands, Egypt, Iran, North Africa, Red Sea and Arabian Gulf areas, and in Asia the Caucasus and Himalayas	Sandflies	Fever, headache, myalgia, arthralgia, neck stiffness	IgM, IgG by ELISA, HI, and N tests
Toscana virus (TOSV)	Italy, Portugal, France, Spain	Sandflies	—	Serology test, RT-PCR
Genus: Nairoviruses				
Congo–Crimean haemorrhagic fever virus (CCHFV)	Eastern Europe, Turkey, Greece, north-western China, central Asia, southern Europe, Africa, Middle East, Indian subcontinent	Ticks	Headache, high fever, back pain, joint pain, stomach pain, vomiting, red eyes, red throat, petechiae (red spots) on the palate are common	Serological test and viral antigen detection by IHC
Reoviridae				
Genus: Coltivirus				
Colorado tick fever virus (CTFV)	Western USA, Canada	Ticks	Fever, headache, myalgia, arthralgia, neck stiffness, retro-orbital pain, photophobia, maculopapular or petechial rash	IgM antibody, RT-PCR

[a] CF test, complement fixation test; EIA, enzyme immunoassay; ELISA, enzyme-linked immunosorbent assay; HI, haemagglutination test; IFA, indirect fluorescent antibody; IgG, immunoglobulin G, IgM, immunoglobulin M; IHC, immunohistochemistry; N test, neutralization test; NAAT, nucleic acid amplification test; PCR, polymerase chain reaction; PRNT, plaque reduction neutralization test; qPCR, quantitative polymerase chain reaction; RT-PCR, reverse transcriptase polymerase chain reaction.

includes macrophages, monocytes, dendritic cells, and mast cells. Arboviruses have also been reported to infect hepatocytes and endothelial cells (ECs).[4] The life cycle of arboviruses is confined to the cytoplasm and endosomal compartments that recognize and interact with viral RNA and virus-infected cells. This results in an effective immune response through two major events: (i) viral sensing by pattern recognition receptors (PRRs) and (ii) activation of protein signalling cascades that regulate the secretion of type I interferons (IFNs). Toll-like receptors (TLRs), retinoic acid-inducible gene I (RIG-I)-like receptors (RLRs), and cyclic GMP–AMP synthase (cGAS) are the three main groups of PRRs involved in the effective detection of viruses.[5] For instance, TLR3 and TLR7 were found to promote central nervous system protection and prevent West Nile virus (WNV) replication.[6] Interestingly, a study reported that knocking out TLR3 from macrophages enhanced the susceptibility of dengue virus (DENV), while knockdown of TLR3 in skin fibroblasts increased viral RNA without any alteration in the total mRNA levels of IFN-I.[7] More recently, a study reported the involvement of RIGs and melanoma differentiation-associated protein 5 in the production of an antiviral immune response against DENV and ZIKV upon recognizing the 5′ end of the viral RNA.[8] An *in vivo* study showed a higher level of DENV RNA and delayed production of IFN in a mitochondrial antiviral signalling protein knockout mouse model.[9] Similarly, an *in vitro* investigation demonstrated enhanced viral replication and reduced host resistance against flaviviruses (ZIKV and DENV).[10]

Each of the three viral nucleic acid sensors (TLR, RLR, and cGAS) causes the transcriptional upregulation of numerous immune components, including IFNα, IFNβ and other IFN subtypes, thereby creating an antiviral environment by activating interferon-stimulating genes (ISGs) against the viruses.[11–13] To that end, we recently investigated the role of two ISGs known as guanylate-binding proteins (GBP1 and GBP2) as biomarkers for DENV infection, and also their associations with oxidative stress markers and dengue severity symptoms (platelet count and haematocrit).[12] In this perspective, we discussed the crucial role of oxidative stress in dengue disease pathogenesis[14,15] and proposed that antioxidant therapy at an early stage of infection could be an effective strategy to alleviate the disease severity of an RNA virus infection.[16]

On the other hand, circulating molecules released by activated immune cells such as monocytes, macrophages, dendritic cells, and ECs are thought to be involved in arbovirus pathogenesis. For instance, an *in vitro* investigation showed that direct exposure of DENV to monocyte-derived macrophages results in the excessive production of cytokines (TNF, IFNα, IL-1, CXCL8, IL-12, CCL3, and CCL5) and results in apoptosis.[17] Previous research in our laboratories indicated the role of CD163 and macrophage-released ferritins as potential markers for severity prediction.[18,19] A study examined the role of various cytokines (IL-1b, IL-1ra, IL-2, IL-4, IL-5, IL-6, IL-9, IL-10, IL-12, IL-13, IL-15, IL-17, IFN-c, MIF, TNF-α, and TNF-β), chemokines (CCL2, CCL3, CCL4, CCL5, CCL11, CXCL10, CXCL12, and IL-8), and growth factors (FGF-2, G-CSF, GM-CSF, HGF, IL-7, PDGF, and VEGF) and their implication in dengue disease severity.[20] Moreover, levels of chymase, tryptase, and VEGF have been reported to be very high in dengue patients, specifically in those with plasma leakage.[21] In line with this, we reported the complete dynamics of two endothelial-released proteins (endoglin and syndecan-1) and their efficacy in predicting disease severity using a machine learning model.[22] We also described a set of potential endothelial biomolecules that could serve as biomarkers for the early prediction of coronavirus disease 2019 (COVID-19) outcomes.[23] Thus, the identification of specific diagnostic (viral detection) and prognostic (early prediction of clinical outcome) biomarkers will pave the way for better treatment and triage. Since endothelial dysfunction is one of the crucial complications of DENV and other RNA virus infections, we recently proposed the use of mesenchymal stem cells in restoring endothelial barrier function.[24] Thus, there are several potential biomarkers that are required to be validated for their clinical applications in the form of devices. Further, the identification of inhibitors against selected endothelial molecules could be a novel therapeutic approach for the management of arbovirus infections.

15.3 Scope of Nanotechnology in Viral Detection

Nanotechnology involves the management of information, the reorganization of matter, and matter-at-a-nanometre scale to develop materials with novel functions and properties.[25] Nanotechnology research has revolutionized various fields such as engineering, biology, chemistry, robotics, communications, and even medicine. Rapid outbreaks of epidemics and pandemics are mostly due to viral genome mutations. A single nucleic acid alteration has a significant negative impact on the human population. For instance, the unexpected emergence and rapid spread of highly lethal viruses may cause a high fatality rate. Therefore, a quick and accurate diagnostic platform could be an effective strategy against sudden pandemics.[26] In general, conventional molecular diagnostic approaches are employed to confirm any pathological condition. These are based on the amplification of nucleic acids (DNA or RNA), such as target amplification (PCR, RT-PCR, and strand displacement amplification), signal amplification (branched DNA assays and hybrid capture), probe amplification (cleavage-invader and ligase chain reaction), or post-amplification analysis (*e.g.* sequencing the amplified products or melting curve analysis). However, there is a minor chance of sample contamination, which could result in a false-negative report.[27]

The high cost and short shelf life of key reagents (such as enzymes and DNA primers) limit the applicability of conventional detection techniques in most developing countries. For instance, current technologies (such as PCR and ELISA) involve various sample preparation steps and lengthy readout periods, which results in delayed reactions and disease containment despite their sensitivity, which limits their application in clinical diagnosis. Additionally, many of these techniques are not feasible in crowded areas such as food courts and airports.[28] This calls for an effective global surveillance system for emerging viruses and novel methods for viral diagnosis and treatment. The past few decades have seen notable advances in nanotechnology. For instance, nanomaterials (NMs) such as fibres, particles, or grains with a size of less than 100 nm have piqued researchers' interest in nanotechnology, particularly in the advancement of disease diagnosis, prevention, and therapy.[29] In addition, sensitivity, multiplexing, and cost-effectiveness could be improved as a result of nanodiagnostics.

Owing to their exceptional qualities, such as high surface area-to-volume ratio, high thermal conductivity, fast signal transduction, improved pharmacokinetic profile, longer drug retention time, biocompatibility, and lack of side effects, NMs have attracted a lot of attention for therapeutic applications. The drawbacks of traditional PCR-based diagnostic methods such as low yield, complexity in the amplification of long, GC-rich genomic DNA, low specificity, fidelity, and efficiency, and laborious standardization procedures may be overcome by integrating conventional diagnostic techniques with nano-based methodologies.[27]

15.4 Recent Advances in Nano-based Viral Diagnosis

15.4.1 Nanomaterials for Viral RNA Extraction

Nucleic acids integrated with NMs play a versatile role in biosensing. Nanomaterials are able to sense and enhance the signals of molecular recognition events owing to their unique physicochemical features.[30] Further, the efficient interaction of magnetic nanoparticles (MNPs) with biomolecules such as nucleic acids (DNA and RNA), proteins, and enzymes, and also their extraction and purification, have been documented in recent years. A schematic representation of the extraction of viral nucleic acids coupled with MNPs is shown in Figure 15.2. Generally, surface-functionalized MNPs draw a charged nucleic acid and separate it from the lysis solution when an external magnetic field is applied. Since the nucleic acid separates from the MNPs on desorption and is collected in the eluate, this technology is easier and faster compared with conventional techniques such as RT-PCR.[31] Lee *et al.* developed a unique method for the preparation of iron oxide silica particles (Fe_3O_4@silica) and demonstrated the isolation of nucleic acids from a ZIKV sample and further analysis by PCR. The cost of the extraction utilizing iron oxide silica particles is 100 times lower than in the standard method, and it exhibits an efficiency comparable to that of the latter.[32]

15.4.2 Nanopore Sequencing

High-throughput sequencing technologies have replaced various traditional sequencing techniques and have made a positive impact on the field of biomedical research.[33] One of the recently advanced fourth-generation

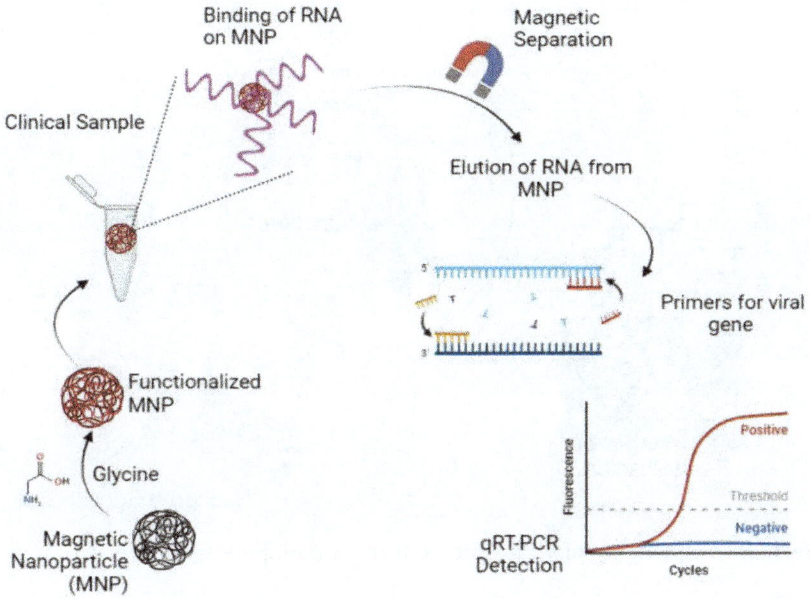

Figure 15.2 Magnetic nanoparticle-based viral RNA extraction.

sequencing methods is the recognition of single molecules using nanopores.[34] Compared with the conventional Sanger sequencing approach, it is the fastest and least expensive.[35] The basic idea is that a single DNA molecule can be recognized as it passes through a tiny, narrow nanopore chamber. Nanopores are classified into two categories, biological and solid-state, and their main benefits include high throughput, label-free operation, extremely long readings (10^4–10^6 bases), and low material requirements.[36] Briefly, the particles comprising the electrolyte suspension are pumped into the tiny tube that runs between the two reservoirs. When a molecule moves through the channel-like passage, it increases the electrical resistance of the channel, and when an electric field is applied the particles that move through the passage are observed as a decrease in current. As a result, nanopore sequencing is one of the best approaches for cheaply sequencing genomes because it can distinguish each nucleotide by analysing the change in electrical conductivity as molecules flow through the channel.[37] A diagrammatic illustration of the application of nanopores in detecting viral nucleic acids is shown in Figure 15.3.

Several companies have proposed different strategies for nanopore sequencing, including NanoTag sequencing (Genia, Roche) and Bayley sequencing (MinION, Oxford Nanopore), and to date independent genomics laboratories have successfully used the MinION, the first commercially accessible nanopore sequencer.[38,39] This portable sequencer can produce very long reads of up to 60 kilobases, and it has been successfully used to sequence

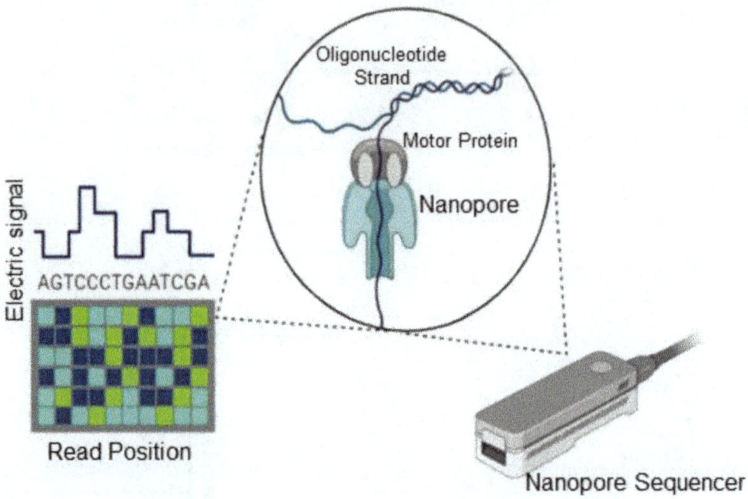

Figure 15.3 Sequencing of viral genome using a nanopore sequencer.

the whole genomes of poxviruses, *Escherichia coli*, and lambda bacteriophages. With this technology, reading data for up to tens of kilobases is possible without any limitations. It has an advantage over other sequencing techniques because it uses nanopores to sequence a single molecule per pore, whereas other techniques rely on DNA cluster analysis.[40] In addition, it is superior to other technologies owing to its fast and simple sample preparation, portability, small footprint, long reads, and flexible run time for data generation. By combining MinION nanopore sequencing with a Web-based pipeline for *in silico* analysis, Greninger *et al.* detected chikungunya virus (CHIKV), Ebola virus (EBOV), and hepatitis C virus (HCV) from four blood samples using the unbiased metagenomic detection platform.[41] A hybrid RNA/cDNA sequence analysis was performed on Venezuelan equine encephalitis virus (VEEV) strain TC-83 and EBOV. In 3 h, the sequence analysis successfully distinguished VEEV TC-83 from the wild type. Additionally, this sequencing technique was used to develop strain-level detection for TC-83.[42]

15.4.3 Virus Sensing *Via* the CRISPR-Cas12A/Cas13a System

15.4.3.1 Based on Different Signal Readouts

15.4.3.1.1 Virus Detection *Via* Fluorescent Signals. The first two sensing systems, SHERLOCK and DETECTOR, were reported based on CRISPR-Cas12A/Cas13a, which saw further modifications with improved sensitivity for virus detection.[43–46] For instance, the signal-to-noise (S/N) ratio, ease of use, and lower background are often benefits of fluorescence-based sensing. In addition to the microplate reader, other devices such as a smartphone,[46] an LED blue light illuminator,[47] a portable instrument,[48] and a microfluidic chip[49] were developed to record the fluorescence signals and enhance the point-of-care (POC) test capability.[50]

Gootenberg *et al.* developed a highly sensitive and single-base resolution (attomolar, aM = 10^{-18} M) SHERLOCK detection system that was capable of detecting ZIKV and DENV viruses.[51] In addition, Gootenberg *et al.* used a paper-based SHERLOCK called SHERLOCKv2 (SHERLOCK version 2) to detect ZIKV RNA using freeze-dried Cas13a to accommodate field deployment. The sensitivity of the paper-based SHERLOCK system's mock ZIKV detection had a limit of detection (LOD) as low as 20 aM.[44] In addition, several other viruses have also been identified using the CRISPR-Cas12a/Cas13a-based platform as depicted in Figure 15.4, including EBOV and Lassa virus (LASV),[49,52] hepatitis B virus (HBV),[53] Japanese encephalitis virus (JEV),[54] and severe acute respiratory syndrome coronavirus 2 (SARS-CoV-2).[55,56] Further, by utilizing particular crRNAs, the CRISPR-Cas13a-based approach was also able to discriminate between drug-resistance mutations in human immunodeficiency virus (HIV)[57] and human brucellosis bacteria.[53]

15.4.3.1.2 Virus Detection *Via* Colorimetric Signals. Owing to their obvious colour changes, gold nanoparticle (AuNP)-based colorimetric tests are more advantageous than those using other NPs. The aggregation and

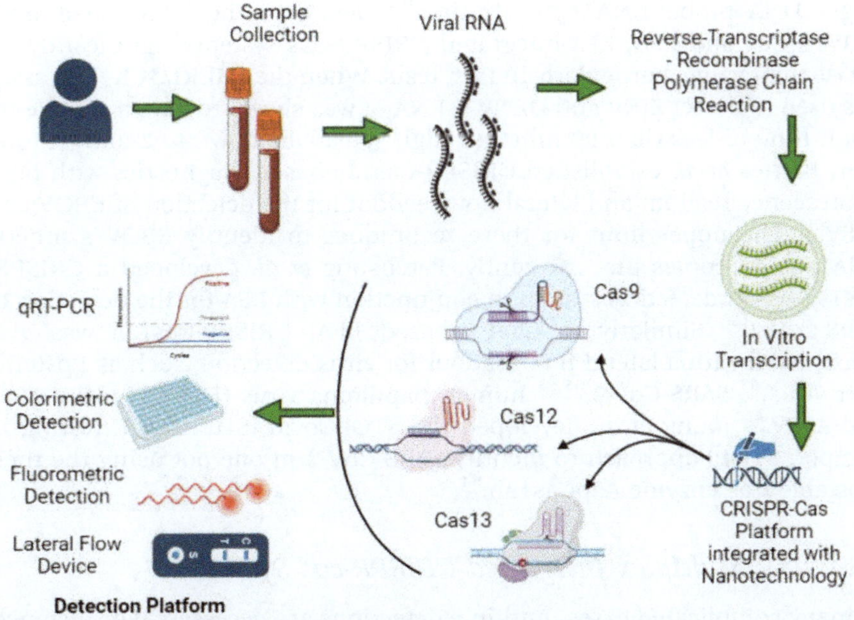

Figure 15.4 CRISPR-Cas platform integrated with nanomaterials for the identification of viruses using various detection systems.

dispersion behaviours of AuNPs display various distance-dependent optical characteristics that may be seen with the naked eye.[58] The colour of the colloidal solution changes from red to purple as the AuNPs aggregate, and the absorption peak shifts to longer wavelengths along with it. After centrifugation, the colloidal solution appears colourless.[59] Yuan *et al.* demonstrated a colorimetric gene-sensing platform with CRISPR-Cas12a/Cas13a systems using this crucial property of AuNPs.[60] A linker ssDNA or RNA that can hybridize with the AuNP–DNA probes was created for this platform. When the target is present, the trans-cleavage activity of Cas12a or Cas13a is activated, which causes the linker ssDNA or ssRNA to degrade and causes the AuNPs to disaggregate. The African swine fever virus (ASFV) was detected with the naked eye in less than 1 h using this platform.

Magnetic pull-down assisted colorimetric diagnosis (M-CDC) was developed by Hu *et al.* based on the CRISPR-Cas12a technology.[61] The M-CDC approach successfully detected the ASFV with 100% sensitivity and 100% specificity. Additionally, an M-CDC improved by horseradish peroxidase was developed, which catalyses tetramethylbenzidine oxidation to produce a colour reaction, enabling visual detection of pathogens. Colorimetric sensors have a high sensitivity in addition to being practical for visual detection.[61]

15.4.3.2 Lateral Flow Assay-integrated CRISPR-cas System

The lateral flow assay (LFA) uses conjugated gold, carbon, or coloured latex NPs within the conjugate pad to simulate antigen–antibody interactions or target DNA–probe DNA hybridization.[62] Owing to the benefits of both CRISPR-Cas and LFA, LFA-integrated CRISPR-Cas systems significantly increase their value, particularly in POC tests. When the SHERLOCK–LFA assay was used to detect ZIKV and DENV ssRNA, it was shown to have a sample-to-result time of less than 90 min and high sensitivity down to 2 aM.[44] Moreover, Barnes *et al.* established CRISPR-Cas13a-based diagnostics with both fluorescence readout and lateral flow readout for the detection of EBOV and LASV.[52] The upper limit for these techniques to identify EBOV synthetic DNA was 10 copies μL^{-1}. Recently, Patchsung *et al.* developed a CRISPR-Cas13a-based detection system in conjunction with LFA for the detection of SARS-CoV-2.[56] Similarly to Cas13a-based LFA, CRISPR-Cas12a was also constructed with a lateral flow readout for virus detection, such as Epstein–Barr virus,[63] SARS-CoV-2,[55,64] human papilloma virus (HPV) 16, HPV 18,[65] and ASFV.[66] Joung *et al.* developed the STOPCovid (SHERLOCK testing in one pot, STOP) approach to identify SARS-CoV-2 in one pot using the thermostable Cas enzyme AapCas12b.[64]

15.4.3.3 Multiplex Integrated CRISPR-cas System

In many complicated cases, multiple detections are necessary. For instance, SARS-CoV-2 infection causes respiratory symptoms similar to those with influenza. For a precise diagnosis, it is crucial to detect two or more pathogens in a single test. For the multiplex detection of various pathogens, Ackerman *et al.* established the CARMEN (combinatorial arrayed reactions for multiplexed evaluation of nucleic acids) platform.[57] This platform is based on a CRISPR-Cas13a detection mixture combined with nanolitre droplets holding amplified samples in a microwell array. A multiplexed detection method using the CARMEN platform was developed that could quickly integrate an extra crRNA and distinguish 169 human-associated viruses while detecting SARS-CoV-2. Additionally, CARMEN-Cas13 has been demonstrated to be effective in subtyping influenza A strains and identifying a number of mutations associated with HIV drug resistance. Moreover, Jiao *et al.* used the CRISPR-Cas12a system to identify RNA viruses *via* multiplex recombinase polymerase amplification and separate LbCas12a/AuNP visual tests.[67]

15.4.3.4 *Virus Detection* Via *Single Nucleotide Polymorphism*

Single nucleotide mutation detection is essential for molecular diagnosis and personalized therapy. A single nucleotide mutation can have a significant impact on the subtypes of viruses that have varying levels of virulence and infectivity. Gootenberg *et al.* used base mismatches in the target–crRNA duplex to distinguish between African and American ZIKV strains and DENV strains 1 and 3.[51] Notably, synthetic mismatch crRNAs permitted robust strain identification with a single-base mismatch by detecting the corresponding strains other than the off-target strains with excellent sensitivity and specificity. Using a PCR-coupled CRISPR-Cas13a platform, HBV and the drug resistance mutations were identified. Utilizing a CRISPR-Cas-based lateral flow biosensor, HPV16 and HPV18 were successfully detected by focusing specific crRNAs on the hypervariable loop V of the L1-encoding gene.[65] Moreover, the SARS-CoV-2 D614G mutation was particularly identified using the ligation method and specific Cas13a/crRNA identification.[68]

15.4.3.5 *Direct Identification of Viruses in a Real Sample*

In addition to the detection of synthetic nucleic acids, CRISPR-Cas12a/Cas13a has been used to identify viruses in real samples such as plasma, serum, urine, swabs, and environmental and food samples. Prior to amplification, viral DNA or RNA in the sample was frequently isolated for the purpose of detecting infected samples. For instance, oropharyngeal and nasopharyngeal samples were mostly evaluated for SARS-CoV-2 infection. Fozouni *et al.* used a CRISPR-Cas13a-based technology to detect SARS-CoV-2 RNA directly from nasal swabs without amplifying it.[46] Some investigations reported the detection of viral nucleic acid straight from the raw sample matrix without nucleic acid extraction to streamline the process.[52,69,70] Myhrvold *et al.* developed HUDSON (heating unextracted diagnostic samples to obliterate nucleases), which identifies DENV from patient samples straight from body fluids (plasma, serum, and urine).[70] Since viruses are ubiquitous in nature, CRISPR-Cas12a/Cas13a systems have been used to identify viruses from a wide variety of sources, including apple leaves, shrimp, canine intestines, and animal blood.[67,71–73] Currently, researchers are exploring the possibilities of detecting viruses in saliva using a highly sensitive POC platform.[74,75]

15.4.4 Plasmonic-based Viral Diagnosis

15.4.4.1 Surface Plasmon Resonance (SPR)-based Sensor

SPR, also known as surface plasmon polarization, is a popular optical technique for observing how the refractive index (RI) of a sensor layer changes after binding a target molecule.[76] It refers to the electromagnetic (EM) resonance of the oscillations of free electrons associated with a plasmonic metal–dielectric semi-infinite interface (silver and gold for the visible spectrum), which generates a surface EM field along the metal–dielectric interface. This field can be employed as a sensing layer to implement SPR-based sensors because it is particularly sensitive to changes in the RI of the dielectric layer.[77,78]

A list of plasmonic-based biosensors for the detection of various arboviruses is provided in Table 15.2. These chip-based plasmonic sensors have a number of benefits, such as (i) label-free detection, (ii) dynamic measurement of binding–unbinding kinetics to observe the mechanism of the reaction occurring over the sensing surface, and (iii) high sensitivity.[76,79] For example, using the DENV serotype as a receptor over a gold chip, Jahanshahi *et al.* developed a rapid immunoglobulin M (IgM) dengue antibody detection SPR technique. The sensing device showed 100% selectivity and a response time of 10 min when using human serum samples.[80] Similarly, another study reported an SPR-based approach for dengue detection using IgM antibodies as a ligand and the DENV E-protein as the target. The linear range of the sensor was determined to be 0.0001–0.01 nM, with a sensitivity of $39.96°\ nM^{-1}$, and the working range was 0.0001–10 nM.[81] Nanograting-based SPR and loop-mediated isothermal amplification techniques have been demonstrated for the detection of HBV. The sensor was capable of detecting viral concentrations as low as five copies per 25 μL with a reaction time of 30 min.[82]

An SPR imaging mode was reported in another study to show the capability of SPR for detecting viral RNA. The sensor chip was created by covalently functionalizing DNA segments corresponding to both tobacco mosaic virus (TMV) and non-TMV RNA. The study demonstrated how DNA capture agents over the gold layer might be used to continuously monitor RNA binding.[83]

15.4.4.2 Localized Surface Plasmon Resonance (LSPR)-based Sensor

LSPR, a kind of SPR phenomenon in which the resonant EM field is restricted to the metallic nanostructure and sensitive to RI changes of the

Table 15.2 List of plasmonic-based biosensors for virus detection.

Virus	Type of biosensor	Detection	Limit of quantification or detection	Ref.
DENV	Electrochemical	Dengue-specific IgM antibody	0.45 μA	95
DENV	SPR optical	DENV2-E protein	1 pM	96
DENV	Optical/electrochemical	NS1 antigen	0.38–0.48 ng mL^{-1}	97
DENV	LFIA integrated with super-paramagnetic NPs	DENV1	0.25 ng mL^{-1}	98
		DENV2	0.1 ng mL^{-1}	
		DENV3	0.25 ng mL^{-1}	
		DENV4	1.0 ng mL^{-1}	
DENV	Lateral flow biosensor coupled with AuNPs	DENV1 RNA	0.01 μM	99
DENV	LSPR immunosensor	NS1 antigen	0.07 μg mL^{-1}	100
ZIKV and DENV	Multiplex electrochemical	NS1 antigen	—	101
ZIKV	Aptamer conjugated with AuNPs	ZIKV	10 ng	102
ZIKV	Electrochemical biosensor based on surface-imprinted polymers	NS1 antigen	2×10^{-4} PFU mL^{-1}	103
ZIKV	Nitrogen-doped porous carbon-based fluorescence sensor	ZIKV RNA	0.23 nM	104
Ebola	Electrochemical	EBOV glycoprotein	2.4 pg mL^{-1}	105
CHIKV	MoS_2 nanosheet-based disposable biosensor (electrochemical detection)	CHIKV DNA	3.4 nmol L^{-1}	106
CHIKV	DNA biosensor using gold shell-coated magnetic nanocubes	CHIKV DNA	0.1 nmol L^{-1}	106

medium surrounding it only within a few tens of nanometres, is another promising candidate to implement plasmonic biosensing with great sensitivity. This type of sensor was developed using metal nanostructures such as nanospheres, nanorods, nanoshells, nanowires, and nanoprisms that were fabricated along with a layer covering the sensing film.[84] In addition to using colloidal particles, advances in nanolithography techniques have made it possible to employ LSPR-based sensors on chip-based substrates with miniaturized sizes, high sensitivity and repeatability, and the ability to integrate with other sensing elements such as microfluidics, *etc.*[85]

In comparison with SPR, LSPR has several advantages, including a high aspect ratio that allows for a larger interaction surface area to immobilize the sensing elements, a miniaturized probe to create compact devices, and broad applicability and compatibility with numerous phenomena, including fluorescence, Raman, and IR spectroscopy, among others. However, the larger aspect ratio makes it easier for biomolecules to fit at the sensing surface than the metal NP, leading to high biomolecule sensitivity.[86] On the other hand, LSPR is less sensitive than traditional SPR in terms of RI sensitivity.[87] To attain the highest sensor performance, SPR is occasionally paired with LSPR in research.[88,89]

Tsang *et al.* developed a novel method for diagnosing EBOV that combined LSPR and luminescence to produce luminescence resonance energy transfer. In a nutshell, EBOV oligonucleotides conjugated with AuNPs were used to achieve LSPR, whereas oligonucleotides coupled to BaGdF5: Yb/Er up-conversion NPs were used for luminescence. Utilizing this approach, the EBOV was detected with LOD values at the pM level.[90] Another study revealed that multicoloured AgNPs may be used for multiplexed viral detection *via* a lateral flow-based system, demonstrating the application of single-channel multiplex analysis without the aid of an external light source.[91] Basso *et al.* demonstrated the specific detection of DENV serotypes using an LSPR-based immunosensor with an aptamer configuration of γ-Fe_2O_3@3-mercaptopropionic acid at AuNPs.[92]

The integration of LSPR with the plasmonic photothermal effect has been reported as another versatile technique for the clinical diagnosis of COVID-19. The technique involved immobilizing the complementary DNA receptors for SARS-CoV-2 onto 2D Au nanoislands for the sensitive detection of specific virus sequences using nucleic acid hybridization. The sensor demonstrated accurate viral detection with an LOD of 0.22 pM.[93]

15.4.4.3 Surface-enhanced Fluorescence (SEF)-based Sensor

The phenomenon of increased fluorescence intensity of a fluorophore material utilizing a plasmonic NM is known as surface-enhanced fluorescence (SEF), sometimes called metal-enhanced fluorescence or plasmon-enhanced fluorescence. This is accomplished by placing the fluorophore close enough to a metallic nanostructure to couple with the fluorophore electrons. Thus, the fluorescence experiences a higher electric field, resulting in improved emission and enhanced fluorescence intensity.[85,94]

SEF is frequently used to set up various POC devices for a range of analyte sensing applications. For instance, plasmonic NPs are used to increase the fluorescence intensity in a lateral flow kit where a fluorescent material is employed as a label to identify binding. Also, they are used for molecular material characterization, single molecule-level spectroscopic analysis, dynamic response of DNA hybridizations, cellular imaging, *etc.* However, the energy-quenching effect might occasionally limit its reach. This can be avoided by choosing the right nanostructures, increasing emission intensity, utilizing radiative decay, *etc.*[79]

Takemaura *et al.* presented a rapid and accurate approach for influenza virus identification utilizing an LSPR-induced fluorescence immunosensor. The sensor has a detection limit for viruses of 0.03 $pg\,mL^{-1}$ and could quickly identify them in 10 min.[107] Inci *et al.* established a unique approach for the detection and quantification of numerous HIV viral serotypes that is appropriate for POC applications. It provided a 70 min test time, effective whole blood sample detection, and a low detection limit (39 copies mL^{-1}).[108] A multifunctional nanosphere (RNs@Au, a combination of red fluorescent nanosphere (RN) quantum dots and Au NPs) was employed as the signal reporter in a combination of SEF and the lateral flow approach to detect the EBOV glycoprotein.[109] In another investigation, just 1 μL of serum or whole blood was utilized for the detection of ZIKV IgG, IgA, and IgM antibodies and also DENV IgG antibodies over a multiplexed plasmonic gold substrate.[110]

15.4.4.4 Surface-enhanced Raman Spectroscopy (SERS)-based Sensor

SERS has become established as a highly selective tool and pre-eminent analytical approach in the field of diagnostic applications over the past few decades, among a number of biosensing transducers. It demonstrates a wide range of benefits: (i) a potential distinct fingerprint signature of the analyte that could result in high selectivity; (ii) a simple sample preparation method; (iii) no signal interference from the analyte medium, which is typically water based; (iv) single-molecule detection; (v) the possibility of multiplexed sensing with a single laser beam; (vi) high throughput; and (vii) POC applicability by using portable Raman probes.[85] The commercial SERS technology uses the optical and chemical characteristics of the surrounding plasmonic NM to strengthen the naturally weak Raman signal.[111,112] Plasmonic metallic nanostructures, which have a confined EM field due to LSPR, might alter the Raman signal by increasing the Raman scattering cross-section if a material is close to or in the vicinity of plasmonic NPs. This improvement is the result of a combination of both EM enhancement and chemical enhancement.[113]

SERS was used for the identification of DENV nucleic acid sequences on a bimetallic nanowave chip. Briefly, polystyrene nanospheres were used to create a wave-like structure for the sensor probe, which was then covered with coatings of AgNPs and AuNPs. The DENV complementary ssDNA was then thiolated and immobilized over the AuNP. It was easy and cost-effective to use since SERS signals were gathered and measured following a single response on the chip's surface without any washing steps. The threshold limit for the detection of DENV DNA was 6 aM.[114] Using SERS-based biosensing and substrates coated with Au@PMPs, the RNA genomes of the WNV and the RVFV were found in 2012. By conjugating the viruses' acquired DNAs with the appropriate Raman reporters, malachite green and erythrosin B, simultaneous detection for both targets was made possible. The operational range for the detection of both viruses was 20–100 nM.[115]

15.4.4.5 *Surface-enhanced Infrared Absorption Spectroscopy (SEIRA)-based Sensor*

The SEIRA phenomenon is employed to boost the IR absorption signal of a target material. The IR signal of a substance is obtained by its atomic vibrations and is often examined using a Fourier transform IR spectrometer. By combining IR spectroscopy and localized plasmonic resonance, SEIRA spectroscopy has an impressively improved sensitivity (by several orders of magnitude). As a result, it functions as a specific biomarker and is widely used in molecular diagnostics.[116,117] SEIRA is accomplished by placing the target material close to a plasmonic NM similar to SEF and SERS. The enhancement factor is lower (10–1000) compared with other surface-enhanced techniques (such as SERS and SEF), but the interaction cross-section for IR absorption is much larger than that of SERS and SEF, making the moderate enhancement factor suitable for a variety of applications.[118]

Using scattering scanning near field optical microscopy (s-SNOM), Brehm *et al.* devised a method to obtain the fingerprint IR spectra of a cylindrical TMV with a diameter of 18 nm. In order to determine the surface topography, IR amplitude, and phase contrast, single viral NPs were cast from an Au/Si substrate and mapped using s-SNOM. Even at a very small illuminated probe volume 10^{-20} I (where I refers to amide I band), IR amplitude and phase contrast were observable, and topography was discernible at around 16 nm of the height of the virus NPs.[119] Another investigation, virus/protein mapping, was carried out using a nanometre-order sharp tip with a lateral resolution of 30 nm.[120]

15.5 Conclusion

The global emergency due to the COVID-19 pandemic has crippled both developed and underdeveloped countries and emphasized the need for continuous bio-surveillance of emerging and re-emerging viral diseases. However, the lack of knowledge of the molecular pathogenesis of viral diseases, specific treatment options, and sensitive detection tools necessitates the vital development of reliable, perceptive, and sophisticated detection technologies for arboviral diseases. Several groups are currently in the process of developing high-throughput methods for the rapid and accurate detection of viral and bacterial diseases. Recently, studies have been carried out on the use of NMs in the detection and identification of viruses. A nano-based sensor can augment routine clinical diagnostic methods for accurate viral detection. Therefore, nanotechnology can be readily applied to modify the current diagnostic platforms to increase the specificity and sensitivity of the clinical diagnosis and prognosis of vector-borne viral infections.

Abbreviations

ASFV	African swine fever virus
AuNP	Gold nanoparticle
CARMEN	Combinatorial arrayed reactions for multiplexed evaluation of nucleic acids
CF test	Complement fixation test
cGAS	Cyclic GMP–AMP synthase
COVID-19	Coronavirus disease 2019
DENV	Dengue virus
EC	Endothelial cell
EIA	Enzyme immunoassay
ELISA	Enzyme-linked immunosorbent assay
EM	Electromagnetic
GBP	Guanylate-binding protein
HI	Haemagglutination test
HIV	Human immunodeficiency virus
HUDSON	Heating unextracted diagnostic samples to obliterate nucleases
IFA	Indirect fluorescent antibody
IFN	Interferon
IgG	Immunoglobulin G
IgM	Immunoglobulin M
IHC	Immunohistochemistry
ISG	Interferon-stimulating gene
LFA	Lateral flow assay
LOD	Limit of detection
LSPR	Localized surface plasmon resonance
M-CDC	Magnetic pull-down assisted colorimetric diagnosis
MNP	Magnetic nanoparticle
NAAT	Nucleic acid amplification test
NM	Nanomaterial
NP	Nanoparticle
N test	Neutralization test
PCR	Polymerase chain reaction
POC	Point-of-care

PRNT	Plaque reduction neutralization test
PRR	Pattern recognition receptor
RI	Refractive index
RIG-I	Retinoic acid inducible gene I
RLR	Retinoic acid inducible gene I-like receptor
RN	Red fluorescent nanosphere
RT-PCR	Reverse transcriptase polymerase chain reaction
SARS-CoV-2	Severe acute respiratory syndrome coronavirus 2
SEF	Surface-enhanced fluorescence
SEIRA	Surface-enhanced infrared absorption spectroscopy
SERS	Surface-enhanced Raman spectroscopy
SPR	Surface plasmon resonance
s-SNOM	Scattering scanning nearfield optical microscopy
STOPCovid	SHERLOCK, testing in one spot COVID
TLR	Toll-like receptor
TMV	Tobacco mosaic virus
WNV	West Nile virus
ZIKV	Zika virus

Acknowledgements

We thank the Mahatma Gandhi Medical Advanced Research Institute (MGMARI), Sri Balaji Vidyapeeth (Deemed to be University), for providing infrastructure facilities for the preparation of this chapter.

References

1. Á Giménez-Richarte, M. I. Ortiz de Salazar, M.-P. Giménez-Richarte, M. Collado, P. L. Fernández and C. Clavijo, *et al.*, Transfusion-transmitted arboviruses: Update and systematic review, *PLoS Neglected Trop. Dis.*, 2022, **16**, e0010843.
2. S. V. Mayer, R. B. Tesh and N. Vasilakis, The emergence of arthropod-borne viral diseases: A global prospective on dengue, chikungunya and zika fevers, *Acta Trop.*, 2017, **166**, 155–163.
3. J. D. Beckham and K. L. Tyler, Arbovirus Infections, *Continuum*, 2015, **21**, 1599–1611.
4. M. S. Diamond and T. C. Pierson, Molecular Insight into Dengue Virus Pathogenesis and Its Implications for Disease Control, *Cell*, 2015, **162**, 488–492.
5. F. McNab, K. Mayer-Barber, A. Sher, A. Wack and A. O'Garra, Type I interferons in infectious disease, *Nat. Rev. Immunol.*, 2015, **15**, 87–103.
6. J. R. Wilson, P. F. de Sessions, M. A. Leon and F. Scholle, West Nile virus nonstructural protein 1 inhibits TLR3 signal transduction, *J. Virol.*, 2008, **82**, 8262–8271.

7. A. M. A. Nasirudeen, H. H. Wong, P. Thien, S. Xu, K.-P. Lam and D. X. Liu, RIG-I, MDA5 and TLR3 synergistically play an important role in restriction of dengue virus infection, *PLoS Neglected Trop. Dis.*, 2011, **5**, e926.
8. M. Chazal, G. Beauclair, S. Gracias, V. Najburg, E. Simon-Lorière and F. Tangy, *et al.*, RIG-I Recognizes the 5' Region of Dengue and Zika Virus Genomes, *Cell Rep.*, 2018, **24**, 320–328.
9. S. T. Perry, T. R. Prestwood, S. M. Lada, C. A. Benedict and S. Shresta, Cardif-mediated signaling controls the initial innate response to dengue virus *in vivo*, *J. Virol.*, 2009, **83**, 8276–8281.
10. L. D. Aarreberg, K. Esser-Nobis, C. Driscoll, A. Shuvarikov, J. A. Roby and M. Gale, Interleukin-1β Induces mtDNA Release to Activate Innate Immune Signaling *via* cGAS-STING, *Mol. Cell*, 2019, **74**, 801–815.e6.
11. S. W. van Leur, T. Heunis, D. Munnur and S. Sanyal, Pathogenesis and virulence of flavivirus infections, *Virulence*, 2021, **12**, 2814–2838.
12. V. Mariappan, S. Adikari, L. Shanmugam, J. M. Easow and A. Balakrishna Pillai, Differential expression of interferon inducible protein: Guanylate binding protein (GBP1 & GBP2) in severe dengue, *Free Radical Biol. Med.*, 2023, **194**, 131–146.
13. M. Robinson, T. E. Sweeney, R. Barouch-Bentov, M. K. Sahoo, L. Kalesinskas and F. Vallania, *et al.*, A 20-Gene Set Predictive of Progression to Severe Dengue, *Cell Rep.*, 2019, **26**, 1104–1111.e4.
14. R. Soundravally, S. L. Hoti, S. A. Patil, C. C. Cleetus, B. Zachariah and T. Kadhiravan, *et al.*, Association between proinflammatory cytokines and lipid peroxidation in patients with severe dengue disease around defervescence, *Int. J. Infect. Dis.*, 2014, **18**, 68–72.
15. A. B. Pillai, K. R. Muthuraman, V. Mariappan, S. S. Belur, S. Lokesh and S. Rajendiran, Oxidative stress response in the pathogenesis of dengue virus virulence, disease prognosis and therapeutics: an update, *Arch. Virol.*, 2019, **164**, 2895–2908.
16. A. B. Pillai, A. R. JeanPierre, V. Mariappan, P. Ranganadin and S. R. Rao, Neutralizing the free radicals could alleviate the disease severity following an infection by positive strand RNA viruses, *Cell Stress Chaperones*, 2022, **27**, 189–195.
17. K. L. Wong, W. Chen, T. Balakrishnan, Y. X. Toh, K. Fink and S.-C. Wong, Susceptibility and response of human blood monocyte subsets to primary dengue virus infection, *PLoS One*, 2012, **7**, e36435.
18. S. S. Sarayu Gopal, A. B. Pillai, V. S. Ramachandrappa, T. Kadhiravan, R. Dhodapkar and J. Kah, *et al.*, Increased serum levels of macrophage activation marker sCD163 in Dengue patients, *J. Clin. Virol.*, 2017, **86**, 62–67.
19. R. Soundravally, B. Agieshkumar, M. Daisy, J. Sherin and C. C. Cleetus, Ferritin levels predict severe dengue, *Infection*, 2015, **43**, 13–19.
20. A. Rathakrishnan, B. Klekamp, S. M. Wang, T. V. Komarasamy, S. K. Natkunam and J. Sathar, *et al.*, Clinical and immunological markers of dengue progression in a study cohort from a hyperendemic area in Malaysia, *PLoS One*, 2014, **9**, e92021.

21. T. Furuta, L. A. Murao, N. T. P. Lan, N. T. Huy, V. T. Q. Huong and T. T. Thuy, *et al.*, Association of mast cell-derived VEGF and proteases in Dengue shock syndrome, *PLoS Neglected Trop. Dis.*, 2012, **6**, e1505.
22. V. Mariappan, S. Adikari, L. Shanmugam, J. M. Easow and A. Balakrishna Pillai, Expression dynamics of vascular endothelial markers: endoglin and syndecan-1 in predicting dengue disease outcome, *Transl. Res.*, 2021, **232**, 121–141.
23. V. Mariappan, P. S. Manoharan, R. Pajanivel, L. Shanmugam, S. R. Rao and A. B. Pillai, Potential biomarkers for the early prediction of SARS-COV-2 disease outcome, *Microb. Pathog.*, 2021, **158**, 105057.
24. A. Balakrishna Pillai, V. Mariappan, A. R. JeanPierre and S. R. Rao, Restoration of vascular endothelial integrity by mesenchymal stromal/stem cells in debilitating virus diseases, *Hum Cell*, 2022, **35**, 1633–1639.
25. F. Sanchez and K. Sobolev, Nanotechnology in concrete – A review, *Constr. Build. Mater.*, 2010, **24**, 2060–2071.
26. M. Vaculovicova, P. Michalek, S. Krizkova, M. Macka and V. Adam, Nanotechnology-based analytical approaches for detection of viruses, *Anal. Methods*, 2017, **9**, 2375–2391.
27. S. G. Ramakrishnan, B. Robert, A. Salim, P. Ananthan, M. Sivaramakrishnan and S. Subramaniam, *et al.*, Nanotechnology based solutions to combat zoonotic viruses with special attention to SARS, MERS, and COVID 19: Detection, protection and medication, *Microb. Pathog.*, 2021, **159**, 105133.
28. M. Qasim, D.-J. Lim, H. Park and D. Na, Nanotechnology for Diagnosis and Treatment of Infectious Diseases, *J. Nanosci. Nanotechnol.*, 2014, **14**, 7374–7387.
29. L. Zhang and T. J. Webster, Nanotechnology and NMs: Promises for improved tissue regeneration, *Nano Today*, 2009, **4**, 66–80.
30. H. Wang, R. Yang, L. Yang and W. Tan, Nucleic acid conjugated nanomaterials for enhanced molecular recognition, *ACS Nano*, 2009, **3**, 2451–2460.
31. S. B. Somvanshi, P. B. Kharat, T. S. Saraf, S. B. Somwanshi, S. B. Shejul and K. M. Jadhav, Multifunctional nano-magnetic particles assisted viral RNA-extraction protocol for potential detection of COVID-19, *Mater. Res. Innovations*, 2021, **25**, 169–174.
32. A. H. F. Lee, S. F. Gessert, Y. Chen, N. V. Sergeev and B. Haghiri, Preparation of iron oxide silica particles for Zika viral RNA extraction, *Heliyon*, 2018, **4**, e00572.
33. C.-S. Ku and D. H. Roukos, From next-generation sequencing to nanopore sequencing technology: paving the way to personalized genomic medicine, *Expert Rev. Med. Devices*, 2013, **10**, 1–6.
34. Y. Feng, Y. Zhang, C. Ying, D. Wang and C. Du, Nanopore-based fourth-generation DNA sequencing technology, *Genomics, Proteomics Bioinf.*, 2015, **13**, 4–16.

35. Y. Yang, R. Liu, H. Xie, Y. Hui, R. Jiao and Y. Gong, *et al.*, Advances in nanopore sequencing technology, *J. Nanosci. Nanotechnol.*, 2013, **13**, 4521–4538.
36. S. Srinivasan and J. Batra, Four Generations of Sequencing- Is it Ready for the Clinic Yet?, *Next-Gener. Sequencing Appl.*, 2014, **1**, 1000107.
37. I. M. Derrington, T. Z. Butler, M. D. Collins, E. Manrao, M. Pavlenok and M. Niederweis, *et al.*, Nanopore DNA sequencing with MspA, *Proc. Natl. Acad. Sci. U. S. A.*, 2010, **107**, 16060–16065.
38. M. Jain, H. E. Olsen, B. Paten and M. Akeson, The Oxford Nanopore MinION: delivery of nanopore sequencing to the genomics community, *Genome Biol.*, 2016, **17**, 239.
39. H. Lu, F. Giordano and Z. Ning, Oxford Nanopore MinION Sequencing and Genome Assembly, *Genomics, Proteomics Bioinf.*, 2016, **14**, 265–279.
40. T. Laver, J. Harrison, P. A. O'Neill, K. Moore, A. Farbos and K. Paszkiewicz, *et al.*, Assessing the performance of the Oxford Nanopore Technologies MinION, *Biomol. Detect. Quantif.*, 2015, **3**, 1–8.
41. A. L. Greninger, S. N. Naccache, S. Federman, G. Yu, P. Mbala and V. Bres, *et al.*, Rapid metagenomic identification of viral pathogens in clinical samples by real-time nanopore sequencing analysis, *Genome Med.*, 2015, **7**, 99.
42. A. Kilianski, P. A. Roth, A. T. Liem, J. M. Hill, K. L. Willis and R. D. Rossmaier, *et al.*, Use of Unamplified RNA/cDNA-Hybrid Nanopore Sequencing for Rapid Detection and Characterization of RNA Viruses, *Emerging Infect. Dis.*, 2016, **22**, 1448–1451.
43. L. Yin, S. Man, S. Ye, G. Liu and L. Ma, CRISPR-Cas based virus detection: Recent advances and perspectives, *Biosens. Bioelectron.*, 2021, **193**, 113541.
44. J. S. Gootenberg, O. O. Abudayyeh, M. J. Kellner, J. Joung, J. J. Collins and F. Zhang, Multiplexed and portable nucleic acid detection platform with Cas13, Cas12a, and Csm6, *Science*, 2018, **360**, 439–444.
45. L. T. Nguyen, B. M. Smith and P. K. Jain, Enhancement of trans-cleavage activity of Cas12a with engineered crRNA enables amplified nucleic acid detection, *Nat. Commun.*, 2020, **11**, 4906.
46. P. Fozouni, S. Son, M. Díaz de León Derby, G. J. Knott, C. N. Gray and M. V. D'Ambrosio, *et al.*, Amplification-free detection of SARS-CoV-2 with CRISPR-Cas13a and mobile phone microscopy, *Cell*, 2021, **184**, 323–333.e9.
47. X. Wang, M. Zhong, Y. Liu, P. Ma, L. Dang and Q. Meng, *et al.*, Rapid and sensitive detection of COVID-19 using CRISPR/Cas12a-based detection with naked eye readout, CRISPR/Cas12a-NER, *Sci. Bull.*, 2020, **65**, 1436–1439.
48. Y. Chen, Y. Shi, Y. Chen, Z. Yang, H. Wu and Z. Zhou, *et al.*, Contamination-free visual detection of SARS-CoV-2 with CRISPR/Cas12a: A promising method in the point-of-care detection, *Biosens. Bioelectron.*, 2020, **169**, 112642.

49. P. Qin, M. Park, K. J. Alfson, M. Tamhankar, R. Carrion and J. L. Patterson, *et al.*, Rapid and Fully Microfluidic Ebola Virus Detection with CRISPR-Cas13a, *ACS Sens.*, 2019, **4**, 1048–1054.
50. X. Ding, K. Yin, Z. Li, R. V. Lalla, E. Ballesteros and M. M. Sfeir, *et al.*, Ultrasensitive and visual detection of SARS-CoV-2 using all-in-one dual CRISPR-Cas12a assay, *Nat. Commun.*, 2020, **11**, 4711.
51. J. S. Gootenberg, O. O. Abudayyeh, J. W. Lee, P. Essletzbichler, A. J. Dy and J. Joung, *et al.*, Nucleic acid detection with CRISPR-Cas13a/C2c2, *Science*, 2017, **356**, 438–442.
52. K. G. Barnes, A. E. Lachenauer, A. Nitido, S. Siddiqui, R. Gross and B. Beitzel, *et al.*, Deployable CRISPR-Cas13a diagnostic tools to detect and report Ebola and Lassa virus cases in real-time, *Nat. Commun.*, 2020, **11**, 4131.
53. S. Wang, H. Li, Z. Kou, F. Ren, Y. Jin and L. Yang, *et al.*, Highly sensitive and specific detection of hepatitis B virus DNA and drug resistance mutations utilizing the PCR-based CRISPR-Cas13a system, *Clin. Microbiol. Infect.*, 2021, **27**, 443–450.
54. L. Li, S. Li, N. Wu, J. Wu, G. Wang and G. Zhao, *et al.*, HOLMESv2: A CRISPR-Cas12b-Assisted Platform for Nucleic Acid Detection and DNA Methylation Quantitation, *ACS Synth. Biol.*, 2019, **8**, 2228–2237.
55. J. P. Broughton, X. Deng, G. Yu, C. L. Fasching, V. Servellita and J. Singh, *et al.*, CRISPR-Cas12-based detection of SARS-CoV-2, *Nat. Biotechnol.*, 2020, **38**, 870–874.
56. M. Patchsung, K. Jantarug, A. Pattama, K. Aphicho, S. Suraritdechachai and P. Meesawat, *et al.*, Clinical validation of a Cas13-based assay for the detection of SARS-CoV-2 RNA, *Nat. Biomed. Eng.*, 2020, **4**, 1140–1149.
57. C. M. Ackerman, C. Myhrvold, S. G. Thakku, C. A. Freije, H. C. Metsky and D. K. Yang, *et al.*, Massively multiplexed nucleic acid detection with Cas13, *Nature*, 2020, **582**, 277–282.
58. W. Zhou, X. Gao, D. Liu and X. Chen, Gold Nanoparticles for In Vitro Diagnostics, *Chem. Rev.*, 2015, **115**, 10575–10636.
59. B. Liu and J. Liu, Freezing Directed Construction of Bio/Nano Interfaces: Reagentless Conjugation, Denser Spherical Nucleic Acids, and Better Nanoflares, *J. Am. Chem. Soc.*, 2017, **139**, 9471–9474.
60. C. Yuan, T. Tian, J. Sun, M. Hu, X. Wang and E. Xiong, *et al.*, Universal and Naked-Eye Gene Detection Platform Based on the Clustered Regularly Interspaced Short Palindromic Repeats/Cas12a/13a System, *Anal. Chem.*, 2020, **92**, 4029–4037.
61. M. Hu, C. Yuan, T. Tian, X. Wang, J. Sun and E. Xiong, *et al.*, Single-Step, Salt-Aging-Free, and Thiol-Free Freezing Construction of AuNP-Based Bioprobes for Advancing CRISPR-Based Diagnostics, *J. Am. Chem. Soc.*, 2020, **142**, 7506–7513.
62. N. Jiang, R. Ahmed, M. Damayantharan, B. Ünal, H. Butt and A. K. Yetisen, Lateral and Vertical Flow Assays for Point-of-Care Diagnostics, *Adv. Healthcare Mater.*, 2019, **8**, e1900244.

63. T. Yuan, O. Mukama, Z. Li, W. Chen, Y. Zhang and J. de Dieu Habimana, *et al.*, A rapid and sensitive CRISPR/Cas12a based lateral flow biosensor for the detection of Epstein-Barr virus, *Analyst*, 2020, **145**, 6388–6394.
64. J. Joung, A. Ladha, M. Saito, N.-G. Kim, A. E. Woolley and M. Segel, *et al.*, Detection of SARS-CoV-2 with SHERLOCK One-Pot Testing, *N. Engl. J. Med.*, 2020, **383**, 1492–1494.
65. O. Mukama, T. Yuan, Z. He, Z. Li, J. de Dieu Habimana and M. Hussain, *et al.*, A high fidelity CRISPR/Cas12a based lateral flow biosensor for the detection of HPV16 and HPV18, *Sens. Actuators, B*, 2020, **316**, 128119.
66. X. Wang, P. Ji, H. Fan, L. Dang, W. Wan and S. Liu, *et al.*, CRISPR/Cas12a technology combined with immunochromatographic strips for portable detection of African swine fever virus, *Commun. Biol.*, 2020, **3**, 62.
67. J. Jiao, K. Kong, J. Han, S. Song, T. Bai and C. Song, *et al.*, Field detection of multiple RNA viruses/viroids in apple using a CRISPR/Cas12a-based visual assay, *Plant Biotechnol. J.*, 2021, **19**, 394–405.
68. Y. Wang, Y. Zhang, J. Chen, M. Wang, T. Zhang and W. Luo, *et al.*, Detection of SARS-CoV-2 and Its Mutated Variants *via* CRISPR-Cas13-Based Transcription Amplification, *Anal. Chem.*, 2021, **93**, 3393–3402.
69. B. Schermer, F. Fabretti, M. Damagnez, V. Di Cristanziano, E. Heger and S. Arjune, *et al.*, Rapid SARS-CoV-2 testing in primary material based on a novel multiplex RT-LAMP assay, *PLoS One*, 2020, **15**, e0238612.
70. C. Myhrvold, C. A. Freije, J. S. Gootenberg, O. O. Abudayyeh, H. C. Metsky and A. F. Durbin, *et al.*, Field-deployable viral diagnostics using CRISPR-Cas13, *Science*, 2018, **360**, 444–448.
71. T. Chaijarasphong, T. Thammachai, O. Itsathitphaisarn, K. Sritunyalucksana and R. Suebsing, Potential application of CRISPR-Cas12a fluorescence assay coupled with rapid nucleic acid amplification for detection of white spot syndrome virus in shrimp, *Aquaculture*, 2019, **512**, 734340.
72. H. Khan, A. Khan, Y. Liu, S. Wang, S. Bibi and H. Xu, *et al.*, CRISPR-Cas13a mediated nanosystem for attomolar detection of canine parvovirus type 2, *Chin. Chem. Lett.*, 2019, **30**, 2201–2204.
73. T. J. Sullivan, A. K. Dhar, R. Cruz-Flores and A. G. Bodnar, Rapid, CRISPR-Based, Field-Deployable Detection Of White Spot Syndrome Virus In Shrimp, *Sci. Rep.*, 2019, **9**, 19702.
74. J. Song, M. G. Mauk, B. A. Hackett, S. Cherry, H. H. Bau and C. Liu, Instrument-Free Point-of-Care Molecular Detection of Zika Virus, *Anal. Chem.*, 2016, **88**, 7289–7294.
75. M. Humaidi, W. P. Tien, G. Yap, C. R. Chua and L. C. Ng, Non-Invasive Dengue Diagnostics—The Use of Saliva and Urine for Different Stages of the Illness, *Diagnostics*, 2021, **11**, 1345.
76. S. Roh, T. Chung and B. Lee, Overview of the characteristics of micro- and nano-structured surface plasmon resonance sensors, *Sensors*, 2011, **11**, 1565–1588.

77. I. Abdulhalim, M. Zourob and A. Lakhtakia, Surface Plasmon Resonance for Biosensing: A Mini-Review, *Electromagnetics*, 2008, **28**, 214–242.
78. A. K. Sharma, R. Jha and B. D. Gupta, Fiber-Optic Sensors Based on Surface Plasmon Resonance: A Comprehensive Review, *IEEE Sens. J.*, 2007, **7**, 1118–1129.
79. A. M. Shrivastav, U. Cvelbar and I. Abdulhalim, A comprehensive review on plasmonic-based biosensors used in viral diagnostics, *Commun. Biol.*, 2021, **4**, 70.
80. P. Jahanshahi, E. Zalnezhad, S. D. Sekaran and F. R. M. Adikan, Rapid immunoglobulin M-based dengue diagnostic test using surface plasmon resonance biosensor, *Sci. Rep.*, 2014, **4**, 3851.
81. N. A. S. Omar, Y. W. Fen, J. Abdullah, Y. Mustapha Kamil, WMEMM Daniyal and A. R. Sadrolhosseini, *et al.*, Sensitive Detection of Dengue Virus Type 2 E-Proteins Signals Using Self-Assembled Monolayers/Reduced Graphene Oxide-PAMAM Dendrimer Thin Film-SPR Optical Sensor, *Sci. Rep.*, 2020, **10**, 2374.
82. C.-W. Lin, J.-H. Lee, N.-F. Chiu, S.-Y. Lee, K.-C. Liu, F.-Y. Tsai *et al.* Hybrid nano plasmonics for integrated biosensor, *2009 Asia Communications and Photonics Conference and Exhibition (ACP)*, Shanghai, 2009, pp. 1–6.
83. B. H. Garcia and R. M. Goodman, Use of surface plasmon resonance imaging to study viral RNA: protein interactions, *J. Virol. Methods*, 2008, **147**, 18–25.
84. P. K. Jain, K. S. Lee, I. H. El-Sayed and M. A. El-Sayed, Calculated absorption and scattering properties of gold nanoparticles of different size, shape, and composition: applications in biological imaging and biomedicine, *J. Phys. Chem. B*, 2006, **110**, 7238–7248.
85. M. Li, S. K. Cushing and N. Wu, Plasmon-enhanced optical sensors: a review, *Analyst*, 2015, **140**, 386–406.
86. K. M. Mayer and J. H. Hafner, Localized surface plasmon resonance sensors, *Chem. Rev.*, 2011, **111**, 3828–3857.
87. J. Cao, E. K. Galbraith, T. Sun and K. T. V. Grattan, Comparison of Surface Plasmon Resonance and Localized Surface Plasmon Resonance-based optical fibre sensors, *J. Phys.: Conf. Ser.*, 2011, **307**, 012050.
88. A. M. Shrivastav, S. K. Mishra and B. D. Gupta, Localized and propagating surface plasmon resonance based fiber optic sensor for the detection of tetracycline using molecular imprinting, *Mater. Res. Express*, 2015, **2**, 035007.
89. A. M. Shrivastav, S. P. Usha and B. D. Gupta, A localized and propagating SPR, and molecular imprinting based fiber-optic ascorbic acid sensor using an *in situ* polymerized polyaniline–Ag nanocomposite, *Nanotechnology*, 2016, **27**, 345501.
90. M.-K. Tsang, W. Ye, G. Wang, J. Li, M. Yang and J. Hao, Ultrasensitive Detection of Ebola Virus Oligonucleotide Based on Upconversion Nanoprobe/Nanoporous Membrane System, *ACS Nano*, 2016, **10**, 598–605.

91. C.-W. Yen, H. de Puig, J. O. Tam, J. Gómez-Márquez, I. Bosch and K. Hamad-Schifferli, *et al.*, Multicolored silver nanoparticles for multiplexed disease diagnostics: distinguishing dengue, yellow fever, and Ebola viruses, *Lab Chip*, 2015, **15**, 1638–1641.
92. C. R. Basso, B. P. Crulhas, M. Magro, F. Vianello and V. A. Pedrosa, A new immunoassay of hybrid nanomater conjugated to aptamers for the detection of dengue virus, *Talanta*, 2019, **197**, 482–490.
93. G. Qiu, Z. Gai, Y. Tao, J. Schmitt, G. A. Kullak-Ublick and J. Wang, Dual-Functional Plasmonic Photothermal Biosensors for Highly Accurate Severe Acute Respiratory Syndrome Coronavirus 2 Detection, *ACS Nano*, 2020, **14**, 5268–5277.
94. S. M. Fothergill, C. Joyce and F. Xie, Metal enhanced fluorescence biosensing: from ultra-violet towards second near-infrared window, *Nanoscale*, 2018, **10**, 20914–20929.
95. O. Parkash, M. A. Abdullah, C. Y. Yean, S. D. Sekaran and R. H. Shueb, Development and Evaluation of an Electrochemical Biosensor for Detection of Dengue-Specific IgM Antibody in Serum Samples, *Diagnostics*, 2020, **11**, 33.
96. N. A. S. Omar, Y. W. Fen, I. Ramli, A. R. Sadrolhosseini, J. Abdullah and N. A. Yusof, *et al.*, An Optical Sensor for Dengue Envelope Proteins Using Polyamidoamine Dendrimer Biopolymer-Based Nanocomposite Thin Film: Enhanced Sensitivity, Selectivity, and Recovery Studies, *Polymers*, 2021, **13**, 762.
97. P. Kanagavalli and M. Veerapandian, Opto-electrochemical functionality of Ru(II)-reinforced graphene oxide nanosheets for immunosensing of dengue virus non-structural 1 protein, *Biosens. Bioelectron.*, 2020, **150**, 111878.
98. B. T. Thanh, N. Van Sau, H. Ju, M. J. K. Bashir, H. K. Jun and T. B. Phan, *et al.*, Immobilization of Protein A on Monodisperse Magnetic Nanoparticles for Biomedical Applications, *J. Nanomater.*, 2019, **2019**, 1–9.
99. F. M. Yrad, J. M. Castañares and E. C. Alocilja, Visual Detection of Dengue-1 RNA Using Gold Nanoparticle-Based Lateral Flow Biosensor, *Diagnostics*, 2019, **9**, 74.
100. S. Farooq, F. Wali, D. M. Zezell, R. E. de Araujo and D. Rativa, Optimizing and Quantifying Gold Nanospheres Based on LSPR Label-Free Biosensor for Dengue Diagnosis, *Polymers*, 2022, **14**, 1592.
101. I. Sampaio, F. D. Quatroni, J. N. Yamauti Costa and V. Zucolotto, Electrochemical detection of Zika and Dengue infections using a single chip, *Biosens. Bioelectron.*, 2022, **216**, 114630.
102. A. Bosak, N. Saraf, A. Willenberg, M. W. C. Kwan, B. W. Alto and G. W. Jackson, *et al.*, Aptamer–gold nanoparticle conjugates for the colorimetric detection of arboviruses and vector mosquito species, *RSC Adv.*, 2019, **9**, 23752–23763.
103. C. Tancharoen, W. Sukjee, C. Thepparit, T. Jaimipuk, P. Auewarakul and A. Thitithanyanont, *et al.*, Electrochemical Biosensor Based on Surface Imprinting for Zika Virus Detection in Serum, *ACS Sens.*, 2019, **4**, 69–75.

104. J. Li, K. Yang, Z. Wu, X. Li and Q. Duan, Nitrogen-doped porous carbon-based fluorescence sensor for the detection of ZIKV RNA sequences: fluorescence image analysis, *Talanta*, 2019, **205**, 120091.
105. X. Jin, H. Zhang, Y.-T. Li, M.-M. Xiao, Z.-L. Zhang and D.-W. Pang, *et al.*, A field effect transistor modified with reduced graphene oxide for immunodetection of Ebola virus, *Microchim. Acta*, 2019, **186**, 223.
106. C. Singhal, A. Dubey, A. Mathur, C. S. Pundir and J. Narang, Paper based DNA biosensor for detection of chikungunya virus using gold shells coated magnetic nanocubes, *Process Biochem.*, 2018, **74**, 35–42.
107. K. Takemura, O. Adegoke, N. Takahashi, T. Kato, T.-C. Li and N. Kitamoto, *et al.*, Versatility of a localized surface plasmon resonance-based gold nanoparticle-alloyed quantum dot nanobiosensor for immunofluorescence detection of viruses, *Biosens. Bioelectron.*, 2017, **89**, 998–1005.
108. F. Inci, O. Tokel, S. Wang, U. A. Gurkan, S. Tasoglu and D. R. Kuritzkes, *et al.*, Nanoplasmonic quantitative detection of intact viruses from unprocessed whole blood, *ACS Nano*, 2013, **7**, 4733–4745.
109. J. Hu, Y.-Z. Jiang, L.-L. Wu, Z. Wu, Y. Bi and G. Wong, *et al.*, Dual-Signal Readout Nanospheres for Rapid Point-of-Care Detection of Ebola Virus Glycoprotein, *Anal. Chem.*, 2017, **89**, 13105–13111.
110. B. Zhang, B. A. Pinsky, J. S. Ananta, S. Zhao, S. Arulkumar and H. Wan, *et al.*, Diagnosis of Zika virus infection on a nanotechnology platform, *Nat. Med.*, 2017, **23**, 548–550.
111. T. J. Moore, A. S. Moody, T. D. Payne, G. M. Sarabia, A. R. Daniel and B. Sharma, In Vitro and In Vivo SERS Biosensing for Disease Diagnosis, *Biosensors*, 2018, **8**, 46.
112. P. Li, F. Long, W. Chen, J. Chen, P. K. Chu and H. Wang, Fundamentals and applications of surface-enhanced Raman spectroscopy–based biosensors, *Curr. Opin. Biomed. Eng.*, 2020, **13**, 51–59.
113. R. Kizek, L. Krejcova, P. Michalek, M. Merlos Rodrigo, Z. Heger and S. Krizkova, *et al.*, 7Nanoscale virus biosensors: state of the art, *Nanobiosens. Dis. Diagn.*, 2015, 47.
114. H. T. Ngo, H.-N. Wang, A. M. Fales, B. P. Nicholson, C. W. Woods and T. Vo-Dinh, DNA bioassay-on-chip using SERS detection for dengue diagnosis, *Analyst*, 2014, **139**, 5655–5659.
115. H. Zhang, M. H. Harpster, W. C. Wilson and P. A. Johnson, Surface-enhanced Raman scattering detection of DNAs derived from virus genomes using Au-coated paramagnetic nanoparticles, *Langmuir*, 2012, **28**, 4030–4037.
116. X. Yang, Z. Sun, T. Low, H. Hu, X. Guo and F. J. García de Abajo, *et al.*, Nanomaterial-Based Plasmon-Enhanced Infrared Spectroscopy, *Adv. Mater.*, 2018, **30**, 1704896.
117. G. Doria, J. Conde, B. Veigas, L. Giestas, C. Almeida and M. Assunção, *et al.*, Noble metal nanoparticles for biosensing applications, *Sensors*, 2012, **12**, 1657–1687.

118. K. Ataka and J. Heberle, Biochemical applications of surface-enhanced infrared absorption spectroscopy, *Anal. Bioanal. Chem.*, 2007, **388**, 47–54.
119. M. Brehm, T. Taubner, R. Hillenbrand and F. Keilmann, Infrared spectroscopic mapping of single nanoparticles and viruses at nanoscale resolution, *Nano Lett.*, 2006, **6**, 1307–1310.
120. I. Amenabar, S. Poly, W. Nuansing, E. H. Hubrich, A. A. Govyadinov and F. Huth, *et al.*, Structural analysis and mapping of individual protein complexes by infrared nanospectroscopy, *Nat. Commun.*, 2013, **4**, 2890.

CHAPTER 16

A Brief Appraisal of Nanomaterials for COVID-19

ISHITA CHAKRABORTY,[a] INDIRA GOVINDARAJU,[a] N. POOJA,[a] RAJIB BISWAS[b] AND NIRMAL MAZUMDER*[a]

[a] Manipal School of Life Sciences, Manipal Academy of Higher Education, Manipal, Karnataka 576104, India; [b] Department of Physics, Tezpur University, Napaam, Assam 784028, India
*Email: nirmaluva@gmail.com

16.1 Introduction

Coronaviruses are enclosed non-segmented positive-sense RNA viruses having a broad host range in humans and other mammals. They are classified as members of the family Coronaviridae and the order Nidovirales. Hepatic, pulmonary, neurological, and gastrointestinal problems can be brought on by coronaviruses. The extremely pathogenic coronavirus has been identified in the human population, but no effective antiviral treatment was available.[1] By using ventilators, oxygen therapy, and other methods, the major objectives of treatments were to lessen symptoms and support breathing. Systems based on nanotechnology have so far shown promise in preclinical studies against viral illnesses such as respiratory viruses, herpes simplex, human papillomavirus, and HIV. Some examples of how nanotechnology-based methods can be used to combat pandemics include developing a sensitive, rapid, and precise COVID-19 diagnostic tool, using nanomaterials to deliver antiviral agents, enhancing contact tracing tools, coating nanomaterial surfaces to inactivate the virus, and developing effective environmental disinfectants.[2] The severe acute respiratory syndrome

Nanoscience & Nanotechnology Series No. 62
Smart Nanomaterials for Infectious Diseases
Edited by Suvardhan Kanchi, Nidhi Sharotri, Rajasekhar Chokkareddy, Deepali Sharma and Falah H. Hussein

Published by the Royal Society of Chemistry, www.rsc.org

coronavirus 2 (SARS-CoV-2) virus is an enveloped, particle-like virus with a diameter of 60–140 nm. Due to its resemblance in physical makeup, it has the capacity to accurately mimic the virus and interact with its proteins. Hence strategies based on nanoparticles (NPs) that aim to tackle this disease have immense potential.[3]

16.2 Nanomaterials for COVID-19

Owing to their special qualities, including their high surface area-to-volume ratio and their capacity to interact with biological systems, nanomaterials have been researched as a potential weapon in the fight against COVID-19.

16.2.1 Detection

The current COVID-19 tests are conducted on viral genetic material taken from nasopharyngeal swabs using the reverse transcription polymerase chain reaction (RT-PCR) method. Alternatively, an enzyme-linked immunosorbent assay (ELISA) is performed to identify the specific antibodies or antigens in a patient's blood. Unfortunately, both methods require expensive equipment, specialized laboratories, and personnel who have received advanced training.[4] Recent advances in nanotechnology and the use of nanomaterials in the production of biosensors have greatly increased their performance.[5] Nanomaterials have significantly increased the efficiency and sensitivity of biosensors because of their remarkable photoelectrochemical properties, high conductivity, and potential for platform downsizing. Metal NPs (MNPs) are used as indicators in colorimetric methods as effective tools to detect and identify target analytes because of their color change at a specific characteristic wavelength.[6]

In biosensing applications for viral infections, gold nanoparticles (AuNPs) are the most often used noble metal MNPs owing to their extraordinary optical/electrical properties, excellent biocompatibility, catalytic properties, and relatively simple fabrication procedure. To detect SARS-CoV-2 immunoglobulin G (IgG) in a patient's plasma, a colorimetric serological test was developed using short antigenic epitopes connected to AuNPs. Four immunodominant linear B-cell epitopes found in the spike (S) and nucleocapsid (N) proteins of SARS-CoV-2 were investigated for their IgG binding affinities and used as highly specific biological motifs on the NP to identify target antibodies.[7] Around 30 min after SARS-CoV-2 antibody injection, a particular bivalent interaction between SARS-CoV-2 antibodies and epitope-functionalized AuNPs promotes NP aggregation, which is seen as a distinct optical transition in the plasmon properties of the AuNPs. Many AuNPs, each with a diameter of around 20 nm, were coated on each viral particle.[8] The binding interactions caused AuNPs to group and the absorbance peak of the colloidal solution to move from 520 to 560 nm. Quantitative evaluations were also made possible by a link between the viral load and the absorbance at 560 nm. Nasal and throat swab samples from 45 SARS-CoV-2-positive patients and 49 SARS-CoV-2-negative patients were used for the real-time polymerase chain reaction (q-PCR) assays that validated this approach. The results showed 98% specificity and 96% sensitivity. The conjugation of AuNPs with *N*-acetylneuraminic acid, a glycan moiety that binds only to the SARS-CoV-2 S protein, was another example (glycan-AuNPs).

Additionally, NPs can be used to find SARS-CoV-2 viral proteins. In one study, polyclonal antibodies directed against the S protein, membrane protein, and envelope protein of SARS-CoV-2 were functionalized with AuNPs.[9] Anti-human IgM-AuNPs were conjugated with AuNPs and utilized to recognize SARS-CoV-2-specific IgM in a lateral flow immunoassay (LFIA) strip in a related experiment. The patient's serum sample, which included IgM specific to SARS-CoV-2, was placed on the sample pad of the strip. Moving down the strip, the sample reached the conjugation pad that held the anti-human IgM-AuNPs. The conjugation pad is where the SARS-CoV-2-specific IgM from the sample attaches to anti-human IgM-AuNPs to form the complex (SARS-CoV-2-IgM)–(anti-human IgM)-AuNP.

16.2.2 Therapeutics

Polymeric NPs are colloidal particles with a size range 10–999 nm that can be produced using organic or synthetic polymers. Common challenges to conventional drug development can be overcome by employing specificity, adjustable release kinetics, and multimodal drug compositions thanks to the adaptability of polymeric-based NP engineering.[10] Because of this, these particles are commonly utilized as drug delivery agents to treat a variety of ailments, including cancer. Owing to their inherent competitive advantages, polymeric-based NPs offer tremendous potential to fight the COVID-19 epidemic.

A different approach to treating COVID-19 is host mimicking. For biomedical applications, such as the most recent use of NPs for viral neutralization, NPs can be coated with cell membranes as a top-down technique to mimic host cells. Several challenges must be overcome in the application of cell membrane-coated NPs against SARS-CoV-2.[11] For instance, cell membrane-coated NPs must be generated in huge quantities to meet the need for therapeutic doses. Large-scale cell farming and a high yield of cell membrane isolation are essential in this regard. It is important to assess carefully the storage stability of formulations containing cell membrane-coated nanoparticles. Clinically standardized lyophilization, cold storage, and reconstitution techniques can be used for this purpose. However, additional research is still required to evaluate the toxicity of NPs wrapped in cell membranes. Cell membrane-coated NPs will become a dependable platform against SARS-CoV-2 by addressing these challenges.[11]

Since they may easily activate the immune system using cover supplied by antigen-introducing cells, AuNPs have attracted particular interest in the development of vaccines. In genetically engineered cells used to make decoy NPs and in human monocytes, which have a large number of cytokine receptors, the SARS-CoV-2 receptor angiotensin-converting enzyme 2 (ACE2) is stably expressed. By out-competing host cells, these nanodecoys efficiently absorb viruses and inflammatory cytokines such as interleukin 6 (IL-6) and granulocyte–macrophage colony-stimulating factor (GM-CSF).[12] These two characteristics permit the effective treatment of viral infection and the corresponding immune dysfunction, providing a potential therapeutic strategy for COVID-19 and other potential epidemics.

Targeting the immune system and creating COVID-19 treatments using NPs have shown promise. Immune cells such as macrophages, dendritic cells, and T cells, which are essential for the immunological response to COVID-19, can be selectively targeted by NPs. These NPs have the potential to increase treatment effectiveness and reduce negative effects by directly delivering medications or other therapeutic substances to immune cells. Moreover, COVID-19 vaccinations can be administered using NPs, either as a solitary shot or a booster shot to heighten the immune response. These NPs can enhance the vaccine's efficiency and stability while also delivering controlled release of the antigen, which may result in a more potent and durable immune response. An excessive immune response brought on by COVID-19 may result in tissue damage and inflammation. By reducing the inflammatory response or enhancing the antiviral response, NPs can be employed to modify the immune response. For instance, anti-inflammatory drugs such as dexamethasone and tocilizumab can be delivered by NPs to patients with COVID-19 to reduce inflammation.[13]

The cellular attachment of viral particles to the alveoli may be prevented by the use of NPs that target viral S protein or ACE2 receptors. Other strategies, such as the application of neutralizing NPs or photocatalytic nanomaterials, can also be used to systemically inactivate viral particles.

16.2.3 Prevention

Inactivated pathogens, which often exhibit poor immunogenicity, or live attenuated pathogens, which carry a risk of reversion to virulent strains, are the two main components of conventional vaccines. This prompted the creation of next-generation subunit vaccines, such as those using viral antigens encoded in RNA or DNA, which have the potential to circumvent these restrictions but have limited immunogenicity.[14] Successful vaccinations are still elusive for many infectious diseases because all of these proteins are easily broken down in the body. Antigens, adjuvants, and immunoregulatory agents have been delivered specifically and released sustainably using platforms based on nanotechnology because they can be controlled for ineffective immune stimulation, loss of bioactivity of immunoreactive agents during circulation, and off-target side effects.[15]

Nanoparticles are being used by pharmaceutical companies to create and administer vaccines. However, for successful vaccine delivery to the host body, Novavax, a nanotechnology-based company, coupled the S protein of SARS-CoV-2 onto the surface of virus-like particles with a nanoscale size.

Owing to their high surface-to-volume ratio, NPs readily bind to either the viral replication system (DNA/RNA) or the viral membrane proteins, which ultimately reduces their ability to infect cells.[16]

Gold, silver, and copper NPs have all been cleverly employed to stop virus proliferation. Owing to their antimicrobial qualities, silver- and copper-based NPs have been utilized in surface coatings and paints for a long time. By including these MNPs in a different matrix, one can take advantage of the production of reactive oxygen species (ROS) and viral cell killing. By altering the permeability of the viral membrane and/or electrostatically attaching the virus to the cationic surface, polyethylenimine (PEI)-coated model surfaces with AgNPs and CuNPs successfully lowered the viral titer.[17] An AgNP-based composite with mesoporous silica protection was created as an effective disinfectant with regulated Ag^{+} release and simple magnetic removal after treatment.

Adjuvants for vaccines that are clinically useful, such as NPs made of aluminum, have been investigated for their ability to cross-present to dendritic cells (DCs) with the consequent activation of cellular immunity. Aluminum adjuvants are known to strongly encourage T2 helper cell development and antibody production through DCs but are unable to elicit an immunological response from T1 helper cells.[18] The cross-presentations by DCs can be improved by using aluminum NPs in conjunction with toll-like receptor ligands.

Several disorders may one day be treated with nucleic acids as medications. Lipid nanoparticles (LNPs) can transport mRNA and also other nucleic acids. Two mRNA-based vaccines, BNT162b2 (Pfizer-BioNTech) and mRNA-1273 (Moderna), were created utilizing LNPs and have received emergency use permission (EUA) from the US Food and Drug Administration (FDA) to prevent COVID-19, which is caused by the SARS-CoV-2 virus.[19] This article by Moura *et al.* focuses on the possible use of LNPs in the creation and administration of COVID-19 mRNA vaccines.[19] Ivermectin, a Food and Drug Administration approved antiparasitic drug, additionally also demonstrated effective antiviral properties. One study reported the effectiveness of ivermectin loaded in nanoparticles (PLGA-*b*-PEG-MAL nanoparticles) to reduce the expression of the viral S protein and its receptor ACE2, both of which are essential for reducing disease transmission rates. Ivermectin was packaged in an orally administrable NP form that could deliver a potent therapeutic antiviral dose.[20]

16.2.4 Nanomaterials in Protective Equipment

Masks, gloves, and gowns are only a few examples of protective gear for protection against COVID-19 that have been enhanced with nanomaterials. Below are some instances of applications of nanomaterials. To increase the filtration effectiveness of N95 masks, nanofiber filters have been employed. Electrospun polymer fibers, which are significantly smaller than the fibers used in conventional masks, are utilized to make these filters. This makes it possible for them to collect smaller particles, such as the coronavirus.[21]

Moreover, antimicrobial coatings for masks, gloves, and other equipment have been made using AgNPs and CuNPs. By eradicating germs and viruses on contact, these coatings can lower the risk of infection. Nanocoatings can also be used to make equipment surfaces that automatically clean themselves, such as touchscreens and doorknobs. To reject dirt and bacteria, these coatings feature nanoscale structures, which reduce the risk of transfer and make cleaning them easier.[22]

Moreover, materials can have NPs added to them to increase their UV resistance. Outdoor gear, such as tents and protective clothing, might benefit from this to block sunlight and lower the possibility of deterioration. Various methods have been implemented to combat the COVID-19 pandemic (Figure 16.1).

Zinc oxide nanoparticles (ZnONPs) have been studied and have demonstrated remarkable antibacterial characteristics. They are widely used as safe ingredients in food and cosmetic applications. Melt-blown polypropylene and nylon–cotton face mask fabrics, and also other textile and face mask materials, can now contain ZnONPs thanks to the creation of a unique nanocomposite (Figure 16.2).[24–27]

Overall, NPs have the potential to improve the COVID-19 resistance capabilities of protective equipment. To understand completely their efficacy and any potential hazards linked to their use, more research is necessary.

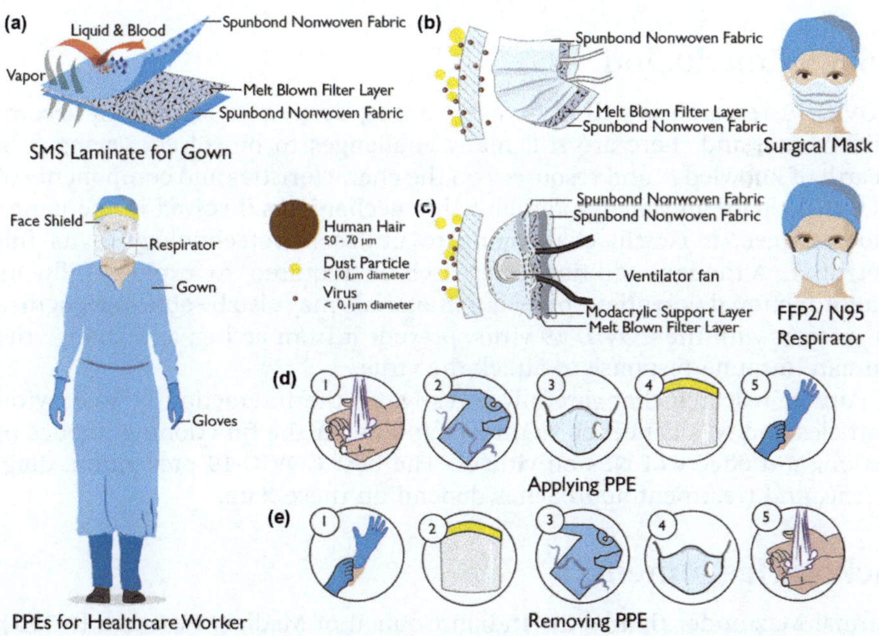

Figure 16.1 Types and applications of protective equipment against COVID-19. Reproduced from ref. 23, https://doi.org/10.3390/nano12050783, under the terms of the CC BY 4.0 license https://creativecommons.org/licenses/by/4.0/.

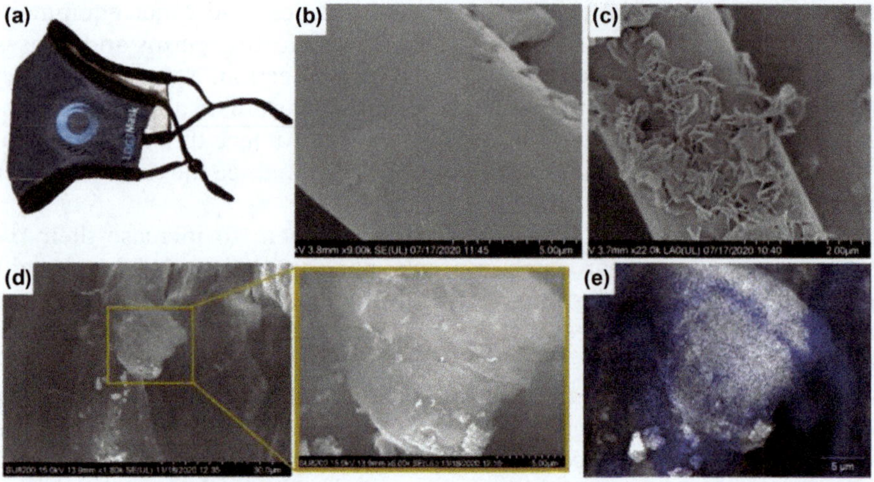

Figure 16.2 (a) A face mask designed by Claros Technologies; (b) polypropylene textile; (c) zinc–polypropylene nanocomposite textile; (d) polyester–cotton fabric with internal zinc particles; (e) energy-dispersive X-ray spectroscopy (EDX) image of the cross-section with zinc marked in blue. Reproduced from ref. 25, https://doi.org/10.1038/s41598-021-03771-1, under the terms of the CC BY 4.0 license https://creativecommons.org/licenses/by/4.0/.

16.3 Conclusion

COVID-19 research is still in its early stages, despite its significant accomplishments, and there are still many challenges to be solved. Owing to a dearth of knowledge and resources on the characteristics and components of COVID-19 pathophysiology and also the mechanisms involved in the nano–bio interface, it is still challenging to utilize nanotechnology to its full potential. Although additional research is required to explore fully its multifunctional potential, some nanomaterials may also be able to recognize or interact with the COVID-19 virus, prevent it from acting, and change the human immune response to attack the virus.

Further research that carefully investigates the interaction between viral particles and NPs is needed to understand better the functioning, modes of action, and effects of NPs on viruses. The best COVID-19 prevention, diagnostic, and treatment approaches depend on these data.

Acknowledgements

Nirmal Mazumder thanks the Indian Council of Medical Research (ICMR), Government of India, for financial support (Project Number ITR/Ad-hoc/43/2020-21, ID No. 2020-3286). Ishita Chakraborty and Pooja N thank MAHE, Manipal, Karnataka, India, for Dr TMA Pai PhD Scholarships. Ishita Chakraborty thanks the Department of Science and Technology (DST),

Government of Karnataka, India, for a PhD Fellowship (Award No. DST/KSTePS/Ph.D. Fellowship/LIF-12:2021-22/1024).

References

1. M. Pal, G. Berhanu, C. Desalegn and V. Kandi, *Cureus*, 2020, **12**, e7423.
2. C. Weiss, M. Carriere, L. Fusco, I. Capua, J. A. Regla-Nava, M. Pasquali, J. A. Scott, F. Vitale, M. A. Unal, C. Mattevi and D. Bedognetti, *ACS Nano*, 2020, **14**, 6383.
3. R. Medhi, P. Srinoi, N. Ngo, H. V. Tran and T. R. Lee, *ACS Appl. Nano Mater.*, 2020, **3**, 8557.
4. L. J. Carter, L. V. Garner, J. W. Smoot, Y. Li, Q. Zhou, C. J. Saveson, J. M. Sasso, A. C. Gregg, D. J. Soares, T. R. Beskid and S. R. Jervey, *ACS Cent. Sci.*, 2020, **6**, 591.
5. X. Huang, Y. Zhu and E. Kianfar, *J. Mater. Res. Technol.*, 2021, **12**, 1649.
6. Z. Li, J. R. Askim and K. S. Suslick, *Chem. Rev.*, 2018, **119**, 231.
7. R. Antiochia, *Microchim. Acta*, 2020, **187**, 639.
8. T. T. Lew, K. M. Aung, S. Y. Ow, S. N. Amrun, L. Sutarlie, L. F. Ng and X. Su, *ACS Nano*, 2021, **15**, 12286.
9. H. Sekimukai, N. Iwata-Yoshikawa, S. Fukushi, H. Tani, M. Kataoka, T. Suzuki, H. Hasegawa, K. Niikura, K. Arai and N. Nagata, *Microbiol. Immunol.*, 2020, **64**, 33.
10. A. Zielińska, F. Carreiró, A. M. Oliveira, A. Neves, B. Pires, D. N. Venkatesh, A. Durazzo, M. Lucarini, P. Eder, A. M. Silva and A. Santini, *Molecules*, 2020, **25**, 3731.
11. L. E. Charelli, G. C. de Mattos, A. de Jesus Sousa-Batista, J. C. Pinto and T. A. Balbino, *J. Nanopart. Res.*, 2022, **24**, 12.
12. L. Mao, Z. Chen, Y. Wang and C. Chen, *J. Inorg. Biochem.*, 2021, **219**, 111454.
13. G. Guerrini, D. Magri, S. Gioria, D. Medaglini and L. Calzolai, *Nat. Nanotechnol.*, 2022, **17**, 570.
14. A. S. Lauring, J. O. Jones and R. Andino, *Nat. Biotechnol.*, 2010, **28**, 573.
15. Y. Zheng, Q. Xie, H. Wang, Y. Hu, B. Ren and X. Li, *Int. J. Biol. Macromol.*, 2020, **165**, 2668.
16. M. D. Shin, S. Shukla, Y. H. Chung, V. Beiss, S. K. Chan, O. A. Ortega-Rivera, D. M. Wirth, A. Chen, M. Sack, J. K. Pokorski and N. F. Steinmetz, *Nat. Nanotechnol.*, 2020, **15**, 646.
17. J. K. Crane, *Immunol. Invest.*, 2020, **49**, 794.
18. X. Wang, W. Sun, W. Yang, S. Gao, C. Sun and Q. Li, *Nanoscale Adv.*, 2019, **1**, 840.
19. I. A. de Moura, A. J. Silva, L. S. de Macêdo, M. D. Invenção, M. M. de Sousa and A. C. de Freitas, *Pathogens*, 2022, **11**, 1444.
20. B. Surnar, M. Z. Kamran, A. S. Shah and S. Dhar, Clinically approved antiviral drug in an orally administrable nanoparticle for COVID-19, *ACS Pharmacol. Transl. Sci.*, 2020, 3(6), 1371–1380.
21. S. Mallakpour, V. Behranvand and C. M. Hussain, *Polym. Bull.*, 2023, **80**, 165.

22. S. P. Deshmukh, S. M. Patil, S. B. Mullani and S. D. Delekar, *Mater. Sci. Eng., C*, 2019, **97**, 954.
23. M. Souri, M. Chiani, A. Farhangi, M. R. Mehrabi, D. Nourouzian, K. Raahemifar and M. Soltani, *Nanomaterials*, 2022, **12**, 783.
24. M. Z. Rahman, M. E. Hoque, M. R. Alam, M. A. Rouf, S. I. Khan, H. Xu and S. Ramakrishna, *Polymers*, 2022, **14**, 1296.
25. A. Gonzalez, H. A. Aboubakr, J. Brockgreitens, W. Hao, Y. Wang, S. M. Goyal and A. Abbas, *Sci. Rep.*, 2021, **11**.
26. C. S. H. Hwang, S. Lee, S. Lee, H. Kim, T. Kang, D. Lee and K-H. Jeong, *ACS Appl. Mater. Interfaces*, 2022, **14**, 54550.
27. T. S. Muzata, A. Gebrekrstos and S. S. Ray, *ACS Omega*, 2021, **6**, 28463.

Subject Index